Foundations of
GEOMETRY

FOUNDATIONS OF
GEOMETRY

Euclidean, Bolyai-Lobachevskian, and Projective Geometry

KAROL BORSUK
WANDA SZMIELEW

Translated by
ERWIN MARQUIT

DOVER PUBLICATIONS, INC.
MINEOLA, NEW YORK

Bibliographical Note

This Dover edition, first published in 2018, is an unabridged and slightly corrected republication of the work originally published by North-Holland Publishing Co., Amsterdam, and Interscience Publishers, New York, in 1960.

Library of Congress Cataloging-in-Publication Data

Names: Borsuk, Karol, author. | Szmielew, Wanda, author.
Title: Foundations of geometry : Euclidean and Bolyai-Lobachevskian geometry projective geometry / by Karol Borsuk and Wanda Szmielew ; translated by Erwin Marquit.
Other titles: Podstawy geometrii. English
Description: Dover edition. | Mineola, New York : Dover Publications, Inc., 2018. | Series: Dover Books on Mathematics | In English; translated from Polish. | Originally published: Amsterdam : North-Holland Publishing Co. ; New York : Interscience Publishers, 1960.
Identifiers: LCCN 2018022377 | ISBN 9780486828091 | ISBN 0486828093
Subjects: LCSH: Geometry—Foundations.
Classification: LCC QA681 .B633 2018 | DDC 516—dc23
LC record available at https://lccn.loc.gov/2018022377

Manufactured in the United States by LSC Communications
82809301   2018
www.doverpublications.com

CONTENTS

INTRODUCTION

Part One

EUCLIDEAN AND BOLYAI-LOBACHEVSKIAN GEOMETRY

INTRODUCTION TO PART ONE

CHAPTER I. AXIOMS OF INCIDENCE AND ORDER

CHAPTER III. AXIOM OF CONTINUITY

CHAPTER IV. MODELS OF ABSOLUTE GEOMETRY

CHAPTER V. EUCLIDEAN GEOMETRY

Part Two

PROJECTIVE GEOMETRY

PREFACE TO THE POLISH EDITION†

This book was planned, at first, to be a textbook in the foundations of geometry, specifically adapted to the present program of studies in this subject in Polish universities. In the process of writing, however, the conception of the work underwent some change in order that the material included constitute a consistent, closed entity which—as it seems to the authors—may be of value independently of any possible changes in the program of the course. With this aim in mind some portions of the book (e.g. Bolyai-Lobachevskian geometry) have been treated much more extensively than in the original outline while some others—dealing with rather marginal problems—have been entirely omitted. The presentation of the material is quite elementary. Even in those portions of Bolyai-Lobachevskian geometry in which the apparatus of differential geometry is usually applied the authors have used exclusively the most elementary notions and methods of the calculus. On the other hand the authors considered it purposeful to introduce the general topological notions at a very early stage of the discussion. The reader who is not familiar with topology will find the necessary information in the Introduction (Section 9).

The book is organized as follows.

In Part I the authors develop Euclidean and Bolyai-Lobachevskian geometry on the basis of an axiom system due, in principle, to Hilbert. It should be noted at once, however, that the authors develop these geometries, in principle, as far as necessary to be able to prove them categorical, i.e., to show that the Cartesian space known from analytic geometry is up to isomorphism the only model of Euclidean geometry, and Klein space (constructed with the help of notions known from the analytic geometry of projective space) is up to isomorphism the only model of Bolyai-Lobachevskian geometry. In this way two aims are achieved. First, it is shown that each of the theories constitutes a uniquely determined

† The Polish original of this work was published under the title *Podstawy Geometrii* by Panstwowe Wydawnictwo Naukowe (Warsaw, 1955).

scheme embracing the geometrical properties of physical space. Secondly, the course in geometry is developed to the point where, by the introduction of coordinates (rectangular coordinates in Euclidean geometry and Beltrami coordinates in Bolyai-Lobachevskian geometry), it becomes possible to employ analytic methods.

Besides the full proof of categoricity, Part I also contains proofs of the consistency of both geometries and—as an example of independence proofs—a proof of the independence of the Axiom of Continuity.

In Part II the authors develop projective geometry on the basis of an axiom system also due, in principle, to Hilbert. In projective geometry use is made of the theorems of Euclidean geometry, this being done by an interpretation of a part of Euclidean geometry in projective geometry. The organization of the material is the same as in Part I. The theory is developed only so far as the introduction of homogeneous projective coordinates in space. This gives the possibility of proving categoricity, which, in this case, reduces to showing that the projective space known from analytic geometry is up to isomorphism the only model of projective geometry.

In spite of such a restricted aim the book is rather extensive. This is so because the authors have set themselves the task of achieving a presentation which does not contain any essential gaps. In this way the reader, after finishing the book, can be confident that the axiom systems adopted are indeed a sufficient basis for the construction of the entire geometry. Such an approach has somewhat encumbered the book, forcing the authors—especially in the initial chapters—to give rigorous proofs of theorems generally well known and frequently trivial. The reader well-versed in the axiomatic treatment of elementary geometry is advised to glance lightly through the first two chapters, and devote attention principally to the conceptional aspects of the work.

It is worth noting that the authors have not taken up in this book the problem of constructing arithmetic on the basis of geometry. Their task was made easier by assuming the arithmetic of real numbers as a theory preceding geometry. Analytic geometry of Cartesian space and projective space is regarded as a branch of arithmetic.

In conclusion, the authors take pleasure in cordially thanking Professor Adam Bielecki for his careful study of the manuscript and for making many valuable and keen remarks which have enabled them to eliminate several errors.

PREFACE TO THE ENGLISH EDITION

The preparation of the English edition of Foundations of Geometry has given the authors the opportunity of introducing essential changes and additions (as well as minor corrections).

In the Polish edition absolute geometry (the common part of Euclidean and Bolyai-Lobachevskian geometries) was based on the Hilbert system of primitive notions and axioms; in the English edition the Hilbert system has been modified in several respects. Lines and planes are now treated as sets of points; therefore the primitive relation of incidence is replaced by the set-theoretical membership relation and does not appear as a primitive notion of geometry. This change has led to a considerable simplification in the logical structure of the discussion of models and categoricity. Furthermore, the primitive relation of congruence of segments is treated in the present edition as a four-termed relation among points; the relation of congruence of angles has been entirely removed from the system of primitive notions, and this has necessitated certain changes in the axioms of congruence. Since the authors wished to avoid any radical changes in the arrangement of the material, they did not avail themselves of the possibility of further limiting the system of primitive notions.

Other essential changes were made to bring out the role of the Axiom of Continuity and the Archimedean Postulate. The authors endeavored to transfer as much material as possible to absolute geometry.

The most essential change introduced in Bolyai-Lobachevskian geometry involves the introduction of the natural basic segment and the natural measure of segments which goes with it. This permits the calculation of the numerical value of the constant $\varkappa$ which appears in a number of formulas of this geometry.

In projective geometry, just as in absolute geometry, lines and planes are now regarded as sets of points, whereas in the original edition, following the traditional approach, the points, the lines, and the planes are treated as three fundamental domains of discourse with equal status in the logical structure of the theory. It should be noticed that the traditional approach, as opposed to the present one, leads to a much simpler and more convenient formulation of the duality law; hence this approach would have been advantageous if the authors had attempted a systematic development of projective geometry. In this work, however, the authors are primarily interested in carrying through simple and precise proofs of categoricity, and from this point of view the new approach is much preferable.

We wish to express our deep gratitude to Professor Alfred Tarski (University of California, Berkeley) for his penetrating comments and criticism of the Polish edition of Foundations of Geometry at the time it appeared. To a large extent these remarks contributed to the changes introduced in the English edition. During the final revision of the English manuscript, Professor Tarski's warm advice has helped us again at every turn. To Professor Stanislaw Jaśkowski (University of Toruń) we are indebted for a remark which led to some simplification of the axiom system for projective geometry.

We are very grateful to Mr. Erwin Marquit (University of Warsaw) for his effort and care in preparing the English translation. We also sincerely appreciate the help extended to us by Professors Henry Helson (University of California, Berkeley), Leon Henkin (University of California, Berkeley) and Steven Orey (University of Minnesota), in connection with the final stylistic revision of the English text.

The final revision of the manuscript of this book, both in material and formal respects, was carried through during the academic year 1957–58. The work was performed partly at the University of Warsaw, Poland, by Karol Borsuk, and partly at the University of California, Berkeley, U.S.A., by Wanda Szmielew, who was then engaged in a research project on the foundations of mathematics sponsored by the U.S. National Science Foundation.

University of Warsaw, Poland   KAROL BORSUK

Institute for Basic Research in Science, University   WANDA SZMIELEW
of California, Berkeley, USA
May, 1958

FOUNDATIONS OF
GEOMETRY

Introduction

1. Geometry before Euclid

The foundations of geometry have attained their present clarity only as a result of a very long process of development which began in very ancient times and concluded in the 20th century. The need for such geometrical notions as segment, line, angle, triangle, circle, length, area, volume, etc. appeared as early as civilization. Even in early antiquity we can find systematic attempts to establish relations between these notions, and this is the beginning of geometry. The famous Rhind papyrus, a copy of which has been preserved from the Hyksos epoch (about 1700 B.C.) testifies to the fact that at that time geometry in Egypt already stood at a rather high level, nevertheless was, in principal, limited to empirically found instructions for calculating the area of a plane figure or the volume of a solid. But only among the Greeks we find a conscious striving to give geometry the form of a science in today's meaning of the word. Beginning with THALES of Milet and PYTHAGORAS of Samos (6th century B.C.) and ending with the Greek mathematicians in Alexandria in the period of the decline of the Roman empire (e.g. PAPPUS of Alexandria, 4th century A.D.), a long list of outstanding Greek mathematicians contributed in an essential way to the development of geometry.

2. Elements of Euclid

One of the most important events in the development of geometry was the systematic treatment of geometry in the form of a uniform deductive system in the work of EUCLID entitled *Elements* ($\sigma\tau o\iota\chi\varepsilon\tilde{\iota}\alpha$) written in Alexandria about 300 B.C. If the value of a scientific work can be measured by the length of time during which it maintains its importance then *Elements* of Euclid is the most valuable scientific book of all time. It has appeared in innumerable editions, and through the entire period of two thousand years has been generally considered a model of rigor and clarity of presentation, and richness of content. Euclid set himself the task of presenting geometry in the form of a system based on a small number of sentences, some of which were called *definitions*, others *axioms*, and still others *postulates*. The remaining statements of his work were to be logical consequences of these three types of initial sentences.

Euclid placed definitions at the beginning of each of the 13 books of his *Elements*. Thus, e.g., the definitions with which he begins the first book are as follows:†

A point is that which has no part. A line is a (breadthless) length. The extremities of a line are points. A straight line is a line which lies evenly with the points on itself. And so on.

As may be seen from these examples, the definitions given by Euclid are not definitions in today's meaning of the word and cannot be used in the construction of a theory. They are rather only explanations of the notions introduced, expressed in imprecise colloquial language and intended to create in the mind of the reader certain intuitive pictures.

The difference between axioms and postulates is not further explained by Euclid. The sentences he called axioms—such as *the whole is greater than the part*—have the character of statements about objects of some very general unspecified kind. The postulates, however, concern specific figures and in character are like the sentences called *axioms* in modern deductive theories. In modern theories the sentences corresponding to Euclid's axioms do not occur, and the terms *axiom* and *postulate* are used interchangeably.

In establishing his system of definitions, axioms, and postulates, Euclid believed that he was creating a sufficient foundation for the construction of geometry, i.e. for the introduction of geometrical notions and for the deduction of their properties. It should be noted that this is not the case. Thus, e.g., Euclid speaks about a point lying between two other points despite the fact that such a relation between points cannot be defined on the basis of his geometry; he makes essential use of the continuity of the space despite the fact that this property cannot be established in his system of geometry. Nevertheless, we are indebted to him for the first attempt known to us to construct an axiomatic theory.

3. Elementary Geometry After Euclid. Euclid's Critics and Commentators

Several years after Euclid the famous Syracusan mathematician ARCHIMEDES supplemented the system of axioms of Euclid by a further system of five axioms which he needed in connection with investigations on the length of curves, the area of surfaces, and the volume of solids. Four of these axioms directly involve the notions of length and area; they are superfluous if these notions are introduced by way of appropriate

---

† All quotations from Euclid in this book are based on T. C. HEATH's *The Thirteen Books of Euclid's Elements*, 2nd ed. (Cambridge 1926).

definitions. On the other hand, the fifth (we shall refer to it as the *Postulate of Archimedes*) is an explicit formulation of a property of segments on a line, a property which was already involved in the considerations of Euclid.

The axiomatic treatment of elementary geometry by Euclid was the subject of investigations by many mathematicians over a period of many centuries. Particular interest was attached to the problem of whether the *parallel postulate* (sometimes known as *Euclid's Fifth Postulate*) is necessary for the construction of elementary geometry. This postulate was formulated as follows:

If a straight line falling (in a plane) on two straight lines makes the interior angles on the same side less than two right angles, the two straight lines, if produced indefinitely, meet on that side on which are the angles less than the two right angles.

For a period of over two thousand years many attempts were made to prove that this postulate is a logical consequence of the remaining assumptions, and therefore that it may be omitted with no loss to the theory. There arose a very extensive literature which contains, among other results, various proofs of the parallel postulate using the remaining assumptions of Euclid together with some further assumption. For example, it suffices to add the assumption that there exists at least one rectangle; or that there exists at least one triangle the sum of whose angles is equal to two right angles. Up to the 19th century it was not settled whether there is a proof of the parallel postulate based only on the remaining assumptions of Euclid.

The most penetrating of the commentators on Euclid, the Italian mathematician Gerolamo SACCHERI (1667–1733) and the Swiss mathematician J. H. LAMBERT (1728–1777), consistently developed the domain of geometry which did not employ the parallel postulate. In particular, they drew attention to the theorem concerning the sum of the angles in a triangle. Without the aid of the parallel postulate they proved that this sum cannot be greater than two right angles. A little later the French mathematician A. M. LEGENDRE proved this fact again. SACCHERI also tried to prove, by deriving a contradiction, that the sum of the angles in a triangle cannot be smaller than two right angles, and he believed that in this way he would be able to obtain a proof of the parallel postulate. Lambert tried to show that the negation of the parallel postulate would lead to conclusions departing too greatly from our picture of space. By systematically investigating the logical consequences of the negation of the parallel postulate, both these mathematicians followed a path which subsequently led to the discovery of what we now call *non-Euclidean geometry*.

4. Bolyai-Lobachevskian Geometry

Only in the 19th century was the parallel postulate shown to be independent of the remaining axioms and postulates of Euclid. A decisive step in this direction was made by the Russian mathematician Nikolai Ivanovich LOBACHEVSKI (1793–1856). In 1829 he published in Kasan his work entitled *O načalah geometrii*, which contained an exposition of a new geometry based precisely on the negation of the parallel postulate. In the history of human thought we often meet with the phenomenon that great discoveries are made simultaneously and independently by several people when the state of science and technology reaches the point where it is ready for these developments. In the field of mathematics one may cite as examples the discovery of differential and integral calculus by NEWTON and LEIBNIZ, and of analytic geometry by FERMAT and DESCARTES. Another typical case is the discovery of non-Euclidean geometry. Simultaneously with Lobachevski and completely independently of him, the Hungarian mathematician Janos BOLYAI (1802–1860) arrived at similar conceptions; the work embodying his ideas, *Appendix scientiam spatii absolute veram exhibens*, appeared in 1832.

It should be mentioned that still earlier the eminent German mathematician Karl Friedrich GAUSS (1777–1855) had arrived at the idea of non-Euclidean geometry, but he never published the results of his investigations. He feared the criticism which might be evoked by an idea which departed so far from the ideas then accepted and sanctified by the tradition of many centuries. Through such an attitude, Gauss lost priority for the discovery of non-Euclidean geometry. Gauss's fears, however, were not without basis. The works of Lobachevski and Bolyai did not receive recognition in the lifetime of their creators. On the contrary, they were regarded as eccentric and pathological; a Russian mathematician well known in this period went so far as to call the work of Lobachevski a satire directed against mathematicians.

At the beginning of the 19th century the idea of non-Euclidean geometry appears quite plainly also in the works of the German mathematicians F. K. SCHWEIKART (1780–1859) and F. A. TAURINUS (1794–1874). But only Lobachevski and Bolyai have made the systematic study of what we now call *Bolyai-Lobachevskian* (or *hyperbolic*) *geometry*.

5. Consistency of Geometry

Lobachevski was convinced that there were no inconsistencies in his geometry. To show the consistency he pointed out that between the formulas of his trigonometry and the formulas of the spherical trigonometry a one-to-one correspondence can be established. By means of this correspondence the problem of consistency of his geometry can, in principle, be reduced to the problem of consistency of spherical geometry which can be treated as a part of arithmetic. However a rigorous proof of consistency was an impossible task at that time since, on the one hand, the foundations of geometry were not yet sufficiently established, and on the other, such general notions as axiomatic theory and consistency were not yet precisely formulated and investigated. (These notions belong to the methodology of deductive sciences, the systematic development of which began only in the final years of the 19th century.)

In the period preceding the final axiomatic approach to geometry the basic idea of the precise proof of the consistency of Bolyai-Lobachevskian geometry was given in 1871 by the German mathematician Felix KLEIN (1849–1925), who, on the basis of the earlier ideas of the Italian mathematician Eugenio BELTRAMI (1835–1900), constructed in his work *Über die sogenannte Nicht Euklidische Geometrie*, Mathematische Annalen 4, the arithmetic model of Bolyai-Lobachevskian geometry.

It was only in the year 1899, however, that David HILBERT, in the work *Grundlagen der Geometrie*, gave the system of primitive notions and axioms of Euclidean geometry and a full proof of the consistency of this axiom system (under the assumption of the consistency of arithmetic). In 1903, in the work *Neue Begründung der Bolyai-Lobatchefskyschen Geometrie*, Mathematische Annalen 57, he proved the consistency of Bolyai-Lobachevskian geometry in a similar manner. Thus the two geometries, Euclidean and Bolyai-Lobachevskian, are equally correct from the standpoint of logic. The question of whether Euclidean or Bolyai-Lobachevskian geometry better describes real space can be settled, if at all, only by way of experiment. It seems, however, that experiment could at most confirm Bolyai-Lobachevskian geometry, but not Euclidean geometry. This is because Euclidean geometry is, in a sense, the limiting case of Bolyai-Lobachevskian geometry, and by means of experiments based only on approximate measurements, we cannot distinguish the limiting case from a very close approximation.

In the same period, and independently of Hilbert, investigations were made in Italy on the foundations of geometry. In particular Mario PIERI in his works *Della geometria elementare come sistema ipotetico-deduttivo*: *Monografia del punto e del moto*, Memorie della R. Accademia delle Scienze di Torino, 1899, and *La geometria elementare instituita sulle nozioni di "punto" e "sfera"*, Memorie di Matematica e di Fisica della Societa Italiana delle Scienze, ser. 3,15, 1908, published two axiomatic systems of Euclidean geometry each of which is based upon only one primitive notion.

6. Riemann Spaces

Simultaneously with the foundations of elementary and Bolyai-Lobachevskian geometry, two other branches of geometry, *differential geometry* and *projective geometry* were cultivated. The development of differential geometry led the German mathematician Bernhard RIEMANN (1826–1866) to introduce (in 1854) a very general class of spaces now called *Riemann spaces*.

Among the general Riemann spaces there stand out, in particular, *spaces with constant curvature* embracing the *parabolic type*, corresponding to the Euclidean space, the *hyperbolic type*, corresponding to the Bolyai-Lobachevskian space, and finally, the *elliptic type*, corresponding to the projective space with suitably chosen metric.

7. Axiomatic Theory

The purpose of axiomatic theory is to approach reality in an abstract form so as to permit the highest possible degree of rigor. In constructing such a theory, the following procedure is adopted:

First of all, a certain system of *primitive notions* is chosen. It is desirable that these notions have the clearest possible intuitive sense. It is taken as a principle that one may employ other notions only when they are defined in terms of the primitive notions, either directly or indirectly by means of previously defined notions.

Next, a certain system of sentences, called *axioms*, is chosen in which there are formulated some properties of the primitive notions. The axioms should state in abstract form some relations holding between the real objects from which the primitive notions were abstracted. It is desirable that the intuitive sense of the axioms not give rise to any doubts. The *theorems* of the theory are the axioms, and those sentences which are logical consequences of the axioms, definitions, and the theorems previously proved.

The geometry in this book will be based on a system of primitive notions and axioms which is a modification of the Hilbert system.

In constructing an axiomatic theory T we usually make use of other axiomatic theories, which are *presupposed* in the following sense: all the primitive notions of those presupposed theories are included in the system of primitive notions of T, and all the axioms of those theories are included in the axiom system of T. Mathematical theories presuppose as a rule mathematical logic and usually also set theory (to a larger or smaller extent). In developing geometry in this book we presuppose mathematical logic, set theory and the arithmetic of real numbers (which can either be treated as an independent theory or can be constructed as a portion of set theory). An axiomatic treatment of these three theories can be found in various special works. We shall not list here the primitive notions and axioms of these theories. In the next section, however, we shall discuss briefly all the basic set-theoretical notions which are relevant to our discussion.

8. Sets and Relations

The basic notion of set theory is that of *membership*. The membership symbol $\in$ occurs in formulas like $x \in X$, which is read *x is an element* (or *a member*) *of the set X*, or *x belongs to the set X*, or, finally, *set X contains x (as an element)*. The set with no elements will be denoted by 0 and will be called the *empty set*. The set consisting of elements $x_1, x_2, \ldots, x_n$ will be denoted by $\{x_1, x_2, \ldots, x_n\}$. If the elements of a set X are sets themselves we refer to set X also as a *family of sets* or a *class*.

If every element of a set X is also an element of a set Y we write $X \subset Y$ and we say that *set X is included in set Y* or *set Y includes set X (as a part)* or *set X is a subset of set Y*. We have:

$$X \subset X,$$

$$\text{if } X \subset Y \text{ and } Y \subset Z, \text{ then } X \subset Z,$$

$$\text{if } X \subset Y \text{ and } Y \subset X, \text{ then } X = Y.$$

For any class $\mathfrak{X}$, the *set-theoretical sum* (or *union*) of the sets of class $\mathfrak{X}$ is the set consisting of all elements which belong to at least one of the sets of $\mathfrak{X}$, while the *set-theoretical product* (or *intersection* or *common part*) of the sets of class $\mathfrak{X}$ is the set consisting of all elements which belong to each of the sets of $\mathfrak{X}$. If $\mathfrak{X} = \{X, Y\}$, then the sum of the sets of class $\mathfrak{X}$ is denoted by $X \cup Y$, and their product by $X \cap Y$. In the case $X \cap Y = 0$ we say that sets X and Y are *disjoint*. If to each natural number n there corresponds a set X_n, then the sum of sets $X_1, X_2, \ldots$ is denoted by $\overset{\infty}{\underset{n=1}{\cup}} X_n$ and the product by $\overset{\infty}{\underset{n=1}{\cap}} X_n$. By the *difference* $X - Y$ we understand the set composed of all elements belonging to X, but not belonging to Y. In case z is an element (but not a set) we shall write $X \cup z$ instead of $X \cup \{z\}$, $X - z$ instead of $X - \{z\}$, and $X \cap Y = z$ instead of $X \cap Y = \{z\}$.

Let us now assume that all the sets under consideration are subsets of some fixed set E, and let us denote by X' the set $E - X$ (called the *complement* of set X to set E). We then have:

$$X \cup 0 = X, \qquad\qquad X \cap E = X,$$
$$X \cup X' = E, \qquad\qquad X \cap X' = 0,$$
$$X \cup Y = Y \cup X, \qquad\qquad X \cap Y = Y \cap X,$$
$$X \cup (Y \cup Z) = (X \cup Y) \cup Z, \qquad X \cap (Y \cap Z) = (X \cap Y) \cap Z,$$
$$X \cup (Y \cap Z) = (X \cup Y) \cap (X \cup Z), \quad X \cap (Y \cup Z) = (X \cap Y) \cup (X \cup Z),$$

and

$$X \cup X = X, \qquad\qquad X \cap X = X,$$
$$X \cup E = E, \qquad\qquad X \cap 0 = 0,$$
$$(X \cup Y)' = X' \cap Y', \qquad (X \cap Y)' = X' \cup Y',$$
$$X'' = X,$$
$$\text{if } X \subset Y, \text{ then } Y' \subset X'.$$

Somewhat less obvious is the following statement:

$$\text{if } X \cap Y \cap Z = 0, \text{ then } X \cup Y \cup Z = (X \cap Y') \cup (Y \cap Z') \cup (Z \cap X').$$

We adopt the convention that the symbol $\cap$ binds more strongly than the symbol $\cup$. Thus, from now on, the parentheses in expressions of the form $(X \cap Y) \cup Z$ and $X \cup (Y \cap Z)$ will always be omitted.

We assume that for any two elements x and y we can construct an *ordered pair* (x,y) which, as opposed to non-ordered pair $\{x,y\}$, uniquely determines its *first term* x and its *second term* y; thus if $(x,y) = (x',y')$, then $x = x'$ and $y = y'$. (We can, e.g., define (x,y) by putting $(x,y) = \{\{x\}, \{x,y\}\}$.)

By a *two-termed* (or *binary*) relation we understand an arbitrary set of ordered pairs. If R is a two-termed relation, instead of $(x,y) \in R$ we write xRy and read *the relation R holds between x and y* or x *is in the relation R to y*. We shall write $x \sim R y$ to express the fact that x *is not in relation R to y*, and $x_1, x_2\, Ry$ instead of $x_1 Ry$ and $x_2 Ry$. The set of those x for which there is a y such that xRy is called the *domain* of relation R, and the set of those y for which there is an x such that xRy is called the *counter-domain* or *range* of relation R. The sum of the domain and the counter-domain is called the *field* of relation R.

For every relation R there exists the *inverse* relation $\breve{R}$ defined by the condition

$$x\breve{R}y \text{ if and only if } yRx.$$

Let the set X be included in the field of relation R. The relation R is said to be *reflexive* in set X if for every element $x \in X$ we have xRx. The relation R is said to be *symmetric* in set X if for every two elements

$x,y \in X$ it follows from xRy that yRx, and *antisymmetric* in set X if for every two elements $x,y \in X$ it follows from xRy that $y \sim Rx$. We say that relation R is *transitive* in set X if for every three elements $x,y,z \in X$ it follows from xRy and yRz that xRz. Relation R is said to be *connected* in set X if for every two distinct elements $x,y \in X$ we have xRy or yRx.

Relations which are reflexive, symmetric and transitive in set X are called *equivalences* in X. If R is an equivalence in X, then X can be uniquely represented as the sum of pairwise disjoint subsets in such a way that two elements $x,y \in X$ belong to the same subset if and only if xRy. These subsets are called *equivalence classes* of set X with respect to relation R.

We say that relation R *partially orders* set X if it is antisymmetric and transitive in X. If relation R is, in addition, connected in X, we say that R *orders* (or *simply orders*) set X. If relation R orders set X, then every element $x_0 \in X$ such that x_0Ry for every $y \in X - x_0$ is called the *first element* of X, and every element $y_0 \in X$ such that xRy_0 for every $x \in X - y_0$ is called the *last element* of X with respect to the relation R.

Let us assume that relation R orders set X. An ordered pair of sets (X_1, X_2) we call a *Dedekind cut* of set X if it satisfies the following conditions:

$$X = X_1 \cup X_2, \quad X_1, X_2 \neq 0,$$

and for any arbitrary elements x_1 and x_2 of set X,

$$\text{if } x_1 \in X_1 \text{ and } x_2 \in X_2, \text{ then } x_1Rx_2.$$

We say that set X has the *Dedekind property* if for every Dedekind cut (X_1, X_2) of set X just one of the following two conditions is satisfied: Either there exists in set X_1 a last element with respect to relation R or there exists in set X_2 a first element with respect to relation R.

We say that a set Y included in set X is *dense* in X with respect to relation R if for every two elements $x,z \in X$ such that xRz there exists an element $y \in Y$ such that xRy and yRz. If relation R orders set X and set X is dense in itself with respect to R, then we say that *relation R densely orders set X*.

If relation R with the domain X and the counterdomain Y has the property that for every element $x \in X$ there exists just one element $y \in Y$ such that xRy, then we call relation R a *function* defined in set X. Instead of xRy, we then write $y = R(x)$ and call y the *value* of function R for the *argument* x or the *R-image* of argument x. We also call X the *set of arguments*, Y the *set of values* of function R and we say that R *maps X onto Y*. If Y is a subset of a set Z, then we say that function R *maps X into Z*.

Instead of the term *function*, we shall also use the term *transformation*.

If X_0 is a subset of the domain X of a function f, then by function f with the *domain restricted to* X_0 we understand the function f_0 defined only in set X_0 by the equality

$$f_0(x) = f(x) \text{ for every } x \in X_0.$$

It is sometimes convenient to extend a function f, defined in a set X, to the class of all subsets of X, in which case, for any $Z \subset X$ we take $f(Z)$ to be the set of all $f(x)$ for $x \in Z$.

If a function f always assigns to different arguments x_1 and x_2 different values $y_1 = f(x_1)$ and $y_2 = f(x_2)$, then we say that it is *univalent*, or *one-to-one*, or *biunique*. In this case the *inverse function* f^{-1}, which assigns the argument x to every value $y = f(x)$, is defined. Let X and Y be the domain and range of a univalent function f; we then say that f *maps in one-to-one way X onto Y*, or that f *establishes a one-to-one correspondence between the elements of X and the elements of Y*.

If a function f defined in a set X takes on the values belonging to a set Y, and a function g defined in set Y takes on the values belonging to a set Z, then the formula

$$h(x) = g(f(x))$$

defines in set X a function h with values belonging to set Z. The function h is called the *superposition* of functions f and g. We shall denote it by gf.

A non-empty set of one-to-one transformations of a set X onto itself forms a *group of transformations* if (*i*) together with transformation f, it contains the inverse transformation f^{-1}, and (*ii*) together with transformations f and g, it contains the superposition gf.

Two sets X and Y are of the *same power* if there exists a function which maps X onto Y in one-to-one way. In particular, sets of the same power as the set of all natural numbers are called *denumerable*. Sets which are not finite nor denumerable are called *non-denumerable*. In particular, the set of all real numbers is non-denumerable; the sets which are of the same power as the set of all real numbers are said to have the *power of the continuum*.

We may regard *n-termed sequences* $(y_1, y_2, \ldots, y_n)$ and *infinite sequences* $(y_1, y_2, \ldots, y_k, \ldots)$ as special cases of functions. The n-termed sequence $(y_1, y_2, \ldots, y_n)$ can be regarded as a function f defined by the formula $f(k) = y_k$ in the set $\{1, 2, \ldots, n\}$, the infinite sequence $(y_1, y_2, \ldots, y_k, \ldots)$ — as a function defined by the same formula in the set of all natural numbers. For brevity, the infinite sequence $(y_1, y_2, \ldots, y_k, \ldots)$ will sometimes be denoted by (y_k).

By generalizing the notion of a two-termed relation we introduce that of an *n-termed relation*. In fact, we define an n-termed relation as an arbitrary set of n-termed sequences (and thus, we identify ordered pairs with two-termed sequences). We shall write $R(x_1,x_2,\ldots,x_n)$ to express the fact that the relation R holds among the elements $x_1,x_2,\ldots,x_n$. The set of those elements x_1 for which there are elements $x_2,\ldots,x_n$ such that $R(x_1,x_2,\ldots,x_n)$ is called the *first domain* of the relation R. The *second, third,...*, *nth domain* of the relation R are defined analogously. The sum of all these domains is called the *field* of relation R.

Let R be an $(n+1)$-termed relation. Let us assume for simplicity that each of the first n domains of R is identical with a set X, and let the $(n+1)$th domain of R be a set Y. If for any elements $x_1,x_2,\ldots,x_n \in X$ there is just one element $y \in Y$ such that $R(x_1,x_2,\ldots,x_n,y)$, then relation R is called a *function of n variables* defined in set X. We then call y the *value of the function R for the arguments* $x_1,x_2,\ldots,x_n$ and instead of $R(x_1,x_2,\ldots,x_n,y)$ we write $y = R(x_1,x_2,\ldots,x_n)$.

9. Topological Space

Topology is the discipline dealing with the notion of a *topological space*. This notion can be conceived in a more or less general way. For our purposes the following definition, originating essentially with HAUS-DORFF, proves to be convenient. We say that *a set E (whose elements are called points) is a topological space with the base* $\mathfrak{U}$ *of neighborhoods*, if $\mathfrak{U}$ is a family of subsets of E satisfying the following two conditions:

(I) If $x_1,x_2 \in E$ and $x_1 \neq x_2$, then there are sets $U_1,U_2 \in \mathfrak{U}$ such that $x_1 \in U_1$, $x_2 \in U_2$, and $U_1 \cap U_2 = 0$.

(II) If $x \in U_1 \cap U_2$ and $U_1,U_2 \in \mathfrak{U}$, then there is a set $U_0 \in \mathfrak{U}$ such that $x \in U_0$ and $U_0 \subset U_1 \cap U_2$.

Under the same conditions we say that $\mathfrak{U}$ *defines a topology in E*.

With the help of the notion of the base $\mathfrak{U}$ of neighborhoods we can define topological notions in space E. In particular:

Every subset of set E containing at least one set $U \in \mathfrak{U}$ such that $x \in U$ is called a *neighborhood* of the point $x \in E$.

A set $G \subset E$ which is a neighborhood for each of its points is said to be *open*. A set $F \subset E$ such that $E-F$ is an open set is said to be *closed*. In particular, the empty set 0 and the entire space E are both open and closed. The sum of any class of open sets is an open set, and the product of any class of closed sets is a closed set. The sum of a finite number of closed sets is a closed set, and the product of a finite number of open sets is an open set.

A point $a \in E$ is called a *point of accumulation* of a set $X \subset E$ if each of the neighborhoods of point a contains infinitely many points of set X. By adding to the set X all its points of accumulation, we obtain a set $\bar{X}$ called the *closure* of set X. The closure $\bar{X}$ of set X can also be defined as the product of all closed sets containing set X. In order for set X to be closed it is necessary and sufficient that $X = \bar{X}$. For every set $X \subset E$

$$X \subset \bar{X}, \quad \bar{\bar{X}} = \bar{X};$$

for every two sets $X, Y \subset E$

$$\text{if } X \subset Y, \text{ then } \bar{X} \subset \bar{Y},$$

$$\overline{X \cup Y} = \bar{X} \cup \bar{Y}, \quad \overline{X \cap Y} \subset \bar{X} \cap \bar{Y}.$$

The set $\omega(X) = X - \overline{E-X}$ is called the *interior* of set $X \subset E$, and its points—*interior points* of set X. For every $X \subset E$

$$\omega(X) \subset X, \quad \omega(\omega(X)) = \omega(X);$$

for every two sets $X, Y \subset E$

$$\omega(X \cup Y) \supset \omega(X) \cup \omega(Y), \quad \omega(X \cap Y) = \omega(X) \cap \omega(Y).$$

Given a set $X \subset E$, we call the set $\bar{X} \cap \overline{E-X}$ the *boundary* of set X.

A set $X \subset E$ is said to be *connected* if for every decomposition into the sum of two non-empty sets $X = X_1 \cup X_2$, we have

$$\bar{X}_1 \cap X_2 \cup X_1 \cap \bar{X}_2 \neq 0.$$

For example, the set of all real numbers with the class of all open intervals as the base of neighborhoods is a connected set. If set X is a connected set and $X \subset Y \subset \bar{X}$, then set Y is also a connected set.

Given a set $X \subset E$, we say that X *is dense in E* if there is a point of set X in every open and non-empty set $G \subset E$.

We say that the sequence (x_n) of points of space E *tends to the limit* $x_0 \in E$ —in symbols, $x_0 = \lim\limits_{x \to \infty} x_n$— if for every neighborhood U of point x_0 the condition $x_n \in U$ is satisfied for almost all indices n. If all points of the sequence (x_n) having the limit x_0 belong to a closed set X, then $x_0 \in X$.

Let E_1 and E_2 be two topological spaces and let a function f map space E_1 into space E_2. We say that function f is *continuous at the point* $x_0 \in E_1$ if for every neighborhood V of the point $f(x_0)$ (in space E_2) there exists a neighborhood U of point x_0 (in space E_1) such that the values which function f assigns to arguments belonging to U belong to V. A function that is continuous at each of its arguments is said, simply, to be *continuous*. We define now the continuity of a function of n variables. Let us assume for simplicity that all the arguments $x^{(1)}, x^{(2)}, \ldots, x^{(n)}$ of function f lie in the same space E_1 and the values, in the space E_2. We say that function f is *continuous* at $(x_0^{(1)}, x_0^{(2)}, \ldots, x_0^{(n)})$ if for every neighborhood V of point $f(x_0^{(1)}, x_0^{(2)}, \ldots, x^{(n)})$ in space E_2 there exist neighborhoods U_i of points $x_0^{(i)}$ ($i = 1, 2, \ldots, n$) in space E_1 such that the values which function f takes on for arguments $x^{(1)}, x^{(2)}, \ldots, x^{(n)}$, belonging respectively to neighborhoods $U_1, U_2, \ldots, U_n$, lie in neighborhood V.

Given a biunique function f mapping a topological space E_1 onto a topological space E_2, if functions f and f^{-1} are both continuous, we say that f is a *bicontinuous* function or a *homeomorphism*. We say that *spaces E_1 and E_2 are homeomorphic*, if there is a homeomorphism mapping E_1 onto E_2. A set homeomorphic with a closed (or open) interval of real numbers is called a *closed* (or *open*) arc.

It is plain that all the topological notions introduced depend on the way the base of neighborhoods is chosen. We say that *two bases of neighborhoods* in *set E, base $\mathfrak{U}$ and base $\mathfrak{B}$, are equivalent* if the identity transformation f ($f(x) = x$ for every $x \in E$) maps homeomorphically the topological space determined in E by $\mathfrak{U}$ onto the topological space determined in E by $\mathfrak{B}$. In order for bases $\mathfrak{U}$ and $\mathfrak{B}$ to be equivalent it is necessary and sufficient that for every set $U \in \mathfrak{U}$ and for every point $x \in U$ there exists a set $V \in \mathfrak{B}$ such that $x \in V \subset U$, and conversely, for every set $V \in \mathfrak{B}$ and for every point $x \in V$ there exists a set $U \in \mathfrak{U}$ such that $x \in U \subset V$.

If $\mathfrak{U}$ is a base of neighborhoods for space E and if $E_0 \subset E$, then the class $\mathfrak{U}_0$ consisting of all sets $U_0 = U \cap E_0$, where $U \in \mathfrak{U}$, constitutes the base of neighborhoods in set E_0. We say that the topology defined in set E_0 by the base $\mathfrak{U}_0$ is *induced* by the topology defined in set E by the base $\mathfrak{U}$.

A set E is called a *metric space* if to each pair of elements $x, y \in E$ is assigned a real number $\varrho(x,y) \geqslant 0$, called the *distance* between x and y, satisfying the following three conditions:

(i) $\varrho(x,y) = 0$ *if and only if* $x = y$,

(ii) $\varrho(x,y) = \varrho(y,x)$,

(iii) $\varrho(x,y) + \varrho(y,z) \geqslant \varrho(x,z)$,

for all elements $x, y, z \in E$.

We call the function ϱ the *metric* in space E.

Given a set $X \subset E$; by the *diameter* of set X we understand the upper bound of the numbers $\varrho(x,y)$, where x and y are any arbitrary points of set X.

Given a point x_0 of a metric space E, we shall refer to the set of all points $x \in E$ for which $\varrho(x,x_0) < \varepsilon$ as the *ε-neighborhood* of point x_0. The class of ε-neighborhoods for all $\varepsilon > 0$ constitutes a base of neighborhoods. In speaking about the topology of metric space E we shall always have in mind the topology defined by the base of ε-neighborhoods. The condition $\lim_{n \to \infty} x_n = x_0$ is then equivalent to the condition $\lim_{n \to \infty} \varrho(x_n, x_0) = 0$.

In order that a function f transforming metric space E_1 with a metric ϱ_1 onto metric space E_2 with a metric ϱ_2 be continuous at point x_0, it is necessary and sufficient that the following *condition of Cauchy* be satisfied:

For every $\varepsilon > 0$ there exists an $\eta > 0$ such that if $\varrho_1(x, x_0) < \eta$, then $\varrho_2(f(x), f(x_0)) < \varepsilon$.

This condition is equivalent to the following *condition of Heine*:

If $\lim_{n \to \infty} x_n = x_0$, then $\lim_{n \to \infty} f(x_n) = f(x_0)$.

If a function f mapping metric space E_1 with metric ϱ_1 onto metric space E_2 with metric ϱ_2 has the property that $\varrho_1(x, y) = \varrho_2(f(x), f(y))$ for any two elements $x, y, \in E_1$, then we say that function f is an *isometric transformation* or an *isometry*. We say that *spaces E_1 and E_2 are isometric* if there exists an isometry mapping space E_1 onto space E_2. All isometric transformations of a metric space E onto itself form a group of transformations.

Let J be any closed arc in a metric space E. It is clear that for every $\varepsilon > 0$ there exists a decomposition of arc J into the sum of a finite number of closed arcs with diameters $< \varepsilon$.

Let J again be any closed arc in a metric space E and let f be a homeomorphism mapping a closed interval of real numbers $<\alpha, \beta>$ onto J. Let us consider all finite sequences $Z = (f(t_0), f(t_1), \ldots, f(t_n))$ where $\alpha = t_0 < t_1 < \ldots < t_n = \beta$. The least upper bound of the numbers

$$|Z| = \sum_{i=1}^{n} \varrho(f(t_{i-1}), f(t_i))$$

will be called the *length of arc J* and will be denoted by $|J|$. The number $|J|$ is positive (it may also be equal to $+\infty$) and does not depend on the way the interval $<\alpha, \beta>$ and the homeomorphism f are chosen, but only on the arc J. If arcs J_1 and J_2 are isometric, then $|J_1| = |J_2|$. If arc J is the sum of two arcs J_1 and J_2, then $|J| \leqslant |J_1| + |J_2|$; if, moreover, arcs J_1 and J_2 have disjoint interiors, then $|J| = |J_1| + |J_2|$. If the number $|J|$ is finite, we say that arc J is *rectifiable*. In this case every arc $J' \subset J$ is also rectifiable. In particular, if $\alpha < t < \beta$, then the closed arc J_t, being the f-image of the interval $<\alpha, t>$, is rectifiable. The length $|J|$ of arc J depends continuously on the number t, hence also on the point $p = f(t)$.

10. Analytic Geometry

A close connection between geometry and arithmetic of real numbers is clearly exhibited in analytic geometry. Analytic geometry is sometimes treated as that part of axiomatic geometry in which properties of the space are studied by means of a coordinate system (i.e., by means of a one-to-one correspondence between points of the space and finite sequences of numbers). Analytic geometry could also be constructed quite independently of axiomatic systems of geometry and, in fact, as a theory in which finite sequences of numbers are regarded as points and all geometrical notions are defined in a purely arithmetic way. Analytic geometry so conceived becomes simply a part of arithmetic. In a book devoted to the foundations of geometry one of the principle tasks is to give a proof of the consistency, or more accurately, to reduce the problem of the consistency of geometry to the problem of the consistency of arithmetic. In this proof it is convenient to employ analytic geometry regarded as a branch of arithmetic. Such an approach to analytic geometry can be found, for instance, in the book by K. Borsuk, *Geometria analityczna*, Warsaw, 1950 (in Polish).

Following the terminology used in this book, we shall understand *n-dimensional Cartesian space* $\mathbf{C}_n$ (we restrict ourselves here to the cases $n = 1,2,3$ only) to be the metric space whose points are all *n*-termed sequences $(x_1, x_2, \ldots, x_n)$ of real numbers and in which the distance ρ between two points $x = (x_1, x_2, \ldots, x_n)$ and $y = (y_1, y_2, \ldots, y_n)$ is defined by the formula

$$\rho(x,y) = \sqrt{(x_1 - y_1)^2 + (x_2 - y_2)^2 + \ldots + (x_n - y_n)^2}.$$

By the *analytic geometry of space* $\mathbf{C}_n$ we shall understand the theory of invariants of isometric transformations of this space onto itself.

By the *n-dimensional projective space* $\mathbf{P}_n$ (we restrict ourselves here to the cases $n=1,2,3$ only) we understand the space whose points are equivalence classes of the set of all $n+1$-termed sequences $(x_0, x_1, \ldots, x_n)$, provided $x_0, x_1, \ldots, x_n$ run over real numbers not equal simultaneously to zero, and the equivalence relation is that of proportionality. The point with *homogeneous coordinates* $x_0, x_1, \ldots, x_n$, i.e. the equivalence class containing the sequence $(x_0, x_1, \ldots, x_n)$, will be denoted by $[x_0, x_1, \ldots, x_n]$. By the *analytic geometry of space* $\mathbf{P}_n$ we shall understand the theory of invariants of linear transformations of space $\mathbf{P}_n$ onto itself, that is transformations mapping the point $[x_0, x_1, \ldots, x_n]$ onto the point $[y_0, y_1, \ldots, y_n]$, where

$$y_i = \alpha_{i,0} x_0 + \alpha_{i,1} x_1 + \ldots + \alpha_{i,n} x_n \text{ for } i = 0,1,\ldots,n$$

and where the coefficients $\alpha_{i,j}$ are real constants whose determinant $|\alpha_{i,j}|$ $(i,j = 0,1,\ldots,n)$ differs from zero.

We do not introduce a metric in space P_n, but we regard P_n as a topological space in which the base of neighborhoods coincides with the class of all sets U consisting of points $[x_0, x_1, \ldots, x_n]$ such that x_i $(i = 1, 2, \ldots, n)$ runs over an open interval I_i of real numbers, provided at least one of the intervals $I_0, I_1, \ldots, I_n$ does not contain the number 0.

In Chapter IV we shall present a brief sketch of the theory of the spaces C_n and P_n. The reader will find there all the elements of analytic geometry employed in this book.

PART ONE

EUCLIDEAN AND BOLYAI-LOBACHEVSKIAN
GEOMETRY

Introduction to Part One

Primitive Notions and Axioms

As the primitive notions of geometry we take a set **S**, two classes
𝔖𝔏 and 𝔓𝔏 of subsets of **S**, and two relations, a three-termed relation **B**
and a four-termed relation **E**, among elements of **S**.

Set **S** will be called *space*, the elements of **S** — *points*. We shall denote
points by the letters *a, b, c, d, e, p, q, r, s*. Instead of writing *a* ∈ **S**, we
shall also write: *a is a point*.

Elements of class 𝔖𝔏 will be called *straight lines*, or, for the sake of
brevity, simply *lines*. We shall denote lines by the letters *K, L, M, N*.
Henceforth, instead of *L* ∈ 𝔖𝔏, we shall write: *L is a line*. Elements
of class 𝔓𝔏 will be called *planes*. We shall denote planes by the letters
P, Q. As in the case of lines, instead of *P* ∈ 𝔓𝔏 we shall write: *P is a plane*.

The formula *p* ∈ *L*, which says that point *p* belongs to line *L*, may
also be read: *point p lies on line L* or *line L passes through point p*.
Similarly, the formula *p* ∈ *P* may be read: *point p lies on plane P* or
plane P passes through point p.

Sets of points will be called *figures*. Thus, lines and planes are special
cases of figures. The formula *F* ⊂ *L*, which says that figure *F* is included
in line *L*, will also be read: *figure F lies on line L*. Similarly, the formula
F ⊂ *P* will be read: *figure F lies on plane P*. The formula *L* ⊂ *P* will
also be read: *plane P passes through line L*.

The relation **B** is called the *betweenness relation*. The formula **B**(*a,b,c*)
is to be read: *point b lies between points a and c*.

The relation **E** is called the *equidistance relation*. The formula
E(*a,b,c,d*) is to be read: *point a is just as far from point b as point c is
from point d*.

We divide the system of axioms of Euclidean geometry as well as the
system of axioms of Bolyai-Lobachevskian geometry into five groups,
four of which are common to both geometries. The first group consists
of nine *axioms of incidence* I1—I9; these express the set-theoretical
relations between points, lines and planes. The second group consists
of nine *axioms of order* O1—O9; these concern the relation **B**. The third
group consists of eight *axioms of congruence* C1—C8; these establish the
fundamental properties of the relation **E**. The *Axiom of Continuity* Co

constitutes the fourth group. The fifth group of axioms also consists of one axiom. In the case of Euclidean geometry this is the Axiom E, which in this book we shall call the *Axiom of Euclid*, since it is equivalent to Euclid's Fifth Postulate. In the case of Bolyai-Lobachevskian geometry, it will be Axiom BL, which in this book will be called the *Axiom of Bolyai-Lobachevski;* it is the negation of the Axiom of Euclid.

The theory based on the first four groups of axioms will be referred to as *absolute geometry*. Thus, absolute geometry consists of all theorems which hold for both Euclidean and Bolyai-Lobachevskian geometry.

In the study of geometry as an axiomatic theory, we are interested not only in the geometrical theorems themselves, but also in the means necessary to prove them. We are interested, for example, in seeing which theorems of Euclidean geometry belong to absolute geometry, and which, therefore, can be proved without the Axiom of Euclid, or which theorems of absolute geometry can be proved without the Axiom of Continuity. In order to make this evident, we shall begin with the theorems which result from the axioms of incidence only. Then, having included the axioms of order, we develop that part of the geometry which is based on the axioms of incidence and axioms of order (Chapter I). In this way by gradually strengthening the axiom system by new groups of axioms, we come, in turn, to that part of absolute geometry which does not involve the Axiom of Continuity (Chapter II), to absolute geometry (Chapter III), and finally to Euclidean and Bolyai-Lobachevskian geometry (Chapter V and VI).

I

Axioms of Incidence and Order

1. Axioms of Incidence

We begin with several definitions. We shall say that points a, b, c are *collinear* if there exists a line L such that $a, b, c \in L$; if this is not the case, we shall say that the points a, b, c are *non-collinear*. Analogously, we shall say that the points a, b, c, d are *coplanar* if there exists a plane P such that $a, b, c, d \in P$; if this is not the case, we shall say that the points a, b, c, d are *non-coplanar*.

The axioms of incidence have the following form:

AXIOM I1. *For any line L there exist (two) distinct points a and b such that $a, b \in L$.*

AXIOM I2. *For any points a and b there exists at least one line L such that $a, b \in L$.*

AXIOM I3. *If points a and b are distinct, then there exists at most one line L such that $a, b \in L$.*

AXIOM I4. *For any plane P there exist (three) non-collinear points a, b, c such that $a, b, c \in P$.*

AXIOM I5. *For any points a, b, c there exists at least one plane P such that $a, b, c \in P$.*

AXIOM I6. *If points a, b, c are non-collinear, then there exists at most one plane P such that $a, b, c \in P$.*

AXIOM I7. *For any line L and any plane P, if there exist two distinct points a and b such that $a, b \in L$ and $a, b \in P$, then $L \subset P$.*

AXIOM I8. *For any planes P and Q, if there exists a point a such that $a \in P$ and $a \in Q$, then there exists a point b distinct from a and such that $b \in P$ and $b \in Q$.*

AXIOM I9. *There exist (four) non-coplanar points a, b, c, d.*

The meaning of Axioms I1—I9 does not require explanation. Axioms I1—I4 express facts well-known from planimetry, and Axioms I5—I9, from stereometry. We shall call Axioms I1—I4 the *plane axioms*, and Axioms I5—I9 the *space axioms*.

2. Three Non-Collinear Points and Four Non-Coplanar Points

The following theorems establish some relations between distinct, non-collinear, and non-coplanar points.

THEOREM 1. *If points a, b, c are non-collinear, then any two of them are distinct.*

PROOF. Let us assume that points a, b, c are non-collinear. Now suppose that two of them are identical, for example, $a = b$. By Axiom I2, there exists a line L such that $b, c \in L$. Since $a = b$, then $a \in L$. As a result, points a, b, c are collinear, which contradicts our assumption. Therefore points a and b must be distinct.

In an analogous manner, the following theorem results from Axiom I5:

THEOREM 2. *If points a, b, c, d are non-coplanar then any two of them are distinct.*

This theorem can be strengthened as follows:

THEOREM 3. *If points a, b, c, d are non-coplanar, then any three of them are non-collinear.*

PROOF. Let us assume that points a, b, c, d are non-coplanar, and let us suppose that three points, say a, b, c are collinear. Then there exists a line L such that $a, b, c \in L$. On the other hand, by Axiom I5 there exists a plane P such that $a, b, d \in P$. As a result, $a, b \in L$ and $a, b \in P$, and, by Theorem 2, points a and b are distinct. Thus because of Axiom I7, we have $L \subset P$ and, in particular, $c \in P$. Therefore, points a, b, c, d are coplanar, in contradiction to our assumption. Thus points a, b, c must be non-collinear.

From Axioms I3 and I6 there result the following sufficient conditions that three points be non-collinear and that four points be non-coplanar.

THEOREM 4. *Given on a line L two distinct points a and b, if $c \sim \in L$, then points a, b, c are non-collinear.*

THEOREM 5. *Given on a plane P three non-collinear points a, b, c, if $d \sim \in P$, then points a, b, c, d are non-coplanar.*

3. Lines and Planes

From Axioms I2 and I3 we obtain at once

THEOREM 6. *Just one line passes through two distinct points.*

The line determined by points a and b ($a \neq b$) will be referred to as *line ab* and will be denoted in the formulas by $\mathbf{L}(ab)$.

Analogously, from Axioms I5 and I6 we obtain

THEOREM 7. *Just one plane passes through three non-collinear points.*

The plane determined by (non-collinear) points a, b, c will be referred to as *plane abc* and will be denoted in the formulas by $\mathbf{P}(abc)$.

The following two theorems give additional ways of determining a plane:

THEOREM 8. *Just one plane passes through a line L and a point c not lying on L.*

PROOF. By Axiom I1, there are two distinct points a and b lying on line L. The plane we are seeking must therefore pass through points a, b and c. Because of Theorem 4, points a, b, c are non-collinear, and by Theorem 7, there exists exactly one plane P such that a, b, $c \in P$. Since points a and b lie on plane P, then, by Axiom I7, so does the entire line L.

We shall call the plane determined by line L and point c (not lying on L) the *plane Lc* and denote it in formulas by $\mathbf{P}(Lc)$.

THEOREM 9. *Just one plane passes through two intersecting and distinct lines.*

PROOF. Let us assume that lines K and L intersect each other and are distinct. There then exists a point a such that $a \in K \cap L$. By Axiom I1, a point c, distinct from a, lies on line L. The plane we are seeking must pass through line K and through point c. But since $K \neq L$, we have $c \sim \in K$, and, by Theorem 8, there exists only one plane P such that $K \subset P$ and $c \in P$. Since along with points a and c the entire line L lies on plane P, this is the plane we are seeking.

We shall call the plane determined by lines K and L ($K \neq L$) *plane KL* and denote it in formulas by $\mathbf{P}(KL)$.

4. Fundamental Existence Theorems

By Axiom I9 and with the help of Theorems 2 and 3, it follows that
(i) there exist at least four distinct non-coplanar points a, b, c, d (Fig. 1);
(ii) there exist at least six distinct lines: $\mathbf{L}(ab)$, $\mathbf{L}(ac)$, $\mathbf{L}(ad)$, $\mathbf{L}(bc)$, $\mathbf{L}(bd)$, $\mathbf{L}(cd)$; and
(iii) there exist at least four distinct planes: $\mathbf{P}(abc)$, $\mathbf{P}(abd)$, $\mathbf{P}(acd)$, $\mathbf{P}(bcd)$.

On the other hand, it can easily be shown that the existence of five distinct points does not follow from the axioms of incidence. Let x, y, z, t denote any four distinct objects (e.g. the natural numbers 1, 2, 3, 4).

Let us regard the objects x, y, z, t as points; all pairs $\{x,y\}$, $\{x,z\}, \dots$, as lines; all triples $\{x,y,z\}$, $\{x,y,t\}, \dots$, as planes. It is easy to show that all the axioms of incidence are satisfied by such an interpretation of points, lines, and planes. Therefore, if the existence of five distinct points were to follow from the axioms of incidence, then the set $\{x,y,z,t\}$ would also have to contain at least five elements. Indeed, this is not the case. †

Further existence theorems have a somewhat different character.

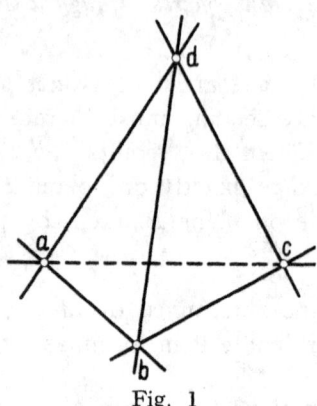

Fig. 1

They state, namely, that for any arbitrarily given points, lines, and planes there exist other points, lines, and planes related to them in some definite manner.

THEOREM 10. (I) *For any point $p \in L$ there exists a point $q \in L$ distinct from p.*

(II) *For any two distinct points p, $q \in P$ there exists a point $r \in P$ non-collinear with points p and q.*

(III) *For any three non-collinear points p, q, r there exists a point s non-coplanar with points p, q, r.*

PROOF. (I) By Axiom I1, two distinct points a and b lie on line L. At least one of them must be distinct from point p.

(II) By Axiom I4, three non-collinear points a, b, c lie on plane P. At least one of them, e.g. point c, does not lie on line pq. By Theorem 4, points p, q, c are non-collinear.

In a similar manner, the last part of the theorem can be proved by using Axiom I9 and Theorem 5.

THEOREM 11. (I) *For any point $p \in P$ there exist points $q, r \in P$ such that points p, q, r are non-collinear.*

---

† This argument does not belong to geometry but to what is called *metageometry*. The notion of *interpretation* is discussed in Chapter IV, Section 1.

(II) *For any two distinct points p and q there exist points r and s such that points p, q, r, s are non-coplanar.*

PROOF. (I) We note that, by Axiom I8 (for $Q = P$), a point $q \neq p$ lies on plane P. Now, making use of Theorem 10 (II), we obtain a point $r \in P$ non-collinear with points p and q.

(II) We produce any plane P through points p and q. By Theorem 10 (II), a point r non-collinear with points p and q lies on it. Making use of Theorem 10 (III), we obtain a point s non-coplanar with points p, q, r.

With the help of Theorems 10 and 11, the reader may easily prove the following existence theorems:

THEOREM 12. (I) *For any point $p \in P$ there exists a line $L \subset P - p$.*

(II) *Two distinct lines $K, L \subset P$ and a line $M \sim \subset P$ pass through any point $p \in P$.*

(III) *Three distinct lines not lying on one plane pass through any point.*

(IV) *A plane not including a line L passes through any point $p \in L$.*

(V) *For any line $L \subset P$ there exists a point $p \in P - L$.*

(VI) *For any line L there exists a skew line (i.e. a line not lying in the same plane as L).*

(VII) *Two distinct planes pass through any line.*

5. Intersections of Lines and Planes

As an immediate consequence of Axiom I3, we obtain

THEOREM 13. *The intersection of two distinct lines consists at most of one point.*

As an immediate consequence of Axiom I7, we have

THEOREM 14. *The intersection of a plane with a line not lying on it consists at most of one point.*

We shall prove

THEOREM 15. *Two distinct planes are either disjoint or their intersection is a line.*

PROOF. Let P and Q be two distinct planes. Suppose that $P \cap Q \neq 0$. There then exists a point a such that $a \in P \cap Q$. On the basis of Axiom I8, we then conclude that there exists a point $b \neq a$ such that $b \in P \cap Q$. Consider line ab. Obviously, $\mathbf{L}(ab) \subset P \cap Q$. We shall show that, conversely, $P \cap Q \subset \mathbf{L}(ab)$. Certainly, if for some point $c \in P \cap Q$ we were to have $c \sim \in \mathbf{L}(ab)$, then, by Theorem 8, we would have $P = Q$, contrary to our hypothesis. From $\mathbf{L}(ab) \subset P \cap Q$ and $P \cap Q \subset \mathbf{L}(ab)$, it follows that $P \cap Q = \mathbf{L}(ab)$.

6. Linear Axioms of Order

We now give the first eight axioms of order characterizing the primitive relation $\mathbf{B}$. We recall that the formula $\mathbf{B}(a,b,c)$ reads: *point b lies between points a and c.*

AXIOM O1. *If* $\mathbf{B}(a,b,c)$, *then points a, b, c are collinear and distinct.*

AXIOM O2. *If* $\mathbf{B}(a,b,c)$, *then* $\mathbf{B}(c,b,a)$.

AXIOM O3. *If* $\mathbf{B}(a,b,c)$, *then* $\sim \mathbf{B}(b,a,c)$.

AXIOM O4. *If points a, b, c are collinear and distinct, then* $\mathbf{B}(a,b,c)$ *or* $\mathbf{B}(b,c,a)$ *or* $\mathbf{B}(c,a,b)$.†

AXIOM O5. *If points a and b are distinct, then there exists a point c such that* $\mathbf{B}(a,b,c)$.

AXIOM O6. *If points a and b are distinct, then there exists a point c such that* $\mathbf{B}(a,c,b)$.

AXIOM O7. *If* $\mathbf{B}(a,b,c)$ *and* $\mathbf{B}(b,c,d)$, *then* $\mathbf{B}(a,b,d)$.

AXIOM O8. *If* $\mathbf{B}(a,b,d)$ *and* $\mathbf{B}(b,c,d)$, *then* $\mathbf{B}(a,b,c)$.

Axioms O1—O8 concern the properties of the relation of betweenness for points lying on a line. That is why we call them *linear* axioms of order. In Section 19 we shall add a *plane* axiom of order to these linear axioms.

We supplement Axiom O7 by

THEOREM 16. *If* $\mathbf{B}(a,b,c)$ *and* $\mathbf{B}(b,c,d)$, *then* $\mathbf{B}(a,c,d)$.

PROOF. From $\mathbf{B}(a,b,c)$ and $\mathbf{B}(b,c,d)$ it follows, by Axiom O2, that

(1) $$\mathbf{B}(c,b,a) \quad \text{and} \quad \mathbf{B}(d,c,b).$$

Substituting in Axiom O7 the variables d, c, b, a for the variables a, b, c, d, respectively, we see that (1) implies $\mathbf{B}(d,c,a)$, from which it follows, by Axiom O2, that $\mathbf{B}(a,c,d)$.

We supplement Axiom O8 by

---

† In colloquial language the connective *or* is used in both the exclusive and the non-exclusive sense. In this book the phrase *either... or* will always be used in the exclusive sense while *or* alone (not preceeded by *either*) will be used in the non-exclusive sense. Thus, e.g., by saying $\mathbf{B}(a,b,c)$ or $\mathbf{B}(b,c,a)$ or $\mathbf{B}(c,a,b)$ in Axiom O4 we merely express the fact that at least one of the conditions involved does hold; if we wished to express the fact that just one of these three conditions holds, we should say *either* $\mathbf{B}(a,b,c)$ *or* $\mathbf{B}(b,c,a)$ *or else* $\mathbf{B}(c,a,b)$.

THEOREM 17. *If* $\mathbf{B}(a,b,d)$ *and* $\mathbf{B}(b,c,d)$, *then* $\mathbf{B}(a,c,d)$.

PROOF. By Axiom O8, from $\mathbf{B}(a,b,d)$ and $\mathbf{B}(b,c,d)$ it follows that $\mathbf{B}(a,b,c)$, which, together with $\mathbf{B}(b,c,d)$ and by Theorem 16, gives $\mathbf{B}(a,c,d)$.

We shall prove one more theorem concerning four points on a line.

THEOREM 18. *Given any four collinear points* a, b, c, d, *if* $\mathbf{B}(a,b,c)$, $\mathbf{B}(a,d,c)$, *and* $b \neq d$, *then either* $\mathbf{B}(a,d,b)$ *or* $\mathbf{B}(b,d,c)$.†

PROOF. Let us assume that $\mathbf{B}(a,d,b)$. Applying Axiom O8 to the formulas $\mathbf{B}(a,b,c)$ and $\mathbf{B}(a,d,b)$, we obtain $\mathbf{B}(c,b,d)$, from which, by Axiom O3, it follows that $\sim\mathbf{B}(b,d,c)$.

Let us next assume that $\sim\mathbf{B}(a,d,b)$. Then, by Axiom O4,

$$\mathbf{B}(d,b,a) \text{ or } \mathbf{B}(b,a,d).$$

By applying Axiom O7 to formulas $\mathbf{B}(b,a,d)$ and $\mathbf{B}(a,b,c)$, we conclude that $\mathbf{B}(c,a,d)$, which is in contradiction to $\mathbf{B}(a,d,c)$. Hence $\mathbf{B}(d,b,a)$, which together with $\mathbf{B}(a,d,c)$, gives $\mathbf{B}(b,d,c)$. This concludes the proof.

7. Segments. Open Segments

The non-ordered pair $\{a,b\}$ formed by distinct points a and b will be called the *segment ab* and denoted by ab or ba. Thus $ab = ba$. We call points a and b the *end points of segment ab*. The set of all points lying between points a and b will be called the *open segment ab* or *segment* (ab) and will be denoted in formulas by (ab). We call a and b the *end points of segment* (ab). Obviously,

$$\text{if } a,b \in L, \quad \text{then} \quad (ab) \subset L.$$

By adding to segment (ab) the end points a and b, we obtain the figure

$$\langle ab \rangle = (ab) \cup ab,$$

which is called the *closed segment ab* or the *segment* $\langle ab \rangle$.

The sets

$$(ab\rangle = (ab) \cup b \text{ and } \langle ab) = a \cup (ab)$$

will be called *open-closed* or *closed-open* segments.

We call the points a and b the *end points of the segments* $\langle ab \rangle$, $(ab\rangle$, and $\langle ab)$.

By Axiom O6, segment (ab) is a non-empty set. We shall now prove

THEOREM 19. *An open segment is an infinite set.*

† See footnote on page 26.

PROOF. Consider any segment (ab). It is sufficient to indicate an infinite sequence of points (p_n) such that (i) for every natural n the point p_n belongs to segment (ab), and (ii) $p_m \neq p_n$ for $m \neq n$.

We define the sequence p_n by induction: p_1 is any point lying between a and b (see Axiom O6); assuming that point p_n has already been specified we take for p_{n+1} any point lying between a and p_n (Fig. 2).

$$a \qquad p_{n+1} \quad p_n \qquad\qquad p_2 \qquad p_1 \qquad\qquad b$$

Fig. 2

We then have

(1) $$\mathbf{B}(a, p_1, b)$$

and

(2) $$\mathbf{B}(a, p_{n+1}, p_n) \quad \text{for} \quad n > 0.$$

We shall prove by induction that for each n

(3) $$\mathbf{B}(a, p_n, b).$$

In fact, because of (1), condition (3) is satisfied for $n = 1$, and from formulas (3) and (2) it follows that $\mathbf{B}(a, p_{n+1}, b)$.

In order to show that any two points of the sequence (p_n) are distinct, it is sufficient to prove that

(4) $$m < n \quad \text{implies} \quad \mathbf{B}(a, p_n, p_m).$$

We shall prove this again by induction with respect to n. For $n = 1$ condition (4) is satisfied vacuously, since there are no indices m smaller than n. Let us now assume that we have already proved (4) for some $n \geqslant 1$ and let $m < n + 1$. If $m = n$, then it follows from (2) that $\mathbf{B}(a, p_{n+1}, p_m)$; and if $m < n$, then, by (4), we obtain $\mathbf{B}(a, p_n, p_m)$, which, together with formula (2), again gives $\mathbf{B}(a, p_{n+1}, p_m)$.

Therefore the theorem is proved.

If a figure containing any two points a and b also contains the entire segment (ab), it will be called a *convex* figure. The line and plane are therefore convex figures. Obviously, the common part of any number of convex figures is again a convex figure.

8. Open Segments on a Line

We shall first of all consider on a line L two open segments having one common end point. By Theorem 17, we obtain at once the following sufficient condition that one segment be contained in another:

THEOREM 20. *For any points a, b, c lying on a line L, if* **B**(a,b,c) *then* $(ab) \subset (ac)$.

We now give a necessary and sufficient condition that segments having a common end point be disjoint.

THEOREM 21. *For any three distinct points a, b, c ∈ L,*

$$(ab) \cap (bc) = 0 \text{ if and only if } \mathbf{B}(a,b,c).$$

PROOF. Assuming **B**(a,b,c), we at once conclude, with the help of Theorem 18, that $(ab) \cap (bc) = 0$.

Let us next assume that $(ab) \cap (bc) = 0$ and suppose that at the same time $\sim$ **B**(a,b,c). Then, by Axiom O4, one of the formulas **B**(b,a,c) or **B**(a,c,b) must hold. By Theorem 20, **B**$b,a,c)$ implies $(ab) \subset (bc)$, and **B**(a,c,b) implies $(bc) \subset (ab)$. Therefore, in both cases $(ab) \cap (bc) \neq 0$, which contradicts our assumption. Thus **B**(a,b,c).

As a direct consequence of Theorem 21, we obtain

THEOREM 22. *For any three distinct points a,b,c ∈ L,*

$$(ab) \cap (ac) \cap (bc) = 0.$$

We now give a necessary and sufficient condition that one segment be contained in another.

THEOREM 23. *Let a,b,c,d ∈ L and a ≠ b, c ≠ d. Then* $(ab) \subset (cd)$ *if and only if a,b ∈ ⟨cd⟩.*

PROOF. Let $a,b, \in \langle cd \rangle$. Let us examine three cases:

Case 1. $ab = cd$. Then, obviously, $(ab) \subset (cd)$.

Case 2. Only one of the two points a and b is identical with one of the other two points c and d; e.g. let $a = c$ and **B**(c,b,d). Then, by Theorem 20, we have $(ab) = (cb) \subset (cd)$.

Case 3. Both points a and b belong to segment (cd); i.e. **B**(c,a,d) and **B**(c,b,d). Then, by Theorem 18, **B**(a,b,c) or **B**(a,b,d) and thus

$$(ab) \subset (ac) \subset (cd) \quad \text{or} \quad (ab) \subset (ad) \subset (cd).$$

We next assume that $(ab) \subset (cd)$. Now suppose that one of the two points a and b does not belong to segment $\langle cd \rangle$, e.g. $a \sim \in \langle cd \rangle$ and

$$(1) \qquad\qquad\qquad \mathbf{B}(a,c,d).$$

Then, from Theorem 21 we obtain $(ca) \cap (cd) = 0$, and therefore, $(ca) \cap (ab) = 0$, which, in turn, by Theorem 21, gives

$$(2) \qquad\qquad\qquad \mathbf{B}(b,a,c).$$

From (1) and (2) we see that $\mathbf{B}(b,a,d)$, from which it follows that $(ab) \cap (ad) = 0$. But formula (1) implies that $(cd) \subset (ad)$, and, therefore, $(ab) \cap (cd) = 0$, in contradiction to our assumption. Hence both points a and b must be in segment $\langle cd \rangle$.

With the aid of Theorem 23, we shall show that segment uniquely determines its own end points a and b, i.e. segment ab. In fact, we have

THEOREM 24. *Let $a,b,c,d \in L$ and $a \neq b$, $c \neq d$. If $(ab) = (cd)$, then $ab = cd$.*

PROOF. Suppose that points a and b are each distinct from points c and d. Because of Theorem 23, we then have

(3) $\mathbf{B}(c,a,d)$ and $\mathbf{B}(c,b,d)$,

as well as

(4) $\mathbf{B}(a,c,b)$ and $\mathbf{B}(a,d,b)$.

But, by (3), it follows that $\mathbf{B}(a,b,c)$ or $\mathbf{B}(a,b,d)$, which contradicts (4). Therefore point a or b coincides with either point c or d. For example, let

(5) $a = c$.

Suppose that $b \neq d$. We then have $\mathbf{B}(c,b,d)$ and $\mathbf{B}(a,d,b)$, i.e. $\mathbf{B}(c,b,d)$ and $\mathbf{B}(c,d,b)$, which is impossible. Therefore $b = d$, which, together with (5), gives $ab = cd$.

9. Division of an Open Segment

We shall prove

THEOREM 25. *If $a,b,c \in L$ and points a,b,c are distinct, then*

$$(ac) - b \subset (ab) \cup (bc).$$

PROOF. From $\mathbf{B}(a,c,b)$ or $\mathbf{B}(b,a,c)$ or $\mathbf{B}(a,b,c)$ we obtain, with the help of Theorems 20 and 18,

$$(ac) \subset (ab) \quad \text{or} \quad (ac) \subset (bc) \quad \text{or} \quad (ac) - b \subset (ab) \cup (bc),$$

respectively, from which the required inclusion follows.

In particular, if point b lies between points a and c, then there will also be the inverse inclusion $(ab) \cup (bc) \subset (ac) - b$. We thus have

THEOREM 26. *Let $a,b,c \in L$. If $\mathbf{B}(a,b,c)$, then*

$$(ac) - b = (ab) \cup (bc).$$

Hence, by removing an arbitrary point, we divide an open segment into two disjoint (by Theorem 21) open segments.

10. Common Part of Two Open Segments

THEOREM 27. *The common part of two open segments lying on the same line is either an open segment or an empty set.*

PROOF. Consider on line L two arbitrary segments (ab) and (cd).
If $(ab) \subset (cd)$, then $(ab) \cap (cd) = (ab)$.

If segments (ab) and (cd) have just one common end point, e.g. $a = c$, then $b \neq d$ and we have one of three possible cases:

$$\mathbf{B}(b,a,d), \text{ or } \mathbf{B}(a,b,d), \text{ or } \mathbf{B}(a,d,b).$$

Accordingly, we have from Theorems 21 and 20 one of the following:

$$(ab) \cap (cd) = (ab) \cap (ad) = 0,$$
$$(ab) \subset (ad) = (cd) \quad \text{and hence} \quad (ab) \cap (cd) = (ab),$$
$$(cd) = (ad) \subset (ab) \quad \text{and hence} \quad (ab) \cap (cd) = (cd).$$

Now, we shall investigate the product $(ab) \cap (cd)$ in case

(1) $$(ab) \sim \subset (cd)$$

and the segments (ab) and (cd) have no common end point, i.e. points a, b, c, d are all distinct. It follows from (1), on the basis of Theorem 23, that end point a or b of segment (ab) does not belong to segment (cd), e.g. $\sim \mathbf{B}(c,a,d)$ and

(2) $$\mathbf{B}(a,c,d).$$

Three cases are now possible:
Case 1. $\mathbf{B}(c,b,d)$; with (2) this gives $\mathbf{B}(a,c,b)$. Then, by Theorem 26,

$$(ab) \cap (cd) = ((ab) - c) \cap ((cd) - b) = ((ac) \cup (bc)) \cap ((bc) \cup (bd)) = (bc),$$

since

$$(ac) \cap (bc) = 0, \ (ac) \cap (bd) \subset (ac) \cap (cd) = 0, \ (bc) \cap (bd) = 0.$$

Case 2. $\mathbf{B}(b,c,d)$; with (2) this gives

$$(cd) \subset (ad) \cap (bd),$$

from which it follows, by Theorem 22, that

$$(ab) \cap (cd) \subset (ab) \cap (ad) \cap (bd) = 0.$$

Case 3. $\mathbf{B}(b,d,c)$; with (2) this gives $\mathbf{B}(a,c,b)$. Then,

$$(cd) \subset (bc) \subset (ab) \quad \text{and hence} \quad (ab) \cap (cd) = (cd).$$

This concludes the proof.

11. Topology on the Line

Let L denote any line. We denote by $\mathfrak{U}$ the class of all open segments included in L. The class $\mathfrak{U}$ has the following properties:

(I) *For any two distinct points* $p,q \in L$, *there exist open segments* $U_1, U_2 \in \mathfrak{U}$ *such that* $p \in U_1$, $q \in U_2$, *and* $U_1 \cap U_2 = 0$.

In fact, it suffices to choose on line L any point b such that $\mathbf{B}(p,b,q)$ and any points a and c satisfying the formulas $\mathbf{B}(a,p,b)$ and $\mathbf{B}(b,q,c)$ (Fig. 3). Then open segments $U_1 = (ab)$ and $U_2 = (bc)$ satisfy the required conditions.

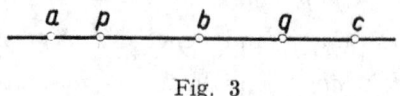

Fig. 3

(II) *If* $U_1, U_2 \in \mathfrak{U}$ *and* $p \in U_1 \cap U_2$, *then there exists an open segment* $U_0 \in \mathfrak{U}$ *such that* $p \in U_0$ *and* $U_0 \subset U_1 \cap U_2$.

In fact, by Theorem 27, the product $U_1 \cap U_2$ is an example of such a segment U_0.

As a result of properties (I) and (II), we have

THEOREM 28. *The class* $\mathfrak{U}$ *of all open segments on line* L *constitutes a base of neighborhoods for line* L.

Thus the topology on line L is defined. Henceforth by the *topology on line L* we shall understand precisely that topology which is determined by the base $\mathfrak{U}$ of all open segments.

For line L we can now define, in turn, all topological notions, in particular, the notions of a *point of accumulation*, the *closure* and *interior* of an arbitrary figure $F \subset L$, the notions of *closed*, *open*, and *connected* sets, the notion of the *limit* of a sequence of points on line L, and the notion of a *continuous function of a point* whose values are real numbers or points, and of a *continuous function of a real variable* whose values are points on the line (all the notions are discussed in Introduction, Section 9). Further, for the topological space L all theorems concerning general topological spaces can be proved; for example, we can prove the following:

THEOREM 29. *If a real function φ is defined and continuous on some connected figure F lying on a line L, and if, for some points a,b ∈ F, we have*

$$\varphi(a) = \alpha, \quad \varphi(b) = \beta \qquad (\alpha < \beta),$$

then for every real number x such that α < x < β, there exists a point p ∈ F such that φ(p) = x.

We can also prove a special theorem relating to space L:

THEOREM 30. *Let (ab) ⊂ L. Then the end points a and b are accumulation points of segment (ab).*

Fig. 4

PROOF. Let us take any neighborhood $U = (pq)$ of point a (Fig. 4). We have to show that

$$(pq) \cap (ab) \neq 0.$$

Suppose that, on the contrary, $(pq) \cap (ab) = 0$. We then, of course, would have

$$(ap) \cap (ab) = 0 \text{ and } (aq) \cap (ab) = 0,$$

from which it would follow that **B**(b,a,p) and **B**(b,a,q). Since, in addition, **B**(p,a,q), then $(bp) \cap (bq) \cap (pq) \neq 0$, in contradiction to Theorem 22. This completes the proof.

Segment (ab), as an element of the base of the neighborhoods, is an open set (in the topological sense). We suggest that the reader check for himself that the closure of the set (ab) is the set $\langle ab \rangle$, and the interior of the set $\langle ab \rangle$ is the set (ab). These facts justify the names *open segment* for the set (ab) and *closed segment* for the set $\langle ab \rangle$.

12. Half-Lines

Let us fix a point a on a line L. We shall now investigate the relation R defined for any two points p and q of the set $L-a$ by the condition:

$$pRq \quad \text{if and only if} \quad \sim \textbf{B}(p,a,q).$$

The formula pRq will be taken to read: *points p and q lie on the same side of point a (on line L).*

On the basis of Axiom O1, for an arbitrary point $p \in L-a$, we have $\sim \textbf{B}(p,a,p)$, that is, pRp. Thus relation R is reflexive.

Furthermore, by Axiom O2, for any two points $p,q \in L-a$, it follows from pRq that qRp and therefore relation R is symmetric.

Assume that pRq and qRr for some distinct points $p,q,r \in L-a$. Then $\sim \mathbf{B}(p,a,q)$, that is, $\mathbf{B}(a,p,q)$ or $\mathbf{B}(a,q,p)$, which, together with $\mathbf{B}(p,a,r)$, would give $\mathbf{B}(q,a,r)$, the latter being contrary to qRr. Therefore $\sim \mathbf{B}(p,a,r)$, that is, pRr. We have shown in this way that relation R is also transitive.

Relation R (of lying on line L on the same side of point a), as a reflexive, symmetric, and transitive relation uniquely determines the partition of the figure $L-a$ into equivalence classes with respect to relation R (see Introduction, Section 8). There are at least two of these classes, since there exists a point $p_0 \in L-a$, and, according to Axioms O5, there exists a point q_0 such that

$$(1) \qquad\qquad \mathbf{B}(p_0,a,q_0),$$

that is, $p_0 \sim Rq_0$. Let us now consider any arbitrary point $r \in L-a$ distinct from points p_0 and q_0. If at the same time we were to have both $r \sim Rp_0$ and $r \sim Rq_0$, then

$$\mathbf{B}(p_0,a,r) \text{ and } \mathbf{B}(q_0,a,r),$$

which, together with formula (1), would give us $a \in (p_0q_0) \cap (p_0r) \cap (q_0r)$, in contradiction to Theorem 22. Therefore one of the conditions rRp_0 or rRq_0 must be satisfied, which means that point r belongs either to the equivalence class determined by point p_0 or to the equivalence class determined by point q_0. We have shown in this manner that the set $L-a$ splits up into exactly two equivalence classes with respect to relation R. Hence:

THEOREM 31. *For any line L and any point $a \in L$, the set $L-a$ can be uniquely represented as the sum of two non-empty and disjoint sets*

$$L-a = A_1 \cup A_2$$

which satisfy the following three conditions:

  (i) *if $p,q \in A_1$, then $\sim \mathbf{B}(p,a,q)$,*
  (ii) *if $p,q \in A_2$, then $\sim \mathbf{B}(p,a,q)$,*
  (iii) *if $p \in A_1$ and $q \in A_2$, then $\mathbf{B}(p,a,q)$.*

We call figures A_1 and A_2 *half-lines* determined on line L by point a; we call point a the *origin* of half-lines A_1 and A_2. We say that half-lines A_1 and A_2 are *complementary* to one another.

We shall denote half-lines by the letters A, B, C, D. The half-line complementary to A will be denoted by A^*. Thus, if half-line A with origin a lies on line L, then

$$L = A \cup a \cup A^*.$$

It is clear that half-line A determines the line L on which it lies; we shall denote this line by $L(A)$. The reader may readily show that if points $p,q \in L$ are distinct, then on line L every half-line with origin p is distinct from every half-line with origin q. Therefore half-line A also determines its origin a and consequently also its complementary half-line $A^* = L - A - a$.

Half-line A is uniquely determined by its origin a and any of its points b. We shall also refer to half-line A as *half-line ab* and denote it in formulas by $\mathbf{H}(ab)$. We shall then denote the complementary half-line by $\mathbf{H}^*(ab)$. It follows at once from the definition of the half-line that for any two distinct points a and b

(2) $p \in \mathbf{H}(ab)$ if and only if $p = b$ or $\mathbf{B}(a,p,b)$ or $\mathbf{B}(a,b,p)$,

(3) $p \in \mathbf{H}^*(ab)$ of and only if $\mathbf{B}(p,a,b)$.

We could take equivalences (2) or (3) as the definitions of a half-line with origin a. More accurately, we may say that the half-lines with origin a are the sets $\mathbf{H}(ab)$ defined (for $b \neq a$) by equivalence (2), or that the half-lines with origin a are the sets $\mathbf{H}^*(ab)$ defined (for $b \neq a$)by equivalence (3). The second of these possible definitions is particularly simple. Besides, it is independent of Theorem 31. That is why in metamathematical investigations it is sometimes more convenient to employ this definition of a half-line instead of the original one.

It is seen at once that we have the following:

THEOREM 32. *For any two distinct points a and p, we have $(ap) \subset \mathbf{H}(ap)$ and $\mathbf{H}(ap) \sim \subset (ap)$.*

Point a as a point of accumulation of segment (ap) (see Theorem 30) is also a point of accumulation of the entire half-line ap. As a result, we have:

THEOREM 33. *The origin a of a half-line $A \subset L$ is a point of accumulation of A.*

It is easily shown that the closure of a half-line A with origin a is the set $\bar{A} = a \cup A$. We shall call the set $\bar{A}$ the *closed half-line.*

Using Theorem 32, the reader may readily prove that the following is a necessary and sufficient condition that an open segment be contained in a half-line:

THEOREM 34. *Let $A \subset L$ and $p,q \in L$, where $p \neq q$. Then $(pq) \subset A$ if and only if $p,q \in \bar{A}$.*

13. Half-Lines on a Line

The basic theorem here is:

THEOREM 35. *Two distinct points a and b are given on a line L. Let*

$$L = A \cup a \cup A^* = B \cup b \cup B^*,$$

where A, A^* and B, B^* are half-lines determined on line L by points a and b, respectively. If $a \in B$ and $b \in A^*$, then

$$A \subset B, \quad B^* \subset A^*, \quad A \cap B^* = 0, \quad A^* \cap B = (ab).$$

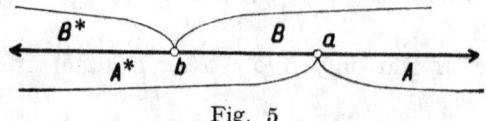

Fig. 5

PROOF. If $p \in A$ (Fig. 5), then $\mathbf{B}(p, a, b)$, and therefore $\sim \mathbf{B}(a, b, p)$, from which it follows that $p \in B$. Hence $A \subset B$. Thus we obtain at once $B^* \subset A^*$ and $A \cap B^* = 0$.

Now, if $p \in A^* \cap B$, then $\sim \mathbf{B}(b, a, p)$ and $\sim \mathbf{B}(a, b, p)$, from which it follows that $\mathbf{B}(a, p, b)$, that is, $p \in (ab)$. Conversely, if $p \in (ab)$, then $\mathbf{B}(a, p, b)$, and therefore $\sim \mathbf{B}(b, a, p)$ and $\sim \mathbf{B}(a, b, p)$, from which it follows that $p \in A^* \cap B$. Hence, $A^* \cap B = (ab)$.

As a simple conclusion from Theorem 35, we obtain the following necessary and sufficient condition that one half-line be contained in another:

THEOREM 36. *If half-lines A and B are determined on line L by points a and b, respectively, where $a \neq b$, then $A \subset B$ if and only if $a \in B$ and $b \sim \in A$.*

PROOF. Four cases are possible:

Case 1. $a \in B$ and $b \sim \in A$; then $A \subset B$.

Case 2. $a \in B^*$ and $b \in A$; then $B \subset A$, which, since $A \neq B$, gives $A \sim \subset B$.

Case 3. $a \in B^*$ and $b \in A^*$; then $A \cap B = 0$, which, since $A \neq 0$, gives $A \sim \subset B$.

Case 4. $a \in B$ and $b \in A$; then $A \cap B = (ab)$. Thus, if it were true that $A \subset B$, then it would follow that $A = (ab)$, which contradicts Theorem 32.

Another consequence of Theorem 35 concerns the division of a half-line:

THEOREM 37. *On a given line L consider two half-lines A and B with distinct origins a and b. If $A \subset B$, then*

$$B - a = (ab) \cup A \quad and \quad (ab) \cap A = 0.$$

PROOF. We have

$$B-a = (B-a) \cap (A^* \cup A) = (ab) \cup A, \quad \text{and} \quad (ab) \subset A^*,$$

from which it follows that $(ab) \cap A = 0$.

By Theorem 35, the common part of two half-lines lying on the same line is either a half-line, or an open segment, or else an empty set. In particular, we have:

THEOREM 38. *For any two distinct points* $a, b \in L$

$$\mathbf{H}(ab) \cap \mathbf{H}(ba) = (ab).$$

PROOF. If we let $A^* = \mathbf{H}(ab)$ and $B = \mathbf{H}(ba)$, then $a \in B$ and $b \in A^*$, from which it follows, by Theorem 35, that $A^* \cap B = (ab)$.

14. Half-lines on a Given Half-Line

THEOREM 39. *Given half-lines* $A, B,$ *and* C *on a line* L, *if* $B \subset A$ *and* $C \subset A$, *then* $B \subset C$ *or* $C \subset B$.

PROOF. This theorem is obvious if any two of the half-lines are identical. Hence we may assume that A, B, C are three distinct half-lines. Let points a, b, c be the origins of half-lines A, B, C, respectively. From our assumption it readily follows that no two of the half-lines are complementary. Hence a, b, c are three distinct points. By Theorem 36, we have

(1) $$b \in A \quad \text{and} \quad c \in A,$$

and, by Theorem 37, it follows that

(2) $$A - b = (ab) \cup B, \quad (ab) \cap B = 0$$

and

(3) $$A - c = (ac) \cup C, \quad (ac) \cap C = 0.$$

Because of (1), either $\mathbf{B}(a,b,c)$ or $\mathbf{B}(a,c,b)$. Assuming that $\mathbf{B}(a,b,c)$, it follows from (2) and (3) that $c \in B$ and $b \sim \in C$, i.e. $C \subset B$. If, however, we assume that $\mathbf{B}(a,c,b)$, we obtain, in a similar manner, $B \subset C$.

15. Orientations of a Line. Axes

Given a line L. We shall investigate the relation $\Re$ defined for any two half-lines A and B on line L by the condition:

$$A \Re B \text{ if and only if } A \subset B \text{ or } B \subset A.$$

Formula $A \Re B$ will read: *half-line* A *has the same orientation as half-line* B.

Relation $\Re$ so defined is obviously reflexive and symmetric. We shall show that relation $\Re$ is also transitive. Let us assume that for three given half-lines $A, B, C \subset L$ we have $A \, \Re \, B$ and $B \, \Re \, C$. Four cases are possible:

Case 1. $A \subset B$ and $B \subset C$; then $A \subset C$ and therefore $A \, \Re \, C$.

Case 2. $B \subset A$ and $C \subset B$; then $C \subset A$ and again $A \, \Re \, C$.

Case 3. $A \subset B$ and $C \subset B$; then, by Theorem 39, $A \subset C$ or $C \subset A$, i.e. $A \, \Re \, C$.

Case 4. $B \subset A$ and $B \subset C$; then $A^* \subset B^*$ and $C^* \subset B^*$ and, by theorem 39, $A^* \subset C^*$ or $C^* \subset A^*$, i.e. $C \subset A$ or $A \subset C$, and consequently $A \, \Re \, C$.

Relation $\Re$, as a reflexive, symmetric, and transitive relation, uniquely defines the partition of the family $\mathfrak{F}$ of all half-lines on line L into equivalence classes with respect to relation $\Re$. It is readily noted that there are just two such classes. In fact, if $L = A \cup a \cup A^*$, then $A \sim \Re A^*$. Now take any half-line $B \subset L$ distinct from half-lines A and A^* and having the origin b. Then $b \neq a$, and one of the following four cases holds:

Case 1. $a \in B$ and $b \in A$; hence $A^* \subset B$ (by Theorem 36).

Case 2. $a \in B$ and $b \in A^*$; hence $A \subset B$.

Case 3. $a \in B^*$ and $b \in A$; hence $A^* \subset B^*$, i.e. $B \subset A$.

Case 4. $a \in B^*$ and $b \in A^*$; hence $A \subset B^*$, i.e. $B \subset A^*$.

Therefore $B \, \Re \, A$ or $B \, \Re \, A^*$.

We can now state the following:

THEOREM 40. *For any line L the family $\mathfrak{F}$ of all half-lines of line L can be uniquely represented as the sum of two non-empty and disjoint classes*

$$\mathfrak{F} = \mathfrak{D}_1 \cup \mathfrak{D}_2$$

satisfying the following two conditions:

(i) *if $A, B \in \mathfrak{D}_1$ or $A, B \in \mathfrak{D}_2$, then half-lines A and B have the same orientation,*

(ii) *if $A \in \mathfrak{D}_1$ and $B \in \mathfrak{D}_2$, then half-lines A and B do not have the same orientation.*

We call each of the classes $\mathfrak{D}_1$ and $\mathfrak{D}_2$ the *orientation* or *sense* of line L. We say that the orientations $\mathfrak{D}_1$ and $\mathfrak{D}_2$ are *opposite* to one another.

Instead of saying that half-line A belongs to the orientation $\mathfrak{D}_1$ (or $\mathfrak{D}_2$), we shall also say that half-line A *has* the orientation $\mathfrak{D}_1$ (or $\mathfrak{D}_2$).

We shall call a line with an orientation (strictly speaking, an ordered pair consisting of a line and one of its orientations) an *oriented line* or *axis*. Thus, from line L we obtain two distinct axes:

$$\mathfrak{L}_1 = (L, \mathfrak{O}_1) \quad \text{and} \quad \mathfrak{L}_2 = (L, \mathfrak{O}_2).$$

Since the sense of line L is determined uniquely by any of its half-lines, then to orient line L it is sufficient to select on it some half-line A.

We shall denote axes by the German letters $\mathfrak{K}, \mathfrak{L}, \mathfrak{M}, \mathfrak{N}$. If $\mathfrak{L}$ is one of the axes of line L, then the other axis of line L will be denoted by $\mathfrak{L}*$.

Let $\mathfrak{L}$ be an axis of line L. We shall say that *point p lies on axis $\mathfrak{L}$* — in symbols, $p \in \mathfrak{L}$ — if $p \in L$. We shall say that *figure F lies on axis $\mathfrak{L}$* if $F \subset L$. Let $p \in L$. Of the two half-lines determined on line L by point p, just one, say A, has the orientation of axis $\mathfrak{L}$. We express this by saying that *point p determines on axis $\mathfrak{L}$ the half-line A*.

16. Order of Points on an Axis

Given an axis $\mathfrak{L}$ of line L. For any point $p \in \mathfrak{L}$ let us denote by A_p the half-line determined on axis $\mathfrak{L}$ by point p. Let $a, b \in \mathfrak{L}$. If $A_a \supset A_b$ and $A_a \neq A_b$, then we say that *point a precedes point b* (on axis $\mathfrak{L}$), and we write $a \prec b$.

We see at once that the relation $\prec$ satisfies, for arbitrary $a, b, c \in \mathfrak{L}$, the following two conditions:

(i) if $a \neq b$, then either $a \prec b$ or $b \prec a$;

(ii) if $a \prec b$ and $b \prec c$, then $a \prec c$.

Hence the relation $\prec$ is connected, antisymmetric, and transitive in set L. We therefore have:

THEOREM 41. *The relation $\prec$ orders the point set L of the axis $\mathfrak{L}$.*

Henceforth we shall understand by axis $\mathfrak{L}$ the point set L ordered by the relation $\prec$.

Instead of $a \prec b$, we shall also write $b \succ a$, which we read: *point b follows point a* (on axis $\mathfrak{L}$).

It is readily noted that if point a determines half-line A on axis $\mathfrak{L}$, then half-line A consists of all points $p \succ a$ and the complementary half-line $A*$ consists of all points $p \prec a$.

Let us now take two distinct points $a, b \in \mathfrak{L}$ and let $a \prec b$. Then the orientation of half-line ab is the same as the orientation of axis $\mathfrak{L}$, and the orientation of half-line ba is opposite to the orientation of axis $\mathfrak{L}$. Since $(ab) = \mathbf{H}(ab) \cap \mathbf{H}(ba)$ (see Theorem 38), then segment (ab) consists of all points p such that $a \prec p \prec b$. Thus, in particular, we obtain a simple connection between the relations $\prec$ and $\mathbf{B}$:

THEOREM 42. *Given any three points $a, b, c \in \mathfrak{L}$, we have $\mathbf{B}(a, b, c)$ if and only if $a \prec b \prec c$ or $c \prec b \prec a$.*

With the help of Theorem 42 we obtain further properties of the relation $\prec$:

(iii) for each point $b \in \mathfrak{L}$ there exist points $a,c \in \mathfrak{L}$ such that $a \prec b \prec c$;

(iv) if for any points $a,c \in \mathfrak{L}$ we have $a \prec c$, then there exists a point $b \in \mathfrak{L}$ such that $a \prec b \prec c$.

Hence

THEOREM 43. *The point set L of the axis $\mathfrak{L}$ with respect to the ordering relation $\prec$ is dense in itself without the first and last points.*

Let us consider a line L. The order of points on line L depends on the orientation chosen. By changing the orientation to its opposite, we change the order to its converse. To establish the order of points on line L, it is sufficient to fix which of two arbitrarily chosen distinct points $a,b \in L$ shall be regarded as being the earlier one. If, e.g. we let $a \prec b$, then by doing so we give line L the orientation of the half-line ab, and this already defines the order. We say that this is the order *from point a to point b*.

Consider any half-line A with origin a. Half-line A defines a certain orientation, and therefore also a certain ordering relation $\prec$, on line $L \supset A$ and thus also on every figure F included in line L. In particular, open half-line A and closed half-line $a \cup A$ may be regarded as ordered sets of points. Origin a is then the first point on half-line $a \cup A$.

Let us consider an arbitrary real function φ defined for points of some figure F lying on axis $\mathfrak{L}$. We shall say that, in the set F, function φ is (i) *increasing*, (ii) *decreasing*, (iii) *non-increasing*, and (iv) *non-decreasing* if, for any two points $p,q \in F$, it follows from $p \prec q$ that (i) $\varphi(p) < \varphi(q)$, (ii) $\varphi(p) > \varphi(q)$, (iii) $\varphi(p) \geqslant \varphi(q)$, (iv) $\varphi(p) \leqslant \varphi(q)$. All functions of these four types are embraced by the term *monotonic* functions in set F.

Similar notions could have equally well been introduced for the function f of a real variable x whose values are points of some axis $\mathfrak{L}$.

Using the topology introduced in Section 11, we shall prove

THEOREM 44. *If a real function φ defined on the points of an axis $\mathfrak{L}$ is monotonic and its range is connected, then φ is continuous.*

PROOF. We know from the arithmetic of real numbers that each connected (and non-empty) set of real numbers either consists of a single point or is identical with a numerical interval (closed, open, open-closed or closed-open, bounded or unbounded). In case φ is a constant function, the theorem is obvious. In the remaining cases we assume that the range of the function φ is the interval of real numbers with the end points α and β, where $\alpha < \beta$ (in particular, we may have $\alpha = -\infty$ or $\beta = +\infty$). Take any point p of axis $\mathfrak{L}$ and a number ε such that

$$0 < \varepsilon < \min (\beta - \varphi(p), \varphi(p) - \alpha).$$

Then

$$\alpha < \varphi(p) - \varepsilon < \beta \quad \text{and} \quad \alpha < \varphi(p) + \varepsilon < \beta.$$

Pick on axis $\mathfrak{L}$ two points q_1 and q_2 such that

$$\varphi(q_1) = \varphi(p) - \varepsilon \quad \text{and} \quad \varphi(q_2) = \varphi(p) + \varepsilon.$$

Since function φ is monotonic, it follows from the inequality

$$\varphi(p) - \varepsilon < \varphi(p) < \varphi(p) + \varepsilon$$

that

$$q_1 < p < q_2 \quad \text{or} \quad q_1 > p > q_2.$$

Let $U = (q_1 q_2)$. Then U is a neighborhood of point p and for any point $p' \in U$ we have the inequality

$$\varphi(p) - \varepsilon \leqslant \varphi(p') \leqslant \varphi(p) + \varepsilon,$$

from which it follows

$$|\varphi(p') - \varphi(p)| \leqslant \varepsilon.$$

This concludes the proof of the theorem.

17. Betweenness Relation for a Line and Two Points

In the preceding sections we developed a linear geometry. We shall now develop a plane geometry.

Consider on a plane P a line K and points a and c. We shall say that *line K lies between points a and c* —in symbols, $\mathbf{B}(a,K,c)$— if and only if $a,c \sim \in K$ and $\mathbf{B}(a,b,c)$ for some point $b \in K$ (Fig. 6).

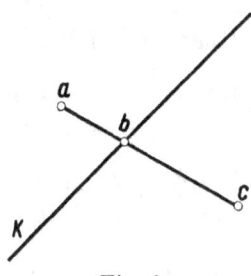

Fig. 6

The following theorem, being a consequence of Axiom O1 and Theorem 13, establishes the connection between the relations $\mathbf{B}(a,b,c)$ and $\mathbf{B}(a,K,c)$:

THEOREM 45. *If a line K intersects a line ac, distinct from K, at a point b, then* $\mathbf{B}(a,K,c)$ *if and only if* $\mathbf{B}(a,b,c)$. *If lines K and ac are disjoint, then* $\sim\mathbf{B}(a,K,c)$.

From the properties of the relation $\mathbf{B}(a,b,c)$ formulated in Axioms O1, O2, O5, and O6, the analogous properties of the relation $\mathbf{B}(a,K,c)$ follow:

THEOREM 46. (I) *If* $\mathbf{B}(a,K,c)$, *then points a and c are distinct.*
(II) *If* $\mathbf{B}(a,K,c)$, *then* $\mathbf{B}(c,K,a)$.

THEOREM 47. (I) *Given a plane P, if* $a \in P$, $K \subset P$, *and* $a \sim \in K$, *then there is a point* $c \in P$ *such that* $\mathbf{B}(a,K,c)$.
(II) *Given a plane P, if* $a,c \in P$ *and* $a \neq c$, *then there is a line* $K \subset P$ *such that* $\mathbf{B}(a,K,c)$.

18. Plane Axiom of Order

The axioms of order for the line were given in Section 6. We now supplement them by the *plane axiom of order* (Fig. 7).

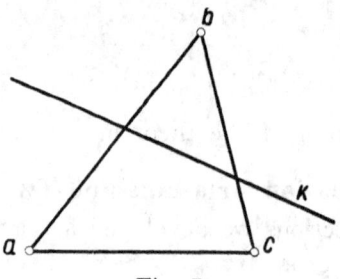

Fig. 7

AXIOM O9. *Given an a plane P three non-collinear points a, b, c and a line K, if* $\mathbf{B}(a,K,b)$ *and* $c \sim \in K$, *then* $\mathbf{B}(b,K,c)$ *or* $\mathbf{B}(a,K,c)$.

Axiom O9 can be strengthened as follows:

THEOREM 48. *Given on a plane P three non-collinear points a,b,c and a line K, if* $\mathbf{B}(a,K,b)$ *and* $c \sim \in K$, *then either* $\mathbf{B}(b,K,c)$ *or* $\mathbf{B}(a,K,c)$.

PROOF. We assume that $\mathbf{B}(a,K,b)$ and, $c \sim \in K$ (Fig. 8). Let us suppose that both $\mathbf{B}(b,K,c)$ and $\mathbf{B}(a,K,c)$ hold. Putting

$$K \cap L\,(ab) = p, \quad K \cap L\,(bc) = q, \quad \text{and} \quad K \cap L\,(ac) = r,$$

we have, by Theorem 45,

(1) $\mathbf{B}(a,p,b)$, $\mathbf{B}(b,q,c)$,   and $\mathbf{B}(a,r,c)$.

Points p, q, r lie on line K and are distinct; because of the symmetry of our assumptions, we can assume that $\mathbf{B}(p,q,r)$. Then, by formula (1), we have simultaneously

$$\mathbf{B}(p,\mathbf{L}(bc),r), \quad \sim \mathbf{B}(a,\mathbf{L}(bc),p), \quad \text{and} \quad \sim \mathbf{B}(a,\mathbf{L}(bc),r),$$

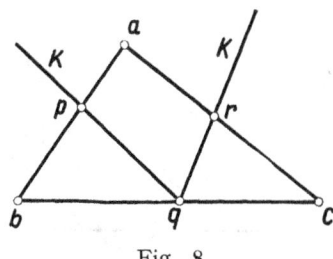

Fig. 8

in contradiction to Axiom O9. We have thus shown that at most one of the conditions $\mathbf{B}(b,K,c)$ and $\mathbf{B}(a,K,c)$ holds. Because of Axiom O9, this concludes the proof.

19. Half-Planes

Let us fix a line K on a plane P. By Theorem 12(V), the set $P-K$ is non-empty. We shall investigate the relation R defined for any two points p and q of set $P-K$ by the condition:

$$pRq \quad \text{if and only if} \quad \sim \mathbf{B}(p,K,q).$$

Formula pRq will be read: *points p and q lie on the same side of line K* (on plane P).

From Theorem 46 it follows at once that the relation R is reflexive and symmetric. We shall now show that the relation R is also transitive. Thus, we take three points $p,q,r \in P-K$ and assume that pRq and qRr. We shall consider two possibilities:

Case 1. Points p,q,r lie on some line L (Fig. 9). By Theorem 45, if line L does not intersect line K, then $\sim \mathbf{B}(p,K,r)$, and if line L intersects line K in a point a, then from $\sim \mathbf{B}(p,K,q)$ and $\sim \mathbf{B}(q,K,r)$ it follows that $\sim \mathbf{B}(p,a,q)$ and $\sim \mathbf{B}(q,a,r)$, which, because of the transitivity of the relation of lying on the line L on the same side of point a (see Section 12), gives $\sim \mathbf{B}(p,a,r)$, that is, $\sim \mathbf{B}(p,K,r)$. Therefore in both cases pRr.

Case 2. Points p,q,r are non-collinear (Fig. 10). Then, by Axiom O9, it immediately follows from $\sim \mathbf{B}(p,K,q)$ and $\sim \mathbf{B}(q,K,r)$ that $\sim \mathbf{B}(p,K,r)$, that is pRr.

We shall now find the number of equivalence classes into which relation R (as a reflexive, symmetric, and transitive relation) divides the set $P-K$. By Theorems 12(V) and 47(I) there exist two points $p_0, q_0 \in P-K$ such that $\mathbf{B}(p_0, K, q_0)$, and hence there are at least two equivalence classes. Let $L = \mathbf{L}(p_0 q_0)$ and $L \cap K = a$. Let us take any point $r \in P-K$. Two cases are possible:

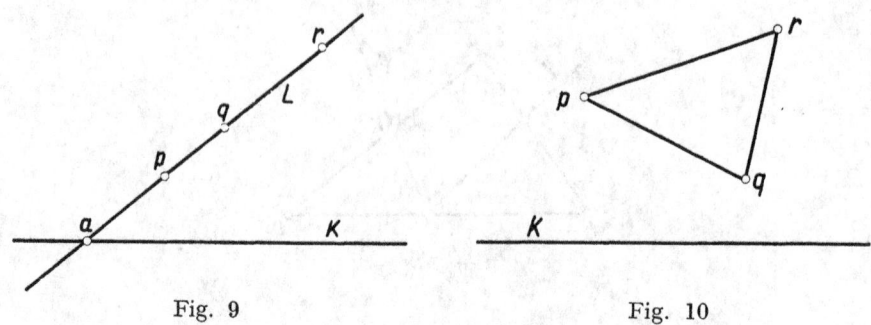

Fig. 9   Fig. 10

Case 1. $r \in L$ (Fig. 11). Since $\mathbf{B}(p_0, a, q_0)$ and since the relation of lying on line L on the same side of point a divides the set $L-a$ into two equivalence classes, it follows that $\sim\mathbf{B}(p_0, a, r)$ or $\sim\mathbf{B}(q_0, a, r)$, that is, $\sim\mathbf{B}(p_0, K, r)$ or $\sim\mathbf{B}(q_0, K, r)$.

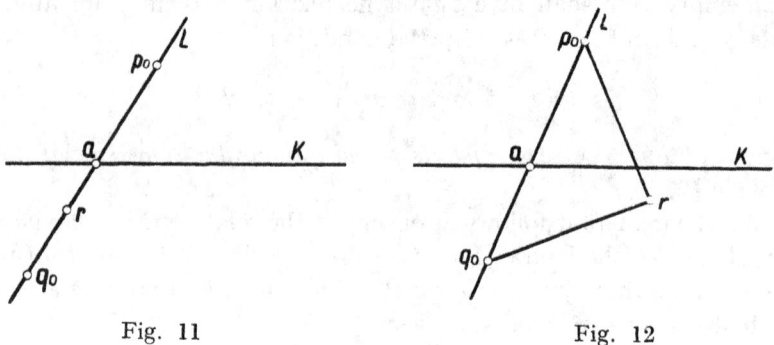

Fig. 11   Fig. 12

Case 2. $r \sim \in L$ (Fig. 12). From Theorem 48 it follows that either $\sim\mathbf{B}(p_0, K, r)$ or $\sim\mathbf{B}(q_0, K, r)$.

Therefore, in both cases, we have either rRp_0 or rRq_0 and hence there are just two equivalence classes. Thus we have

THEOREM 49. *For any line K on a given plane P, the set $P-K$ can be uniquely represented as the sum of two non-empty and disjoint sets*

$$P-K = W_1 \cup W_2$$

satisfying the following two conditions:

(i) *if $p,q \in W_1$ or $p,q \in W_2$, then $\sim \mathbf{B}(p,K,q)$;*

(ii) *if $p \in W_1$ and $q \in W_2$, then $\mathbf{B}(p,K,q)$.*

We call the figures W_1 and W_2, the *half-planes* determined on plane P by line K; we call line K the *boundary of half-planes* W_1 and W_2. We say that half-planes W_1 and W_2 are *complementary* to one another.

A half-plane will be denoted by the letter W. The half-plane complementary to W will be denoted by W^*. Therefore, if a half-plane W with a boundary K lies on plane P, then

$$P = W \cup K \cup W^*.$$

It is clear that half-plane W determines the plane P on which it lies. The reader may easily confirm that if the lines $M,N \subset P$ are distinct, then on plane P each of the half-planes with the boundary M is distinct from each of the half-planes with the boundary N. Hence half-plane W determines its boundary K and therefore also determines its complementary half-plane $W^* = P - W - K$.

Half-plane W is uniquely defined by its boundary K and any of its points b. We shall also call half-plane W *half-plane Kb* and denote it in the formulas by $\mathbf{HP}(Kb)$. We shall then denote the complementary half-plane by $\mathbf{HP}^*(Kb)$. From the definition of the half-plane it follows at once that for any line K and for any point $b \sim \in K$

(1) $p \in \mathbf{HP}^*(Kb)$ if and only if $\mathbf{B}(p,K,b)$.

This equivalence can be adopted as the definition of a half-plane with boundary K. More accurately, we can say that half-planes with boundary K are the sets $\mathbf{HP}^*(Kb)$ defined (for $b \sim \in K$) by equivalence (1). Such a definition of a half-plane would be independent of Theorem 49. That is why in metamathematical investigations it is sometimes more convenient to employ this definition of a half-plane instead of the origin one.

Let $P = W \cup K \cup W^*$. If a figure $F_1 \neq 0$ lies in half-plane W and a figure F_2 lies in half-plane W^* (Fig. 13), then we write $\mathbf{B}(F_1,K,F_2)$, which we read: *line K lies between figures F_1 and F_2*, or *figures F_1 and F_2 lie* (on plane P) *on opposite sides of line K*.

The following obvious theorem establishes a sufficient condition that a half-line lie on a half-plane:

THEOREM 50. *Given a half-plane $W \subset P$ with boundary K, if $a \in K$ and $p \in W$, then $\mathbf{H}(ap) \subset W$.*

With the help of Theorem 50 and 32 we obtain a necessary and sufficient condition that an open segment be contained in a half-plane:

THEOREM 51. *Given in a plane P a half-plane W with boundary K and an open segment (pq), then $(pq) \subset W$ if and only if $pq \subset K \cup W$ and $pq \sim \subset K$.*

We recall that by removing a point from a line, we divided the line into two half-lines and that by removing a line from a plane, we divided

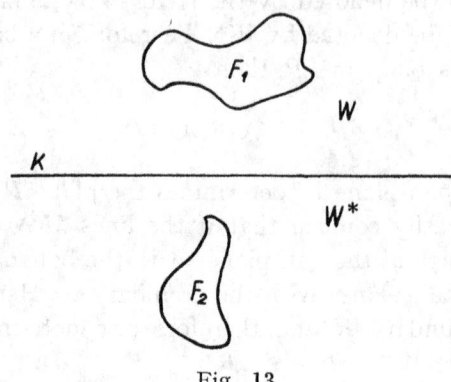

Fig. 13

the plane into two half-planes. Similarly, by removing a plane P from the space S we could divide the space S into two *half-spaces* with the *boundary* P. For this purpose we should introduce the relation **B**

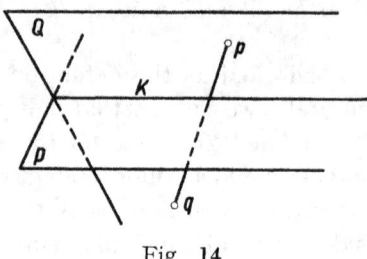

Fig. 14

defined for any points $p,q \in S$ and for any plane P by the condition: **B** (p,P,q) if and only if $p,q \sim \in P$ and there exists a point $a \in P$ such that **B** (p,a,q). The half-spaces with the boundary P are the equivalence classes with respect to the relation R defined for $p,q \in S-P$ by the condition: pRq if and only if $\sim$**B** (p,P,q).

The following theorem gives a simple connection between the relations **B** (p,P,q) and **B** (p,K,q) (Fig. 14):

THEOREM 52. *If a plane Q intersects a plane P along a line K, then for any two points $p,q \in Q$ we have* **B**(p,P,q) *if and only if* **B**(p,K,q).

20. Pencils and Half-Pencils of Half-Lines

Given a point p on a plane P (Fig. 15). The family of all half-lines with origin p and lying in plane P will be called the *pencil* $\mathfrak{P}$ determined by point p (on plane P). We shall call point p the *vertex of the pencil* $\mathfrak{P}$.

Let us now take two complementary half-lines A and A^* in pencil $\mathfrak{P}$. The line $K = A \cup p \cup A^*$ determines on plane P two half-planes W and W^*. We divide the class of half-lines $\mathfrak{P}-\{A,A^*\}$ into two disjoint classes $\mathfrak{H}$ and $\mathfrak{H}^*$ by including into class $\mathfrak{H}$ those half-lines of pencil $\mathfrak{P}$

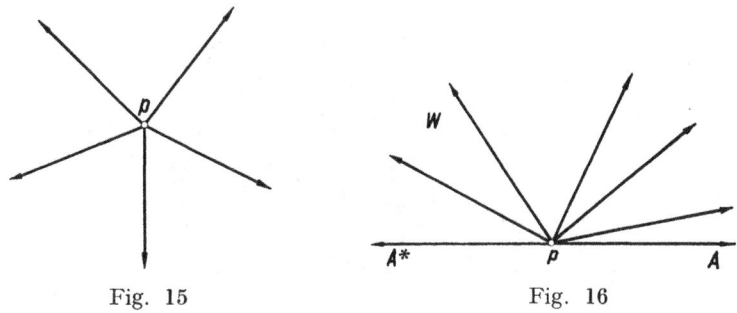

Fig. 15   Fig. 16

which lie in half-plane W (Fig. 16) and into class $\mathfrak{H}^*$ those half-lines of pencil $\mathfrak{P}$ which lie in half-plane W^*. Classes $\mathfrak{H}$ and $\mathfrak{H}^*$ will be called *half-pencils* of pencil $\mathfrak{P}$ with *end half-lines* A and A^*. We call point p the *vertex of half-pencils* $\mathfrak{H}$ and $\mathfrak{H}^*$.

From the definition of half-pencils $\mathfrak{H}$ and $\mathfrak{H}^*$ it is immediately seen that, for any half-pencil $\mathfrak{H}$ of pencil $\mathfrak{P}$,

$$\text{if } B \in \mathfrak{H}, \quad \text{then} \quad B^* \in \mathfrak{H}^*.$$

If half-lines $A_1, A_2, \ldots, A_n \in \mathfrak{P}$ belong to some half-pencil $\mathfrak{H} \subset \mathfrak{P}$, then we say that they are *co-half-pencilar*. We shall now prove a theorem which allows us to establish a necessary and sufficient condition that half-lines be co-half-pencilar.

THEOREM 53. *Given a half-pencil* $\mathfrak{H}$, *if* $A_1, A_2, \ldots, A_n \in \mathfrak{H}$ *and* A *is an end half-line of half-pencil* $\mathfrak{H}$, *then a half-line* B *may be produced from any point* $a \in A$ *to intersect each of the half-lines* $A_1, A_2, \ldots, A_n$ *in only one point.*

PROOF. For $n = 1$ the theorem is obvious. Let us now assume that there exists a half-line B with origin $a \in A$ which intersects each of the half-lines A_i ($i = 1, 2, \ldots, n$) at a certain point a_i (Fig. 17). Let W be the half-plane with boundary $L(A)$ in which the half-pencil $\mathfrak{H}$ lies. By Theorem 50, we have $B \subset W$. Without restricting the generality of the proof, we may assume that the last of the points $a_1, a_2, \ldots, a_n$ on the half-line $a \cup B$ is point a_n. Therefore

$$a \prec a_i \prec a_n \quad \text{for} \quad i = 1, 2, \ldots, n-1,$$

from which, by Theorem 42, it follows that

$$(1) \qquad\qquad \mathbf{B}(a, a_i, a_n) \quad \text{for} \quad i = 1, 2, \ldots, n-1.$$

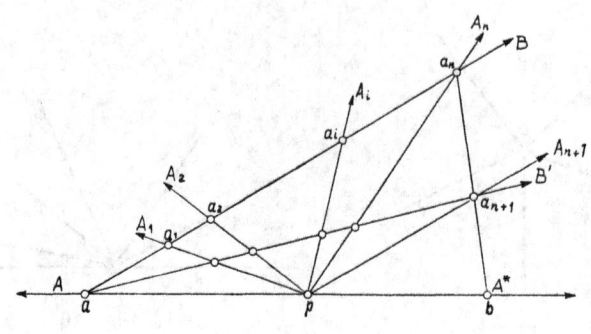

Fig. 17

Let us add to the half-lines $A_1, A_2, \ldots, A_n \in \mathfrak{H}$ a new half-line $A_{n+1} \in \mathfrak{H}$ and let us assume that

$$(2) \qquad\qquad B \cap A_{n+1} = 0.$$

We shall now seek a half-line B' with origin a intersecting each of the half-lines $A_1, A_2, \ldots, A_n, A_{n+1}$. For this purpose we select on half-line A^* any point b. Denoting by p the vertex of the half-pencil $\mathfrak{H}$, we thus have

$$(3) \qquad\qquad \mathbf{B}(a, p, b).$$

By Theorem 51, we have $(ba_n) \subset W$. Applying Axiom O9 to points a, b, a_n we see from (2) and (3) that half-line A_{n+1} intersects the segment (ba_n) in some point a_{n+1}, where

$$(4) \qquad\qquad \mathbf{B}(b, a_{n+1}, a_n).$$

Applying Theorem 48, we obtain from (1) and (3)

$$5) \qquad\qquad A_i \cap (ba_n) = 0 \quad \text{for} \quad i = 1, 2, \ldots, n-1.$$

Let $B' = \mathbf{H}(aa_{n+1})$. Then $B' \subset W$ and half-line B' intersects half-line

A_{n+1}. We shall show that half-line B' also intersects each of the half-lines $A_1, A_2, \ldots, A_n$. Obviously, by applying Axiom O9 to the points a, b, a_{n+1}, we obtain, because of (3), (4), and (5),

$$A_i \cap (aa_{n+1}) \neq 0 \quad \text{for} \quad i = 1, 2, \ldots, n,$$

from which it follows that $A_i \cap B' \neq 0$ for $i = 1, 2, \ldots, n$. This concludes the proof.

We shall now prove:

THEOREM 54. *Half-lines $A_1, A_2, \ldots, A_n$ of a given pencil $\mathfrak{P}$ are co-half-pencilar if and only if there exists a line L intersecting each of these half-lines in only one point.*

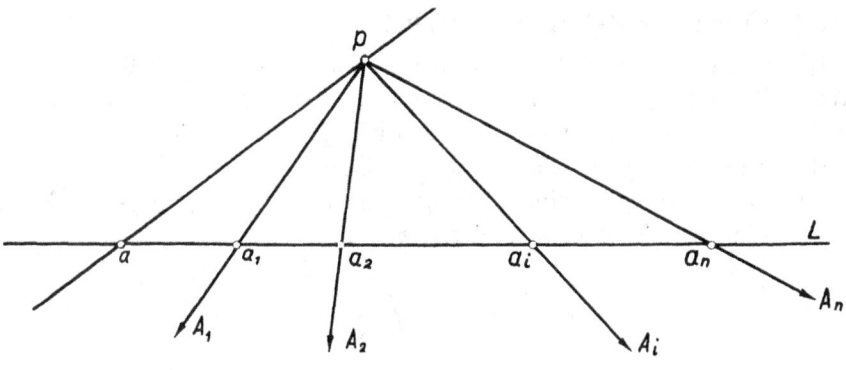

Fig. 18

PROOF. The first part of the theorem follows directly from Theorem 53. Let us now assume that a line L intersects half-lines $A_1, A_2, \ldots, A_n \in \mathfrak{P}$ respectively in points $a_1, a_2, \ldots, a_n$ (Fig. 18). Let us choose an axis $\mathfrak{L}$ of line L; on axis $\mathfrak{L}$ let e.g. the point a_1 be the first of the points $a_1, a_2, \ldots, a_n$. We now take on axis $\mathfrak{L}$ any point $a \prec a_1$ (see Theorem 43). Then

$$\sim \mathbf{B}(a_1, a, a_i) \quad \text{for} \quad i = 2, 3, \ldots, n,$$

from which it follows that all points $a_1, a_2, \ldots, a_n$, and therefore all half-lines $A_1, A_2, \ldots, A_n$, lie in the half-plane $\mathbf{L}(pa)a_1$, provided p is the vertex of pencil $\mathfrak{P}$. Thus half-lines $A_1, A_2, \ldots, A_n$ are co-half-pencilar.

As a direct consequence of Theorems 53 and 54 we obtain:

THEOREM 55. *Given a half-pencil $\mathfrak{H}$, if $A_1, A_2, \ldots, A_n \in \mathfrak{H}$ and A is an end half-line of half-pencil $\mathfrak{H}$, then half-lines $A, A_1, A_2, \ldots, A_n$ are co-half-pencilar.*

As a consequence of Theorem 55 we have:

THEOREM 56. *Given a pencil $\mathfrak{P}$ and half-lines $A,B,C \in \mathfrak{P}$, if half-lines A,B,C are not co-half-pencilar and if lines $\mathbf{L}(A)$, $\mathbf{L}(B)$, $\mathbf{L}(C)$ are distinct, then*

$$\mathbf{B}(A,\mathbf{L}(B),C) \ \text{and} \ \mathbf{B}(B,\mathbf{L}(C),A) \ \text{and} \ \mathbf{B}(C,\mathbf{L}(A),B).$$

PROOF. Let us assume that lines $\mathbf{L}(A)$, $\mathbf{L}(B)$, $\mathbf{L}(C)$ are distinct. Consider line $\mathbf{L}(B)$. If half-lines A and C were to lie on the same side of line $\mathbf{L}(B)$, then half-lines A and C would belong to the half-pencil with the end half-line B, and therefore they would be co-half-pencilar. This would contradict our hypothesis. Hence $\mathbf{B}(A,\mathbf{L}(B),C)$. Cyclic permutation of variables A, B, C gives us $\mathbf{B}(B,\mathbf{L}(C),A)$ and $\mathbf{B}(C,\mathbf{L}(A),B)$.

21. Betweenness Relation for Half-Lines

Given a pencil $\mathfrak{P}$. Let us take three half-lines $A,B,C \in \mathfrak{P}$ (Fig. 19). We shall say that *half-line B lies between half-lines A and C* — in symbols, $\mathbf{B}(A,B,C)$ — if the following two conditions are satisfied:

(I) half-lines A, B, C are co-half-pencilar;

(II) for every line L intersecting half-lines A, B, C, respectively, in some points a, b, c the formula $\mathbf{B}(a,b,c)$ holds.

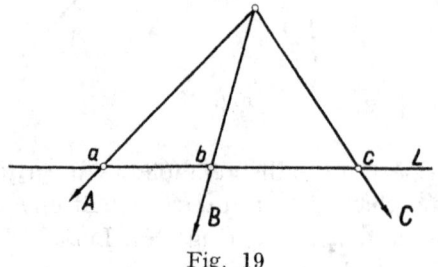

Fig. 19

Since, by Condition (I) and Theorem 54, there exists at least one line L intersecting half-lines A, B, C, then we have at once:

THEOREM 57. *Given a pencil $\mathfrak{P}$ and three half-lines $A,B,C \in \mathfrak{P}$, if $\mathbf{B}(A,B,C)$, then $\mathbf{B}(A,\mathbf{L}(B),C)$ and half-lines A and B lie in one half-plane with boundary $\mathbf{L}(C)$.*

As a further conclusion from the definition, we have:

THEOREM 58. *Given a pencil $\mathfrak{P}$ and three half-lines $A,B,C \in \mathfrak{P}$, we have:*
(I) *if $\mathbf{B}(A,B,C)$, then half-lines A, B, C are distinct;*
(II) *if $\mathbf{B}(A,B,C)$, then $\mathbf{B}(C,B,A)$;*
(III) *if $\mathbf{B}(A,B,C)$, then $\sim\mathbf{B}(B,A,C)$.*

Hence, the properties of the relation **B** for points on a line, as formulated in Axioms O1, O2, O3, apply to the relation **B** for half-lines of a pencil $\mathfrak{P}$.

The following theorem establishes a sufficient condition that **B**(A,B,C):

THEOREM 59. *Given a pencil* $\mathfrak{P}$, *three distinct co-half-pencilar half-lines* $A,B,C \in \mathfrak{P}$, *and a line* L *intersecting half-lines* A,B,C *in points* a, b, c, *if* **B**(a,b,c), *then* **B**(A,B,C).

PROOF. Let us take any line $L' \neq L$ intersecting half-lines A,B,C in points a', b', c', respectively. We have to show that **B**(a',b',c').

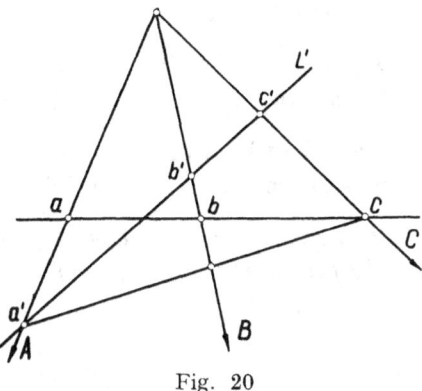

Fig. 20

Let us assume that $a \neq a'$, $c \neq c'$ (Fig. 20). Applying Axiom O9 to points c, a, a', we obtain $B \cap (a'c) \neq 0$; applying Axiom O9 to points a', c, c', we obtain $B \cap (a'c') \neq 0$, i.e. **B**(a',b',c'). If $a = a'$, then $c \neq c'$ and it suffices to investigate the points a, c, c' only. Similarly, if $c = c'$, then $a \neq a'$, and it suffices to investigate the points c, a, a' only.

We now restrict ourselves to half-lines of some half-pencil $\mathfrak{H}$. Then every finite number of half-lines can be intersected by one line, and, by Theorem 59, all the properties of the relation of betweenness for points on a line established in Section 6 at once apply to the relation of betweenness for half-lines of a half-pencil.

Hence from Axiom O4 we obtain:

THEOREM 60. *If half-lines* A,B,C *belonging to a half-pencil* $\mathfrak{H}$ *are distinct, then*

$$\mathbf{B}(A,B,C) \quad \text{or} \quad \mathbf{B}(B,C,A) \quad \text{or} \quad \mathbf{B}(C,A,B).$$

From Axioms O5 and O6 we have:

THEOREM 61. *Given a half-pencil* $\mathfrak{H}$, *we have:*

(I) *If* $A,B \in \mathfrak{H}$ *and* $A \neq B$, *then there exists a half-line* $C \in \mathfrak{H}$ *such that* **B**(A,B,C).

(II) *If* $A,B \in \mathfrak{H}$ *and* $A \neq B$, *then there exists a half-line* $C \in \mathfrak{H}$ *such that* $\mathbf{B}(A,C,B)$.

Corresponding to Axioms O7 and O8 and Theorems 16 and 17, we have:

THEOREM 62. *For any four half-lines* $A,B,C,D \in \mathfrak{H}$ *we have:*
(I)   *if* $\mathbf{B}(A,B,C)$ *and* $\mathbf{B}(B,C,D)$, *then* $\mathbf{B}(A,B,D)$;
(II)   *if* $\mathbf{B}(A,B,D)$ *and* $\mathbf{B}(B,C,D)$, *then* $\mathbf{B}(A,B,C)$;
(III) *if* $\mathbf{B}(A,B,C)$ *and* $\mathbf{B}(B,C,D)$, *then* $\mathbf{B}(A,C,D)$;
(IV)   *if* $\mathbf{B}(A,B,D)$ *and* $\mathbf{B}(B,C,D)$, *then* $\mathbf{B}(A,C,D)$.

Corresponding to Theorem 18, we have:

THEOREM 63. *For any half-lines* $A,B,C,D \in \mathfrak{H}$ *it follows from* $\mathbf{B}(A,B,C)$, $\mathbf{B}(A,D,C)$, *and* $B \neq D$ *that either* $\mathbf{B}(A,D,B)$ *or* $\mathbf{B}(B,D,C)$.

Let A be one of the end half-lines of half-pencil $\mathfrak{H}$. We shall extend the field of the relation $\mathbf{B}$ to the class $\{A\} \cup \mathfrak{H}$. We then have:

THEOREM 64. *If half-line* A *is an end half-line of half-pencil* $\mathfrak{H}$, *then for any two distinct half-lines* $B,C \in \mathfrak{H}$, *we have either* $\mathbf{B}(A,B,C)$ *or* $\mathbf{B}(A,C,B)$.

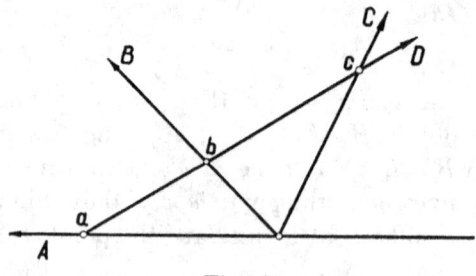

Fig. 21

PROOF. From any point a on half-line A we produce a half-line D (Fig. 21) intersecting half-lines B and C in points b and c, respectively. Then either $\mathbf{B}(a,b,c)$ or $\mathbf{B}(a,c,b)$, and therefore either $\mathbf{B}(A,B,C)$ or $\mathbf{B}(A,C,B)$.

Next, we extend the field of the relation $\mathbf{B}$ to the class $\{A,A^*\} \cup \mathfrak{H}$. We shall prove:

THEOREM 65. *Let* $\mathfrak{H}$ *be a half-pencil with end half-lines* A *and* A^*. *Then, for any half-lines* $B,C \in \mathfrak{H}$, *if* $\mathbf{B}(A,B,C)$, *then* $\mathbf{B}(B,C,A^*)$.

PROOF. Assume that $\mathbf{B}(A,B,C)$. We supplement the construction from the proof of Theorem 64 by choosing on half-line A^* any point a_1 (Fig. 22). Applying Axiom O9 to the points a,b,a_1, we readily conclude that half-line C intersects the segment (ba_1) in some point c_1. Hence line ba_1 intersects half-lines B, C, A^* at points b, c_1, a_1, respectively, where $\mathbf{B}(b,c_1,a_1)$, i.e., $\mathbf{B}(B,C,A^*)$.

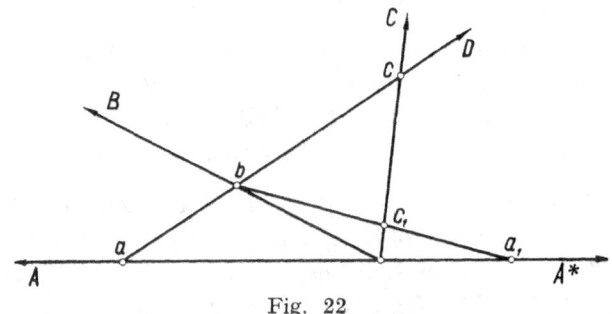

Fig. 22

We shall now give a few theorems concerning the relation $\mathbf{B}$ for half-lines of the entire pencil $\mathfrak{P}$.

THEOREM 66. *Given a pencil $\mathfrak{P}$ and three co-half-pencilar half-lines $A,B,C \in \mathfrak{P}$, if $\mathbf{B}(A,B,C)$, we then have:*

(I) *Half-lines B and C belong to the half-pencil with end half-lines A and A^*.*

(II) *Half-line B belongs to every half-pencil $\mathfrak{H} \subset \mathfrak{P}$ containing half-lines A and C.*

PROOF. Assume that $\mathbf{B}(A,B,C)$. Let a line L intersect half-lines A, B, C in points a, b, c, respectively (Fig. 23). Then $\mathbf{B}(a,b,c)$. Hence

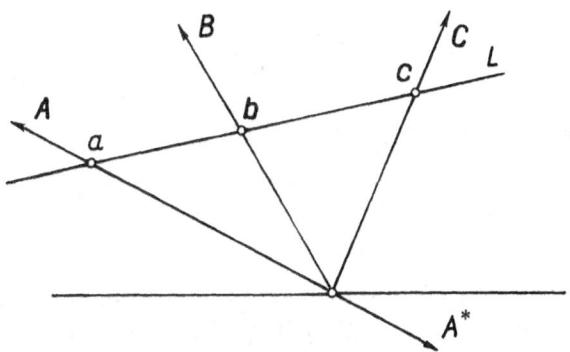

Fig. 23

points b and c, and therefore also the entire half-lines B and C, lie in the same half-plane with boundary $\mathbf{L}(A)$. Therefore half-lines B and C belong to the half-pencil with end half-lines A and A^*.

Let us now take any half-pencil $\mathfrak{H} \subset \mathfrak{P}$ containing half-lines A and C, and let W be the half-plane of half-pencil $\mathfrak{H}$. Then $a,c \in W$, from which, by Theorem 51, it follows that $b \in W$. Therefore the entire half-line B lies in half-plane W, and consequently $B \in \mathfrak{H}$.

THEOREM 67. *Given a pencil $\mathfrak{P}$ and three distinct half-lines $A,B,C \in \mathfrak{P}$, if half-lines A, B, C are co-half-pencilar and $\mathbf{B}(A,B,C)$, then:*

(I)   *Half-lines A, B^*, C are not co-half-pencilar.*

(II) *Half-lines A^*, B^*, C are co-half-pencilar and $\mathbf{B}(C,A^*,B^*)$.*

PROOF. (I) By Theorem 66 (II), each half-pencil $\mathfrak{H} \subset \mathfrak{P}$ containing half-lines A and C, also contains half-line B, and therefore does not contain half-line B^*. Thus half-lines A, B^*, C are not co-half-pencilar.

(II) From $\mathbf{B}(A,B,C)$ it follows, by Theorem 66 (I), that half-lines B and C belong to the half-pencil with end half-lines A and A^*. Applying Theorem 65 we obtain $\mathbf{B}(B,C,A^*)$. Hence, repeating the argument, we conclude that half-lines C and A^* belong to the half-pencil with end half-lines B and B^* (and therefore, by Theorem 55, half-lines A^*, B^*, C are co-half-pencilar) and that $\mathbf{B}(C,A^*,B^*)$.

Finally, we give a necessary and sufficient condition that $\mathbf{B}(A,B,C)$.

THEOREM 68. *Given a pencil $\mathfrak{P}$, three co-half-pencilar half-lines $A,B,C \in \mathfrak{P}$ and points $a \in A$ and $c \in C$, we have $\mathbf{B}(A,B,C)$ if and only if $B \cap (ac) \neq 0$.*

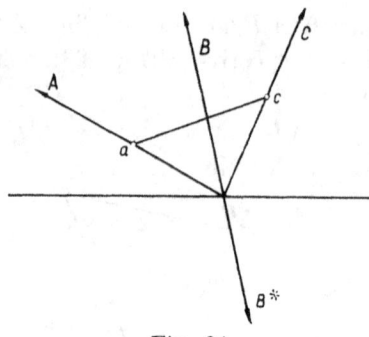

Fig. 24

PROOF. From $B \cap (ac) \neq 0$ it follows immediately that $\mathbf{B}(A,B,C)$. Let us next assume that $\mathbf{B}(A,B,C)$ (Fig. 24). Then, by Theorem 57, we have $\mathbf{B}(A,\mathbf{L}(B),C)$, and therefore

$$\mathbf{L}(B) \cap (ac) \neq 0.$$

On the other hand, by Theorem 67(I), half-lines A, B^*, C are not co-half-pencilar, and as a result

$$B^* \cap (ac) = 0.$$

Thus $B \cap (ac) \neq 0.$

22. Ordering of a Half-Pencil

Let us consider any half-pencil $\mathfrak{H}$ and let us pick one of its two end half-lines, say A. Let $B,C \in \mathfrak{H}$. If $B \neq C$ and $\mathbf{B}(A,B,C)$ (Fig. 25), then we say that *half-line B precedes half-line C* and we write $B \prec C$. By Theorem 64, the relation $\prec$ is connected and anti-symmetric, and from Theorem 62(IV) it follows that the relation $\prec$ is transitive. Hence:

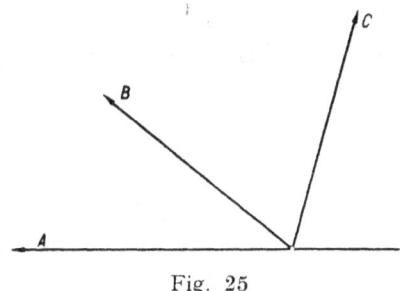

Fig. 25

THEOREM 69. *The relation $\prec$ orders the half-pencil $\mathfrak{H}$.*

We agree to say, more specifically, that *the relation $\prec$ orders half-pencil $\mathfrak{H}$ from the end half-line A to the end half-line A**.

Instead of $B \prec C$ we shall also write $C \succ B$ and read: *half-line C follows half-line B.*

Let us now take three half-lines $B,C,D \in \mathfrak{H}$ and let $\mathbf{B}(B,C,D)$. By Theorem 62(II)(III), if $B \prec D$, then $B \prec C \prec D$, and if $D \prec B$, then $D \prec C \prec B$. Hence we have a simple connection between the relations $\prec$ and $\mathbf{B}$: For any half-lines $B,C,D \in \mathfrak{H}$ we have $\mathbf{B}(B,C,D)$ if and only if $B \prec C \prec D$ or $D \prec C \prec B$. As a result of this equivalence and because of Theorem 61, it follows at once that the order of a half-pencil $\mathfrak{H}$ is dense without the first and last half-line.

If we single out the end half-line $A*$ instead of the end half-line A, then, by Theorem 65, we change the order of half-lines in $\mathfrak{H}$ to its inverse. To fix the order of half-lines in a half-pencil $\mathfrak{H}$, it is sufficient to determine which of any two arbitrarily chosen half-lines $B,C \in \mathfrak{H}$ we shall regard as being the earlier one. For instance, by assuming $B \prec C$ we single out the end half-line A for which $\mathbf{B}(A,B,C)$. We say that this is the order *from half-line B to half-line C.*

23. Angles

Given a pencil of half-lines $\mathfrak{P}$ with vertex p. Let us consider in pencil $\mathfrak{P}$ two co-half-pencilar half-lines A and B. Thus half-lines A and B have a common origin p and do not lie on one line. We shall call the non-ordered pair $\{AB\}$ the *angle AB* and denote it in formulas by AB or BA. Half-lines A and B are called the *sides of angle AB*, and point p is called its *vertex*. If $a \in A$ and $b \in B$, then we denote angle AB also by $\not< apb$ or $\not< bpa$.

In analogy to the notion of an open segment ab (consisting of all points c lying between points a and b) we introduce the notion of an *open angle AB* consisting of all half-lines $C \in \mathfrak{P}$ lying between half-lines A and B (Fig. 26). The open angle AB will be referred to in the text as *angle* (AB) and denoted in formulas by (AB). We also introduce the notation: $\langle AB)$ for the class $\{A\} \cup (AB)$ and $(AB\rangle$ for the class $(AB) \cup \{B\}$.

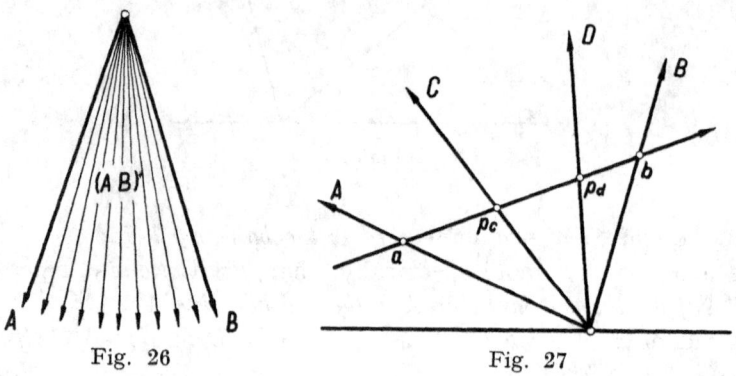

Fig. 26   Fig. 27

Consider any half-pencil $\mathfrak{H} \subset \mathfrak{P}$ such that $A,B \in \mathfrak{H}$ (Fig. 27). Then, by Theorem 66(II), we have $(AB) \subset \mathfrak{H}$. In half-pencil $\mathfrak{H}$, and consequently in angle (AB), let us fix the order from half-line A to half-line B. Further, let us take any points $a \in A$ and $b \in B$ and fix the order on segment (ab) (just as on a part of the line ab) as being from point a to point b. By Theorem 68, each half-line $C \in (AB)$ intersects segment (ab) in some point p_C. It is readily shown that for any half-lines $C,D \in (AB)$

$$C \prec D \quad \text{if and only if} \quad p_C \prec p_D.$$

Thus, the order of the angle (AB) from half-line A to half-line B does not depend on the choice of half-pencil $\mathfrak{H}$.

The sum (Fig. 28) of all half-lines $C \in (AB)$ (i.e. the set of points c for which there exists a half-line C such that $\mathbf{B}(A,C,B)$ and $c \in C$) will be called the *inner domain of angle AB*.

We shall prove:

THEOREM 70. *Given an angle AB with vertex p, let W_A be the half-plane with boundary $\mathbf{L}(A)$ containing side B, and let W_B be the half-plane with* *boundary* $L(B)$ *containing side* A. *Then the inner domain of angle* AB *coincides with the set* $W_A \cap W_B$.

PROOF. Choose any points $a \in A$ and $b \in B$ (Fig. 29). Now, if $B(A,C,B)$, then, by Theorem 68, there exists a point c such that $C = H(pc)$ and $B(a,c,b)$. We therefore have $\sim B(c,a,b)$ and $\sim B(a,b,c)$, from which it follows that $c \in W_A \cap W_B$ and $C \subset W_A \cap W_B$. Thus we have shown that the inner domain of angle AB is included in set $W_A \cap W_B$.

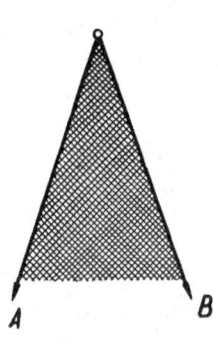

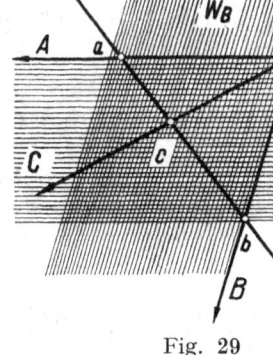

Fig. 28   Fig. 29

Next, suppose that $q \in W_A \cap W_B$. Then $C = H(pq) \subset W_A$ and, by our hypothesis, $B \subset W_A$. It therefore follows from Theorem 55 that there exists a line L intersecting half-lines A, B, C in some points a, b, c. Obviously

(1) $\sim B(c,a,b)$.

On the other hand, $C \subset W_B$ and, by our hypothesis, $A \subset W_B$, from which it follows

(2) $\sim B(a,b,c)$.

From formulas (1) and (2) we obtain $B(a,c,b)$, i.e. $B(A,C,B)$, and therefore point q belongs to the inner domain of angle AB. This is what we were required to prove.

24. Characterization of Lines and Planes in Terms of the Betweenness Relation

From Axioms O1 and O4 there follows at once:

THEOREM 71. *For any two distinct points* a *and* b, *the line* ab *is identical with the set of points* p *satisfying the following condition:*

$$p = a \ or \ p = b \ or \ B(a,b,p) \ or \ B(b,p,a) \ or \ B(p,a,b).$$

We shall now prove an analogous theorem for the plane:

THEOREM 72. *For any three non-collinear points a, b, c, the plane abc is identical with the set of points p satisfying the following condition:*

$$p \in L(bc) \text{ or } p \in L(ca) \text{ or } p \in L(ab)$$

$$\text{or } B(a,L(pb),c) \text{ or } B(b,L(pc),a) \text{ or } B(c,L(pa),b).$$

PROOF. If $p \in L(bc) \cup L(ca) \cup L(ab)$, then, obviously, $p \in P(abc)$. Assume now that $p \neq b$ and $B(a,L(pb),c)$. Thus for some point $b' \in L(pb)$ we have $B(a,b',c)$. Then $b' \neq b$, $p \in L(bb')$, $b' \subset L(ac) \subset P(abc)$, $L(bb') \subset P(abc)$. Hence $p \in P(abc)$. By cyclic permutation of the variables a, b, c we find that each of the conditions $B(b,L(pc),a)$ or $B(c,L(pa),b)$ also leads us to conclusion that $p \in P(abc)$.

Let us assume, conversely, that $p \in P(abc)$, and let

$$p \sim \in L(bc) \cup L(ca) \cup L(ab).$$

We consider the half-lines $A = H(pa)$, $B = H(pb)$, $C = H(pc)$. If half-lines A, B, C are co-half-pencilar, then, by Theorem 60, we have

$$B(A,B,C) \text{ or } B(B,C,A) \text{ or } B(C,A,B),$$

which implies

$$B(A,L(pb),C) \text{ or } B(B,L(pc),A) \text{ or } B(C,L(pa),B).$$

Hence

$$B(a,L(pb),c) \text{ or } B(b,L(pc),a) \text{ or } B(c,L(pa),b).$$

If half-lines A, B, C are not co-half-pencilar, then, by Theorem 56, we have $B(A,L(pb),C)$, and therefore $B(a,L(pb),c)$.

This concludes the proof.

Theorems 71 and 72 indicate the possibility of defining the notions of the line and the plane in terms of the relation B for points. Hence the system of primitive notions of geometry may be limited to the set of points S and to the two relations B and E (see p. 19). Obviously, the system of axioms would then have to undergo some modification.

25. Triangles. Open Triangles

The non-ordered triple $\{a,b,c\}$ of non-collinear points a, b, c will be called the *triangle abc*. We shall call points a, b, c the *vertices*, segments ab, bc, ac the *sides*, and angles bac, abc, acb the *(interior) angles of triangle abc*. These angles will also be called *angle a, angle b, angle c*, respectively, of triangle abc and will be denoted in formulas by the symbols $\sphericalangle a$,

$\measuredangle\, b$, $\measuredangle c$. We say that *angles a and b are adjacent to side ab* and that *angle c lies opposite side ab* or *is included between sides ca and cb.*

Let

$$W_a = \mathsf{HP}(\mathsf{L}(bc)a), \ \ W_b = \mathsf{HP}(\mathsf{L}(ac)b), \ \ W_c = \mathsf{HP}(\mathsf{L}(ab)c).$$

The common part of these three half-planes W_a, W_b, W_c (Fig. 30) will be called the *open triangle abc* or *triangle (abc)* and will be denoted in the formulas by (abc). Therefore

(1) $$(abc) = W_a \cap W_b \cap W_c.$$

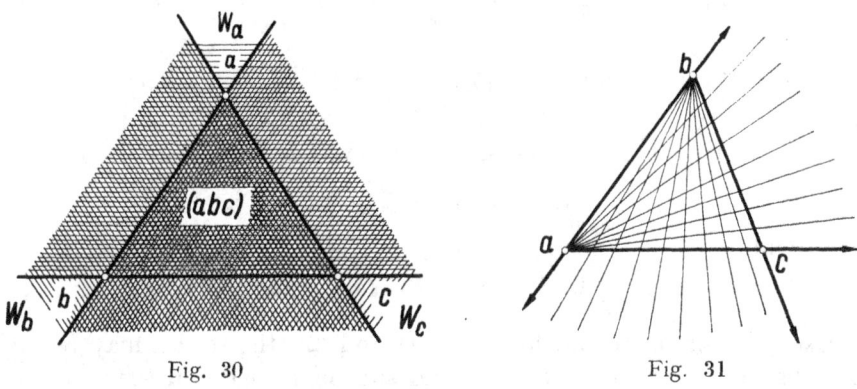

Fig. 30   Fig. 31

We have here an analogy to the notion of segment (ab) which, by Theorem 38, is the common part of half-lines ab and ba.

We shall leave it to the reader to show that

$$W_a \cup W_b \cup W_c = \mathsf{P}(abc),$$

from which it immediately follows that

$$(W_a^* \cup \mathsf{L}\,(bc)) \cap (W_b^* \cup \mathsf{L}\,(ac)) \cap (W_c^* \cup \mathsf{L}\,(bc)) = 0$$

and consequently

(2) $$W_a^* \cap W_b^* \cap W_c^* = 0.$$

It is at once seen from Theorem 70 that triangle (abc) may also be defined as the common part of the inner domains of any two of its angles (Fig. 31). The reader may readily show that an open triangle is a non-empty set.

We call points a, b, c the *vertices*, segments (ab), (bc), (ac) the *sides*, and the figure

$$F = (ab) \cup (bc) \cup (ac) \cup abc$$

the *boundary of triangle (abc)*. If we add to open triangle *abc* its boundary *F*, we obtain the *closed triangle abc*, which we also call *triangle ⟨abc⟩* and denote it in formulas by ⟨*abc*⟩.

It may readily be shown that

(3) $\langle abc \rangle = (W_a \cup L(bc)) \cap (W_b \cup L(ac)) \cap (W_c \cup L(ab)).$

In fact, because of (1)

$$W_a \cap W_b \cap W_c = (abc);$$

then

$$W_a \cap W_b \cap L(ab) = (W_a \cap L(ab)) \cap (W_b \cap L(ab)) = H(ba) \cap H(ab) = (ab);$$

similarly,

$$W_b \cap W_c \cap L(bc) = (bc), \quad W_a \cap W_c \cap L(ac) = (ac);$$

further

$$W_a \cap L(ab) \cap L(ac) = a, \quad W_b \cap L(ab) \cap L(bc) = b,$$

$$W_c \cap L(ac) \cap L(bc) = c;$$

and finally

$$L(ab) \cap L(bc) \cap L(ac) = 0.$$

Using Theorem 51 and formulas (1) and (2), the reader may readily establish the following necessary and sufficient condition that an open triangle contain an open segment:

THEOREM 73. *Given on a plane P a triangle (abc) and a segment (pq), we have (pq) ⊂ (abc) if and only if*

$$pq \subset \langle abc \rangle, \quad pq \sim \subset L(ab), \quad pq \sim \subset L(bc), \quad pq \sim \subset L(ac).$$

We shall now give a characteristic property of points belonging to an open triangle *(abc)*:

THEOREM 74. *Given a triangle (abc) and a point p on a plane P, in order that p ∈ (abc) it is necessary and sufficient that there exist a point q such that q ∈ (bc) and p ∈ (aq).*

PROOF. From Theorem 73 it follows immediately that this is a sufficient condition. We shall now show that this condition is also necessary. Let us suppose, therefore, that $p \in (abc)$. Then p belongs to the inner domain of angle *bac*, i.e., point p belongs to some half-line A lying between half-lines *ab* and *ac* (Fig. 32). By Theorem 68, there exists a point $q \in (bc)$ such that $A = H(aq)$, from which it follows that $\sim B(p,a,q)$. Since $\sim B(a, L(bc), p)$, then also $\sim B(a,q,p)$ and therefore $B(a,p,q)$, that is, $p \in (aq)$.

Triangle *(abc)* is therefore the sum of segments *(aq)* for $q \in (bc)$, that is, the sum of the segments cut off by line bc from the half-lines lying between half-lines ab and ac.

From Theorem 74 and 51 we obtain the following necessary and sufficient condition that a half-plane contain an open triangle:

THEOREM 75. *Given a half-plane W with boundary K and a triangle (abc) in a plane P, we have $(abc) \subset W$ if and only if $a,b,c \in K \cup W$.*

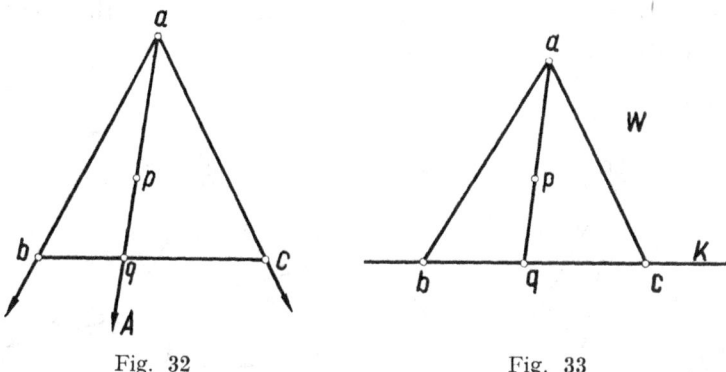

Fig. 32   Fig. 33

As a direct consequence of Theorem 75 we obtain (using formulas (1) and (3)) a necessary and sufficient condition that one open triangle be included in another:

THEOREM 76. *Given triangles (abc) and (pqr) in a plane P, we have $(abc) \subset (pqr)$ if and only if $a,b,c \in \langle pqr \rangle$.*

By means of Theorems 74 and 75, we shall prove

THEOREM 77. *For any point p of a half-plane W there exists a triangle $(abc) \subset W$ such that $p \in (abc)$.*

PROOF. Let K be the boundary of half-plane W. We choose points $b,q,c \in K$ and $a \in W$ such that $\mathbf{B}(b,q,c)$ and $\mathbf{B}(a,p,q)$ (Fig. 33). Then $p \in (abc) \subset W$.

26. The Open Triangle and the Line

Employing the notions introduced in the last section, we can give Axiom O9 a new, more intuitive form:

THEOREM 78. *Consider in a plane P a triangle (abc) and a line L. If line L intersects one of the sides of triangle (abc) and does not pass through any of its vertices, then line L also intersects at least one of the remaining sides of triangle (abc).*

In this form, Axiom O9 appears in the literature as the *Pasch Axiom*,

taking the name of the 19th century German mathematician Moritz
Pasch who first drew attention to the importance of this axiom to the
foundations of geometry.

From Theorem 48 we obtain:

THEOREM 79. *A line cannot simultaneously intersect all three sides of
a triangle (abc).*

We shall now give one more property characterizing the points of a
triangle (abc):

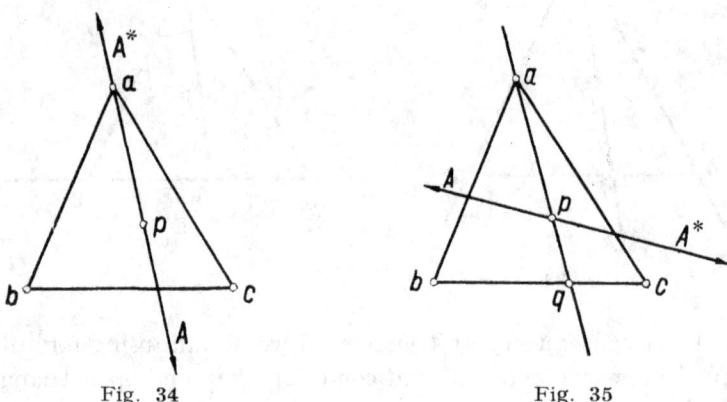

Fig. 34   Fig. 35

THEOREM 80. *Given in a plane P a triangle (abc) with boundary F and
a point p, in order that p $\in$ (abc) it is necessary and sufficient that every
half-line A $\subset$ P with origin p intersect boundary F in just one point.*

PROOF. Let us assume that $p \in (abc)$, and let us take a half-line $A \subset P$
with origin p. It is easy to show (by using the definition of the open triangle
and Theorem 74) that half-line A has at most one point in common with
boundary F. To find this point we shall distinguish between three cases:

Case 1. Half-line A passes through one of the three vertices a, b, c.
Then, obviously, $A \cap F \neq 0$.

Case 2. Half-line A^* passes through one of the vertices, say vertex a
(Fig. 34). Then, by Theorem 74, $A \cap (bc) \neq 0$.

Case 3. Line $L = A \cup p \cup A^*$ does not pass through any of the
vertices a, b, c (Fig. 35). Then, by Theorem 74, there is a point q such
that $\mathbf{B}(b,q,c)$ and $\mathbf{B}(a,p,q)$. Of the two points b and c, only one, for
example b, lies on the same side of line aq as half-line A. Applying
Pasch's Axiom to triangle (abq) and line L, we conclude that

$$A \cap ((ab) \cup (bq)) \neq 0.$$

Thus, in all cases $A \cap F \neq 0$.

Let us now assume that $p \sim \in(abc)$. Two cases are possible:

Case 1. Point p lies on one of the three lines ab, bc, ac, say on line ab. Then one of the half-lines of line ab with origin p has an infinite number of points in common with side (ab).

Case 2. $p \sim \in \mathbf{L}(ab) \cup \mathbf{L}(bc) \cup \mathbf{L}(ac)$. Then one of the three lines ab, bc, ac lies between point p and the vertex opposite this line, for example, $\mathbf{B}(p, \mathbf{L}(ab), c)$ (Fig. 36). Then $\mathbf{H}(ap) \cap F = 0$. Putting $A = \mathbf{H}(ap) — (ap\rangle$, we obtain a half-line with origin p and not intersecting boundary F.

This completes the proof of the theorem.

THEOREM 81. *Given in a plane P a triangle (abc) with boundary F and a point p, in order that $p \in (abc)$ it is necessary and sufficient that every line $L \subset P$ passing through point p intersect boundary F in just two points.*

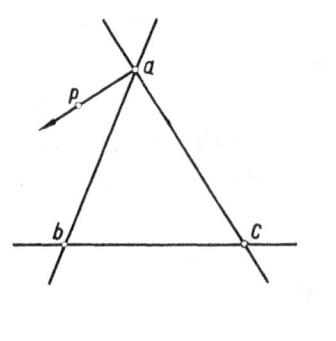

Fig. 36

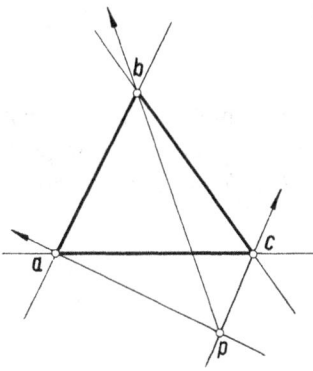

Fig. 37

PROOF. In one direction the theorem follows directly from Theorem 80. We assume now that $p \in P — (abc)$. If point p lies on one of the lines ab, bc, ac, then, obviously, this line has, by Theorem 19, more than two points in common with boundary F. Now, let us assume that $p \sim \in \mathbf{L}(ab) \cup \mathbf{L}(bc) \cup \mathbf{L}(ac)$ (Fig. 37). Then one of the following conditions holds:

$$\mathbf{B}(c, \mathbf{L}(ab), p) \quad \text{or} \quad \mathbf{B}(a, \mathbf{L}(bc), p) \quad \text{or} \quad \mathbf{B}(b, \mathbf{L}(ac), p).$$

For example, let $\mathbf{B}(b, \mathbf{L}(ac), p)$. Then line ac intersects each of the half-lines pa, pb, and pc, and therefore, these half-lines are co-half-pencilar. By Theorems 60 and 58(III), we have just one of the following:

$$\mathbf{B}(\mathbf{H}(pa), \mathbf{H}(pb), \mathbf{H}(pc)) \quad \text{or} \quad \mathbf{B}(\mathbf{H}(pb), \mathbf{H}(pc), \mathbf{H}(pa))$$

$$\text{or} \quad \mathbf{B}(\mathbf{H}(pc), \mathbf{H}(pa), \mathbf{H}(pb)).$$

For example, let $\mathbf{B}(\mathbf{H}(pa), \mathbf{H}(pb), \mathbf{H}(pc))$. Then $\sim \mathbf{B}(\mathbf{H}(pb), \mathbf{H}(pa), \mathbf{H}(pc))$, from which it follows that $\mathbf{L}(pa) \cap (bc) = 0$. Thus it can be seen at once that a is the only point common to line pa and boundary F.

From Theorem 81, with the help of Theorem 73, we obtain:

THEOREM 82. *The common part of any open triangle and a line lying in the plane of the triangle is either an empty set or an open segment.*

As a second conclusion from Theorem 81 we have:

THEOREM 83. *Given in a plane P two open triangles (abc) and (pqr), if $(abc) = (pqr)$, then $abc = pqr$.*

PROOF. Assume that $(abc) = (pqr)$. By Theorem 76,

(1) $$a,b,c \in \langle pqr \rangle$$

and

(2) $$p,q,r \in \langle abc \rangle.$$

Suppose that $p \in (abc)$. Then, by Theorem 81, line pq would intersect one of the sides of triangle (abc), for example, side (ab) (Fig. 38). Then $\mathbf{B}(a,\mathbf{L}(pq),b)$, and therefore, one of the two points a and b would not

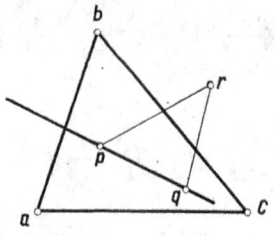

Fig. 38

belong to triangle $\langle pqr \rangle$, in contradiction to (1). Hence p does not belong to triangle (abc). In a similar manner, we may prove that p does not belong to side (ab), (bc), or (ac). Because of formula (2), point p must therefore be one of the three vertices a, b, c; a similar argument may be applied to vertices q and r. Therefore $abc = pqr$.

By Theorem 83, triangle (abc) uniquely determines its vertices a, b, c, its sides, and its boundary.

27. Topology on the Plane

Let P be any plane. Let us denote by $\mathfrak{U}$ the class of all open triangles included in plane P. Class $\mathfrak{U}$ has the following two properties:

(I) *For any two distinct points $p,q \in P$ there exist open triangles $U_1, U_2 \in \mathfrak{U}$ such that $p \in U_1$, $q \in U_2$ and $U_1 \cap U_2 = 0$.*

PROOF. By Theorem 47 (II) there exists on plane P a line K such that $\mathbf{B}(p,K,q)$. Let $P = W \cup K \cup W^*$, $p \in W$, $q \in W^*$. Then, by Theorem 77, there exist open triangles $U_1, U_2 \in \mathfrak{U}$ such that $p \in U_1 \subset W$, $q \in U_2 \subset W^*$. Thus we obviously have $U_1 \cap U_2 = 0$.

(II) *If* $U_1, U_2 \in \mathfrak{U}$ *and* $p \in U_1 \cap U_2$, *then there exists an open triangle* $U_0 \in \mathfrak{U}$ *such that* $p \in U_0$ *and* $U_0 \subset U_1 \cap U_2$.

PROOF. Let us produce two distinct lines K and L through point p. Let

$$I_i^{(K)} = U_i \cap K, \qquad I_i^{(L)} = U_i \cap L \quad (i = 1,2).$$

By Theorem 82, figures $I_1^{(K)}, I_2^{(K)}, I_1^{(L)}, I_2^{(L)}$ are open segments; by Theorem 27 the products

$$I^{(K)} = I_1^{(K)} \cap I_2^{(K)} \quad \text{and} \quad I^{(L)} = I_1^{(L)} \cap I_2^{(L)}$$

are also open segments. Certainly

$$I^{(K)}, \ I^{(L)} \subset U_1 \quad \text{and} \quad I^{(K)}, \ I^{(L)} \subset U_2.$$

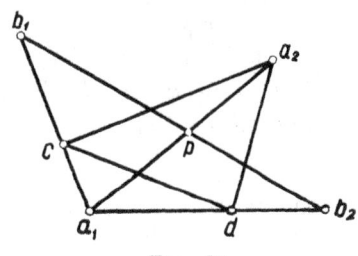

Fig. 39

Let $I^{(K)} = (a_1 a_2)$, $I^{(L)} = (b_1 b_2)$. Now, consider points c and d such that $\mathbf{B}(a_1, c, b_1)$ and $\mathbf{B}(a_1, d, b_2)$ (Fig. 39) and let $U_0 = (a_2 c d)$. It may readily be shown, by means of Theorems 74 and 76, that $p \in U_0$ and $U_0 \subset U_1 \cap U_2$.

As a result of the properties (I) and (II), we have:

THEOREM 84. *The class* $\mathfrak{U}$ *of all open triangles on a plane P constitutes a base of neighborhoods for P.*

Henceforth, by the *topology* on plane P we shall understand precisely that topology which is determined by the base $\mathfrak{U}$ of all open triangles of plane P. This topology induces a certain topology on each line $L \subset P$. It is readily shown that this topology coincides with the topology which we defined earlier on line L (in Section 11). To show this it is sufficient to note that the non-empty intersection of an open triangle with line L

is, by Theorem 82, an open segment, and that each open segment (ab) of line L can be regarded as the common part of line L and some open triangle (pqr) (Fig. 40).

Having the topology on plane P, we can introduce the notions of a *point of accumulation*, the *interior*, the *boundary*, and the *closure* of a set of points, the notions of an *open* set, a *closed* set, the *limit* of a sequence of points (on plane P), a *continuous* function of a point (defined on some set $E \subset P$). In addition, all theorems concerning general topological spaces (see Introduction, Section 9) can be proved for topological space P.

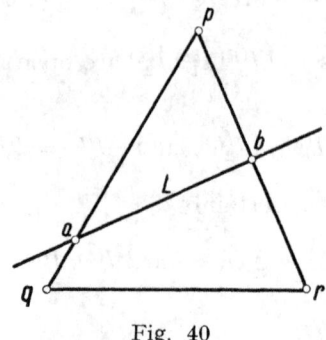

Fig. 40

In further course, we shall need the following special topological theorem concerning the operation of closure (we recall that the closure of a set F is denoted by $\bar{F}$):

THEOREM 85. *Given two figures F and G on a plane P, if G is an open set, then*

$$\overline{F \cap G} = \bar{F} \cap G.$$

PROOF. Certainly $\overline{F \cap G} \subset \bar{F} \cap G$. It is thus sufficient to show that $\bar{F} \cap G \subset \overline{F \cap G}$, which amounts to showing that $\bar{F} \cap G \subset \overline{F \cap G}$. Since $F \cap G \subset \overline{F \cap G}$, then we need only to show that if p is a point of accumulation of set F and $p \in G$, then $p \in \overline{F \cap G}$.

We take any neighborhood U of point p. Since $p \in G$ and G is an open set, then there exists a neighborhood $U' \subset U$ of point p entirely contained in G. Since p is a point of accumulation of set F, then an infinite number of points of set F lie in the neighborhood U'. Therefore an infinite number of points of the set $F \cap G$ lie in the neighborhood U' and hence in neighborhood U of point p, that is, point p is a point of accumulation of the set $F \cap G$. Thus $p \in \overline{F \cap G}$.

We shall now consider some topological properties of plane P. By Theorem 77, we have:

THEOREM 86. *Every half-plane $W \subset P$ is an open set.*

Let

(1) $$P = W \cup K \cup W^*.$$

Since half-plane W is an open set, then its complement $K \cup W^*$ (to plane P) is a closed set.

Take a point $a \in K$ and produce from it any half-line $A \subset W$. By Theorem 33, point a is a point of accumulation of half-line A, and therefore a point of accumulation of half-plane W. Thus $W \cup K \subset \bar{W}$. On the other hand, from the fact that $W \cup K$ is a closed set, it follows that $\bar{W} \subset W \cup K$. Thus

THEOREM 87. *For any half-plane $W \subset P$ with boundary K,*

$$\bar{W} = W \cup K.$$

From Theorem 86 and 87 it follows at once (see Introduction, Section 9) that line K is the boundary, in the topological sense, of each of the two half-planes W and W^*. This accounts for the name given it in Section 19.

Let us return to formula (1). Because of Theorem 86, the set $W \cup W^*$ is open, and therefore line K is a closed set. Therefore we have

THEOREM 88. *Every line $K \subset P$ is a closed set.*

Let us pass on to the triangle. Any triangle (abc) belongs to the base of neighborhoods, and therefore is an open set. This explains its name *open triangle*. Triangle $\langle abc \rangle$ is the product of three closed half-planes (see formula (3) on page 60) and therefore is a closed set. Hence, here too the term *closed triangle* finds justification.

Using methods similar to that for the half-plane, the reader can easily prove the following theorem:

THEOREM 89. *For any triangle $abc \subset P$, triangle $\langle abc \rangle$ is the closure of triangle (abc), and triangle (abc) is the interior of triangle $\langle abc \rangle$.*

It follows from this that the set

$$F = \langle abc \rangle - (abc) = (ab) \cup (bc) \cup (ac) \cup abc$$

is the boundary, in the topological sense, of the triangle (abc). This explains why in Section 25 we called F the boundary of triangle (abc).

Triangle $\langle abc \rangle$ determines triangle (abc) as its interior. Triangle (abc) determines, in turn, by Theorem 83, triangle abc. Thus

THEOREM 90. *Given in a plane P two closed triangles abc and pqr, if $\langle abc \rangle = \langle pqr \rangle$, then $abc = pqr$.*

By Theorem 90, the triangle ⟨abc⟩ determines: the points a, b, c, which we call the *vertices of the triangle* ⟨abc⟩; the segments ⟨ab⟩, ⟨bc⟩, ⟨ac⟩, which we call its *sides*; and finally, the angles a, b, c, which we call its *angles*.

28. Polygonal Lines

We shall call the finite sequence of distinct points $Z = (a_1, a_2, \ldots, a_n)$ a *polygonal line* (Fig. 41). The points $a_1, a_2, \ldots, a_n$ will be called the *vertices*, and the segments

$$a_1 a_2, \ a_2 a_3, \ \ldots, a_{n-1} a_n, \ a_n a_1$$

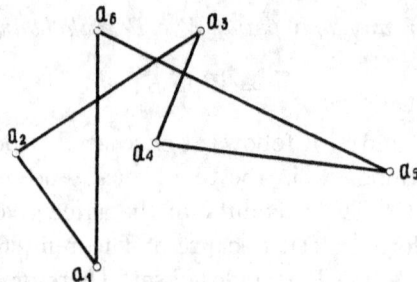

Fig. 41

the *sides of polygonal line* Z†. If pq is a pair of vertices which is not a side, then we shall call line pq a *diagonal* of polygonal line Z.

We say that two polygonal lines Z and Z' are *equivalent* if they have the same vertices and the same sides. Thus if, Z and Z' are equivalent and $Z = (a_1, a_2, \ldots, a_n)$, then $Z' = (a_k, a_{k+1}, \ldots, a_n, a_1, a_2, \ldots, a_{k-1})$ or $Z' = (a_k, a_{k-1}, \ldots, a_1, a_n, a_{n-1}, \ldots, a_{k+1})$ for some k, where $1 \leqslant k \leqslant n$. The relation of the equivalence of polygonal lines is reflexive, symmetric, and transitive. As a result, we can identify equivalent polygonal lines (or more accurately, we can replace every polygonal line by the class of all polygonal lines equivalent to it). Then segment ab and triangle abc can be included under the notion of the polygonal line.

We say that polygonal line $Z = (a_1, a_2, \ldots, a_n)$ is a *plane polygonal line* if points $a_1, a_2, \ldots, a_n$ are coplanar. We say that a plane polygonal line $Z = (a_1, a_2, \ldots, a_n)$ is *convex* if it satisfies the following condition: If segment pq is an arbitrary side of Z, then all vertices $a_i \neq p, q$ lie on the same side of line pq (Fig. 42). This condition may be somewhat weakened by restricting it to sides $pq \neq a_n a_1$:

† In this definition we restrict ourselves to a very special kind of polygonal lines, namely the *closed polygonal lines*, as polygonal lines of other kinds will not be needed in this book.

THEOREM 91. *In order for a plane polygonal line $Z = (a_1, a_2, \ldots, a_n)$ to be convex, it is sufficient that, for $i = 1, 2, \ldots, n-1$, all points $a_1, a_2, \ldots, a_{i-1}, a_{i+2} \ldots, a_n$ lie on the same side of line $a_i a_{i+1}$.*

PROOF. For $n \leqslant 3$ the theorem is obvious. Assume now that we have already proved the theorem for every polygonal line with $n \geqslant 3$ vertices. Let us consider the polygonal line $Z = (a_1, a_2, \ldots, a_{n+1})$ satisfying, for $i = 1, 2, \ldots, n$, the condition:

(1)   points $a_1, a_2, \ldots, a_{i-1}, a_{i+2}, \ldots, a_{n+1}$ lie on the same side of line $a_i a_{i+1}$

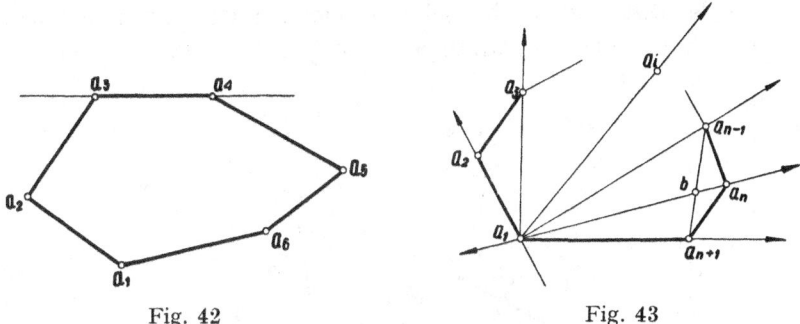

Fig. 42   Fig. 43

(Fig. 43). Setting $i = 1$, we see from condition (1) that all half-lines $a_1 a_i$ for $i = 2, 3, \ldots, n+1$ are co-half-pencilar. Because of our inductive assumption, it follows from condition (1) that polygonal line $Z' = (a_1, a_2, \ldots a_n)$ is convex. Therefore points $a_3, a_4, \ldots, a_{n-1}$ lie in the inner domain of angle $a_2 a_1 a_n$ and as a result

(2) $\mathbf{B}(\mathbf{H}(a_1 a_2), \mathbf{H}(a_1 a_i), \mathbf{H}(a_1 a_n))$ for $i = 3, 4, \ldots, n-1$.

Using condition (1) for $i = n-1$ and for $i = n$, we see that point a_1 lies in the interior of angle $a_{n-1} a_n a_{n+1}$. Therefore half-line $a_n a_1$ intersects segment $(a_{n-1} a_{n+1})$ in a point b which, by condition (1), belongs to segment $(a_1 a_n)$ and hence also to half-line $a_1 a_n$. Thus we have

(3) $\mathbf{B}(\mathbf{H}(a_1 a_{n-1}), \mathbf{H}(a_1 a_n), \mathbf{H}(a_1 a_{n+1}))$.

From formulas (2) and (3) we conclude, with the help of Theorems 62(III) (IV), that

$\mathbf{B}(\mathbf{H}(a_1 a_2), \mathbf{H}(a_1 a_i), \mathbf{H}(a_1 a_{n+1}))$ for $i = 3, 4, \ldots, n$.

Hence all points $a_3, a_4, \ldots, a_n$ lie in the inner domain of angle $a_2 a_1 a_{n+1}$, that is, on the same side of line $a_1 a_{n+1}$ as point a_2. This result and condition (1) lead us to the conclusion that polygonal line Z is convex.

The following is a direct consequence of the theorem we have just proved:

THEOREM 92. *If a polygonal line* $Z = (a_1, a_2, \ldots, a_n)$, *where* $n > 1$, *is convex, then the polygonal line* $Z' = (a_1, a_2 \ldots a_{n-1})$ *is also convex.*

From this theorem it follows, in general, that every sub-sequence of a convex polygonal line is a convex polygonal line.

29. Quadrangles

A polygonal line with four vertices will be called a *quadrangle*.

The reader will prove for himself the following necessary and sufficient condition that a plane quadrangle (a,b,c,d) be convex:

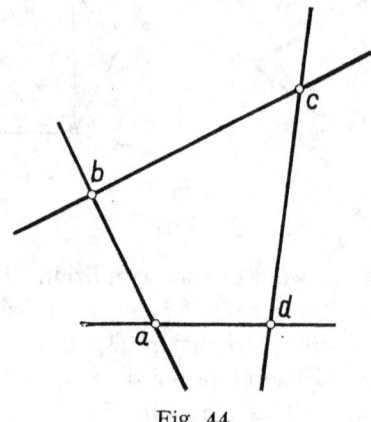

Fig. 44

THEOREM 93. *In order for a plane quadrangle* (a,b,c,d) *to be convex, it is necessary and sufficient that one of the following two conditions be satisfied*:
(I)   *diagonals* ac *and* bd *intersect in a point* $p \in (ac) \cap (bd)$;
(II) $\mathbf{B}(b,\mathbf{L}(ac),d)$ *and* $\mathbf{B}(a,\mathbf{L}(bd),c)$.

By the *angles of a convex quadrangle* (a,b,c,d) we shall understand the following:

$$\sphericalangle a = \sphericalangle dab, \quad \sphericalangle b = \sphericalangle abc, \quad \sphericalangle c = \sphericalangle bcd, \quad \sphericalangle d = \sphericalangle cda.$$

Let us consider a convex quadrangle (a,b,c,d) (Fig. 44). We denote by W_{ab} the half-plane with boundary $\mathbf{L}(ab)$ containing points c and d. We define half-planes W_{bc}, W_{cd}, W_{da} in a similar manner. The figure

(1) $(abcd) = W_{ab} \cap W_{bc} \cap W_{cd} \cap W_{da}$

is, as a product of four open sets (see Theorem 86), also an open set;

we shall call it an *open quadrangle*. We call its closure a *closed quadrangle* and denote it by $\langle abcd \rangle$.

It is readily shown that

(2) $$(abcd) = (abc) \cup (ac) \cup (acd).$$

From formula (2) and from Theorem 89 it follows that

(3) $$\langle abcd \rangle = \overline{(abcd)} = \langle abc \rangle \cup \langle acd \rangle.$$

Therefore

(4) $$\langle abcd \rangle = (abcd) \cup (ab) \cup (bc) \cup (cd) \cup (da) \cup abcd.$$

Proceeding as in the case of the triangle, we can easily show (see Theorem 87) that

$$\overline{W_{ab}} \cap \overline{W_{bc}} \cap \overline{W_{cd}} \cap \overline{W_{da}} = (abcd) \cup (ab) \cup (bc) \cup (cd) \cup (da) \cup abcd.$$

Therefore

(5) $$\langle abcd \rangle = \overline{W_{ab}} \cap \overline{W_{bc}} \cap \overline{W_{cd}} \cap \overline{W_{da}}.$$

By Theorem 51 and 75, we obtain from formula (2)

THEOREM 94. *Given a quadrangle $(abcd)$ and a half-plane W in a plane P, we have $(abcd) \subset W$ if and only if $a,b,c,d \in \bar{W}$.*

The following theorem results from Theorems 51, 75, and 94, on the basis of formulas (1) and (5):

THEOREM 95. *For any quadrangle $(abcd)$ in a plane P, the following holds:*

(I) *For any segment $(pq) \subset P$, we have $(pq) \subset (abcd)$ if and only if $pq \subset \langle abcd \rangle$, $pq \sim \subset L(ab)$, $pq \sim \subset L(bc)$, $pq \sim \subset L(cd)$, and $pq \sim \subset L(da)$.*

(II) *For any triangle $(pqr) \subset P$, we have $(pqr) \subset (abcd)$ if and only if $p,q,r \in \langle abcd \rangle$, and $(abcd) \subset (pqr)$ if and only if $a,b,c,d \in \langle pqr \rangle$.*

The proof is left to the reader.

Any convex polygonal line can be treated in a similar manner as the quadrangle.

30. The Intersection Point of the Diagonals of a Quadrangle

By Theorem 93(I), the diagonals ac and bd of a convex quadrangle (a,b,c,d) intersect in some point p (Fig. 45). If vertices a, b, c, d are variable, point p is a function of these vertices, say $p = f(a,b,c,d)$. Employing the topology introduced for the plane (cf. Section 27), we shall prove that f is a continuous function. The exact formulation of the theorem is as follows:

THEOREM 96. *Let p be the point of intersection of the diagonals ac and bd of a given convex quadrangle (a,b,c,d). For any neighborhood U_p of point p there exist neighborhoods U_a, U_b, U_c, U_d of points a, b, c, d, respectively, such that if $a' \in U_a$, $b' \in U_b$, $c' \in U_c$, $d' \in U_d$, then points a',b',c',d' are distinct, quadrangle (a',b',c',d') is convex, and point of intersection p' of its diagonals $a'c'$ and $b'd'$ belongs to the neighborhood U_p.*

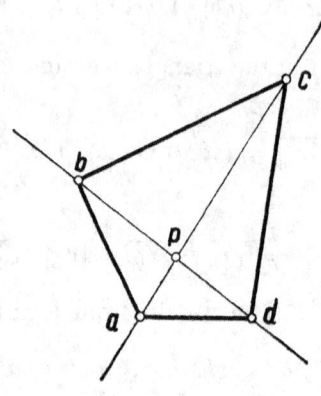

Fig. 45

PROOF. We may assume that the neighborhood U_p is some open triangle included in quadrangle $(abcd)$ (Fig. 46). We shall seek the neighborhoods U_a, U_b, U_c, U_d.

We choose a point e on half-line ca such that $c \prec p \prec a \prec e$. We also choose points c_1 and c_2 on half-line bc such that

(1) $b \prec c_1 \prec c \prec c_2,$

where

(2) $\mathbf{B}(a,\mathbf{L}(ec_2),d).$

(We leave it to the reader to show that this is always possible.) Then $\mathbf{B}(e,\mathbf{L}(bd),c_i)$ $(i = 1,2)$ and therefore half-line bd intersects segment (ec_i) in some point q_i. Obviously

(3) $b \prec q_1 \prec p \prec q_2 \prec d.$

By Theorem 82 line bd intersects open triangle U_p along an open segment I. Since segment I and segment (q_1q_2) both contain point p, their common part is then an open segment I' also containing point

p. Let us take on segment I' two points r_1 and r_2 such that $\mathbf{B}(p,r_1,b)$ and $\mathbf{B}(p,r_2d)$. Then, by formula (3), we have on half-line bd

(4) $$b \prec r_1 \prec p \prec r_2 \prec d.$$

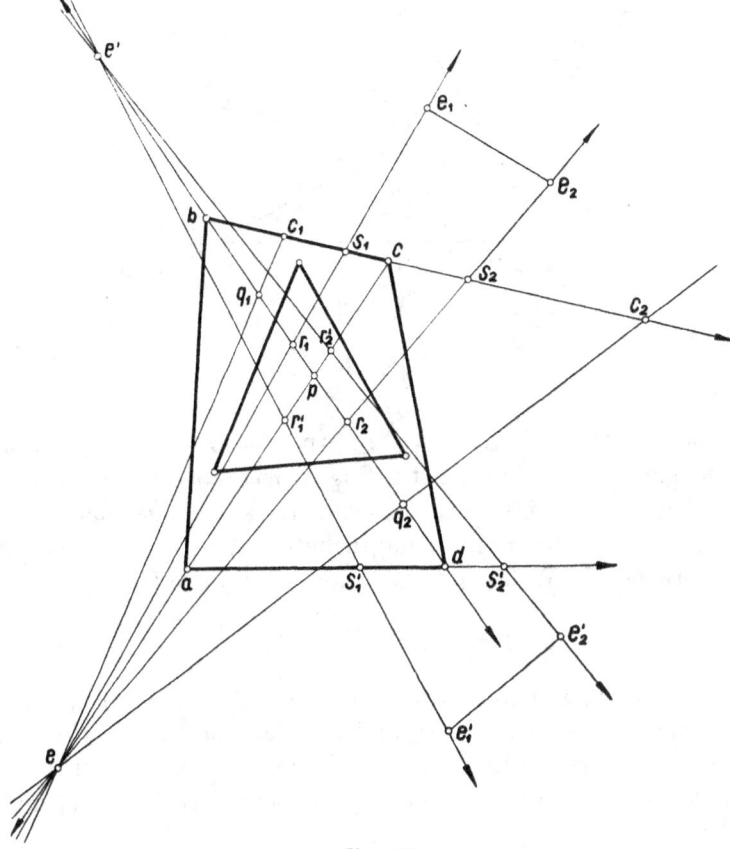

Fig. 46

Line er_i $(i = 1,2)$ intersects segment (cc_i) in some point s_i. On half-line bc, because of formula (1), we have, $b \prec s_1 \prec c \prec s_2$. By formula (2)

(5) $$\mathbf{B}(a,\mathbf{L}(es_2),d).$$

Taking a point e_i on half-line es_i such that $e \prec r_i \prec s_i \prec e_i$, we obtain an open triangle $U = (ee_1e_2)$ constituting a neighborhood of each of the points a and c and intersecting segment (bd) along segment (r_1r_2) the latter containing point p and being included, along with its end points, in neighborhood U_p. In addition, we have as a result of formula (4)

(6) $$\mathbf{B}(b,\mathbf{L}(ee_i),d) \quad \text{for} \quad i = 1,2.$$

Repeating the above construction, we take, on half-line db a point e' such that $d \prec p \prec b \prec e'$ and we find, on half-line ad, points s_1' and s_2' satisfying the condition $a \prec s_1' \prec d \prec s_2'$ and such that by choosing a

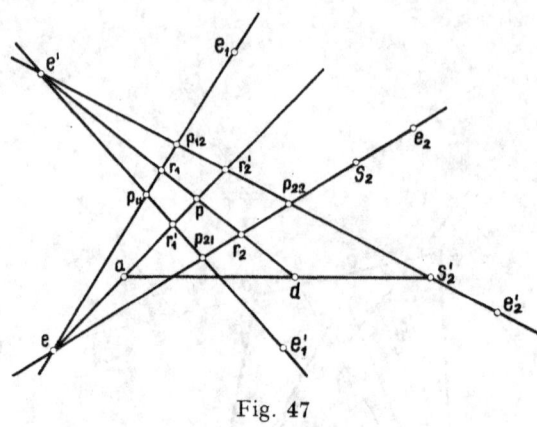

Fig. 47

point e_i' on half-line $e's_i'$ (for $i = 1,2$), where $e' \prec s_i' \prec e_i'$, we obtain an open triangle $U' = (e'e_1'e_2')$ constituting a neighborhood of both points b and d. This triangle intersects segment (ac) along a segment $(r'r_2')$, the latter containing point p and being included, along with its end points, in neighborhood U_p. We have for the sides of triangle U'

(7) $\mathbf{B}(a, \mathbf{L}(e'e_i'), c)$ for $i = 1,2$.

Applying Pasch's Axiom in turn: to triangle $e'pr_1'$ and line ee_1, to triangle $e'pr_2'$ and line ee_1, to triangle epr_2 and line $e'e_1'$, and to triangle $e'ds_2'$ and line ee_2 (Fig. 47), we are led to the result (in the last case, with the help of formula (5)) that lines ee_1 and ee_2 intersect both lines $e'e_1'$ and $e'e_2'$.
We let

$$p_{11} = \mathbf{L}(ee_1) \cap \mathbf{L}(e'e_1'), \qquad p_{12} = \mathbf{L}(ee_1) \cap \mathbf{L}(e'e_2'),$$
$$p_{21} = \mathbf{L}(ee_2) \cap \mathbf{L}(e'e_1'), \qquad p_{22} = \mathbf{L}(ee_2) \cap \mathbf{L}(e'e_2').$$

We may assume that points $p_{11}, p_{12}, p_{21}, p_{22}$ all belong to the neighborhood U_p. For, if e.g. $p_{11} \sim \in U_p$, then we can choose a point $p_{11}^* \in U_p$ (Fig. 48) on segment $(p_{11}r_1)$. Then line $e'p_{11}^*$ intersects segment $(e_1'e_2')$ in some point $\cdot e_1'^*$, and segment $(p_{21}r_2)$ in some point p_{21}^*, it being readily noted that if $p_{21} \in U_p$, then $p_{21}^* \in U_p$. Triangle $(e'e_1'e_2')$ is now replaced by triangle $(e'e_1'^*e_2')$. This operation must be repeated at most four times. We therefore assume that

(8) $p_{11}, p_{12}, p_{21}, p_{22} \in U_p$.

It is readily shown that on half-lines ee_1, ee_2, $e'e_1'$, $e'e_2'$ the points are ordered in the following manner (Fig. 49):

(9)
$$e \prec p_{11} \prec p_{12} \prec e_1, \quad e \prec p_{21} \prec p_{22} \prec e_2,$$
$$e' \prec p_{11} \prec p_{21} \prec e_1', \quad e' \prec p_{12} \prec p_{22} \prec e_2'.$$

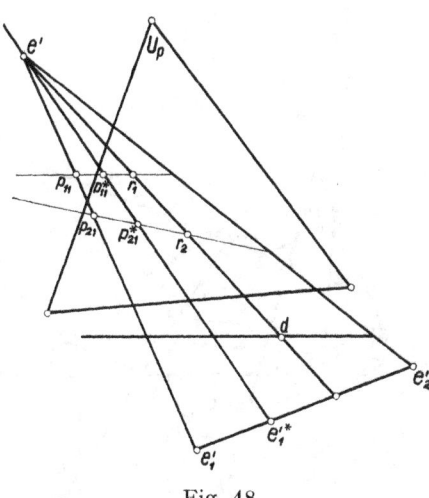

Fig. 48

It thus follows that the quadrangles

$$(p_{11}, p_{12}, p_{22}, p_{21}), \quad (p_{12}, p_{22}, e_2, e_1), \quad (p_{21}, p_{22}, e_2', e_1')$$

are convex and that

(10)
$$\mathbf{L}(ee_1) \cap U' = (p_{11}p_{12}), \quad \mathbf{L}(ee_2) \cap U' = (p_{21}p_{22}),$$
$$\mathbf{L}(e'e_1') \cap U = (p_{11}p_{21}), \quad \mathbf{L}(e'e_2') \cap U = (p_{12}p_{22}).$$

Now, let $V_p = (p_{11}p_{12}p_{22}p_{21})$ and

$$U_a = (ep_{11}p_{21}), \quad U_b = (e'p_{11}p_{12}), \quad U_c = (p_{12}p_{22}e_2e_1),$$
$$U_d = (p_{21}p_{22}e_2'e_1').$$

By using formulas (6)–(10) and Theorems 73, 74, 76, 94 we can easily show that the sets V_p, U_a, U_b, U_c, U_d are disjoint, that they are neighborhoods of points p, a, b, c, d, respectively, that $V_p \subset U_p$ and

(11)
$$U_a, U_c \subset U, \quad U_b, U_d \subset U',$$

and that

(12)
$$\mathbf{B}(U_b, \mathbf{L}(ee_i), U_d), \quad \mathbf{B}(U_a, \mathbf{L}(e'e_i'), U_c) \quad \text{for} \quad i = 1, 2.$$

Let us now take any points $a' \in U_a$, $b' \in U_b$, $c' \in U_c$, $d' \in U_d$. The points a', b', c', d' are obviously distinct. By formulas (10), (11), and (12), line $a'c'$ intersects sides $(p_{11}p_{21})$ and $(p_{12}p_{22})$, and line bd intersects sides $(p_{11}p_{12})$ and $(p_{21}p_{22})$ of quadrangle V_p. It then readily follows that

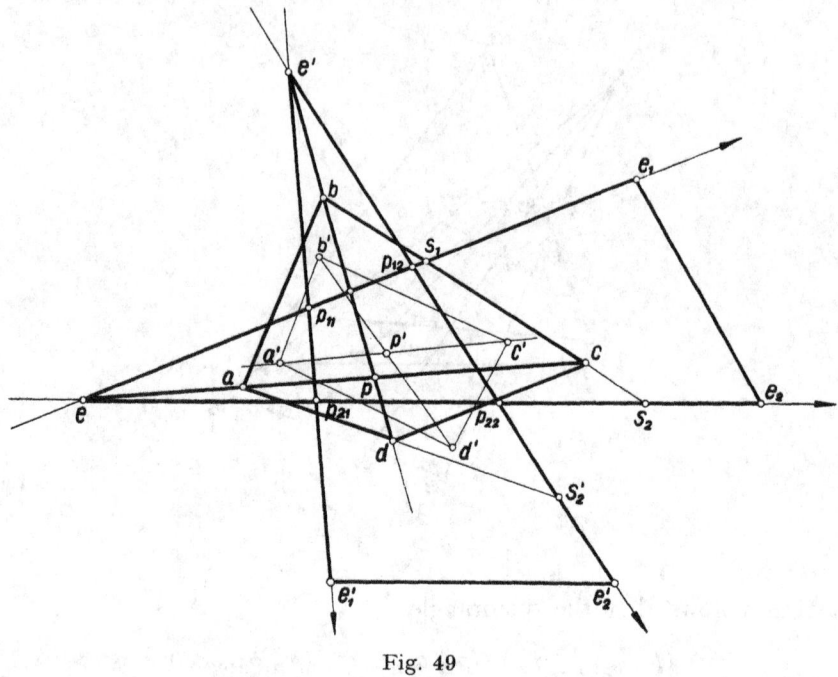

Fig. 49

lines $a'c'$ and $b'd'$ intersect in a point $p' \in V_p$. It is easy to notice that $p' \in (a'c') \cap (b'd')$. Therefore, by Theorem 93(I), rectangle (a',b',c',d') is convex.

This concludes the proof.

31. Polygons and Their Triangulation

A plane figure E which can be represented as a finite sum of closed triangles $T_1, T_2, \ldots T_n$ with disjoint interiors will be called a *polygon*. We call the set $\mathfrak{T} = \{T_1, T_2, \ldots, T_n\}$ a *triangulation*† of polygon E; by the *vertices of triangulation* $\mathfrak{T}$ we understand the vertices of triangles $T_1, T_2, \ldots, T_n$. The empty set, the closed triangle, and the closed quadrangle are special cases of a polygon.

---

† The notion of a triangulation as introduced here differs somewhat from that usually adopted in other fields of mathematics (e.g. algebraic topology). The definition adopted here is, however, more convenient for our purposes.

The following theorem immediately follows from the definition of a polygon:

THEOREM 97. *The finite sum of polygons lying in the same plane and having disjoint interiors is a polygon.*

We shall now investigate various figures and show that they are polygons.

Let us take any triangle (abc) and a half-plane $W \subset \mathbf{P}(abc)$ with boundary K. We shall investigate the figure $F = \overline{(abc) \cap W}$. If $K \cap (abc) = 0$, then it readily follows from the convexity of the open triangle and from Theorem 89 that either $F=0$ or $F=\langle abc \rangle$. Therefore, in this case, F is a polygon.

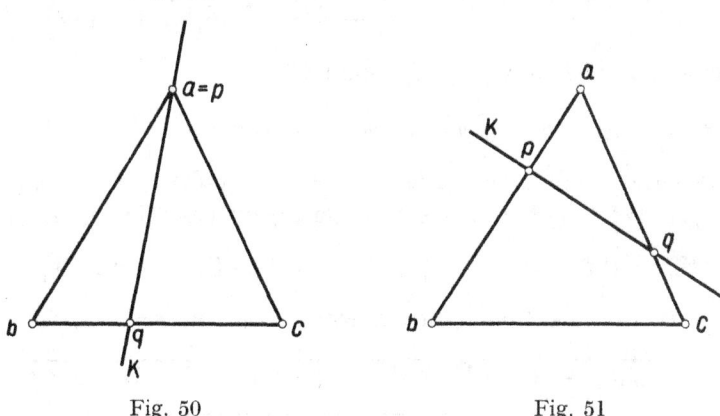

Fig. 50   Fig. 51

Let us next assume that $K \cap (abc) \neq 0$. By Theorem 81, line K then intersects the boundary of triangle (abc) in two distinct points, say p and q. We shall consider two possible cases:

Case 1. Point p is one of the vertices of triangle (abc), and point q lies on the side opposite this vertex, e.g. $p = a$ and $\mathbf{B}(b,q,c)$ (Fig. 50). Then either $F = \langle abq \rangle$ or $F = \langle acq \rangle$.

Case 2. Points p and q lie on two different sides of triangle (abc), e.g. $p \in (ab)$ and $q \in (ac)$ (Fig. 51). Then either $F = \langle apq \rangle$ or $F = \langle bpqc \rangle$. Thus we have proved

LEMMA 1. *The figure $\overline{(abc) \cap W}$ is a polygon.*

It follows from the lemma, by Theorem 85, that:

LEMMA 2. *The figure $\overline{\langle abc \rangle \cap W}$ is a polygon.*

With the help of Theorem 97 we can generalize Lemma 2 by replacing triangle $\langle abc \rangle$ by any polygon E:

THEOREM 98. *For any polygon E in a plane P and for any half-plane $W \subset P$, the figure $\overline{E \cap W}$ is a polygon.*

PROOF. If $\mathfrak{T}$ is a triangulation of polygon E, then

$$\overline{E \cap W} = \overline{(\bigcup_{T \in \mathfrak{T}} T) \cap W} = \overline{\bigcup_{T \in \mathfrak{T}} (T \cap W)} = \bigcup_{T \in \mathfrak{T}} \overline{T \cap W}.$$

We now take two triangles, a closed triangle pqr and an open triangle abc, both lying in a plane P. Triangle (abc) is the common part of three half-planes W_a, W_b, W_c (see page 59). With the help of Theorem 85 we obtain

$$\overline{\langle pqr \rangle \cap (abc)} = \overline{\langle pqr \rangle \cap W_a \cap W_b \cap W_c} =$$

$$= \overline{\overline{\langle pqr \rangle \cap W_a} \cap W_b \cap W_c} = \overline{\overline{\overline{\langle pqr \rangle \cap W_a} \cap W_b} \cap W_c},$$

from which there follows, by Theorem 98,

LEMMA 3. *The figure $\overline{\langle pqr \rangle \cap (abc)}$ is a polygon.*

Consider the figure $\langle pqr \rangle - \langle abc \rangle$. We have $\langle abc \rangle = \bar{W}_a \cap \bar{W}_b \cap \bar{W}_c$. Since $W_a^* \cap W_b^* \cap W_c^* = 0$ (see formula (2) on page 59), then

$$W_a^* \cup W_b^* \cup W_c^* = W_a^* \cap \bar{W}_b \cup W_b^* \cap \bar{W}_c \cup W_c^* \cap \bar{W}_a$$

(see Introduction, Section 8), from which, by Theorem 85, we obtain

$$F = \overline{\langle pqr \rangle - \bar{W}_a \cap \bar{W}_b \cap \bar{W}_c} = \overline{\langle pqr \rangle \cap (W_a^* \cup W_b^* \cup W_c^*)}$$

$$= \overline{\langle pqr \rangle \cap (W_a^* \cap \bar{W}_b \cup W_b^* \cap \bar{W}_c \cup W_c^* \cap \bar{W}_a)}$$

$$= \overline{\overline{\langle pqr \rangle \cap W_a^* \cap W_b} \cup \overline{\langle pqr \rangle \cap W_b^* \cap W_c} \cup \overline{\langle pqr \rangle \cap W_c^* \cap W_a}}.$$

By Theorem 98, each of the components of this sum is a polygon; it may readily be noted that these polygons have disjoint interiors. Hence, by Theorem 97, we obtain

LEMMA 4. *The figure $\langle pqr \rangle - \langle abc \rangle$ is a polygon.*

Lemma 4 can be generalized in the following manner:

LEMMA 5. *If a figure E is a polygon, then the figure $\overline{E - \langle abc \rangle}$ is a polygon.*

Indeed, if $\mathfrak{T}$ is a triangulation of polygon E, then

$$\overline{E - \langle abc \rangle} = \overline{(\bigcup_{T \in \mathfrak{T}} T) - \langle abc \rangle} = \bigcup_{T \in \mathfrak{T}} \overline{(T - \langle abc \rangle)}.$$

On the basis of Lemma 5 we arrive at the general theorem:

THEOREM 99. *If polygons E_1 and E_2 lie in one plane, then the figure $\overline{E_1 - E_2}$ is a polygon.*

PROOF. Let $\mathfrak{T} = \{T_1, T_2, \ldots T_n\}$ be a triangulation of polygon E_2. We use induction with respect to n. If $n = 1$, the theorem is obvious by Lemma 5. In general, for $n > 1$

$$\overline{W_1 - W_2} = \overline{W_1 - \bigcup_{i=1}^{n} T_i} = \overline{\overline{W_1 - T_n} - \bigcup_{i=1}^{n-1} T_i} \ ,$$

and the problem reduces to the case in which polygon E has a triangulation consisting of only $n - 1$ triangles.

Finally, we shall prove a theorem which establishes the relation between two triangulations of the same polygon (we recall that the interior of the set F is denoted by $\omega(F)$).

THEOREM 100. *If $\mathfrak{T}_1$ and $\mathfrak{T}_2$ are any two triangulations of a polygon E, then there exists a triangulation $\mathfrak{T}$ of polygon E such that each closed triangle $T \in \mathfrak{T}_1 \cup \mathfrak{T}_2$ has a triangulation $\mathfrak{T}_T \subset \mathfrak{T}$.*

PROOF. By Theorem 85, we have for any two triangles $T_1 \in \mathfrak{T}_1$ and $T_2 \in \mathfrak{T}_2$

$$\overline{\omega(T_1) \cap T_2} = \overline{\omega(T_1) \cap \omega(T_2)} = \overline{T_1 \cap \omega(T_2)}.$$

We put

$$E(T_1, T_2) = \overline{\omega(T_1) \cap T_2} = \overline{T_1 \cap \omega(T_2)}.$$

By Lemma 3 and Theorem 89, the figure $E(T_1, T_2)$ is a polygon. It is readily noted that if the ordered pair of triangles $T_1 \in \mathfrak{T}_1$ and $T_2 \in \mathfrak{T}_2$ is distinct from the ordered pair of triangles $T_1' \in \mathfrak{T}_1$ and $T_2' \in \mathfrak{T}_2$, then the interiors of polygons $E(T_1, T_2)$ and $E(T_1', T_2')$ are disjoint. Indeed, we have $E(T_1, T_2) \subset T_1 \cap T_2$; thus, $\omega(E(T_1, T_2)) \subset \omega(T_1) \cap \omega(T_2)$; similarly, we have $\omega(E(T_1', T_2')) \subset \omega(T_1') \cap \omega(T_2')$. If, therefore, e.g. $T_1 \neq T_1'$, then $\omega(T_1) \cap \omega(T_1') = 0$, from which it follows that figures $\omega(E(T_1, T_2))$ and $\omega(E(T_1', T_2'))$ are disjoint.

For any triangle $T_1 \in \mathfrak{T}_1$ we have

$$T_1 = \overline{\omega(T_1)} = \overline{\omega(T_1) \cap E} = \bigcup_{T_2 \in \mathfrak{T}_2} \overline{\omega(T_1) \cap T_2} = \bigcup_{T_2 \in \mathfrak{T}_2} E(T_1, T_2).$$

Similarly, for any triangle $T_2 \in \mathfrak{T}_2$, we have

$$T_2 = \overline{\omega(T_2)} = \overline{E \cap \omega(T_2)} = \bigcup_{T_1 \in \mathfrak{T}_1} \overline{T_1 \cap \omega(T_2)} = \bigcup_{T_1 \in \mathfrak{T}_1} E(T_1, T_2).$$

Let $\mathfrak{T}(T_1, T_2)$ be any triangulation of polygon $E(T_1, T_2)$. Summing up all triangulations $\mathfrak{T}(T_1, T_2)$, where $T_1 \in \mathfrak{T}_1$, $T_2 \in \mathfrak{T}_2$, we obtain the sought-for triangulation $\mathfrak{T}$ of polygon E.

Axioms of Congruence

1. Axioms of Congruence. Congruence of Segments. Congruence of Figures

For the time being, we shall give only the first three axioms of congruence which characterize the primitive relation **E**. We recall that the formula **E**$(a,b;c,d)$ is to be read: *point a is just as far from point b as point c is from point d.*

AXIOM C1. *If* **E**$(a,a;p,q)$, *then* $p = q$.

AXIOM C2. **E**$(a,b;b,a)$.

AXIOM C3. *If* **E**$(a,b;p,q)$ *and* **E**$(a,b;r,s)$, *then* **E**$(p,q;r,s)$.

As a simple consequence of these three axioms we obtain:

THEOREM 1. **E**$(a,b;a,b)$.

PROOF. By Axiom C2,

$$\mathbf{E}(a,b;b,a) \quad \text{and} \quad \mathbf{E}(a,b;b,a),$$

from which it follows, by Axiom C3, that **E**$(b,a;b,a)$. Replacing variables "a" and "b" by "b" and "a", respectively, we obtain the required theorem.

THEOREM 2. *If* **E**$(a,b;c,d)$, *then* **E**$(c,d;a,b)$.

PROOF. Assume that **E**$(a,b;c,d)$. Since also **E**$(a,b;a,b)$ (see Theorem 1), we have, by Axiom C3, **E**$(c,d;a,b)$.

THEOREM 3. *If* **E**$(a,b;c,d)$ *and* **E**$(c,d;p,q)$, *then* **E**$(a,b;p,q)$.

PROOF. From **E**$(a,b;c,d)$ it follows, by Theorem 2, that **E**$(c,d;a,b)$. Since also **E**$(c,d;p,q)$, we then have **E**$(a,b;p,q)$.

THEOREM 4. *If* **E**$(a,b;c,d)$, *then* **E**$(b,a;c,d)$.

PROOF. From **E**$(a,b;b,a)$ (see Axiom C2) and **E**$(a,b;c,d)$ it follows, by Axiom C3, that **E**$(b,a;c,d)$.

THEOREM 5. *If* **E**$(a,b;c,d)$, *then* **E**$(a,b;d,c)$.

PROOF. From $E(a,b;c,d)$ it follows, by Theorem 2, that $E(c,d;a,b)$; therefore, by Theorem 4, we have $E(d,c;a,b)$, which implies $E(a,b;d,c)$.

On the basis of Theorems 4 and 5 we can now define the *congruence relation* for segments in the following way: Given two arbitrary segments *ab* and *cd*, we say that *segment ab is congruent to segment cd* —in symbols, $ab \equiv cd$ — if and only if $E(a,b;c,d)$.

From Theorems 1–3 it follows immediately that the relation $\equiv$ has the following properties:

THEOREM 6. *For any segments ab, cd, pq:*
(I) $ab \equiv ab$;
(II)   *if* $ab \equiv cd$, *then* $cd \equiv ab$;
(III) *if* $ab \equiv cd$ *and* $cd \equiv pq$, *then* $ab \equiv pq$.

We shall say that *a figure F_1 is congruent to a figure F_2* — in symbols, $F_1 \equiv F_2$ — if there exists a one-to-one transformation f of figure F_1 onto figure F_2 such that

$$pq \equiv f(p)f(q)$$

for any two distinct points $p,q \in F_1$. We refer to transformation f as a *congruence* function, and we say that *it realizes the congruence of figure F_1 to figure F_2*. From the reflexivity, symmetry, and transitivity of the congruence relation for segments, it follows that the congruence relation for any arbitrary figures is reflexive, symmetric and transitive.

We now give the remaining axioms of congruence. Further properties of the relation $\equiv$ for segments are formulated in these axioms. Of course, they can be formulated in terms of the primitive relation E as well.

AXIOM C4. *If* $B(a_1,b_1,c_1)$, $B(a_2,b_2,c_2)$, $a_1b_1 \equiv a_2b_2$, *and* $b_1c_1 \equiv b_2c_2$, *then* $a_1c_1 \equiv a_2c_2$.

AXIOM C5. *For every half-line A with origin a and for every segment pq there exists just one point $b \in A$ such that* $ab \equiv pq$.

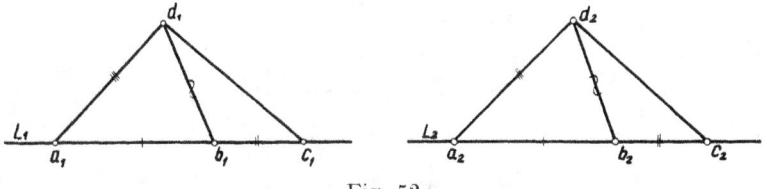

Fig. 52

AXIOM C6. *Given lines L_1 and L_2 and points $a_1,b_1,c_1 \in L_1$, $d_1 \sim \in L_1$, $a_2,b_2,c_2 \in L_2$, $d_2 \sim \in L_2$, if* $B(a_1,b_1,c_1)$, $B(a_2,b_2,c_2)$, $a_1b_1 \equiv a_2b_2$, $b_1c_1 \equiv b_2c_2$, $d_1a_1 \equiv d_2a_2$ *and* $d_1b_1 \equiv d_2b_2$, *then* $d_1c_1 \equiv d_2c_2$ (Fig. 52).

AXIOM C7. *Given a half-plane W with boundary K, a segment $ab \subset K$, and a triangle pqr, if $ab \equiv pq$, then there exists just one point $c \in W$ such that $ac \equiv pr$ and $bc \equiv qr$.*

2. Relations Between Segments on Two Lines

We supplement Axiom C4 by the following:

THEOREM 7. *If* $\mathbf{B}(a_1,b_1,c_1)$, $\mathbf{B}(a_2,b_2,c_2)$, $a_1b_1 \equiv a_2b_2$, *and* $a_1c_1 \equiv a_2c_2$, *then* $b_1c_1 \equiv b_2c_2$.

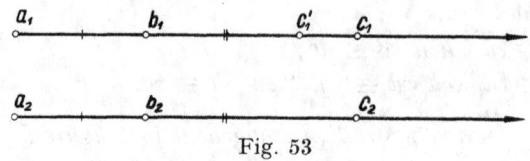

Fig. 53

PROOF. By Axiom C5 there exists a point c_1' on half-line b_1c_1 (Fig. 53) such that

(1) $$b_1c_1' \equiv b_2c_2.$$

From the construction it further results that $\mathbf{B}(a_1,b_1,c_1')$. Because of Axiom C4 we have $a_1c_1' \equiv a_2c_2$, which, by Theorem 6, gives $a_1c_1' \equiv a_1c_1$. It is readily seen that points c_1 and c_1' lie on line a_1b_1 on the same side of point a_1 and, by Theorem 6 (I) and Axiom C5, we have $c_1' = c_1$. Because of formula (1) we thus have $b_1c_1 = b_2c_2$, which was to be proved.

The following theorems, may be said to be the converse of Axiom C4 and Theorem 7:

THEOREM 8. *We assume that* $\mathbf{B}(a_1,b_1,c_1)$.

(I) *If* $a_1c_1 \equiv a_2c_2$, *then there exists a point* b_2 *such that* $\mathbf{B}(a_2,b_2,c_2)$, $a_1b_1 \equiv a_2b_2$, *and* $b_1c_1 \equiv b_2c_2$.

(II) *If* $b_1c_1 \equiv b_2c_2$, *then there exists a point* a_2 *such that* $\mathbf{B}(a_2,b_2,c_2)$, $a_1b_1 \equiv a_2b_2$, *and* $a_1c_1 \equiv a_2c_2$.

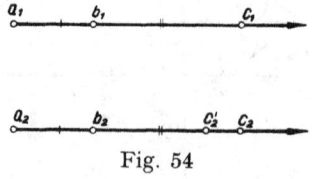

Fig. 54

PROOF. (I) Let us take points b_2 and c_2' on half-line a_2c_2 such that $\mathbf{B}(a_2,b_2,c_2')$, $a_1b_1 \equiv a_2b_2$, and $b_1c_1 \equiv b_2c_2'$ (Fig. 54). By Axiom C4, we have

$a_1c_1 \equiv a_2c_2'$, and hence, by Axiom C5, it follows that $c_2' = c_2$ and b_2 is the point we are seeking.

(II) Let us take a point a_2 such that $\mathbf{B}(a_2,b_2,c_2)$ and $a_1b_1 \equiv a_2b_2$. Then, by Axiom C4, we have $a_1c_1 \equiv a_2c_2$, and therefore a_2 is the point we are seeking.

The last theorem of this section deals with segments on two half-lines.

THEOREM 9. *Let there be given a half-line A_1 with origin p_1, a half-line A_2 with origin p_2, and points $a_1,b_1 \in A_1$, $a_2,b_2 \in A_2$ such that $p_1a_1 \equiv p_2a_2$ and $p_1b_1 \equiv p_2b_2$. Then $a_1 = b_1$ implies $a_2 = b_2$, and $\mathbf{B}(p_1,a_1,b_1)$ implies $\mathbf{B}(p_2,a_2,b_2)$.*

PROOF. If $a_1 = b_1$, then $p_2a_2 \equiv p_1a_1 = p_1b_1 \equiv p_2b_2$. Hence $p_2a_2 \equiv p_2b_2$, from which it follows that $a_2 = b_2$.

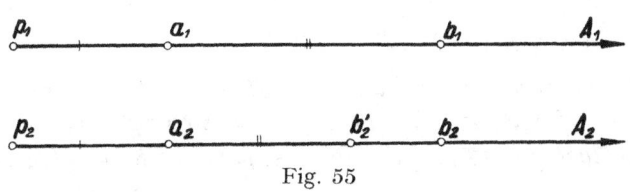

Fig. 55

Let us next assume that $\mathbf{B}(p_1,a_1,b_1)$ (Fig. 55). Since $p_1a_1 \equiv p_2a_2$, then, by Theorem 8(II), there exists a point b_2' such that $\mathbf{B}(p_2,a_2,b_2')$ and $p_1b_1 = p_2b_2'$. It is obvious that $b_2' \in A_2$. Since $p_1b_1 \equiv p_2b_2$, then $b_2' = b_2$, and therefore $\mathbf{B}(p_2,a_2,b_2)$, which was to be proved.

3. Relations Between Segments on Two Planes

We supplement Axiom C6 by the following theorem:

THEOREM 10. *Given lines L_1 and L_2 and points $a_1,b_1,c_1 \in L_1$, $d_1 \sim \in L_1$, $a_2,b_2,c_2 \in L_2$, $d_2 \sim \in L_2$, if $\mathbf{B}(a_1,b_1,c_1)$, $\mathbf{B}(a_2,b_2,c_2)$, $a_1b_1 \equiv a_2b_2$, $b_1c_1 \equiv b_2c_2$, $d_1a_1 \equiv d_2a_2$ and $d_1c_1 \equiv d_2c_2$, then $d_1b_1 \equiv d_2b_2$.*

PROOF. Let $W_1 = \mathbf{HP}(L_1d_1)$ (Fig. 56). Since $a_1b_1 \equiv a_2b_2$, then by Axiom C7 there exists a point $d_1' \in W_1$ such that $d_1'a_1 \equiv d_2a_2$ and $d_1'b_1 \equiv d_2b_2$. Since

(1) $\mathbf{B}(a_1,b_1,c_1)$, $\mathbf{B}(a_2,b_2,c_2)$ and $a_1b_1 \equiv a_2b_2$, $b_1c_1 \equiv b_2c_2$,

then, by Axiom C6, we also have $d_1'c_1 \equiv d_2c_2$. Thus, we have simultaneously

$$d_1'a_1 \equiv d_2a_2, \; d_1'c_1 \equiv d_2c_2 \quad \text{and} \quad d_1a_1 \equiv d_2a_2, \; d_1c_1 \equiv d_2c_2.$$

Then from (1) it follows that $a_1c_1 \equiv a_2c_2$. Therefore, by Axiom C7, point d_1' coincides with point d_1, and hence $d_1b_1 \equiv d_2b_2$, which was to be shown.

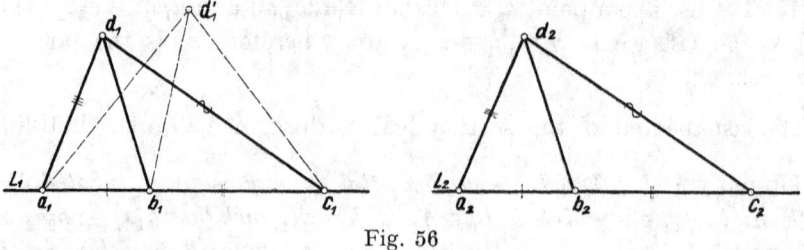

Fig. 56

4. Congruence of Angles

We say that *angle A_1B_1 with vertex p_1 is congruent to angle A_2B_2 with vertex p_2* —in symbols, $A_1B_1 \equiv A_2B_2$ — if there exist points $a_1 \in A_1$, $b_1 \in B_1$, $a_2 \in A_2$, $b_2 \in B_2$ such that $p_1a_1 \equiv p_2a_2$, $p_1b_1 \equiv p_2b_2$ and $a_1b_1 \equiv a_2b_2$ (Fig. 57).

We shall now prove the following:

THEOREM 11. *Given an angle A_1B_1 with vertex p_1 and an angle A_2B_2 with vertex p_2, if $A_1B_1 \equiv A_2B_2$, then, for any points $c_1 \in A_1, d_1 \in B_1, c_2 \in A_2$, $d_2 \in B_2$ such that $p_1c_1 \equiv p_2c_2$ and $p_1d_1 \equiv p_2d_2$, we have $c_1d_1 \equiv c_2d_2$.*

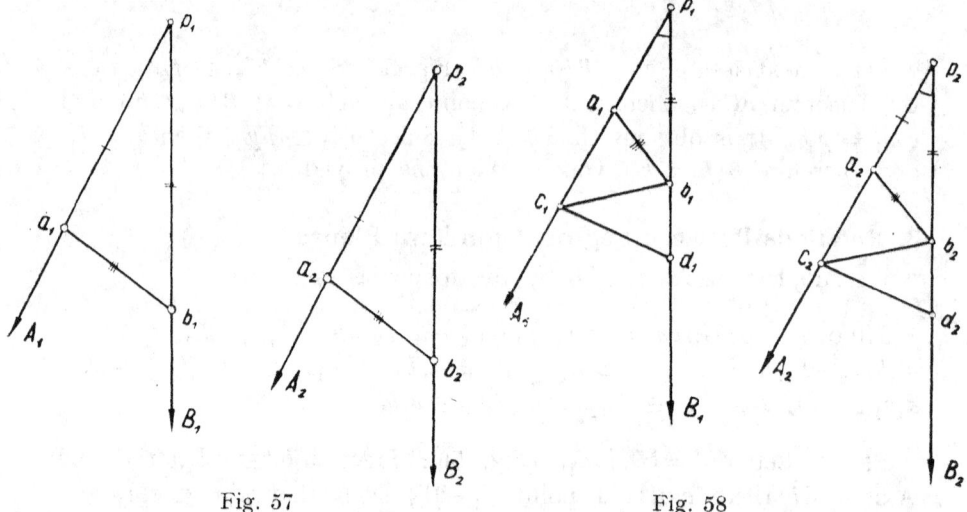

Fig. 57   Fig. 58

PROOF. Let us assume that $A_1B_1 \equiv A_2B_2$. From the definition of the congruence relation for angles, it follows that there exist points $a_1 \in A_1$, $b_1 \in B_1$, $a_2 \in A_2$ and $b_2 \in B_2$ (Fig. 58) such that

(1) $p_1a_1 \equiv p_2a_2$,

(2) $p_1b_1 \equiv p_2b_2$,

(3) $a_1b_1 \equiv a_2b_2$.

We now take any points $c_1 \in A_1$, $d_1 \in B_1$, $c_2 \in A_2$, $d_2 \in B_2$ such that

(4) $$p_1 c_1 \equiv p_2 c_2,$$

(5) $$p_1 d_1 \equiv p_2 d_2.$$

If $c_1 = a_1$ and $d_1 = b_1$, then because of formulas (1), (2), (4), (5), and Theorem 9 we conclude that $c_2 = a_2$ and $d_2 = b_2$, which, by (3), gives $c_1 d_1 \equiv c_2 d_2$.

Let us next assume that at least one of the following two conditions is satisfied: $c_1 \neq a_1$ or $b_1 \neq d_1$. For example, let $a_1 \neq c_1$. Then either $\mathbf{B}(p_1, a_1, c_1)$ or $\mathbf{B}(p_1, c_1, a_1)$. If $\mathbf{B}(p_1, a_1, c_1)$, then by formulas (1), (4), and Theorem 9 we also have $\mathbf{B}(p_2, a_2, c_2)$. Hence by formulas (1), (4), and Theorem 7 we have $a_1 c_1 \equiv a_2 c_2$, and because of formulas (1), (2), and (3) it follows, with the help of Axiom C6, that $c_1 b_1 \equiv c_2 b_2$. If $\mathbf{B}(p_1, c_1, a_1)$, we arrive at the same formula by using the same argument as above, except that instead of Axiom C6 we use Theorem 10. Thus, we always have

(6) $$c_1 b_1 \equiv c_2 b_2.$$

If $d_1 = b_1$, then $d_2 = b_2$ and from (6) it follows that $c_1 d_1 \equiv c_2 d_2$. If $d_1 \neq b_1$, then an argument similar to that which led to formula (6) gives $c_1 d_1 \equiv c_2 d_2$.

This concludes the proof.

From the reflexivity and symmetry of the congruence relation for segments it follows at once that the congruence relation for angles is reflexive and symmetric. From Theorem 11 and from the transitivity of the congruence relation for segments, it at once follows that the congruence relation for angles is transitive. To show this we take three angles: $A_1 B_1$ with vertex p_1, $A_2 B_2$ with vertex p_2, and $A_3 B_3$ with vertex p_3, and we assume that

(7) $$A_1 B_1 \equiv A_2 B_2 \quad \text{and} \quad A_2 B_2 \equiv A_3 B_3.$$

We choose points $a_1 \in A_1$, $b_1 \in B_1$, $a_2 \in A_2$, $b_2 \in B_2$, $a_3 \in A_3$, $b_3 \in B_3$ such that

(8) $$p_1 a_1 \equiv p_2 a_2 \equiv p_3 a_3 \quad \text{and} \quad p_1 b_1 \equiv p_2 b_2 \equiv p_3 b_3.$$

Because of Theorem 11 and by formulas (7) and (8) we have $a_1 b_1 \equiv a_2 b_2$ and $a_2 b_2 \equiv a_3 b_3$, which implies $a_1 b_1 \equiv a_3 b_3$. Then it follows from (8) that $p_1 a_1 \equiv p_3 a_3$ and $p_1 b_1 \equiv p_3 b_3$. Hence, from the definition of the congruence relation for angles we get $A_1 B_1 \equiv A_3 B_3$.

We therefore have:

THEOREM 12. *For any angles* A_1B_1, A_2B_2, A_3B_3:
(I) $A_1B_1 \equiv A_1B_1$;
(II)   *if* $A_1B_1 \equiv A_2B_2$, *then* $A_2B_2 \equiv A_1B_1$;
(III) *if* $A_1B_1 \equiv A_2B_2$ *and* $A_2B_2 \equiv A_3B_3$, *then* $A_1B_1 \equiv A_3B_3$.

We shall now prove for angles a theorem analogous to Axiom C7.

THEOREM 13. *For every half-plane W with boundary K, for every half-line A ⊂ K with origin p, and for every angle CD there exists just one half-line B ⊂ W with origin p such that* $AB \equiv CD$.

PROOF. We denote the vertex of angle CD by q. Let us now take any two points $c \in C$ and $d \in D$ and a point $a \in A$ such that $pa \equiv qc$ (Fig. 59).

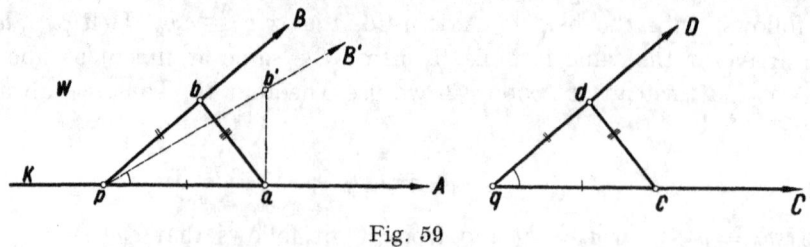

Fig. 59

Because of Axiom C7 there exists a point $b \in W$ such that $pb \equiv qd$ and $ab \equiv cd$. Putting $B = \mathbf{H}(pb)$, we have $AB \equiv CD$. If, besides half-line B, there were to exist still another half-line $B' \subset W$ with origin p such that $AB' \equiv CD$, then by choosing on it a point b' such that $pb' \equiv qd$, we would have, by Theorem 11, $ab' \equiv cd$. But, by Axiom C7, point b' would have to coincide with point b and therefore $B'=B$.

5. Adjacent Angles. Vertical Angles

We say that two angles are *adjacent* if they have one common side and if the two remaining sides are complementary half-lines.

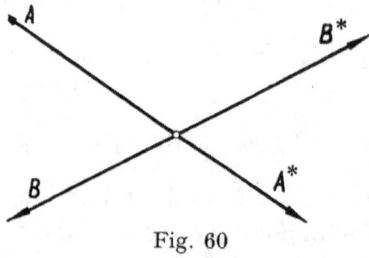

Fig. 60

Thus the angle adjacent to angle AB is the angle $A*B$, as well as the angle $AB*$ (Fig. 60).

THEOREM 14. *Angles adjacent to congruent angles are congruent. In other words, if $A_1B_1 \equiv A_2B_2$, then $A_1^*B_1 \equiv A_2^*B_2$.*

PROOF. Let point p_1 be the vertex of angle A_1B_1 and point p_2 the vertex of angle A_2B_2 (Fig. 61). We select any points $a_1 \in A_1$, $b_1 \in B_1$, $c_1 \in A_1^*$ and we take points $a_2 \in A_2$, $b_2 \in B_2$, $c_2 \in A_2^*$ such that

(1) $$p_1a_1 \equiv p_2a_2,$$

(2) $$p_1b_1 \equiv p_2b_2,$$

(3) $$p_1c_1 \equiv p_2c_2.$$

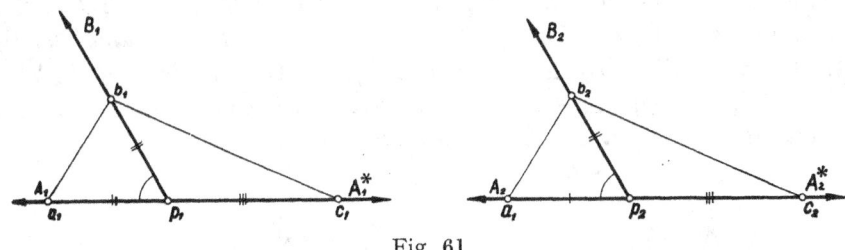

Fig. 61

Because of (1) and (2), it follows from $A_1B_1 \equiv A_2B_2$ that $a_1b_1 \equiv a_2b_2$, which, together with formulas (1)–(3), gives $b_1c_1 \equiv b_2c_2$. Thus we conclude from (2) and (3) that $A_1^*B_1 \equiv A_2^*B_2$.

From Theorem 14 we readily obtain:

THEOREM 15. *Given two adjacent angles A_1B_1 and $A_1^*B_1$, if three half-lines A_2, B_2, C, belonging to a pencil $\mathfrak{P}$, satisfy the following conditions:*

$$A_1B_1 \equiv A_2B_2, \quad A_1^*B_1 \equiv CB_2, \quad \mathbf{B}(A_2,\mathbf{L}(B_2),C),$$

then angles A_2B_2 and CB_2 are adjacent, i.e. $C = A_2^*$.

PROOF. By Theorem 14 it follows from $A_1B_1 \equiv A_2B_2$ that $A_1^*B_1 \equiv A_2^*B_2$ (Fig. 62), and since also $A_1^*B_1 \equiv CB_2$ and half-lines A_2^* and C lie on the

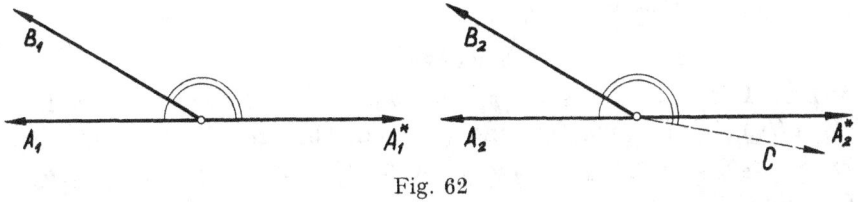

Fig. 62

same half-plane with boundary $\mathbf{L}(B_2)$, then, by Theorem 13, we have $C = A_2^*$, which was to be proved.

We say that two angles are *vertical* angles if the sides of one of these angles are half-lines complementary to the sides of the other angle. Thus angles AB and A^*B^* are vertical angles and so are angles AB^* and A^*B (Fig. 60).

Angles AB and A^*B^* are both adjacent to angle AB^*. Hence, by Theorems 12(I) and 14, it follows that $AB \equiv A^*B^*$. Thus we have proved:

THEOREM 16. *Vertical angles are congruent.*

6. Relations Between Angles of Two Half-Pencils

THEOREM 17. *Given pencils* $\mathfrak{P}_i$ *with vertices* p_i, *half-lines* $A_i, B_i, C_i \in \mathfrak{P}_i$, *and points* $a_i \in A_i$, $b_i \in B_i$, $c_i \in C_i$, *where* $i = 1,2$, *let us assume that* (i) $\mathbf{B}(A_i, \mathbf{L}(B_i), C_i)$, (ii) $A_1B_1 \equiv A_2B_2$ *and* $B_1C_1 \equiv B_2C_2$, *and* (iii) $p_1a_1 \equiv p_2a_2$, $p_1b_1 \equiv p_2b_2$, *and* $p_1c_1 \equiv p_2c_2$. *Under these assumptions, if points* a_1, b_1, c_1 *are collinear, then points* a_2, b_2, c_2 *are also collinear.*

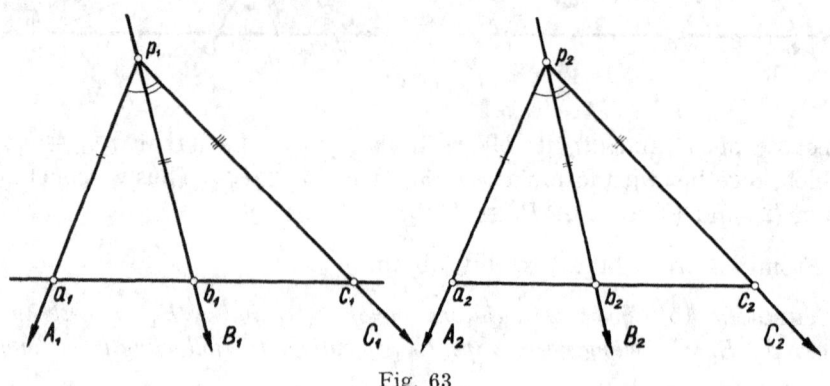

Fig. 63

PROOF. We assume that points a_1, b_1, c_1 (Fig. 63) are collinear. From $A_1B_1 \equiv A_2B_2$, $p_1a_1 \equiv p_2a_2$, $p_1b_1 \equiv p_2b_2$ it follows that $a_1b_1 \equiv a_2b_2$. Hence

(1) $\sphericalangle\, a_1b_1p_1 \equiv \sphericalangle\, a_2b_2p_2$.

Similarly, from $B_1C_1 \equiv B_2C_2$, $p_1b_1 \equiv p_2b_2$, $p_1c_1 \equiv p_2c_2$ it follows that $b_1c_1 \equiv b_2c_2$. Therefore

(2) $\sphericalangle\, p_1b_1c_1 \equiv \sphericalangle\, p_2b_2c_2$.

By $\mathbf{B}(A_1, \mathbf{L}(B_1), C_1)$ angles $a_1b_1p_1$ and $p_1b_1c_1$ are adjacent. Hence, from $\mathbf{B}(A_2, \mathbf{L}(B_2), C_2)$, formulas (1) and (2), and Theorem 15 it follows that angles $a_2b_2p_2$ and $p_2b_2c_2$ are also adjacent. Consequently, points, a_2, b_2, c_2 are collinear.

On the basis of Theorem 54 of Chapter I and as an immediate consequence of Theorem 17 we obtain

THEOREM 18. *Given two pencils $\mathfrak{P}_1$ and $\mathfrak{P}_2$ and half-lines $A_1,B_1,C_1 \in \mathfrak{P}_1$ and $A_2,B_2,C_2 \in \mathfrak{P}_2$, let us assume that (i) $\mathbf{B}(A_1,\mathbf{L}(B_1),C_1)$ and $\mathbf{B}(A_2,\mathbf{L}(B_2),C_2)$, and (ii) $A_1B_1 \equiv A_2B_2$ and $B_1C_1 \equiv B_2C_2$. Under these assumptions, if half-lines A_1, B_1, C_1 are co-half-pencilar, then half-lines A_2, B_2, C_2 are also co-half-pencilar.*

With the aid of this theorem many theorems concerning angles can be proved similarly to the corresponding theorems for segments.

We shall now prove theorems for angles analogous to Axiom C4 and Theorem 7.

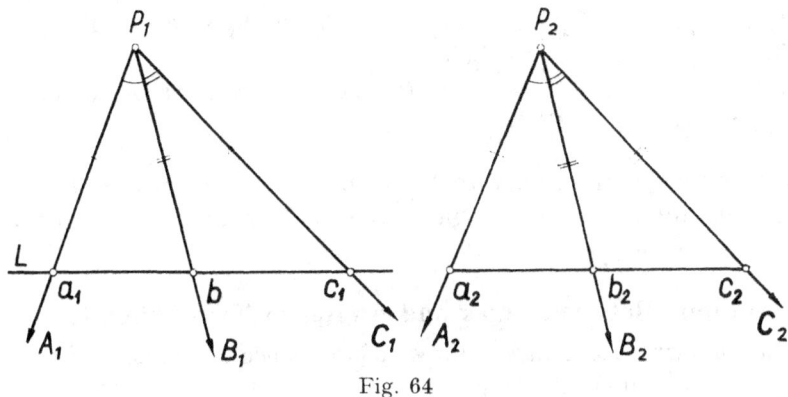

Fig. 64

THEOREM 19. *Given two half-pencils $\mathfrak{H}_1$ and $\mathfrak{H}_2$ and half-lines $A_1,B_1,C_1 \in \mathfrak{H}_1$ and $A_2,B_2,C_2 \in \mathfrak{H}_2$, let us assume that $\mathbf{B}(A_1,B_1,C_1)$ and $\mathbf{B}(A_2,B_2,C_2)$. Then we have:*

(I)   *if $A_1B_1 \equiv A_2B_2$ and $B_1C_1 \equiv B_2C_2$, then $A_1C_1 \equiv A_2C_2$;*
(II) *if $A_1B_1 \equiv A_2B_2$ and $A_1C_1 \equiv A_2C_2$, then $B_1C_1 \equiv B_2C_2$.*

PROOF. Suppose that

(3) $$A_1B_1 \equiv A_2B_2 \text{ and } B_1C_1 \equiv B_2C_2.$$

Let point p_1 be the vertex of half-pencil $\mathfrak{H}_1$ and point p_2 the vertex of half-pencil $\mathfrak{H}_2$. Further, let a line L intersect half-lines A_1, B_1, C_1 in points a_1,b_1,c_1, respectively (Fig. 64). Now take points $a_2 \in A_2$, $b_2 \in B_2$, $c_2 \in C_2$ such that

(4) $$p_1a_1 \equiv p_2a_2, \quad p_1b_1 \equiv p_2b_2, \quad p_1c_1 \equiv p_2c_2.$$

By Theorem 17, points a_2,b_2,c_2 are collinear and, because of $\mathbf{B}(A_1,B_1,C_1)$ and $\mathbf{B}(A_2,B_2,C_2)$, we have

$$\mathbf{B}(a_1,b_1,c_1) \text{ and } \mathbf{B}(a_2,b_2,c_2).$$

By formulas (3) and (4), we also have

$$a_1b_1 \equiv a_2b_2 \text{ and } b_1c_1 \equiv b_2c_2.$$

Hence $a_1c_1 \equiv a_2c_2$ and consequently $A_1C_1 \equiv A_2C_2$.

Part (II) of the theorem is obtained from part (I) with the help of Theorem 18, the proof being analogous to that used in obtaining the corresponding theorem for segments (Theorem 7) from Axiom C4.

The following theorem may be said to be the converse of Theorem 19:

THEOREM 20. *Given half-lines* A_1,B_1,C_1 *such that* $\mathbf{B}(A_1,B_1,C_1)$, *we have*:
(I) *If* $A_1C_1 \equiv A_2C_2$, *then there is a half-line* B_2 *such that* $\mathbf{B}(A_2,B_2,C_2)$, $A_1B_1 \equiv A_2B_2$ *and* $B_1C_1 \equiv B_2C_2$.
(II) *If* $B_1C_1 \equiv B_2C_2$, *then there is a half-line* A_2 *such that* $\mathbf{B}(A_2,B_2,C_2)$, $A_1B_1 \equiv A_2B_2$ *and* $A_1C_1 \equiv A_2C_2$.

We prove this theorem with the help of Theorem 18, the proof being analogous to the proof of the corresponding theorem for segments, namely, Theorem 8.

7. Relations Between Sides and Angles of Two Triangles

The following theorem is a direct consequence of the definition of the congruence of angles:

THEOREM 21. *Given triangles* $a_1b_1c_1$ *and* $a_2b_2c_2$, *if* $a_1b_1 \equiv a_2b_2$, $b_1c_1 \equiv b_2c_2$, *and* $a_1c_1 \equiv a_2c_2$, *then* $\sphericalangle\, a_1 \equiv \sphericalangle\, a_2$, $\sphericalangle\, b_1 \equiv \sphericalangle b_2$, *and* $\sphericalangle\, c_1 \equiv \sphericalangle\, c_2$.

The remaining relations between the angles and sides of two arbitrary triangles are given in the following:

THEOREM 22. *Given two triangles* $a_1b_1c_1$ *and* $a_2b_2c_2$, *we have*:
(I) *If* $a_1b_1 \equiv a_2b_2$, $b_1c_1 \equiv b_2c_2$, *and* $\sphericalangle\, b_1 \equiv \sphericalangle\, b_2$, *then* $a_1c_1 \equiv a_2c_2$, $\sphericalangle\, a_1 \equiv \sphericalangle\, a_2$, *and* $\sphericalangle\, c_1 \equiv \sphericalangle\, c_2$.
(II) *If* $a_1c_1 \equiv a_2c_2$, $\sphericalangle\, a_1 \equiv \sphericalangle\, a_2$, *and* $\sphericalangle\, c_1 \equiv \sphericalangle\, c_2$, *then* $a_1b_1 \equiv a_2b_2$, $b_1c_1 \equiv b_2c_2$, *and* $\sphericalangle\, b_1 \equiv \sphericalangle\, b_2$.

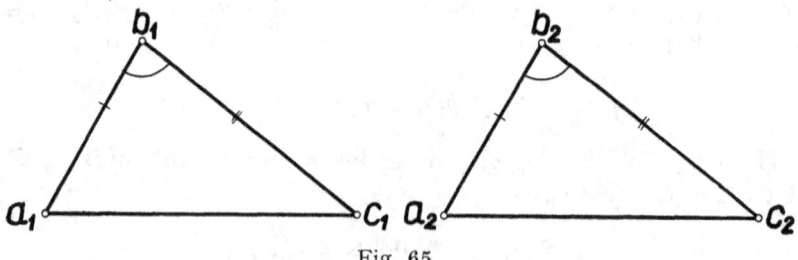

Fig. 65

PROOF. (I) From $a_1b_1 \equiv a_2b_2$, $b_1c_1 \equiv b_2c_2$, and $\sphericalangle\, b_1 \equiv \sphericalangle\, b_2$ (Fig. 65), it follows that $a_1c_1 \equiv a_2c_2$. This, together with $a_1b_1 \equiv a_2b_2$ and $b_1c_1 \equiv b_2c_2$, give us, by Theorem 21, $\sphericalangle\, a_1 \equiv \sphericalangle\, a_2$ and $\sphericalangle\, c_1 \equiv \sphericalangle\, c_2$.

(II) Let us choose on half-line a_1b_1 a point b_1' (Fig. 66) such that

$$(1) \qquad\qquad\qquad a_1b_1' \equiv a_2b_2.$$

Applying part (I) of the theorem to triangles $a_1b_1'c_1$ and $a_2b_2c_2$, we conclude that $\sphericalangle\, a_1c_1b_1' \equiv \sphericalangle\, c_2$. By our hypothesis, $\sphericalangle\, a_1c_1b_1 \equiv \sphericalangle\, c_2$, and from the construction it follows that half-lines c_1b_1' and c_1b_1 lie on the same half-plane with boundary $\mathbf{L}(a_1c_1)$. Hence, according to Theorem 13, half-line c_1b_1' coincides with half-line c_1b_1, from which it follows that $b_1' = b_1$. By formula (1) we therefore have $a_1b_1 \equiv a_2b_2$. Applying part (I) of the theorem to triangles $a_1b_1c_1$ and $a_2b_2c_2$, we obtain $b_1c_1 \equiv b_2c_2$ and $\sphericalangle\, b_1 \equiv \sphericalangle\, b_2$.

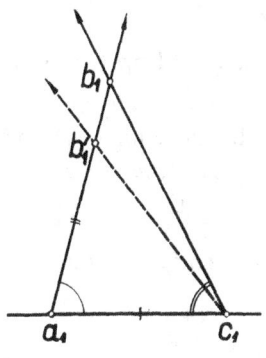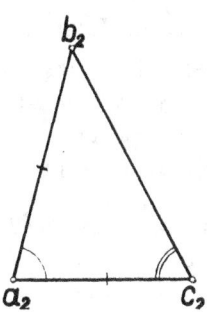

Fig. 66

From Theorem 22 we obtain two sufficient conditions for the congruence of triangles:

1. *If two sides and the included angle of one triangle are congruent respectively to two sides and the included angle of another triangle, then the two triangles are congruent.*

2. *If one side and the two angles adjacent to it in one triangle are congruent respectively to one side and the two angles adjacent to it in another triangle, then the two triangles are congruent.*

8. Relations Between Sides and Angles of a Triangle

THEOREM 23. *In a triangle the angles opposite congruent sides are congruent, and the sides opposite congruent angles are congruent. In other words, in a triangle abc we have ab $\equiv$ bc if and only if $\sphericalangle\, a \equiv \sphericalangle\, c$.*

PROOF. Assume that $ab \equiv bc$. Since $ab = ba$, $bc = cb$ and $\sphericalangle\, abc = \sphericalangle\, cba$, then by Theorems 6 and 12 we have

$$ab \equiv cb, \quad bc \equiv ba, \quad \measuredangle\, abc \equiv \measuredangle\, cba,$$

from which we conclude, by Theorem 22(I), that $\measuredangle\, a \equiv \measuredangle\, c$.

Assume that $\measuredangle\, a \equiv \measuredangle\, c$. Then by Theorems 6 and 12 we have

$$ac \equiv ca, \quad \measuredangle\, bac \equiv \measuredangle\, bca, \quad \measuredangle\, bca \equiv \measuredangle\, bac,$$

from which we conclude by Theorem 22(II) that $ab \equiv bc$.

A triangle in which two sides are congruent is called an *isosceles* triangle. We shall continue to call the two congruent sides the *sides* of the isosceles triangle and the remaining side we shall call the *base*. In particular, it follows from Theorem 23 that the base angles of an isosceles triangle are congruent.

A triangle having all three sides congruent is called an *equilateral* triangle. In particular, it follows from Theorem 23 that all three angles of an equilateral triangle are congruent.

9. The Relations Less-Than and Greater-Than for Segments

We shall say that *a segment a_1b_1 is less than a segment a_2b_2*—in symbols, $a_1b_1 < a_2b_2$ — if there exists a point p such that $\mathbf{B}(a_2,p,b_2)$ and $a_1b_1 \equiv a_2p$.

We shall now prove the so-called *law of extensionality* for the relation *less-than* with respect to the congruence relation.

THEOREM 24. *Given any two segments a_1b_1 and a_2b_2, if $a_1b_1 < a_2b_2$, then:*
(I) $ab \equiv a_1b_1$ *implies* $ab < a_2b_2$;
(II) $ab \equiv a_2b_2$ *implies* $a_1b_1 < ab$.

PROOF. From the definition of the relation less-than for segments it follows that there exists a point p such that

$$(1) \qquad\qquad \mathbf{B}(a_2,p,b_2)$$

and

$$(2) \qquad\qquad a_1b_1 \equiv a_2p.$$

Assume that $ab \equiv a_1b_1$. Then $ab \equiv a_2p$ and hence $ab < a_2b_2$.

Assume that $ab \equiv a_2b_2$. By formula (1) and Theorem 8(I) there then exists a point q such that $\mathbf{B}(a,q,b)$ and $a_2p \equiv aq$, from which it follows, by formula (2), that $a_1b_1 \equiv aq$, that is, $a_1b_1 < ab$.

We next show that the relation $<$ partially orders the family of all segments. To do this we should show that the relation $<$ (for segments) is antisymmetric and transitive.

THEOREM 25. *Given any three segments a_1b_1, a_2b_2, and a_3b_3, we have:*

(I) *if $a_1b_1 < a_2b_2$, then $a_2b_2 \sim < a_1b_1$;*
(II) *if $a_1b_1 < a_2b_2$ and $a_2b_2 < a_3b_3$, then $a_1b_1 < a_3b_3$.*

PROOF. Assume that $a_1b_1 < a_2b_2$. Then there exists a point p (Fig. 67) such that

(3) $$\mathbf{B}(a_2,p,b_2) \quad \text{and} \quad a_1b_1 \equiv a_2p.$$

Suppose $a_2b_2 < a_1b_1$. There then exists a point q such that $\mathbf{B}(a_1,q,b_1)$ and $a_2b_2 \equiv a_1q$. Therefore there exists a point r such that

(4) $$\mathbf{B}(a_2,b_2,r) \quad \text{and} \quad a_1b_1 \equiv a_2r.$$

By formulas (3) and (4) we have $p \neq r$. Thus, there exist on half-line

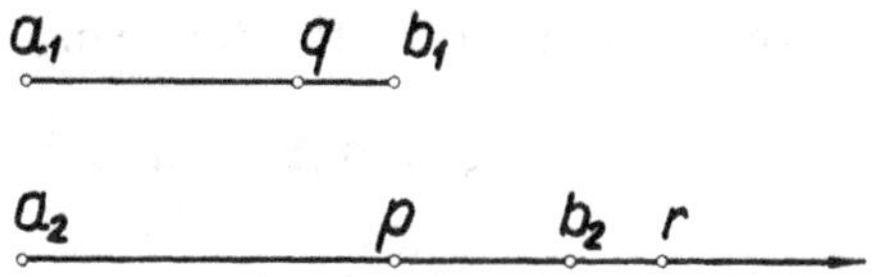

Fig. 67

a_2b_2 two distinct points p and r such that segments a_2p and a_2r are congruent to segment a_1b_1, which is contrary to Axiom C5. Thus $a_2b_2 \sim < a_1b_1$.

Let us next assume that $a_1b_1 < a_2b_2$ and $a_2b_2 < a_3b_3$. Then there exists a point p such that $\mathbf{B}(a_3,p,b_3)$ and $a_2b_2 \equiv a_3p$. By Theorem 24(II) we then have $a_1b_1 < a_3p$, that is, there exists a point q such that $\mathbf{B}(a_3,q,p)$ and $a_1b_1 < a_3q$. Since $\mathbf{B}(a_3,q,b_3)$, then $a_1b_1 < a_3b_3$.

We shall say that *a segment a_1b_1 is greater than a segment a_2b_2*—in symbols, $a_1b_1 > a_2b_2$ — if $a_2b_2 < a_1b_1$. We shall now prove the *law of trichotomy* for segments. This law states that only one of the following three relations: the relation less-than, the relation greater-than, and the relation of congruence, holds between any pair of arbitrarily chosen segments.

THEOREM 26. *For any two segments a_1b_1 and a_2b_2, either $a_1b_1 \equiv a_2b_2$ or $a_1b_1 < a_2b_2$ or else $a_1b_1 > a_2b_2$.*

PROOF. Choose on half-line a_1b_1 a point p such that $a_1p \equiv a_2b_2$. If $p = b_1$, then $a_1b_1 \equiv a_2b_2$. If $\mathbf{B}(a_1,b_1,p)$, then $a_1b_1 < a_1p$, from which, by Theorem 24(II), we obtain $a_1b_1 < a_2b_2$. Finally, if $\mathbf{B}(a_1,p,b_1)$, then $a_2b_2 < a_1b_1$, that is, $a_1b_1 > a_2b_2$. We have therefore shown that

$$a_1b_1 \equiv a_2b_2 \quad \text{or} \quad a_1b_1 < a_2b_2 \quad \text{or} \quad a_1b_1 > a_2b_2.$$

It follows from Axiom C5 and Theorem 25(I) that each of these three possibilities excludes the remaining two.

10. The Relations Less-Than and Greater-Than for Angles

We shall say that *an angle A_1B_1 is less than an angle A_2B_2* — in symbols, $A_1B_1 < A_2B_2$ — if there exists a half-line C such that $\mathbf{B}(A_2,C,B_2)$ and $A_1B_1 \equiv A_2C$. We shall say that *an angle A_1B_1 is greater than an angle A_2B_2* — in symbols, $A_1B_1 > A_2B_2$, — if $A_2B_2 < A_1B_1$.

The relation *less-than (greater-than)* for angles has the same properties as the relation less-than (greater-than) for segments. In particular, the relation less-than for angles is extensional with respect to the relation of congruence, and partially orders the family of all angles. In addition, the relations $<$, $>$, and $\equiv$ for angles are connected by means of the law of trichotomy. We therefore have the following three theorems:

THEOREM 27. *Given any two angles A_1B_1 and A_2B_2, if $A_1B_1 < A_2B_2$, then*

(I) $AB \equiv A_1B_1$ *implies* $AB < A_2B_2$;
(II) $AB \equiv A_2B_2$ *implies* $A_1B_1 < AB$.

THEOREM 28. *For any three angles A_1B_1, A_2B_2, A_3B_3,*
(I)   *if* $A_1B_1 < A_2B_2$, *then* $A_2B_2 \sim\, < A_1B_1$;
(II) *if* $A_1B_1 < A_2B_2$ *and* $A_2B_2 < A_3B_3$, *then* $A_1B_1 < A_3B_3$.

THEOREM 29. *For any two angles A_1B_1 and A_2B_2, either $A_1B_1 \equiv A_2B$ or $A_1B_1 < A_2B_2$ or else $A_1B_1 > A_2B_2$.*

The proofs of Theorems 27–29 are analogous to the proofs of the corresponding theorems for segments, namely, Theorems 24–26.

The last theorem on the relation less-than will deal with adjacent angles.

THEOREM 30. *If $A_1B_1 < A_2B_2$, then $A_1^{*}B_1 > A_2^{*}B_2$.*

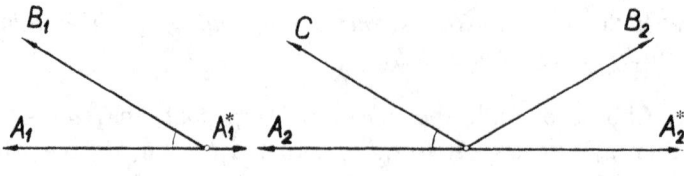

Fig. 68

PROOF. Let $A_1B_1 < A_2B_2$ and take a half-line C such that $\mathbf{B}(A_2,C,B_2)$ and $A_1B_1 \equiv A_2C$ (Fig. 68). Then $\mathbf{B}(C,B_2,A_2^{*})$ and consequently

$A_2^* B_2 < A_2^* C$ and $A_1^* B_1 \equiv A_2^* C$. It then follows by Theorem 27(II) that $A_1^* B_1 > A_2^* B_2$.

11. Midpoint of a Segment

Consider a segment ab. A point $c \neq a,b$ collinear with points a and b and satisfying the condition $ac \equiv bc$ is called a *midpoint of segment ab*.

THEOREM 31. *If a point c is a midpoint of a segment ab, then* $\mathbf{B}(a,c,b)$.

PROOF. Indeed, from $\mathbf{B}(b,a,c)$ it would follow that $ac < bc$, and from $\mathbf{B}(a,b,c)$ it would follow that $bc < ac$. But these two inequalities exclude, by Theorem 26, the congruence $ac \equiv bc$.

We shall now prove

THEOREM 32. *For each segment ab there exists just one midpoint c.*

PROOF. We shall show first that segment ab has at least one midpoint c. Let $P = W \cup \mathbf{L}(ab) \cup W^*$ denote some plane containing segment ab.

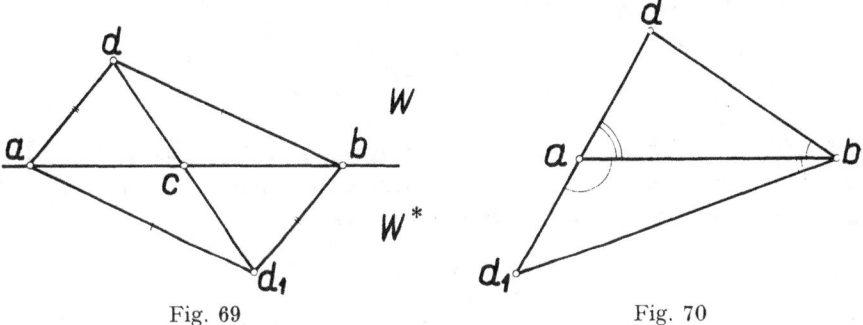

Fig. 69   Fig. 70

Take any point $d \in W$ (Fig. 69). By Axiom C7 there exists in half-plane W^* a point, d_1 such that

(1) $ad_1 \equiv bd$ and $bd_1 \equiv ad$,

from which it follows, by Theorem 21, that

(2) $\sphericalangle bad_1 \equiv \sphericalangle abd$ and $\sphericalangle abd_1 \equiv \sphericalangle bad$.

By Theorem 18 and formula (2) neither points d,a,d_1 nor points d,b,d_1 are collinear (Fig. 70); by formula (1) we have

(3) $\sphericalangle bd_1 d \equiv \sphericalangle add_1$ and $\sphericalangle bdd_1 \equiv \sphericalangle ad_1 d$.

Line ab intersects segment (dd_1) in a point c such that

(4) $\mathbf{B}(d,c,d_1)$.

We shall show that $\boldsymbol{B}(a,c,b)$. To do this we suppose that $\boldsymbol{B}(c,a,b)$ (Fig. 71). Then

$$\boldsymbol{B}(\boldsymbol{H}(dc), \boldsymbol{H}(da), \boldsymbol{H}(db))$$

and

$$\boldsymbol{B}(\boldsymbol{H}(d_1c), \boldsymbol{H}(d_1a), \boldsymbol{H}(d_1b)),$$

from which, by formulas (3) and (4), we obtain

$$\measuredangle\, add_1 = \measuredangle\, adc < \measuredangle\, bdc = \measuredangle\, bdd_1 \equiv \measuredangle\, ad_1d = \measuredangle\, ad_1c < \measuredangle\, bd_1c$$
$$= \measuredangle\, bd_1d \equiv \measuredangle\, add_1.$$

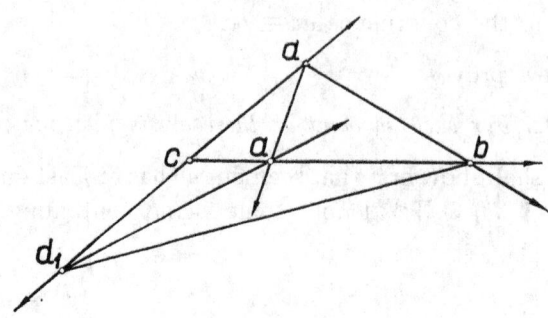

Fig. 71

Therefore, by Theorems 27(I) and 28(II), we have $\measuredangle\, add_1 < \measuredangle\, add_1$, which contradicts Theorem 29. In the same manner we prove that $\boldsymbol{B}(a,b,c)$ cannot hold. Hence we must have

(5) $\boldsymbol{B}(a,c,b).$

From formulas (3) and (2) we obtain, because of formulas (4) and (5),

(6) $\measuredangle\, bd_1c \equiv \measuredangle\, adc$ and $\measuredangle\, cbd_1 \equiv \measuredangle\, cad.$

$$a \quad\quad\quad\quad\quad c \;\; c' \quad\quad\quad b$$

Fig. 72

From formulas (1) and (6) we conclude, by comparing triangles cbd_1 and cad, that $ac \equiv cb$. Thus point c is a midpoint of segment ab.

Now, let us suppose that point $c' \neq c$ is also a midpoint of segment ab (Fig. 72). We then have

$$ac \equiv bc \quad \text{and} \quad ac' \equiv bc',$$

and by Theorem 31 we have $\boldsymbol{B}(a,c,b)$ and $\boldsymbol{B}(a,c',b)$. Then either $\boldsymbol{B}(a,c,c')$ or $\boldsymbol{B}(a,c',c)$. Assume, for example, that $\boldsymbol{B}(a,c,c')$; then $\boldsymbol{B}(b,c',c)$. Hence

$ac < ac' \equiv bc' < bc \equiv ac$, from which we conclude, by Theorem 25(II), that $ac < ac$. But this is impossible. Therefore point c is the only mid-point of segment ab.

12. Bisector of an Angle

Given an angle AB. A half-line $C \neq A,B$ co-half-pencilar with half-lines A and B and satisfying the condition $AC \equiv BC$ is called the *bisector* of angle AB. From Theorem 29 there follows at once:

THEOREM 33. *If a half-line C is the bisector of an angle AB, then* $\mathbf{B}(A,C,B)$.

We now prove:

THEOREM 34. *For every angle AB there exists just one bisector C.*

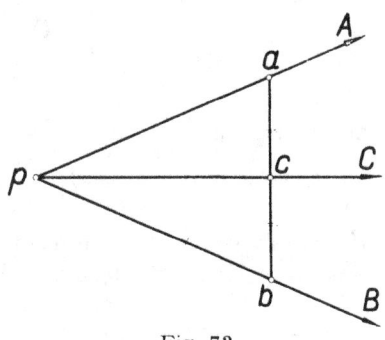

Fig. 73

PROOF. Let point p be the vertex of angle AB (Fig. 73). Take points $a \in A$ and $b \in B$ such that $pa \equiv pb$. Let c be the midpoint of segment ab; we then have $ac \equiv bc$. Therefore $\sphericalangle\, apc \equiv \sphericalangle\, bpc$ and half-line $C = \mathbf{H}\,(pc)$ is the bisector of angle AB.

We can prove that half-line C is the only bisector of angle AB by an argument similar to that by which we established the uniqueness of midpoint c of segment ab in the proof of Theorem 32.

13. External Angles of a Triangle

Consider any triangle abc. An angle adjacent to any of the internal angles: $\sphericalangle\, a$, $\sphericalangle\, b$, $\sphericalangle\, c$, of this triangle is called an *external* angle of triangle abc.

THEOREM 35. *An external angle of a triangle abc is greater than each of the interior angles not adjacent to it.*

PROOF. Take a point b_1 such that

(1) $\mathbf{B}(b,c,b_1)$.

Angle acb_1 is an external angle of triangle abc (Fig. 74).

Case 1. We show that $\angle acb_1 > \angle a$. Let d be the midpoint of side ac (see Theorem 32); we then have

$$(2) \qquad\qquad\qquad ad \equiv cd.$$

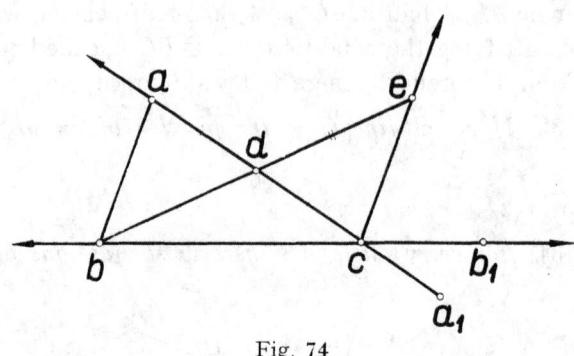

Fig. 74

On half-line bd we choose a point e such that

$$(3) \qquad\qquad\qquad \mathbf{B}(b,d,e) \text{ and } bd \equiv ed.$$

For triangles dab and dce it follows, from formulas (2), (3), and Theorems 16, 22(I), that $\angle a \equiv \angle ace$. Since, by formula (3), we have

$$\mathbf{B}(\mathbf{H}(cb), \mathbf{H}(ca), \mathbf{H}(ce)),$$

then

$$\mathbf{B}(\mathbf{H}(ca), \mathbf{H}(ce), \mathbf{H}(cb_1))$$

and $\angle acb_1 > \angle a$.

Case 2. We show that $\angle acb_1 > \angle b$. On half-line ac take a point a_1 such that $\mathbf{B}(a,c,a_1)$. Since, according to Theorem 16, $\angle acb_1 \equiv \angle bca_1$, Case 2 reduces to Case 1.

14. Two Non-Intersecting Lines on a Plane

We shall first prove two theorems on two non-intersecting half-lines in a half-plane.

THEOREM 36. *Given a half-plane W_1 with boundary K_1 and a half-plane W_2 with boundary K_2, two distinct points $a_1,b_1 \in K_1$, two distinct points $a_2,b_2 \in K_2$, and four half-lines $A_1,B_1 \subset W_1$ and $A_2,B_2 \subset W_2$ with origins a_1, b_1, a_2, b_2, respectively, if $a_1b_1 \equiv a_2b_2$, $\mathbf{H}(a_1b_1)A_1 \equiv \mathbf{H}(a_2b_2)A_2$, and $\mathbf{H}(b_1a_1)B_1 \equiv \mathbf{H}(b_2a_2)B_2$, then $A_1 \cap B_1 = 0$ implies $A_2 \cap B_2 = 0$.*

PROOF. Suppose that half-line A_2 intersects half-line B_2 in some point p_2 (Fig. 75). On half-line B_1 take a point p_1 such that $b_1p_1 \equiv b_2p_2$.

From $a_1b_1 \equiv a_2b_2$ and $\sphericalangle\, a_1b_1p_1 \equiv \sphericalangle\, a_2b_2p_2$ it then follows that $\sphericalangle\, b_1a_1p_1 \equiv \sphericalangle\, b_2a_2p_2$, which, since $\textbf{H}(a_1b_1)A_1 \equiv \sphericalangle\, b_2a_2p_2$, gives us $A_1 = \textbf{H}(a_1p_1)$.

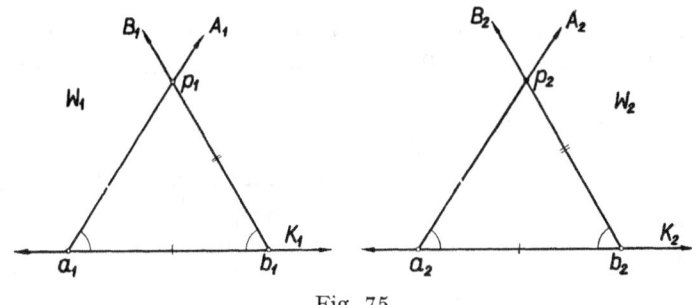

Fig. 75

Therefore $A_1 \cap B_1 \neq 0$. Hence $A_1 \cap B_1 = 0$ implies $A_2 \cap B_2 = 0$.

THEOREM 37. *Given a half-plane W with boundary K, two distinct points $a,b \in K$, and two half-lines $A,B \subset W$ with origins a and b, respectively, if* $\textbf{H}(ab)A \equiv \textbf{H}^*(ba)B$ *or* $\textbf{H}(ab)A > \textbf{H}^*(ba)B$, *then* $A \cap B = 0$.

PROOF. If half-line A were to intersect half-line B in some point p (Fig. 76), then angle $\textbf{H}^*(ba)B$ would be external to triangle pab, and since

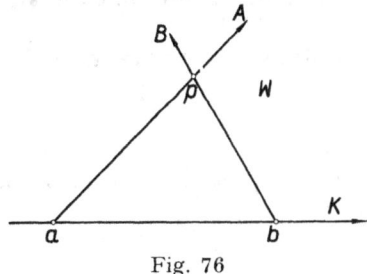

Fig. 76

angle $\textbf{H}(ba)B$ is not adjacent to angle $\textbf{H}(ab)A$ (the latter being identical to angle pab), it would have to be greater than it.

As a consequence of Theorem 37 we have:

THEOREM 38. *Given, on a plane P, a half-plane W with boundary K, two distinct points $a, b \in K$, and two half-lines $A,B \subset W$ with origins a and b, if* $\textbf{H}(ab)A \equiv \textbf{H}^*(ba)B$, *then* $\textbf{L}(A)$ *and* $\textbf{L}(B)$ *are disjoint lines.*

PROOF. We assume that

(1) $$\textbf{H}(ab)A \equiv \textbf{H}^*(ba)B.$$

Then $A \cap B = 0$ (Fig. 77). Further, $A^*, B^* \subset W^*$, and by (1) we have

$\mathsf{H}(ab)A^* \equiv \mathsf{H}^*(ba)B^*$. As a result $A^* \cap B^* = 0$. From $A \cap B = 0$ and $A^* \cap B^* = 0$ it follows that $\mathsf{L}(A) \cap \mathsf{L}(B) = 0$.

As an important consequence of Theorem 38 we have

THEOREM 39. *Let L be a line in a given plane P. For any point $a \in P{-}L$ there is at least one line $K \subset P$ passing through a and not intersecting L.*

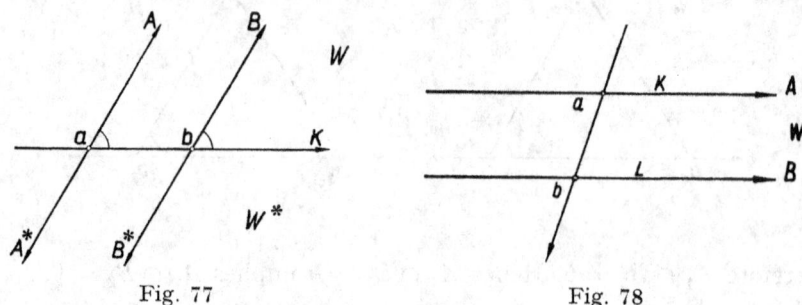

Fig. 77   Fig. 78

PROOF. Take any point $b \in L$ (Fig. 78) and any half-plane $W \subset P$ with boundary $\mathsf{L}(ab)$. Let $B = L \cap W$. From point a produce a half-line $A \subset W$ such that

(1) $$\mathsf{H}(ab)A \equiv \mathsf{H}^*(ba)B.$$

Then line $K = \mathsf{L}(A)$ lies in plane P and lines K and L are disjoint.

15. Relations Between Sides and Angles of a Triangle (Conclusion)

We shall prove:

THEOREM 40. *In a triangle abc the greater angle lies opposite the greater side and the greater side lies opposite the greater angle. In other words, $ab > bc$ if and only if $\angle c > \angle a$.*

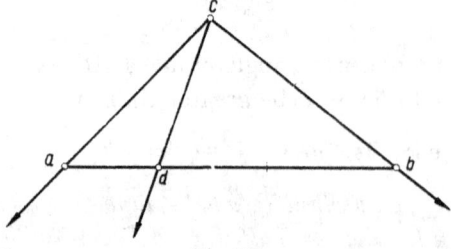

Fig. 79

PROOF. Assume that $ab > bc$ (Fig. 79). Then there exists a point d such that $\mathsf{B}(a,d,b)$ and $bd \equiv bc$. Since $\angle bdc$ is an external angle of triangle acd, then, by Theorem 35, we have $\angle a < \angle bdc$. Further, it follows from $\mathsf{B}(\mathsf{H}(ca), \mathsf{H}(cd), \mathsf{H}(cb))$ that $\angle dcb < \angle c$. Since triangle bcd

is isosceles, we conclude finally, by Theorem 23, that $\angle bdc \equiv \angle dcb$. We therefore have $\angle a < \angle bdc \equiv \angle dcb < \angle c$, from which it follows that $\angle c > \angle a$.

We next assume that $\angle c > \angle a$. By Theorem 26 there are three possible cases: $ab \equiv bc$, or $ab < bc$, or $ab > bc$. But if $ab \equiv bc$, then $\angle c \equiv \angle a$, and from $ab < bc$ it follows, by what was proved above, that $\angle c < \angle a$, which, by Theorem 29, contradicts our assumption. Therefore $ab > bc$.

16. Relations Between Sides and Angles of Two Triangles (Continued)

We shall prove:

THEOREM 41. *Given two triangles* $a_1b_1c_1$ *and* $a_2b_2c_2$, *if* $a_1b_1 \equiv a_2b_2$ *and* $b_1c_1 \equiv b_2c_2$, *then* $\angle b_1 < \angle b_2$ *if and only if* $a_1c_1 < a_2c_2$.

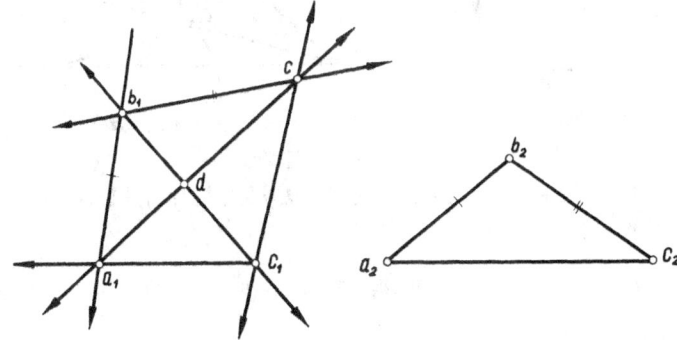

Fig. 80

PROOF. Let $a_1b_1 \equiv a_2b_2$ and $b_1c_1 \equiv b_2c_2$. Let us assume that

(1) $$\angle b_1 < \angle b_2.$$

On the half-plane $\mathbf{L}(a_1b_1)c_1$, we choose a point c (Fig. 80) such that $a_1c \equiv a_2c_2$ and $b_1c \equiv b_2c_2$. Then

(2) $$\angle a_1b_1c \equiv \angle b_2 \quad \text{and} \quad a_1c \equiv a_2c_2.$$

Hence it suffices to show that $a_1c_1 < a_1c$.

We have $b_1c_1 \equiv b_2c_2 \equiv b_1c$, and therefore

(3) $$b_1c_1 \equiv b_1c.$$

Further, by formulas (1) and (2), it readily follows, that

(4) $$\mathbf{B}(\mathbf{H}(b_1a_1), \mathbf{H}(b_1c_1), \mathbf{H}(b_1c)),$$

and therefore half-line b_1c_1 intersects segment (a_1c) in some point d such that

(5) $$\mathbf{B}(a_1,d,c).$$

If $d = c_1$, then $\mathbf{B}(a_1,c_1,c)$ and $a_1c_1 < a_1c$ (Fig. 81).

Let us next assume that $d \neq c_1$. To show that $a_1 c_1 < a_1 c$ it suffices, by Theorem 40, to derive the inequality $\sphericalangle\, a_1 c c_1 < \sphericalangle\, a_1 c_1 c$. To do this we note that from formula (5) it follows that

(6) $$\boldsymbol{B}\,(\boldsymbol{H}(c_1 a_1), \boldsymbol{H}(c_1 d), \boldsymbol{H}(c_1 c)).$$

Let us examine two possibilities:

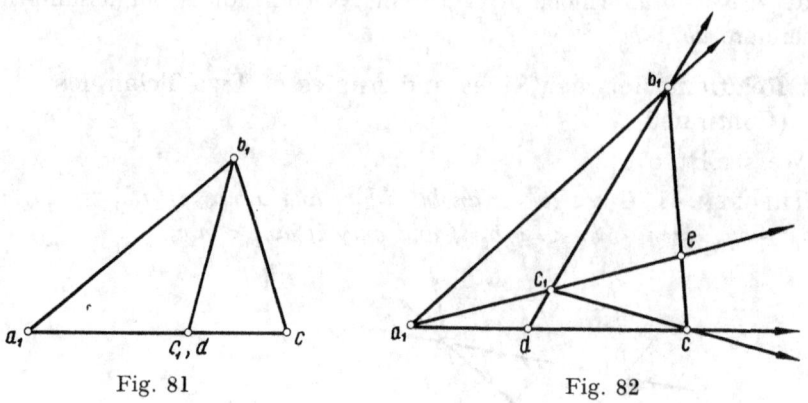

Fig. 81   Fig. 82

Case 1. $\boldsymbol{B}\,(b_1, d, c_1)$, and therefore $\boldsymbol{B}\,(\boldsymbol{H}(cb_1), \boldsymbol{H}(cd), \boldsymbol{H}(cc_1))$. Then, by formulas (5), (3), and (6)

$$\sphericalangle\, a_1 c c_1 = \sphericalangle\, d c c_1 < \sphericalangle\, b_1 c c_1 \equiv \sphericalangle\, b_1 c_1 c = \sphericalangle\, d c_1 c < \sphericalangle\, a_1 c_1 c,$$

from which it follows that $\sphericalangle\, a_1 c c_1 < \sphericalangle\, a_1 c_1 c$.

Case 2. $\boldsymbol{B}\,(b_1, c_1, d)$ (Fig. 82), and therefore $\boldsymbol{B}\,(\boldsymbol{H}(a_1 b_1), \boldsymbol{H}(a_1 c_1), \boldsymbol{H}(a_1 d))$, from which, by formula (5), it follows that $\boldsymbol{B}\,(\boldsymbol{H}(a_1 b_1), \boldsymbol{H}(a_1 c_1), \boldsymbol{H}(a_1 c))$. Thus half-line $a_1 c_1$ intersects segment $(b_1 c)$ at some point e. Thus we have $\boldsymbol{B}\,(\boldsymbol{H}(c_1 b_1), \boldsymbol{H}(c_1 e), \boldsymbol{H}(c_1 c))$ and, by formula (4), $\boldsymbol{B}\,(a_1, c_1, e)$, which, together with formulas (3) and (6), gives us

$$\sphericalangle\, a_1 c c_1 < \sphericalangle\, c c_1 e < \sphericalangle\, c c_1 b_1 \equiv \sphericalangle\, c_1 c b_1 < \sphericalangle\, d c_1 c < \sphericalangle\, a_1 c_1 c,$$

and therefore $\sphericalangle\, a_1 c c_1 < \sphericalangle\, a_1 c_1 c$.

Let us now assume that $a_1 c_1 < a_2 c_2$. There are now three possible cases: $\sphericalangle\, b_1 \equiv \sphericalangle\, b_2$, or $\sphericalangle\, b_1 < \sphericalangle\, b_2$, or $\sphericalangle\, b_1 > \sphericalangle\, b_2$. But from $\sphericalangle\, b_1 \equiv \sphericalangle\, b_2$ it follows that $a_1 c_1 \equiv a_2 c_2$, and from the first part of the theorem, already proved, it follows that if we were to have $\sphericalangle\, b_1 > \sphericalangle\, b_2$ then we would have $a_1 c_1 > a_2 c_2$, which is in contradiction to our assumption. Therefore we must have $\sphericalangle\, b_1 < \sphericalangle\, b_2$, which was to be proved.

17. Free Segments. The Relation Less-than and Greater-than for Free Segments. Addition of Free Segments

By Theorem 6, the relation $\equiv$ for segments is reflexive, symmetric, and transitive. The equivalence classes of the family of all segments with respect to the relation $\equiv$ will be called *free segments*. Individual segments will be now called *bound segments*. We call each bound segment belonging to a free segment the *representative* of the latter. We denote free segments by the small German letters $\mathfrak{a}, \mathfrak{b}, \mathfrak{c}, \mathfrak{d}, \mathfrak{x}, \mathfrak{y}$. A free segment with a representative ab will be denoted by $[ab]$.

We say that *a free segment* $\mathfrak{a} = [a_1b_1]$ *is less than a free segment* $\mathfrak{b} = [a_2b_2]$ — in symbols, $\mathfrak{a} < \mathfrak{b}$ — if $a_1b_1 < a_2b_2$. By Theorem 24 this definition does not depend on the choice of the representatives of free segments $\mathfrak{a}$ and $\mathfrak{b}$. Instead of $\mathfrak{a} < \mathfrak{b}$ we also write $\mathfrak{b} > \mathfrak{a}$ and say that *free segment* $\mathfrak{b}$ *is greater than free segment* $\mathfrak{a}$.

From the properties of the relation less-than for bound segments (see Theorems 25 and 26) there immediately follow the corresponding properties of the relation *less-than* for free segments.

THEOREM 42. *Given any three free segments* $\mathfrak{a}, \mathfrak{b}, \mathfrak{c}$, *we have*: (I) *if* $\mathfrak{a} < \mathfrak{b}$, *then* $\mathfrak{b} \sim < \mathfrak{a}$; (II) *if* $\mathfrak{a} < \mathfrak{b}$ *and* $\mathfrak{b} < \mathfrak{c}$, *then* $\mathfrak{a} < \mathfrak{c}$; (III) *either* $\mathfrak{a} = \mathfrak{b}$, *or* $\mathfrak{a} < \mathfrak{b}$, *or else* $\mathfrak{a} > \mathfrak{b}$.

From (I) and (II) it follows that the relation $<$ is antisymmetric and transitive; from (III) it follows, in particular, that the relation $<$ is also connected. Therefore the relation less-than orders the family of free segments.

We say *a free segment* $\mathfrak{c} = [ab]$ *is the sum of free segments* $\mathfrak{a}$ *and* $\mathfrak{b}$ — in symbols, $\mathfrak{c} = \mathfrak{a} + \mathfrak{b}$ — if there exists a point p such that $\mathbf{B}(a,p,b)$, $\mathfrak{a} = [ap]$ and $\mathfrak{b} = [pb]$. By Theorem 8(I), this definition does not depend on the choice of the representative of free segment $\mathfrak{c}$. It is also readily noted that for any free segments $\mathfrak{a}$ and $\mathfrak{b}$ there exists just one free segment $\mathfrak{c}$ constituting their sum. We shall call the operation leading from two free segments $\mathfrak{a}$ and $\mathfrak{b}$ to their sum $\mathfrak{a} + \mathfrak{b}$ the *addition of free segments*.

An obvious relation between the operation of addition and the relation less-than for free segments is given by the following theorem:

THEOREM 43. *Given any free segments* $\mathfrak{a}$ *and* $\mathfrak{b}$, *we have* $\mathfrak{a} < \mathfrak{b}$ *if and only if there exists a free segment* $\mathfrak{x}$ *such that* $\mathfrak{a} + \mathfrak{x} = \mathfrak{b}$.

The following two theorems establish the basic properties of the addition of free segments.

THEOREM 44. *Given any free segments* $\mathfrak{a}, \mathfrak{b}, \mathfrak{c}, \mathfrak{d}$, *we have*:
(I) $\mathfrak{a} + \mathfrak{b} = \mathfrak{b} + \mathfrak{a}$;
(II) $(\mathfrak{a} + \mathfrak{b}) + \mathfrak{c} = \mathfrak{a} + (\mathfrak{b} + \mathfrak{c})$;

(III) *if* $a < b$, *then* $a + c < b + c$;

(IV) *if* $a < b$ *and* $c < \mathfrak{d}$, *then* $a + c < b + \mathfrak{d}$.

PROOF. Formula (I) is obvious by the definition of the sum. We now prove formula (II). Let

(1) $$a + b = [p_1 p_3].$$

For some point p_2 lying between p_1 and p_3 we then have

(2) $$a = [p_1 p_2],$$

and

(3) $$b = [p_2 p_3].$$

We now take a point p_4 such that $\mathbf{B}(p_1, p_3, p_4)$ and

(4) $$c = [p_3 p_4].$$

From formulas (1) and (4) we obtain

(5) $$(a + b) + c = [p_1 p_4].$$

On the other hand, from $\mathbf{B}(p_2, p_3, p_4)$ and formulas (3) and (4) we get

$$b + c = [p_2 p_4].$$

From $\mathbf{B}(p_1, p_2, p_4)$ and formula (2) it thus follows that

(6) $$a + (b + c) = [p_1 p_4].$$

The required equality results from formulas (5) and (6).

Formula (III) is obtained immediately from Theorem 43 and from formulas (I) and (II); formula (IV) follows from formulas (III) and (I) and Theorem 42 (II).

THEOREM 45. *For any free segment* a *there exists just one free segment* x *such that* $a = x + x$.

PROOF. Let $a = [ab]$ and let c be the midpoint of bound segment ab. By Theorems 31 and 32, segment $[ac]$ is the only free segment satisfying the equation $a = x + x$.

From Theorems 42, 43, and 44 it follows that:

(I) *The relation* $<$ *orders the family of free segments.*

(II) *Addition of free segments is commutative and associative, and for any two distinct free segments* a *and* b *there exists either a free segment* x *such that* $a + x = b$ *or a free segment* $\mathfrak{y}$ *such that* $b + \mathfrak{y} = a$.

(III) *The relation* $<$ *and the operation* $+$ *are connected to each other* (see Theorem 44 (III)) *by the law of monotony.*

Properties I–III can be expressed briefly as follows: *The family of free segments is a commutative half-group (without the zero element) with respect to the operation of addition, ordered by the relation less than.*† This half-group has, by Theorem 45, a special property, namely, each of its elements is divisible by 2, or more accurately, each of its elements may be represented as the sum of two identical elements.

18. Subtraction of Free Segments. Multiplication of Free Segments by Dyadic Numbers

The discussion of this section will have a purely algebraic character. We shall use only the fact that the family of free segments is a commutative half-group with respect to addition, ordered by the relation *less than*, and having the property formulated in Theorem 45; however, the nature of the free segments themselves will not play any role. Since similar arguments are well known from algebra, we shall, in general, by-pass the proofs (which are, in fact, very simple). The theorems will be arranged in such an order that each will be a consequence of the preceding theorems and definitions.

We say that an *open segment* c *is the difference of free segments* a *and* b — in symbols, $c = a - b$ — if $a = b + c$.

The difference c exists only when $a > b$, and is then uniquely determined by free segments a and b. Certainly, the expression $a - b$ has sense only when $a > b$.

We shall call the operation leading from free segments a and b to their difference $a - b$ the *subtraction of free segments*. From the definition of the difference, there immediately follows:

THEOREM 46. (I) *For any free segments* a *and* b *we have* $(a + b) - b = a$ *and* $(a - b) + b = a$ *(if* $a > b$*).*

(II) *If each of free segments* a *and* b *is greater than free segment* c, *then* $a + (b - c) = (a - c) + b$.

(III) *If free segment* a *is greater than free segment* c, *then* $a < b$ *implies* $a - c < b - c$.

(IV) *If free segment* c *is greater than free segment* b, *then* $a < b$ *implies* $c - a > c - b$.

We define the *product* $n \cdot a$ *of a natural number* n *with a free segment* a by induction:

† We are obviously dealing here with a special case of an ordered half-group. If in conditions 1–3, instead of free segments, we were to speak of elements of some non-empty set X in which a certain relation $<$ is defined and a certain operation $+$ is defined and performable, then we would obtain a general definition of a commutative half-group (without the zero element) with respect to operation $+$, ordered by the relation $<$.

$$1 \cdot \mathfrak{a} = \mathfrak{a}, \quad (n+1) \cdot \mathfrak{a} = n \cdot \mathfrak{a} + \mathfrak{a},$$

i.e. the product $n \cdot \mathfrak{a}$ is the sum of n free segments, each of which is equal to $\mathfrak{a}$.

For the product of a natural number with a free segment the following distributive and associative laws hold:

THEOREM 47. *For any natural numbers m and n and for any free segments $\mathfrak{a}$ and $\mathfrak{b}$:*
(I) $n \cdot (\mathfrak{a} + \mathfrak{b}) = n \cdot \mathfrak{a} + n \cdot \mathfrak{b}$; II) $(m+n) \cdot \mathfrak{a} = m \cdot \mathfrak{a} + n \cdot \mathfrak{a}$;
(III) $(m \cdot n) \cdot \mathfrak{a} = m \cdot (n \cdot \mathfrak{a})$.

Let k be any non-negative integer. By Theorem 45, for any free segment $\mathfrak{a}$ there exists just one free segment $\mathfrak{x}$ such that $\mathfrak{a} = 2^k \cdot \mathfrak{x}$.

We shall call this free segment $\mathfrak{x}$ the *quotient* of free segment $\mathfrak{a}$ divided by the number 2^k and denote it by $\dfrac{1}{2^k} \cdot \mathfrak{a}$.

THEOREM 48. *For any integers $k, l \geqslant 0$ and $n > 0$ and for any free segments $\mathfrak{a}$ and $\mathfrak{b}$, we have*

(I) $\dfrac{1}{2^k} \cdot (\mathfrak{a} + \mathfrak{b}) = \dfrac{1}{2^k} \cdot \mathfrak{a} + \dfrac{1}{2^k} \cdot \mathfrak{b}$; (II) $\left(\dfrac{1}{2^k} \cdot \dfrac{1}{2} \right) \cdot \mathfrak{a} = \dfrac{1}{2^k} \cdot \left(\dfrac{1}{2^l} \cdot \mathfrak{a} \right)$;

(III) $2^k \cdot \left(\dfrac{1}{2^k} \cdot \mathfrak{a} \right) = \mathfrak{a}$; (IV) $n \cdot \left(\dfrac{1}{2^k} \cdot \mathfrak{a} \right) = \dfrac{1}{2^k} \cdot (n \cdot \mathfrak{a})$.

Now, let w be any (positive) dyadic number, i.e. $w = n/2^k$, where n and k are integers and $n > 0$, $k \geqslant 0$. We define the *product* $w \cdot \mathfrak{a}$ of a dyadic number w with a free segment $\mathfrak{a}$ as the free segment $n \cdot \left(\dfrac{1}{2^k} \cdot \mathfrak{a} \right)$.

The product $w \cdot \mathfrak{a}$ is independent of the choice of the numbers n and k. Indeed, by Theorem 48(II)–(IV) and Theorem 47(III), we have

$$\frac{n}{2^k} \cdot \mathfrak{a} = n \cdot \left(\frac{1}{2^k} \cdot \mathfrak{a} \right) = 2^l \cdot \left(\frac{1}{2^l} \cdot \left(n \cdot \left(\frac{1}{2^k} \cdot \mathfrak{a} \right) \right) \right) = 2^l \cdot \left(n \cdot \left(\frac{1}{2^l} \cdot \left(\frac{1}{2^k} \cdot \mathfrak{a} \right) \right) \right)$$

$$= (2^l \cdot n) \cdot \left(\frac{1}{2^{k+l}} \cdot \mathfrak{a} \right) = \frac{2^l \cdot n}{2^{k+l}} \cdot \mathfrak{a}.$$

We shall call the operation leading from a dyadic number w and a free segment $\mathfrak{a}$ to their product $w \cdot \mathfrak{a}$ *multiplication of a dyadic number by a free segment*. The following theorem establishes the basic properties of this operation:

THEOREM 49. *For any dyadic numbers w and v and for any free segments $\mathfrak{a}$ and $\mathfrak{b}$:*

(I) $w \cdot (\mathfrak{a} + \mathfrak{b}) = w \cdot \mathfrak{a} + w \cdot \mathfrak{b}$;   (II) $(w + v) \cdot \mathfrak{a} = w \cdot \mathfrak{a} + v \cdot \mathfrak{a}$;

(III) $w \cdot (v \cdot \mathfrak{a}) = (w \cdot v) \cdot \mathfrak{a}$;   (IV)   *if* $\mathfrak{a} < \mathfrak{b}$, *then* $w \cdot \mathfrak{a} < w \cdot \mathfrak{b}$;

(V)   *if* $w < v$ *then* $w \cdot \mathfrak{a} < v \cdot \mathfrak{a}$;   (VI)   *if* $w \cdot \mathfrak{a} < w \cdot \mathfrak{b}$, *then* $\mathfrak{a} < \mathfrak{b}$;

(VII)   *if* $w \cdot \mathfrak{a} < v \cdot \mathfrak{a}$, *then* $w < v$.

19. Triangle Inequality

THEOREM 50. *For any three distinct points a, b, c, we have*

(1) $$[ab] + [bc] \geqslant [ac],$$

and $[ab] + [bc] = [ac]$ *if and only if* $\mathbf{B}(a,b,c)$.

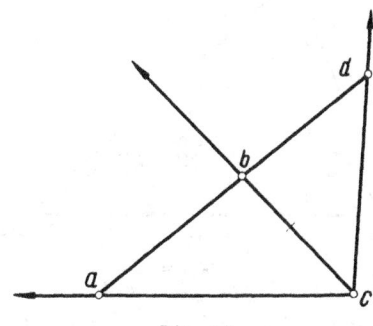

Fig. 83

PROOF. If points a, b, c are non-collinear (Fig. 83), then by taking a point d such that $\mathbf{B}(a,b,d)$ and $bd \equiv bc$, we obtain $\mathbf{B}(\mathbf{HL}(ca), \mathbf{HL}(cb), \mathbf{HL}(cd))$, from which it follows that $\sphericalangle\, acd > \sphericalangle\, bcd \equiv \sphericalangle\, adc$. Therefore $ad > ac$. Then, $[ad] = [ab] + [bc]$. Thus, $[ab] + [bc] > [ac]$.

Assume now that points a, b, c are collinear. If $\mathbf{B}(a,b,c)$, then obviously $[ab] + [bc] = [ac]$. If, however, $\mathbf{B}(a,c,b)$ or $\mathbf{B}(b,a,c)$, then $ab > ac$ or $bc > ac$, from which it follows that $[ab] > [ac]$ or $[bc] > [ac]$, and thus obviously $[ab] + [bc] > [ac]$.

We call inequality (1) the *triangle inequality*. By induction this inequality can readily be generalized to any number of points:

THEOREM 51. *For any distinct points $a_1, a_2, \ldots, a_n$, we have*

$$\sum_{i=1}^{n-1} [a_i a_{i+1}] \geqslant [a_1 a_n].$$

From Theorem 50 it follows in particular that in any triangle abc the sum of two (free) sides is greater than the third (free) side, i.e.

$[ab] + [bc] > [ac]$. This inequality can be generalized in the following manner:

THEOREM 52. *Given a triangle abc and a convex polygonal line* $Z = (p_0, p_1, \ldots, p_n)$, *if* $p_0 = a$, $p_n = c$ *and* $p_i \in (abc)$ *for* $i = 1, 2, \ldots, n-1$, *then*

$$\sum_{i=1}^{n} [p_{i-1} p_i] < [ab] + [bc].$$

PROOF. For $n = 1$, the theorem reduces to Theorem 50. Let us now assume that the theorem is true if the number of vertices of polygonal line Z, distinct from point a, is equal to n ($n \geqslant 1$). Consider any triangle

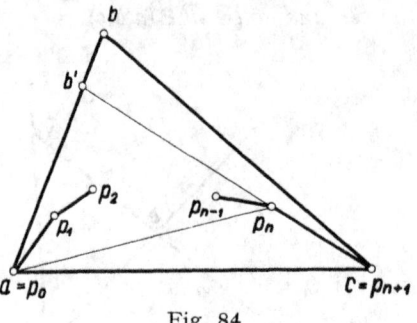

Fig. 84

abc and polygonal line $(p_0, p_1, \ldots, p_n, p_{n+1})$ such that $p_0 = a$, $p_{n+1} = c$ and $p_i \in (abc)$ for $i = 1, 2, \ldots, n$ (Fig. 84). We denote by b' the point of intersection of line $p_n p_{n+1}$ with side (ab). With the help of Theorem 92 of Chapter I it can easily be shown that, for $i = 1, 2, \ldots n-1$, points p_i of the convex polygonal line $(p_0, p_1, \ldots, p_n)$ lie in triangle $(ab' p_n)$. By our inductive assumption we therefore have

$$\sum_{i=1}^{n} [p_{i-1} p_i] < [ab'] + [b' p_n],$$

from which it follows, by Theorems 44(II), 46(I), and 50, that

$$\sum_{i=1}^{n+1} [p_{i-1} p_i] < ([ab'] + [b' p_n]) + [p_n p_{n+1}] = [ab'] + [b'c]$$

$$= ([ab] - [bb']) + [b'c] < ([ab] - [bb']) +$$
$$+ ([bb'] + [bc]) = [ab] + [bc].$$

Hence, the theorem is proved.

From Theorem 71 of Chapter I and from Theorem 50 we obtain at once the following necessary and sufficient condition that three points be collinear:

THEOREM 53. *Three distinct points a, b, c are collinear if and only if one of the segments [ab], [bc], [ac] is the sum of the remaining two.*

20. Free Angles. Calculus of Free Angles

The congruence relation for angles, as a reflexive, symmetric, and transitive relation (see Theorem 12), permits all angles to be divided into disjoint classes of congruent angles. Each such class will be called a *free angle* and each particular angle belonging to this class will be referred to as a *representative* of the class. Individual angles will now be called *bound angles*.

We shall denote free angles by the German capitals $\mathfrak{A}$, $\mathfrak{B}$, $\mathfrak{C}$, $\mathfrak{D}$, $\mathfrak{X}$. We shall denote a free angle with a representative AB by $[AB]$. If $[AB]$ is any free angle, then free angle $[A*B]$ consists, as is readily noted, of angles adjacent to representatives of free angle $[AB]$. We shall call angle $[A*B]$ the free angle *adjacent* to free angle $[AB]$.

We shall introduce the calculus of free angles in a manner analogous to that for free segments. We shall say that a *free angle* $\mathfrak{A} = [A_1B_1]$ *is less than a free angle* $\mathfrak{B} = [A_2B_2]$ —in symbols, $\mathfrak{A} < \mathfrak{B}$ — if $A_1B_1 < A_2B_2$. By Theorem 27, this definition does not depend on the choice of the representatives of angles $\mathfrak{A}$ and $\mathfrak{B}$. Instead of $\mathfrak{A} < \mathfrak{B}$ we also write $\mathfrak{B} > \mathfrak{A}$ and say that *free angle $\mathfrak{B}$ is greater than free angle $\mathfrak{A}$.*

From Theorems 28 and 29 there result the following properties of the relation $<$ for free angles:

THEOREM 54. *Given any free angles $\mathfrak{A}$, $\mathfrak{B}$, and $\mathfrak{C}$, we have:*

(I)   *if $\mathfrak{A} < \mathfrak{B}$, then $\mathfrak{B} \sim < \mathfrak{A}$;*

(II)   *if $\mathfrak{A} < \mathfrak{B}$ and $\mathfrak{B} < \mathfrak{C}$, then $\mathfrak{A} < \mathfrak{C}$;*

(III) *either $\mathfrak{A} = \mathfrak{B}$, or $\mathfrak{A} < \mathfrak{B}$, or else $\mathfrak{A} > \mathfrak{B}$.*

We shall say that a free angle $\mathfrak{C} = [AB]$ is the *sum* of free angles $\mathfrak{A}$ and $\mathfrak{B}$ —in symbols, $\mathfrak{C} = \mathfrak{A} + \mathfrak{B}$ —if there exists a half-line C such that $\mathbf{B}(A,C,B)$, $\mathfrak{A} = [AC]$ and $\mathfrak{B} = [CB]$. By Theorem 20(I) this definition does not depend on the choice of the representative of free angle $\mathfrak{C}$. It is also readily noted that if the sum $\mathfrak{A} + \mathfrak{B}$ exists, then it is uniquely determined by the free angles $\mathfrak{A}$ and $\mathfrak{B}$. In contrast to the addition of free segments, however, the *addition of free angles* is not always performable. This results immediately from the following:

THEOREM 55. *The sum of free angles $\mathfrak{A}$ and $\mathfrak{B}$ exists if and only if free angle $\mathfrak{B}$ is smaller than the free angle adjacent to $\mathfrak{A}$.*

PROOF. Assume that free angle $\mathfrak{B}$ is smaller than the free angle adjacent to $\mathfrak{A}$. Let $\mathfrak{A} = [AB]$, $\mathfrak{B} = [A_1B_1]$. Then $A_1B_1 < A*B$ and thus there exists a half-line C such that $\mathbf{B}(A*,C,B)$ and $BC = A_1B_1$ (Fig. 85). Then

B(A,B,C), and thus $[AC] = \mathfrak{A} + \mathfrak{B}$. Therefore the sum of free angles $\mathfrak{A}$ and $\mathfrak{B}$ exists.

Let us now assume that the sum of free angles $\mathfrak{A}$ and $\mathfrak{B}$ exists. Let $\mathfrak{A} + \mathfrak{B} = [AC]$. There then exists a half-line B such that **B**(A,B,C) and $\mathfrak{A} = [AB]$, $\mathfrak{B} = [BC]$. From **B**(A,B,C) it follows that **B**(B,C,A^*), and therefore $BC < BA^*$, i.e. free angle $\mathfrak{B}$ is smaller than angle $[BA^*]$ adjacent to free angle $\mathfrak{A}$.

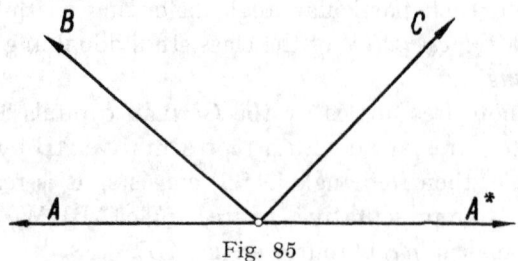

Fig. 85

Using Theorem 55 we can prove all further theorems concerning operations with free angles in a manner completely analogous to that used in the proofs of the corresponding theorems for free segments. Therefore these proofs may be omitted.

Between the relation *less than* and the addition of free angles the following relation holds:

THEOREM 56. *Given any free angles* $\mathfrak{A}$ *and* $\mathfrak{B}$, *we have* $\mathfrak{A} < \mathfrak{B}$ *if and only if there exists a free angle* $\mathfrak{X}$ *such that* $\mathfrak{A} + \mathfrak{X} = \mathfrak{B}$.

The following two theorems establish the basic properties of the addition of free angles:

THEOREM 57. *Given any free angles* $\mathfrak{A}$, $\mathfrak{B}$, $\mathfrak{C}$, $\mathfrak{D}$, *we have*:

(I) *If the sum* $\mathfrak{A} + \mathfrak{B}$ *exists, then the sum* $\mathfrak{B} + \mathfrak{A}$ *also exists, and* $\mathfrak{A} + \mathfrak{B} = \mathfrak{B} + \mathfrak{A}$.

(II) *If the sums* $\mathfrak{A} + \mathfrak{B}$ *and* $(\mathfrak{A} + \mathfrak{B}) + \mathfrak{C}$ *exist, then the sums* $\mathfrak{B} + \mathfrak{C}$ *and* $\mathfrak{A} + (\mathfrak{B} + \mathfrak{C})$ *also exist (and vice versa), and* $(\mathfrak{A} + \mathfrak{B}) + \mathfrak{C} = \mathfrak{A} + (\mathfrak{B} + \mathfrak{C})$.

(III) *If* $\mathfrak{A} < \mathfrak{B}$ *and the sum* $\mathfrak{B} + \mathfrak{C}$ *exists, then the sum* $\mathfrak{A} + \mathfrak{C}$ *also exists, and* $\mathfrak{A} + \mathfrak{C} < \mathfrak{B} + \mathfrak{C}$.

(IV) *If* $\mathfrak{A} < \mathfrak{B}$ *and* $\mathfrak{C} < \mathfrak{D}$, *and the sum* $\mathfrak{B} + \mathfrak{D}$ *exists, then the sum* $\mathfrak{A} + \mathfrak{C}$ *also exists, and* $\mathfrak{A} + \mathfrak{C} < \mathfrak{B} + \mathfrak{D}$.

As an immediate consequence of Theorems 33 and 34 we have:

THEOREM 58. *For any free angle* $\mathfrak{A}$ *there exists just one free angle* $\mathfrak{X}$ *such that* $\mathfrak{A} = \mathfrak{X} + \mathfrak{X}$.

We shall say that *a free angle* $\mathfrak{C}$ *is the difference of free angles* $\mathfrak{A}$ *and* $\mathfrak{B}$ — in symbols, $\mathfrak{C} = \mathfrak{A} - \mathfrak{B}$ —if $\mathfrak{A} = \mathfrak{B} + \mathfrak{C}$.

The difference $\mathfrak{A} - \mathfrak{B}$ exists only if $\mathfrak{A} > \mathfrak{B}$ and is then uniquely determined by free angles $\mathfrak{A}$ and $\mathfrak{B}$. The following theorem results immediately from the definition of the difference:

THEOREM 59. *Given free angles* $\mathfrak{A}$ *and* $\mathfrak{B}$, *if* $\mathfrak{A} > \mathfrak{B}$, *then the sum* $(\mathfrak{A} - \mathfrak{B}) + \mathfrak{B}$ *exists and* $(\mathfrak{A} - \mathfrak{B}) + \mathfrak{B} = \mathfrak{A}$.

As in the case of segments, the *product* $n \cdot \mathfrak{A}$ *of a natural number* n *with a free angle* $\mathfrak{A}$ is defined by induction:

$$1 \cdot \mathfrak{A} = \mathfrak{A},$$

and if the product $n \cdot \mathfrak{A}$ and the sum $n \cdot \mathfrak{A} + \mathfrak{A}$ exist, then

$$(n + 1) \cdot \mathfrak{A} = n \cdot \mathfrak{A} + \mathfrak{A}.$$

Since the addition of free angles is not always performable, then there does not always exist the product of a natural number n with a free angle $\mathfrak{A}$. The following theorem establishes the basic properties of the *multiplication of a free angle by a natural number*:

THEOREM 60. *Given any natural numbers* m *and* n *and any free angles* $\mathfrak{A}$ *and* $\mathfrak{B}$, *if the sums and products standing on the left-hand side of equalities* (I), (II), (III) *exist, then the respective sums and products standing on the right-hand side of equalities* (I), (II), (III) *exist (and vice versa), and*

(I) $n \cdot (\mathfrak{A} + \mathfrak{B}) = n \cdot \mathfrak{A} + n \cdot \mathfrak{B}$,
(II) $(m + n) \cdot \mathfrak{A} = m \cdot \mathfrak{A} + n \cdot \mathfrak{A}$,
(III) $(m \cdot n) \cdot \mathfrak{A} = m \cdot (n \cdot \mathfrak{A})$.

Let k be any non-negative integer. By Theorem 58, for any free angle $\mathfrak{A}$ there exists just one free angle $\mathfrak{X}$ such that $\mathfrak{A} = 2^k \cdot \mathfrak{X}$. We shall call this free angle the *quotient* of free angle $\mathfrak{A}$ divided by the number 2^k and denote it by $\dfrac{1}{2^k} \cdot \mathfrak{A}$.

THEOREM 61. *For any integers* $k, l \geqslant 0$ *and* $n > 0$, *and for any free angles* $\mathfrak{A}$ *and* $\mathfrak{B}$, *we have*:

(I) *If the sum* $\mathfrak{A} + \mathfrak{B}$ *exists, then the sum* $\dfrac{1}{2^k} \cdot \mathfrak{A} + \dfrac{1}{2^k} \cdot \mathfrak{B}$ *also exists,* *and* $\dfrac{1}{2^k} \cdot (\mathfrak{A} + \mathfrak{B}) = \dfrac{1}{2^k} \cdot \mathfrak{A} + \dfrac{1}{2^k} \cdot \mathfrak{B}$.

(II) $\left(\dfrac{1}{2^k} \cdot \dfrac{1}{2^l} \right) \cdot \mathfrak{A} = \dfrac{1}{2^k} \cdot \left(\dfrac{1}{2^l} \cdot \mathfrak{A} \right)$.

(III) *There exists the product* $2^k \cdot \left(\dfrac{1}{2^k} \cdot \mathfrak{A} \right)$, *and* $2^k \cdot \left(\dfrac{1}{2^k} \cdot \mathfrak{A} \right) = \mathfrak{A}$.

(IV) *If the product* $n \cdot \mathfrak{A}$ *exists, then the product* $n \cdot \left(\dfrac{1}{2^k} \cdot \mathfrak{A} \right)$ *also exists,*
and $n \cdot \left(\dfrac{1}{2^k} \cdot \mathfrak{A} \right) = \dfrac{1}{2^k} \cdot (n \cdot \mathfrak{A}).$

Consider any positive dyadic number $w = \dfrac{n}{2^k}$. If the product $n \cdot \left(\dfrac{1}{2^k} \cdot \mathfrak{A} \right)$
exists, then we shall regard this product as the *product of dyadic number*
w with free angle $\mathfrak{A}$ and denote it by $w \cdot \mathfrak{A}$. It is readily shown that
segment $w \cdot \mathfrak{A}$ does not depend on the choice of numbers n and k.

THEOREM 62. *For any dyadic numbers* w *and* v *and for any free*
angles $\mathfrak{A}$ *and* $\mathfrak{B}$ *we have:*

(I) *If the sum* $\mathfrak{A} + \mathfrak{B}$ *and the product* $w \cdot (\mathfrak{A} + \mathfrak{B})$ *exist, then the products*
$w \cdot \mathfrak{A}$, $w \cdot \mathfrak{B}$ *and the sum* $w \cdot \mathfrak{A} + w \cdot \mathfrak{B}$ *also exist (and visa versa), and*
$w \cdot (\mathfrak{A} + \mathfrak{B}) = w \cdot \mathfrak{A} + w \cdot \mathfrak{B}.$

(II) *If the product* $(w + v) \cdot \mathfrak{A}$ *exists, then the products* $w \cdot \mathfrak{A}$, $v \cdot \mathfrak{A}$ *and*
the sum $w \cdot \mathfrak{A} + v \cdot \mathfrak{A}$ *also exist (and vice versa), and* $(w + v) \cdot \mathfrak{A} =$
$w \cdot \mathfrak{A} + v \cdot \mathfrak{A}.$

(III) *If the products* $v \cdot \mathfrak{A}$ *and* $w \cdot (v \cdot \mathfrak{A})$ *exist, then the product* $(w \cdot v) \cdot \mathfrak{A}$
also exists and $(w \cdot v) \cdot \mathfrak{A} = w \cdot (v \cdot \mathfrak{A}).$

(IV) *If the product* $w \cdot \mathfrak{B}$ *exists and* $\mathfrak{A} < \mathfrak{B}$, *then the product* $w \cdot \mathfrak{A}$
also exists and $w \cdot \mathfrak{A} < w \cdot \mathfrak{B}.$

(V) *If the product* $v \cdot \mathfrak{A}$ *exists and* $w < v$, *then the product* $w \cdot \mathfrak{A}$ *also*
exists and $w \cdot \mathfrak{A} < v \cdot \mathfrak{A}.$

(VI) *If the products* $w \cdot \mathfrak{A}$ *and* $w \cdot \mathfrak{B}$ *exist, then* $w \cdot \mathfrak{A} < w \cdot \mathfrak{B}$ *implies*
$\mathfrak{A} < \mathfrak{B}.$

(VII) *If the products* $w \cdot \mathfrak{A}$ *and* $v \cdot \mathfrak{A}$ *exist, then* $w \cdot \mathfrak{A} < v \cdot \mathfrak{A}$ *implies*
$w < v.$

21. Addition of Free Angles in a Pencil

THEOREM 63. *Given a pencil* $\mathfrak{P}$ *and three co-half-pencilar half-lines*
$A,B,C \in \mathfrak{P}$, *if* $\mathbf{B}(A,B,C)$, *then the angle* $[AB] + [CA^*]$ *is adjacent to the*
angle $[BC]$; *in other words*

(1) $$[AB] + [CA^*] = [CB^*].$$

PROOF. Assume that $\mathbf{B}(A,B,C)$ (Fig. 86). Applying Theorem 65 of
Chapter I twice, we obtain $\mathbf{B}(C,A^*,B^*)$. Therefore $[A^*B^*] + [CA^*] = [CB^*]$
which, together with $AB \equiv A^*B^*$ (see Theorem 16), gives formula (1).

22. The Sum of Two Angles of a Triangle

Let us consider any triangle *abc*. The free angles [≮*a*], [≮*b*], [≮*c*] with the respective representatives ≮*a*, ≮*b*, ≮*c* we shall call the *free angles of triangle abc*. The sum of any two of them, say [≮*a*] and [≮*b*], exists, since angle *b* is less than the angle adjacent to angle *a* (see Theorem 55).

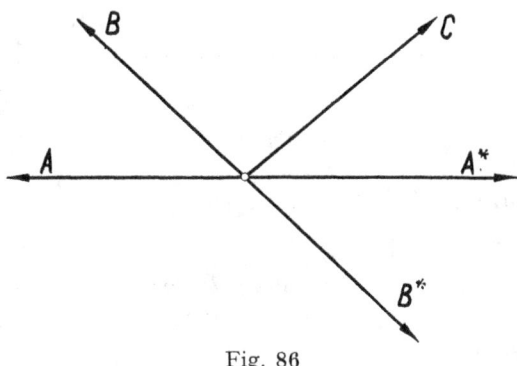

Fig. 86

The triangle inequality (Theorem 50) establishes a certain relation between the sum of two free sides of a triangle and the third side. The question arises of whether there exists a definite relation between the sum of two free angles of a triangle and the third angle. We answer this question in Chapter III; for the time being, we prove only the following implication:†

THEOREM 64. *If for every plane P, every line L ⊂ P, and every point p ∈ P — L there exists only one line K ⊂ P passing through point p and not intersecting line L, then in any triangle abc the sum of any two free angles is identical to the angle adjacent to the third free angle.*

PROOF. Assume that for any plane *P*, any line *L* ⊂ *P*, and any point *p* ∈ *P*−*L* there exists only one line *K* ⊂ *P* passing through point *p* and not intersecting line *L*. Take any triangle *abc* (Fig. 87). Let *Q* = **P** (*abc*) and *W* = **HP**(**L**(*bc*)*a*). Because of our assumption, in plane *Q* we can produce just one line *M* passing through point *c* and not cutting line *ab*. Let *C* = *M* ∩ *W* and *C** = *M* ∩ *W**. Then, by Theorems 38 and 16, we have **H**(*ca*)*C* ≡ ≮*a* and **H**(*cb*)*C** ≡ ≮*b*. Therefore, by Theorem 63, the sum of angles [≮*a*] and [≮*b*] is identical to the angle adjacent to angle [≮*c*].

† This implication is employed in Chapter IV for constructing sentences which, on the basis of absolute geometry, are equivalent to the Axiom of Euclid (sentence E on page 197). For this purpose we shall need also Theorems 26, 27, 28, and 47 from Chapter III.

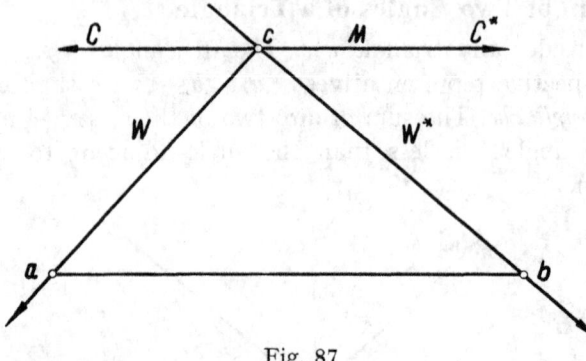

Fig. 87

23. Right, Acute, and Obtuse Angles

We shall prove

THEOREM 65. *Let W be a half-plane determined on a plane P by a line L and let $L = A \cup a \cup A^*$. There exists just one half-line $C \subset W$ with origin a such that $AC \equiv A^*C$.*

PROOF. Consider any half-line $D \subset W$ with origin a. If $AD \equiv A^*D$, the half-line C we are seeking coincides with half-line D. Now assume that

(1) $AD \sim \equiv A^*D$.

We produce from point a a half-line $D_1 \subset W$ (Fig. 88) such that

(2) $AD \equiv A^*D_1$.

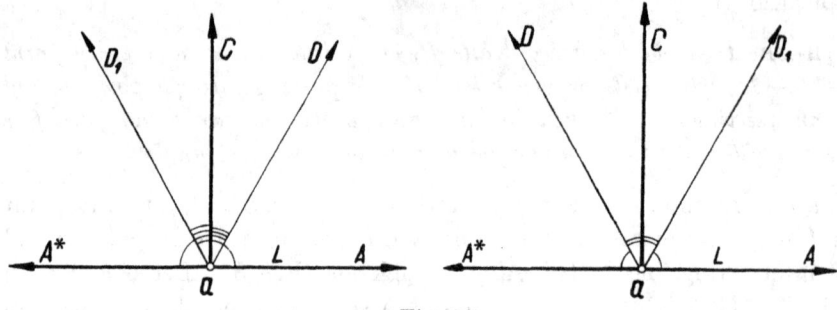

Fig. 88

From formula (1) it follows that $D_1 \neq D$. Let half-line C be the bisector of angle DD_1; therefore

(3) $DC \equiv D_1C$.

Now, there are two possibilities: (i) $\mathbf{B}(A,D,D_1)$, in which case $\mathbf{B}(A^*,D_1,D)$; (ii) $\mathbf{B}(A,D_1,D)$, in which case $\mathbf{B}(A^*,D,D_1)$. Using Theorem 19 we now conclude from formulas (2) and (3) that

(4) $AC \equiv A^*C$.

The proof that in half-plane W there exists just one half-line C satisfying condition (4) is similar to the proof that any angle has just one bisector.

An angle congruent to its adjacent angle is called a *right angle*. From Theorem 65 we obtain at once

THEOREM 66. *There exist right angles.*

We shall prove

THEOREM 67. *An angle congruent to a right angle is again a right angle and any two right angles are congruent. In other words, if angle AB is a right angle, then $AB \equiv A_1B_1$ if and only if angle A_1B_1 is a right angle*

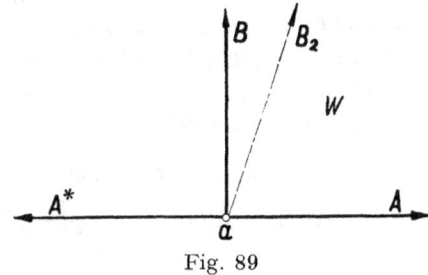

Fig. 89

PROOF. We consider a right angle AB. Then

(5) $$AB \equiv A^*B.$$

We assume that

(6) $$AB \equiv A_1B_1.$$

Then

(7) $$A^*B \equiv A_1^*B_1$$

and from formulas (5), (6), and (7) it follows by Theorem 12 that $A_1B_1 \equiv A_1^*B_1$, that is, angle A_1B_1 is a right angle.

Let us next assume that angle A_1B_1 is a right angle. Let W be that half-plane with boundary $L(A) = A \cup a \cup A^*$ which includes half-line B (Fig. 89). We produce on half-plane W from point a a half-line B_2 such that $AB_2 \equiv A_1B_1$. As we have just proved, angle AB_2 is a right angle, from which, by Theorem 65, we obtain $B_2 = B$. Therefore $AB \equiv A_1B_1$, which was to be proved.

We refer to an angle smaller than any right angle as an *acute angle,* and to an angle greater than any right angle as an *obtuse angle.*

From Theorems 28(II) and 67 there immediately follows:

THEOREM 68. *An angle smaller than any acute angle is again an acute angle. An angle greater than any obtuse angle is again an obtuse angle. Any acute angle is smaller than any obtuse angle.*

It is readily shown that the following theorem holds:

THEOREM 69. *An angle adjacent to a right angle is again a right angle. An angle adjacent to an acute angle is an obtuse angle. An angle adjacent to an obtuse angle is an acute angle.*

24. Right, Acute, and Obtuse Free Angles

By Theorems 66 and 67, all right angles constitute a free angle which we shall call the *right free angle* and denote by the letter $\Re$. Thus, on the basis of absolute geometry, it has been possible to single out, from among all free angles, one free angle, namely, the right free angle.

Free angles smaller than free angle $\Re$ will be called *acute free angles* and free angles greater than free angle $\Re$, *obtuse free angles*.

From Theorem 69 there immediately follows:

THEOREM 70. *Free angle $\Re$ is adjacent to itself.*

By Theorem 54, there exists the sum of free angle $\Re$ and any free acute angle $\mathfrak{X}$, and we have

THEOREM 71. *For any acute free angle $\mathfrak{X}$, angles $\Re + \mathfrak{X}$ and $\Re - \mathfrak{X}$ are adjacent.*

PROOF. Consider any half-pencil $\mathfrak{H}$ with end half-lines A and A^*, and half-lines $B, C \in \mathfrak{H}$ such that $\mathbf{B}(A, B, C)$, $\Re = [AB]$, $\mathfrak{X} = [BC]$ (Fig. 90).

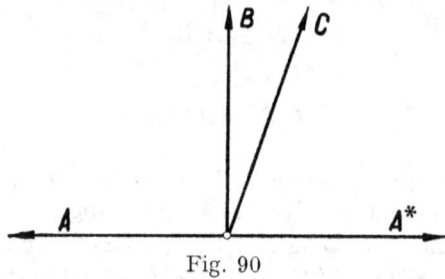

Fig. 90

Then $\mathbf{B}(B, C, A^*)$ and $\Re = [BA^*]$. Thus $\Re + \mathfrak{X} = [AC]$, $\Re - \mathfrak{X} = [A^*C]$, and therefore, angles $\Re + \mathfrak{X}$ and $\Re - \mathfrak{X}$ are adjacent.

For any acute free angle $\mathfrak{X}$, angle $\Re + \mathfrak{X}$ is obtuse. On the other hand, if $\mathfrak{A}$ is an obtuse free angle, then $\mathfrak{A} > \Re$ and, by Theorem 56, there exists a free angle $\mathfrak{X}$ such that $\mathfrak{A} = \Re + \mathfrak{X}$. Since the sum of free angles $\Re$ and $\mathfrak{X}$ exists, free angle $\mathfrak{X}$ must be smaller than the free angle adjacent to $\Re$. Therefore, by Theorem 70, free angle $\mathfrak{X}$ is acute. We thus have:

THEOREM 72. *Every obtuse free angle $\mathfrak{A}$ is the sum of free angle $\mathfrak{R}$ and some acute free angle $\mathfrak{X}$.*

25. Right, Acute, and Obtuse Triangles

We shall prove

THEOREM 73. *In every triangle there are at least two acute angles.*

PROOF. Assume that in a triangle abc angle a is either a right or an obtuse angle. Then, by Theorem 69, the angle adjacent to angle a is either a right or an acute angle; and since, by Theorem 35, angles b and c are smaller than it, therefore, by Theorem 68, they must be acute.

A triangle in which all angles are acute will be called an *acute triangle*; a triangle in which one of the angles is a right angle, a *right triangle*; a triangle in which one of the angles is obtuse, an *obtuse triangle*.

In a right triangle the side opposite the right angle is called the *hypotenuse*, while the remaining sides will be simply referred to as the *sides*. With the help of Theorem 73 and 68 the following theorem is obtained from Theorem 40:

THEOREM 74. (I) *In a right triangle the hypotenuse is greater than each of the sides.*

(II) *In an obtuse triangle the side opposite the obtuse angle is greater than each of the remaining sides.*

Besides the relations between sides and angles of two arbitrary triangles, formulated in Theorem 22, the following three additional relations hold in case they are both right triangles:

THEOREM 75. *Given a right triangle $a_1b_1c_1$ with right angle b_1 and a right triangle $a_2b_2c_2$ with right angle b_2, we have:*

(I) *If $a_1c_1 \equiv a_2c_2$ and $\sphericalangle a_1 \equiv \sphericalangle a_2$, then $a_1b_1 \equiv a_2b_2$, $b_1c_1 \equiv b_2c_2$, and $\sphericalangle c_1 \equiv \sphericalangle c_2$.*

(II) *If $b_1c_1 \equiv b_2c_2$ and $\sphericalangle a_1 \equiv \sphericalangle a_2$, then $a_1b_1 \equiv a_2b_2$, $a_1c_1 \equiv a_2c_2$, and $\sphericalangle c_1 \equiv \sphericalangle c_2$.*

(III) *If $a_1b_1 \equiv a_2b_2$ and $a_1c_1 \equiv a_2c_2$, then $b_1c_1 \equiv b_2c_2$, $\sphericalangle a_1 \equiv \sphericalangle a_2$ and $\sphericalangle c_1 \equiv \sphericalangle c_2$.*

PROOF. (I) Choose a point b_1' on half-line a_1b_1 such that $a_1b_1' \equiv a_2b_2$ (Fig. 91). Suppose $b_1' \neq b_1$. Comparing triangles $a_1b_1'c_1$ and $a_2b_2c_2$ we see that $\sphericalangle a_1b_1'c_1 \equiv \sphericalangle b_2$, and therefore $\sphericalangle a_1b_1'c_1$ is a right angle. Then, as readily seen, triangle $c_1b_1b_1'$ has two right angles, in contradiction to Theorem 73. Therefore we have $b_1' = b_1$ and consequently $a_1b_1 \equiv a_2b_2$, from which it immediately follows that $b_1c_1 \equiv b_2c_2$ and $\sphericalangle c_1 \equiv \sphericalangle c_2$.

(II) Take a point a such that $\textbf{B}(a_1,b_1,a)$ and $ab_1 \equiv a_2b_2$ (Fig. 92). Comparing triangles ab_1c_1 and $a_2b_2c_2$ we see that $\sphericalangle a = \sphericalangle a_2$ and $ac_1 \equiv a_2c_2$.

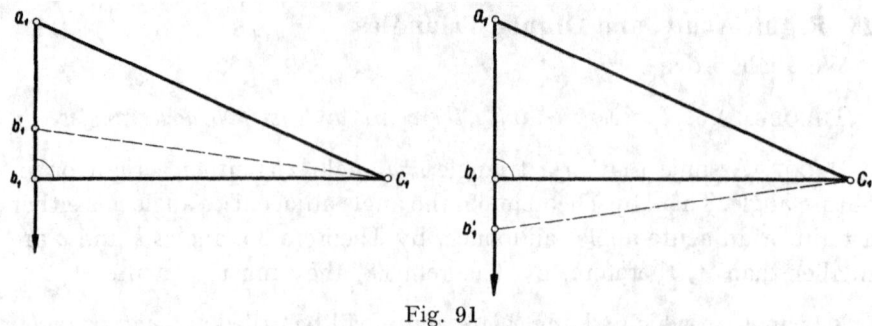

Fig. 91

Therefore $\sphericalangle a_1 \equiv \sphericalangle a$, as a result of which $a_1c_1 \equiv ac_1$. Thus $a_1c_1 \equiv a_2c_2$. By Theorem 75(I), we conclude that also $a_1b_1 \equiv a_2b_2$ and $\sphericalangle c_1 \equiv \sphericalangle c_2$.

(III) Take a point c_1' on half-line b_1c_1 such that $b_1c_1' \equiv b_2c_2$. Suppose that $c_1' \neq c_1$. From triangles $a_1b_1c_1'$ and $a_2b_2c_2$ we see that $a_1c_1' \equiv a_2c_2$, and therefore $a_1c_1' \equiv a_1c_1$. Thus $\sphericalangle a_1c_1'c_1 \equiv \sphericalangle a_1c_1c_1'$. But this is impos-

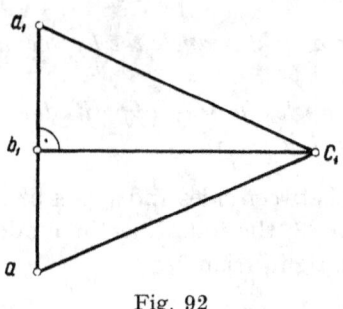

Fig. 92

sible, since if $\textbf{B}(b_1,c_1',c_1)$ (Fig. 93), then angle $a_1c_1c_1'$ is acute and angle $a_1c_1'c_1$, being adjacent to acute angle $a_1c_1'b_1$, is obtuse (see Theorem 69); and if $\textbf{B}(b_1,c_1,c_1')$ (Fig. 94), then angle $a_1c_1'c_1$ is acute and angle $a_1c_1c_1'$,

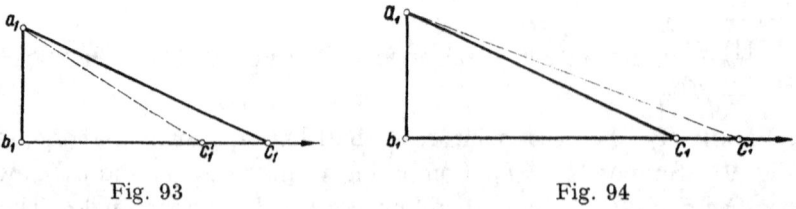

Fig. 93   Fig. 94

being adjacent to acute angle $a_1c_1b_1$, is obtuse. Therefore point c_1' must coincide with point c_1. Hence $b_1c_1 \equiv b_2c_2$, from which it follows that also $\sphericalangle a_1 \equiv \sphericalangle a_2$ and $\sphericalangle c_1 \equiv \sphericalangle c_2$.

26. Perpendicular Lines

We say that *a line K is perpendicular to a line L* —in symbols, $K \perp L$ —
if line K intersects line L in some point p, and there exist half-lines $A \subset K$
and $B \subset L$, both with origin p, such that the angle AB is a right
angle. Then, evidently, all the angles A^*B, AB^*, A^*B^* are also right
angles. Obviously, $K \perp L$ implies $L \perp K$.

From Theorem 65 there follows at once:

THEOREM 76. *Let L be a line on a given plane P. For any point a on L
there is just one line $K \subset P$ passing through point a and perpendicular
to line L.*

A second theorem of this kind is

THEOREM 77. *For any point a not lying on a given line L there exists
just one line K passing through point a and perpendicular to line L.*

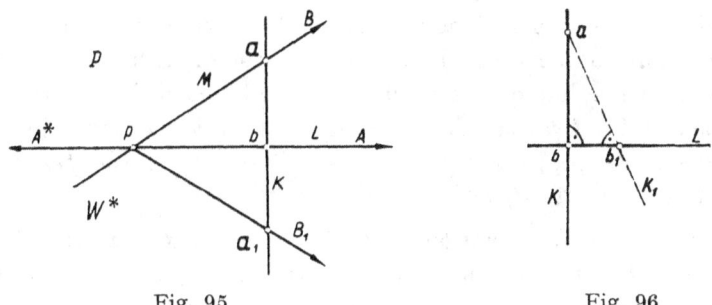

Fig. 95   Fig. 96

PROOF. Let $P = \mathbf{P}(La)$ (Fig. 95). Produce through point a any line M
intersecting line L in some point p. Let $L = A \cup p \cup A^*$, $\mathbf{H}(pa) = B$,
and $\mathbf{H}^*(La) = W^*$. Produce, in half-plane W^*, from point p a half-line
B_1 such that

(1) $$AB \equiv AB_1.$$

Take a point a_1 on half-line B_1 such that $pa \equiv pa_1$. We shall show that
line $K = \mathbf{L}(aa_1)$ is perpendicular to line L. Indeed, if points a,p,a_1 are
collinear, then angles AB and AB_1 are adjacent, and, by formula (1),
they are right angles. We thus conclude that $K \perp L$. If, however, points
a,p,a_1 are non-collinear, then, denoting by b the point in which line L
intersects segment (aa_1), we conclude from triangles abp and a_1bp that
$\sphericalangle abp \equiv \sphericalangle a_1bp$, and hence angle abp is a right angle, i.e. $K \perp L$.

If through point a were to pass another line $K_1 \neq K$ perpendicular
to line L (Fig. 96), then, denoting by b_1 its point of intersection with

line L, we would obtain a triangle abb_1 with two right angles, which, as we know, is impossible.

From Theorems 76 and 77 we obtain at once

THEOREM 78. *Given three lines K, L, and M in a plane P, if lines K and L are distinct and both perpendicular to line M, then they are disjoint.*

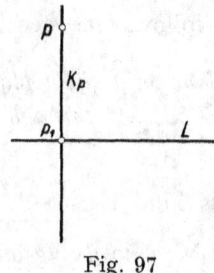

Fig. 97

27. Perpendicular Projection Upon a Line

Consider a line L on a plane P (Fig. 97). As we know, just one line $K_p \subset P$ perpendicular to line L can be produced through each point $p \in P$. The point p_1 in which line K_p intersects line L will be called the *perpendicular projection of point p upon line L*, and the function f correlating to each point $p \in P$ its projection $f(p) = p_1$, will be called the *perpendicular projectivity* (in plane P) *upon line L*.

THEOREM 79. *Given on a plane P two non-perpendicular lines K and L, if f is the perpendicular projectivity upon line L (in plane P), then $\mathbf{B}(p,q,r)$ implies $\mathbf{B}(f(p),f(q),f(r))$, for any three points $p,q,r \in K$.*

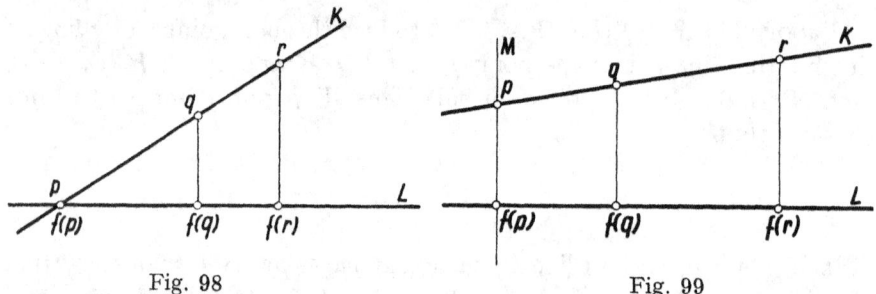

Fig. 98   Fig. 99

PROOF. We may assume that $K \neq L$. From $\mathbf{B}(p,q,r)$ it follows that points p,q,r are distinct, and therefore, since $K \sim \perp L$, we conclude, by Theorem 76, that points $f(p)$, $f(q)$ and $f(r)$ are distinct. By Pasch's Axiom (Theorem 78 of Chapter I), the proof is obvious (with the aid of Theorem 78) if either point p or point r lies on line L (Fig. 98). If, however, $p,r \sim \in L$ (Fig. 99), then $p \neq f(p)$, and putting $M = L(pf(p))$ we see, again with the aid of Theorem 78, that

$$\sim \mathbf{B}(f(q),M,q), \quad \sim \mathbf{B}(q,M,r), \quad \sim \mathbf{B}(r,M,f(r)),$$

from which we conclude, since the relation of lying on the same side of line K is transitive, that $\sim \mathbf{B}(f(q),M,f(r))$, that is, $\sim \mathbf{B}(f(q),f(p),f(r))$. In a similar manner it can be shown that $\sim \mathbf{B}(f(p),f(r),f(q))$. Therefore $\mathbf{B}(f(p),f(q),f(r))$, which concludes the proof of the theorem.

From Theorem 74(I) we obtain immediately the following property of the perpendicular projection upon a line:

THEOREM 80. *Consider on a plane P a line L and a point a not lying on L. Let point a_1 be the perpendicular projection of point a upon line L. Then for any point $p \in L - a_1$ we have $aa_1 < ap$.*

28. Perpendicular Bisector of a Segment

We say that a line K is *perpendicular* to a segment ab —in symbols, $K \perp ab$ or $ab \perp K$ —if it is perpendicular to line ab.

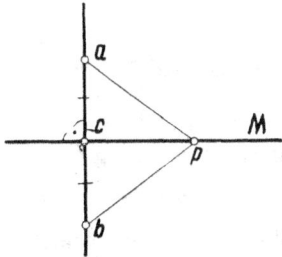

Fig. 100

Let us take any segment ab in a plane P. The line $M \subset P$ passing through the midpoint c of segment ab and perpendicular to ab (Fig. 100) is called the *perpendicular bisector* of segment ab (in plane P). It is readily noted that for each point $p \in M$ we have $pa \equiv pb$, and that for any point $p \in P$ it follows from $pa \equiv pb$ that $p \in M$. We thus have

THEOREM 81. *For any segment ab on a plane P, the perpendicular bisector $M \subset P$ of ab is identical to the set of points $p \in P$ satisfying the condition $pa \equiv pb$.*

29. The Saccheri Quadrangle

If in a convex quadrangle (a,b,c,d) angles b and c are right angles, and sides ab and cd are congruent (Fig. 101), then the quadrangle is called a *Saccheri quadrangle with the lower base bc*. Side ad is called the *upper base*. The line M determined by the midpoint p of the lower base bc and by the midpoint q of the upper base ad is called the *median* of the

Saccheri quadrangle (a,b,c,d). It is readily seen that a Saccheri quadrangle exists. We shall prove two basic properties of a Saccheri quadrangle.

THEOREM 82. *The upper base angles of a Saccheri quadrangle are congruent.*

PROOF. Let (a,b,c,d) be a Saccheri quadrangle with lower base bc (Fig. 102). Comparing triangles abc and dbc we get $ac \equiv bd$; comparing triangles dab and adc we thus obtain $\sphericalangle a \equiv \sphericalangle d$.

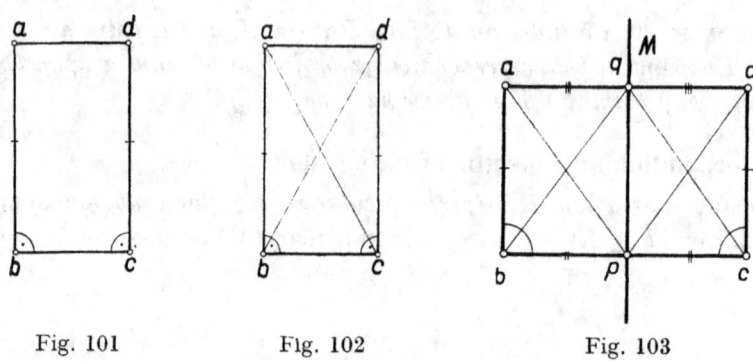

Fig. 101   Fig. 102   Fig. 103

THEOREM 83. *The median of a Saccheri quadrangle is perpendicular to both of its bases.*

PROOF. Let (a,b,c,d) be a Saccheri quadrangle with lower base bc (Fig. 103). We denote by p and q, respectively, the midpoints of bases bc and ad. By Theorem 82, triangles aqb and dqc are congruent and, specifically, $bq \equiv cq$; therefore, by Theorem 81, line $M = \mathbf{L}(pq)$ is the perpendicular bisector of segment bc. Thus $M \perp bc$. Similarly, triangles apb and dpc are congruent and $ap \equiv dp$; therefore line M is the perpendicular bisector of segment ad, from which it follows that $M \perp ad$.

The last theorem gives a relation between two Saccheri quadrangles:

THEOREM 84. *Given a Saccheri quadrangle* (a_1,b_1,c_1,d_1) *with lower base* b_1c_1 *and a Saccheri quadrangle* (a_2,b_2,c_2,d_2) *with lower base* b_2c_2, *if* $a_1b_1 \equiv a_2b_2$ *and* $b_1c_1 \equiv b_2c_2$, *then* $c_1d_1 \equiv c_2d_2$ *and* $d_1a_1 \equiv d_2a_2$.

PROOF. From $a_1b_1 \equiv c_1d_1$, $a_2b_2 \equiv c_2d_2$ and $a_1b_1 \equiv a_2b_2$ it follows at once that $c_1d_1 \equiv c_2d_2$. Furthermore, triangles $a_1b_1c_1$ and $a_2b_2c_2$ are congruent (Fig. 104) and, specifically, $a_1c_1 \equiv a_2c_2$ and $\sphericalangle a_1c_1b_1 \equiv \sphericalangle a_2c_2b_2$. Hence $\sphericalangle a_1c_1d_1 \equiv \sphericalangle a_2c_2d_2$. Thus triangles $a_1c_1d_1$ and $a_2c_2d_2$ are congruent and, specifically, $d_1a_1 \equiv d_2a_2$.

30. Rectangles

If all the angles a, b, c, d of a plane quadrangle (a,b,c,d) are right angles, we call the quadrangle a *rectangle*. From Theorem 78 it at once follows that every rectangle is a convex quadrangle. We shall show that the rectangle is a special case of the Saccheri quadrangle.

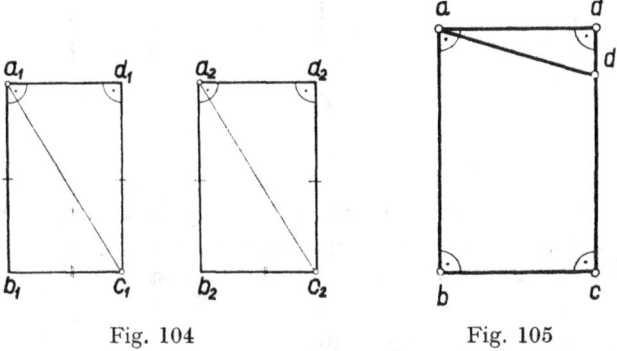

Fig. 104   Fig. 105

THEOREM 85. *If a quadrangle (a,b,c,d) is a rectangle, then $ab \equiv cd$.*

PROOF. Assume that quadrangle (a,b,c,d) is a rectangle. If, for example, side ab were smaller than side cd (Fig. 105), then by taking on segment (cd) a point d' such that $ab \equiv cd'$, we would obtain a Saccheri quadrangle (a,b,c,d') with lower base bc. Angle bad' would be acute, since it is smaller than angle a, and angle $cd'a$ would be obtuse, since it is larger than angle d. But this would contradict Theorem 82.

The following theorem will be used in Chapter III to construct a rectangle with sides that are sufficiently large:

THEOREM 86. *If a quadrangle (a,b,c,d) is a rectangle and points a_1 and b_1 are defined by the conditions*

(1) $\mathbf{B}(a,d,a_1)$, $\mathbf{B}(b,c,b_1)$, $ad \equiv a_1d$, $bc \equiv b_1c$,

then quadrangle (a_1,b_1,c,d) is also a rectangle.

PROOF. Assume that quadrangle (a,b,c,d) is a rectangle (Fig. 106) and that points a_1 and b_1 satisfy condition (1). Then angles a_1dc and b_1cd are obviously right angles. Because of the symmetry of our assumption, it is sufficient to show that angle da_1b_1 is a right angle. Indeed, comparing triangles bcd and b_1cd we obtain

(2) $bd \equiv b_1d$

and $\measuredangle bdc \equiv \measuredangle b_1dc$, from which it follows that

(3) $\measuredangle adb \equiv \measuredangle a_1db_1$.

From (1), (2) and (3) it follows that $\sphericalangle\, a \equiv \sphericalangle da_1b_1$, and thus angle da_1b_1 is a right angle.

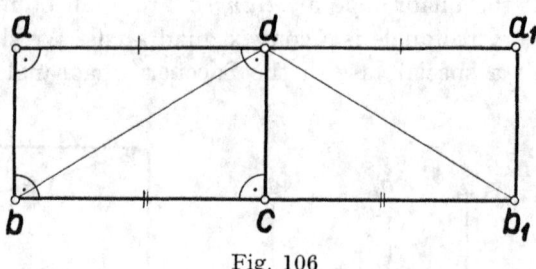

Fig. 106

In each of the last two theorems we have proved some properties of the rectangle under the assumption that it exists. It still remains an open question whether rectangles, in general, exist.

31. Line Perpendicular to a Plane

We say that *a line K is perpendicular to a plane P* —in symbols, $K \perp P$ — or that *plane P is perpendicular to line K* — in symbols $P \perp K$ —if line K intersects plane P in some point a and is perpendicular to every line L lying on plane P and passing through point a.

THEOREM 87. *Let a line K pass through a point a of a plane P. In order that line K be perpendicular to plane P it is sufficient that line K be perpendicular to two distinct lines L_1 and L_2 lying on plane P and passing through point a.*

PROOF. Assume that $L_1, L_2 \subset P$, $L_1 \neq L_2$, $a \in L_1 \cap L_2$, $K \perp L_1$, and $K \perp L_2$ (Fig. 107). Let L be any line lying on plane P, passing through point a, and distinct from lines L_1 and L_2. We have to show that $K \perp L$. We denote by A one of the half-lines determined on line L by point a; further, we denote by B_1 and B_2 those half-lines determined by point a on lines L_1 and L_2 for which $\mathbf{B}(B_1,A,B_2)$. Choose points $b_1 \in B_1$ and $b_2 \in B_2$. Half-line A intersects segment (b_1b_2) in some point c such that

(1) $\mathbf{B}(b_1,c,b_2)$.

On line K, take points p_1 and p_2 on opposite sides of point a such that

(2) $ap_1 \equiv ap_2$.

Comparing right triangles p_1ab_1 and p_2ab_1 we obtain

(3) $p_1b_1 \equiv p_2b_1$

and comparing right triangles p_1ab_2 and p_2ab_2 we obtain

(4) $p_1b_2 \equiv p_2b_2$.

By formulas (3) and (4), triangles $p_1 b_1 b_2$ and $p_2 b_1 b_2$ are congruent and, specifically,

$$(5) \qquad\qquad \sphericalangle p_1 b_1 b_2 = \sphericalangle p_2 b_1 b_2;$$

comparing triangles $p_1 b_1 c$ and $p_2 b_1 c$ we have, by virtue of formulas (3), (5), and (1),

$$(6) \qquad\qquad p_1 c \equiv p_2 c.$$

Finally, comparing triangles $p_1 ac$ and $p_2 ac$ we infer, by formulas (2) and (6), that $\sphericalangle p_1 ac \equiv \sphericalangle p_2 ac$. We thus conclude that angle $p_1 ac$ is a right angle, i.e. $K \perp L$, which was to be proved.

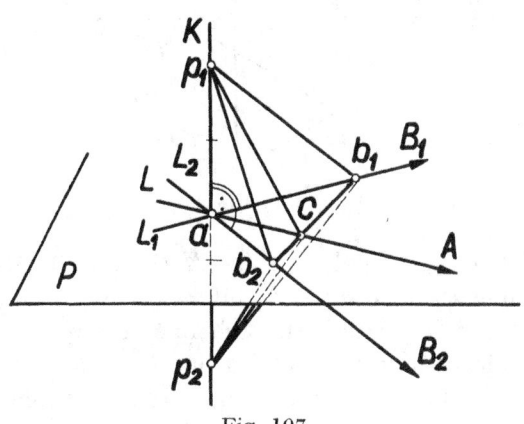

Fig. 107

THEOREM 88. *There exists just one plane P passing through a given point a and perpendicular to a given line K.*

PROOF. Let us examine two cases:

Case 1. Point a lies on line K (Fig. 108). Through line K we produce two distinct planes Q_1 and Q_2 (see Theorem 12 (VII) of Chapter I). Through point a we produce lines $L_1 \subset Q_1$ and $L_2 \subset Q_2$ such that

$$(7) \qquad\qquad L_1 \perp K \text{ and } L_2 \perp K.$$

Obviously, $L_1 \neq L_2$. Let $P = \mathbf{P}(L_1 L_2)$ (see Theorem 9 of Chapter I). It is obvious that $a \in P$, and from Theorem 87 it follows, by formula (7), that $P \perp K$.

If some other plane P' passing through point a were to be perpendicular to line K, then, by choosing a point $p \in P' - P$, we would have on plane Kp two distinct lines both passing through point a and perpendicular to line K, in fact, line ap and line $P \cap \mathbf{P}(Kp)$ (see Theorem 15 of

Chapter I). This, as we know, is impossible. Therefore plane P is the only plane passing through point a and perpendicular to line K.

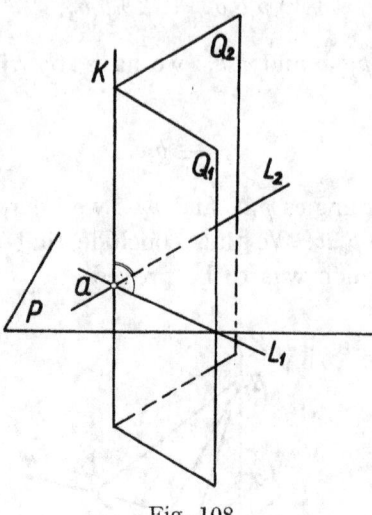

Fig. 108

Case 2. Point a does not lie on line K (Fig. 109). Through point a we produce a line $L_1 \perp K$. Line L_1 intersects line K in some point b. Let $Q_1 = \mathbf{P}(KL_1)$. We produce through line K any plane $Q_2 \neq Q_1$. In plane Q_2 we produce through point b a line $L_2 \perp K$. We have $L_1 \neq L_2$. Let $P = \mathbf{P}(L_1L_2)$. Obviously $a \in P$, and from Theorem 87 it follows that $P \perp K$.

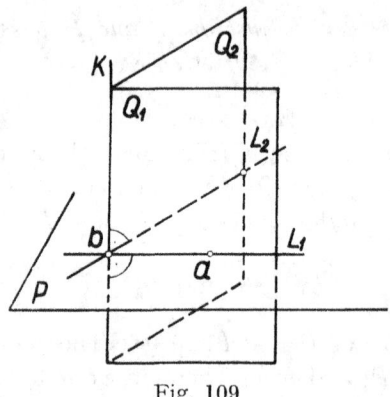

Fig. 109

Suppose that another plane P' passing through point a were to be perpendicular to line K. Let b' be the point in which line K intersects plane P'. If we were to have $b' = b$, then through point b on line K there would pass two distinct planes P and P' both perpendicular to K, which we have already shown to be impossible (Case 1). On the other hand, if $b' \neq b$, then through point a there would pass two distinct lines perpendicular to line K, in fact, $L(ab)$ and $L(ab')$, which, as we know, is also impossible. Therefore plane P is the only plane passing through point a and perpendicular to line K.

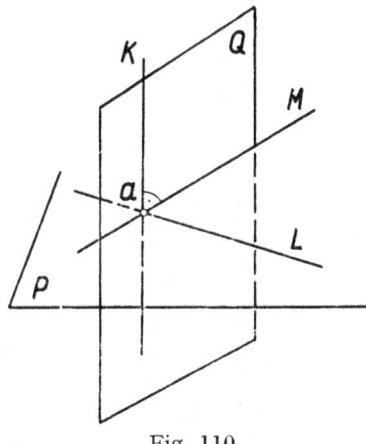

Fig. 110

THEOREM 89. *There exists just one line K passing through a given point a and perpendicular to a given plane P.*

PROOF. Let us examine two cases:

Case 1. Point a lies on plane P (Fig. 110). We produce through point a some line $L \subset P$ and a plane Q such that

(8) $Q \perp L$

(see Theorem 86). The intersection of planes P and Q is a line M (see Theorem 15 of Chapter I). We produce through point a in plane Q a line $K \perp M$. Since, by formula (8), $K \perp L$, and since $L, M \subset P$ and $L \neq M$, then, by Theorem 87, we obtain $K \perp P$.

If some other line K' passing through point a were to be perpendicular to plane P, then the intersection of plane KK' with plane P would be a line N, to which two distinct lines K and K' passing through point a and lying in plane KK' would be perpendicular. As we know, this is impossible. Therefore line K is the only line passing through point a and perpendicular to plane P.

Case 2. Point a does not lie on plane P (Fig. 111). Take any line $L \subset P$; let b denote the perpendicular projection of point a upon line L (in plane La). We thus have

(9) $L(ab) \perp L.$

Through point b we produce a line M in plane P such that

(10) $$M \perp L.$$

Let c denote the perpendicular projection of point a upon line M (in plane Ma). We then have

(11) $$L(ac) \perp M.$$

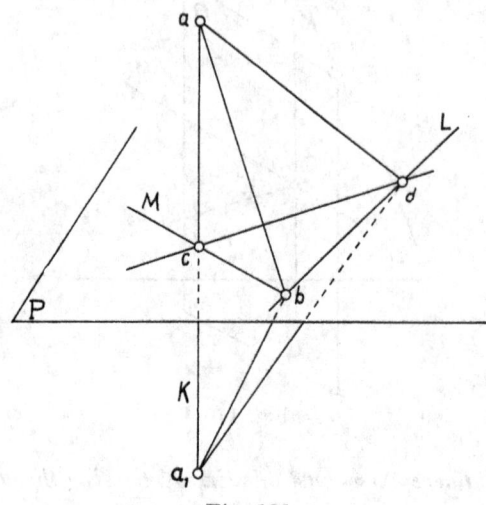

Fig. 111

We shall show that line $K = L(ac)$ is perpendicular to plane P. If $b = c$, then, by Theorem 87, it immediately follows from formulas (9) and (11) that $K \perp P$. If $b \neq c$, then, choosing on line L a point $d \neq b$, we have $L(cd) \neq M$, and, by Theorem 87 and formula (11), it suffices to show that $L(ac) \perp L(cd)$. To do this we take on line ac a point a_1 such that

(12) $$B(a,c,a_1)$$

and

(13) $$ac \equiv a_1c.$$

Since from formulas (9) and (10) it follows that $L \perp P(abc)$, then $L \perp L(a_1b)$, that is, angle a_1bd is a right angle. By formula (13), the right triangles abc and a_1bc are congruent and, specifically,

(14) $$ab \equiv a_1b.$$

Comparing right triangles abd and a_1bd we obtain, by formula (14),

(15) $$ad \equiv a_1d;$$

comparing triangles acd and a_1cd we obtain, because of formulas (13) and (15),

$$\sphericalangle acd \equiv \sphericalangle a_1cd,$$

and since, by formula (12), angles acd and a_1cd are adjacent, then angle acd is a right angle, that is, $\mathbf{L}(ac) \perp \mathbf{L}(cd)$, and therefore $K \perp P$.

If another line K' passing through point a were perpendicular to plane P, then by intersecting plane P by plane KK' we would obtain a line N which would not pass through point a. At the same time, two distinct lines K and K' perpendicular to N could be produced through point a, which, as we know, is impossible. Hence line K is the only line passing through point a and perpendicular to plane P.

32. Perpendicular Planes

We say that *a plane P is perpendicular to a plane Q* — in symbols, $P \perp Q$ — if there exists on plane P a line K perpendicular to plane Q.

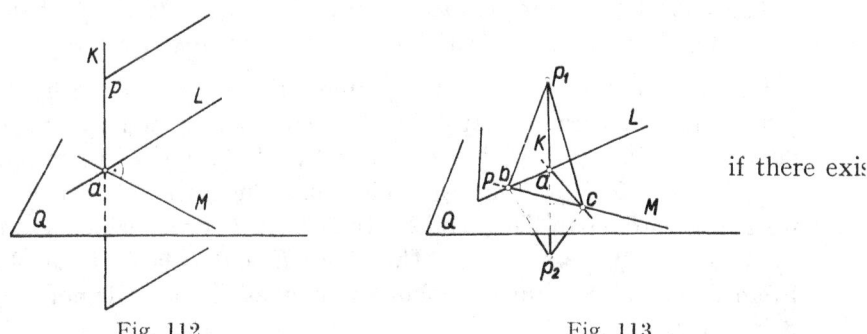

if there exis

Fig. 112   Fig. 113

THEOREM 90. *If $P \perp Q$, then $Q \perp P$.*

PROOF. If $K \subset P$, $K \perp Q$ (Fig. 112), a is the intersection point of line K and plane Q, and L is the intersection line of planes P and Q, then by producing in plane Q through point a a line M perpendicular to L we obtain $M \perp P$. Therefore $Q \perp P$.

THEOREM 91. *Let $P \perp Q$ and let L be the intersection line of planes P and Q. In order that a line K lying in plane P be perpendicular to plane Q it is sufficient that line K be perpendicular to line L.*

PROOF. Assume that $K \subset P$ and $K \perp L$, and let a be the intersection point of lines K and L (Fig. 113). By Theorem 90, we have $Q \perp P$, and therefore on plane Q there exists a line $M \perp P$. Let b be the point of intersection of lines M and L. If $a = b$, then $K \perp L$ and $K \perp M$, from which it follows that $K \perp Q$. Assume now that $a \neq b$, and take on line M a point $c \neq b$. In order to prove that $K \perp Q$ it suffices to prove, since $K \perp \mathbf{L}(ab)$, that $K \perp \mathbf{L}(ac)$. To do this we choose on line K points p_1 and p_2 such that $\mathbf{B}(p_1,a,p_2)$ and $p_1a \equiv p_2a$. Right triangles p_1ab and p_2ab are congruent and, specifically, $p_1b \equiv p_2b$. Thus, right triangles p_1bc

and p_2bc are congruent and $p_1c \equiv p_2c$. Finally, triangles p_1ac and p_2ac are congruent and $\sphericalangle p_1ac \equiv \sphericalangle p_2ac$. Hence $K \perp \mathbf{L}(ac)$, from which it follows that $K \perp Q$, which was to be proved.

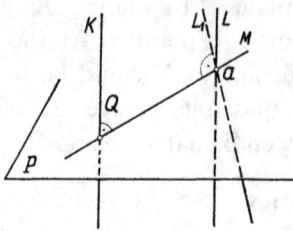

Fig. 114

THEOREM 92. *If lines K and L are distinct and both are perpendicular to a plane P, then lines K and L lie on one plane and are disjoint.*

PROOF. Let a be the point of intersection of line L with plane P (Fig. 114). Furthermore, let Q be the plane determined by line K and point a. Obviously $Q \perp P$. We denote by M the line of intersection of planes P and Q. Producing through point a in plane Q a line $L' \perp M$, we see, according to Theorem 91, that $L' \perp P$, from which it follows, by Theorem 89, that $L = L'$. Thus lines K and L both lie in plane Q. From Theorem 78 it further follows that lines K and L do not have any point in common.

33. Perpendicular Projection Upon a Plane

Consider a plane P. By Theorem 89, just one line K_p perpendicular to P can be produced through every point p (Fig. 115). Point p_1 in which line K_p intersects plane P will be called the *perpendicular projection of point p upon plane P*, and the function f correlating to each point $p \in P$

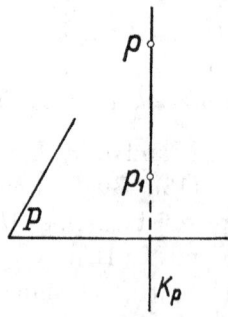

Fig. 115

the projection $f(p) = p_1$ will be called the *perpendicular projectivity upon plane P*.

THEOREM 93. *If f is the perpendicular projectivity upon plane P and points p, q, r are collinear, then the points $f(p)$, $f(q)$, $f(r)$ are also collinear and* $\mathbf{B}(p,q,r)$ *implies* $\mathbf{B}(f(p),f(q),f(r))$ *whenever points p, q, r do not lie on a line perpendicular to P.*

PROOF. Denote by K_p, K_q, K_r, respectively, the lines passing through point p, q, r and perpendicular to plane P (Fig. 116). By Theorem 92, it is readily seen that lines K_p, K_q, K_r all lie in a plane Q, perpendicular to P, and hence points $f(p)$, $f(q)$, $f(r)$ lie on line $P \cap Q$ and are therefore collinear.

The second part of the theorem follows immediately from Theorem 79 and from the fact that, by Theorem 92, the perpendicular projection

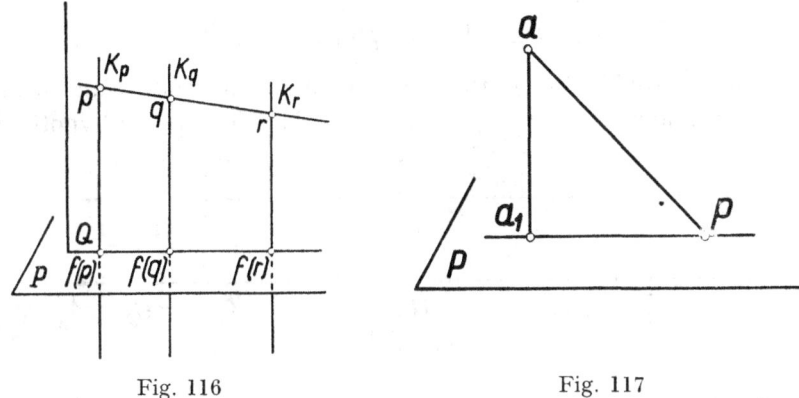

Fig. 116   Fig. 117

upon plane P, restricted to points of plane Q, is the perpendicular projection (in Q) upon line $P \cap Q$.

From Theorem 80 we obtain at once the following property of the perpendicular projection upon a plane:

THEOREM 94. *Assume that* $a \sim \in P$. *Let* a_1 *be the perpendicular projection of point a upon plane P. Then for any point* $p \in P - a_1$ *we have* $aa_1 < ap$.

PROOF. It is sufficient to note that point a_1 is the projection of point a upon line a_1p in plane aa_1p (Fig. 117). We may then use Theorem 80.

34. Congruence of Two Lines and Two Planes. Perfect Homogeneity of the Line and of the Plane

On page 81 we defined the congruence relation for any arbitrary figures. At present, we shall prove some theorems concerning the congruence of lines and planes. We begin with two auxiliary theorems.

THEOREM 95. *Given two lines L_1 and L_2, two distinct points $a_1, b_1 \in L_1$, and two distinct points $a_2, b_2 \in L_2$, if $a_1 b_1 \equiv a_2 b_2$, then for every point $p_1 \neq a_1, b_1$ of line L_1 there is at most one point $p_2 \neq a_2, b_2$ of line L_2 such that*

(1) $$p_1 a_1 \equiv p_2 a_2 \text{ and } p_1 b_1 \equiv p_2 b_2.$$

PROOF. We assume that $a_1 b_1 \equiv a_2 b_2$. We take any point $p_1 \in L_1 - \{a_1, b_1\}$ (Fig. 118). Let $L_2 = A_2 \cup a_2 \cup A_2^*$. The points $p_2 \in L_2 - a_2$ satisfying the condition $p_1 a_1 \equiv p_2 a_2$ are identical with the points q_2 and q_2^* defined by the conditions:

$$q_2 \in A_2 \text{ and } p_1 a_1 \equiv q_2 a_2; \quad q_2^* \in A_2^* \text{ and } p_1 a_1 \equiv q_2^* a_2.$$

Point a_2 is the midpoint of segment $q_2 q_2^*$. If we were to have simultaneously $q_2 \neq b_2$, $q_2^* \neq b_2$ and

$$p_1 b_1 \equiv q_2 b_2 \text{ and } p_1 b_1 \equiv q_2^* b_2,$$

then point b_2 would also be the midpoint of segment $q_2 q_2^*$, and consequently it would coincide with point a_2, in contradiction to our hypothesis.

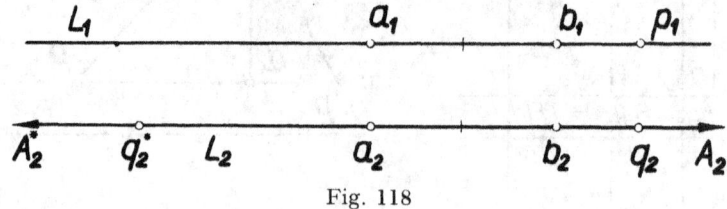

Fig. 118

Therefore at most one of the two points q_2 and q_2^* satisfies condition (1).

THEOREM 96. *Given two planes P_1 and P_2, three non-collinear points $a_1, b_1, c_1 \in P_1$, and three non-collinear points $a_2, b_2, c_2 \in P_2$, if*

(2) $$a_1 b_1 \equiv a_2 b_2, \ b_1 c_1 \equiv b_2 c_2, \ a_1 c_1 \equiv a_2 c_2,$$

then for every point $p_1 \neq a_1, b_1, c_1$ of plane P_1 there is at most one point $p_2 \neq a_2, b_2, c_2$ of plane P_2 such that

(3) $$p_1 a_1 \equiv p_2 a_2, \ p_1 b_1 \equiv p_2 b_2, \ p_1 c_1 \equiv p_2 c_2.$$

PROOF. We assume (2) and we take any point $p_1 \in P_1 - \{a_1, b_1, c_1\}$. We now examine two cases:

Case 1. $p_1 \in \mathbf{L}(a_1 b_1)$. Then point $p_2 \in P_2$ satisfying the conditions $p_2 \neq a_2, b_2$ and

(4) $$p_1 a_1 \equiv p_2 a_2 \text{ and } p_1 b_1 \equiv p_2 b_2$$

must belong to the set $\mathbf{L}(a_2 b_2) - \{a_2, b_2\}$ (see formulas (2), (4) and Theorem 53). By Theorem 95, there exists in this set at most one point p_2 satisfying condition (4).

Case 2. $p_1 \sim \in \mathbf{L}(a_1b_1)$ (Fig. 119). Then $p_2 \sim \in \mathbf{L}(a_2b_2)$. Let us put $P_2 = W_2 \cup \mathbf{L}(a_2b_2) \cup W_2^*$. The points $p_2 \in P_2 - \mathbf{L}(a_2b_2)$ satisfying the conditions $p_1a_1 \equiv p_2a_2$ and $p_1b_1 \equiv p_2b_2$ are identical with the points q_2 and q_2^* defined by the conditions:

$$q_2 \in W_2, \quad p_1a_1 \equiv q_2a_2, \quad p_1b_1 \equiv q_2b_2;$$
$$q_2^* \in W_2^*, \quad p_1a_1 \equiv q_2^*a_2, \quad p_1b_1 \equiv q_2^*b_2.$$

Line a_2b_2 is the perpendicular bisector (in plane P_2) of segment $q_2q_2^*$. If we were to have simultaneously $q_2 \neq c_2$, $q_2^* \neq c_2$ and

$$p_1c_1 \equiv q_2c_2 \text{ and } p_1c_1 \equiv q_2^*c_2,$$

then point c_2 would lie on the perpendicular bisector of segment $q_2q_2^*$, that is, on line a_2b_2, in contradiction to our hypothesis. Therefore at most one of the two points q_2 and q_2^* satisfies condition (3).

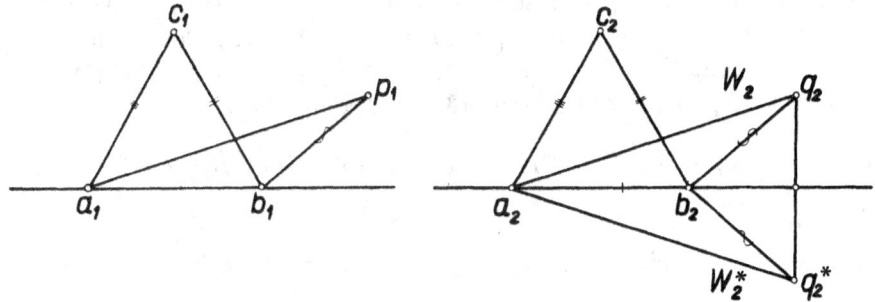

Fig. 119

THEOREM 97. *Given two lines L_1, L_2, and points $a_1,b_1 \in L_1$, $a_2,b_2 \in L_2$, if $a_1b_1 \equiv a_2b_2$, then there exists a congruence function f mapping L_1 onto L_2 and such that $f(a_1) = a_2$ and $f(b_1) = b_2$.*

PROOF. Let $A_i = \mathbf{H}(a_ib_i)$ $(i = 1,2)$. We define a transformation f of line L_1 onto line L_2 in the following way (see Axiom C5):

(5) $f(a_1) = a_2;$

(6)   if $p \in A_1$ then $f(p) \in A_2$, if $p \in A_1^*$ then $f(p) \in A_1^*$,

and

(7) $a_1p \equiv a_2f(p).$

Then

$$f(b_1) = b_2,$$

and, by Theorem 9, for any distinct points $p,q \in L_1 - a_1$,

(8) $\mathbf{B}(p,a_1,q)$ implies $\mathbf{B}(f(p),a_2,f(q))$; $\mathbf{B}(a_1,p,q)$ implies $\mathbf{B}(a_2,f(p),f(q))$.

From formulas (5), (7), (8) (see Axiom C4 and Theorem 7) it is at once seen that the function f realizes the congruence of lines L_1 and L_2.

From Theorem 97 it follows, in particular, that any two lines are congruent to one another, and that in Theorem 95 the phrase *at most one* can be replaced by the phrase *just one*.

THEOREM 98. *Given two planes P_1 and P_2, three non-collinear points $a_1, b_1, c_1 \in P_1$, and three non-collinear points $a_2, b_2, c_2 \in P$, if*

$$a_1 b_1 \equiv a_2 b_2, \ b_1 c_1 \equiv b_2 c_2, \ a_1 c_1 \equiv a_2 c_2,$$

then there exists a congruence function f mapping the plane P_1 onto the plane P_2 such that

(9) $$f(a_1) = a_2, \ f(b_1) = b_2, \ f(c_1) = c_2.$$

PROOF. We put $A_i = \mathbf{H}(a_i b_i)$, $L_i = \mathbf{L}(A_i)$, $W_i = \mathbf{HP}(L_i c_i)$. We define a transformation f of plane P_1 onto plane P_2 in the following manner: The transformation f is defined on line L_1 by formulas (5)–(7). For points $p \in P_1 - L_1$ (see Theorem 13 and Axiom C5)

(10) if $p \in W_1$ then $f(p) \in W_2$, if $p \in W_1^*$ then $f(p) \in W_2^*$,

and

(11) $$A_1 \mathbf{H}(a_1 p) \equiv A_2 \mathbf{H}(a_2 f(p)) \ \text{and} \ a_1 p \equiv a_2 f(p).$$

Thus formula (9) is satisfied.

The function f (with the domain restricted to line L_1) realizes the congruence of lines L_1 and L_2. Therefore, the formula

(12) $$pq \equiv f(p) f(q)$$

holds for every two distinct points $p, q \in L_1$.

Consider now any point $q \in P_1 - L_1$. We put $B_1 \equiv \mathbf{H}(a_1 q)$ and $B_2 = \mathbf{H}(a_2 f(q))$. Then, for every point $p \in \mathbf{L}(a_1 q) - a_1$ (see formula (10) and Theorem 14),

if $p \in B_1$ then $f(p) \in B_2$, if $p \in B_1^*$ then $f(p) \in B_2^*$,

and $a_1 p \equiv a_2 f(p)$. In addition, $f(a_1) = a_2$. Hence, the function f realizes the congruence of lines $a_1 q$ and $a_2 f(q)$.

As a result, formula (12) is satisfied by every two distinct points $p, q \in P_1$ collinear with point a_1.

We next assume that points $a_1, p, q \in P_1$ are not collinear. We shall examine separately three cases.

Case 1. $p \in L_1$, $q \sim \in L_1$ (Fig. 120). Then $p \neq a_1$ and moreover $\sphericalangle pa_1q \equiv \sphericalangle f(p)a_2f(q)$ (see formulas (6), (11), and Theorem 14), from which it follows that $pq \equiv f(p)f(q)$.

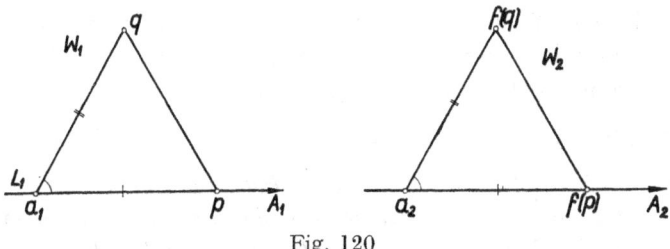

Fig. 120

Case 2. $p,q \in W_1$, or $p,q \in W_1^*$ (Fig. 121). Let $\mathbf{B}(A_1, \mathbf{H}(a_1p), \mathbf{H}(a_1q))$. Then $A_1\mathbf{H}(a_1p) < A_1\mathbf{H}(a_1q)$, from which it follows that $A_2\mathbf{H}(a_2f(p)) < A_2\mathbf{H}(a_2f(q))$. Therefore $\mathbf{B}(A_2, \mathbf{H}(a_2f(p)), \mathbf{H}(a_2f(q)))$. Applying Theorem 19(II), we thus see that $\sphericalangle pa_1q \equiv \sphericalangle f(p)a_2f(q)$, from which it follows that $pq \equiv f(p)f(q)$.

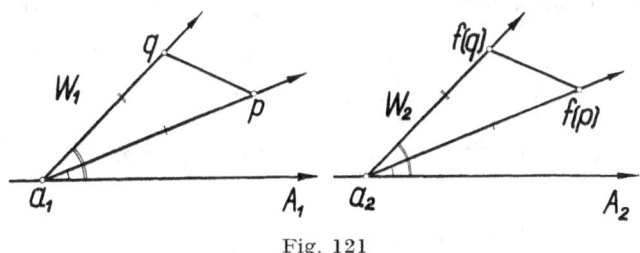

Fig. 121

Case 3. $p \in W_1$, $q \in W_1^*$ (Fig. 122). Line L_1 intersects segment (pq) in some point r and (see formulas 6 and 11)

$$\sphericalangle pa_1r \equiv \sphericalangle f(p)a_2f(r) \quad \text{and} \quad \sphericalangle qa_1r \equiv \sphericalangle f(q)a_2f(r).$$

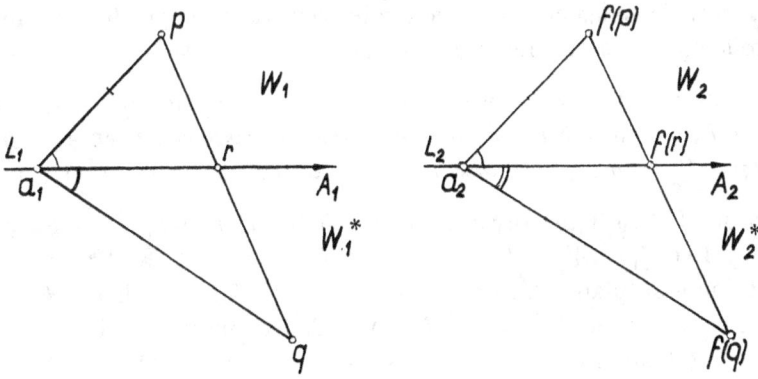

Fig. 122

Applying Theorem 19(I), we thus see that $\sphericalangle pa_1q \equiv \sphericalangle f(p)a_2f(q)$, from which it follows that $pq \equiv f(p)f(q)$.

This concludes the proof.

From Theorem 98 it follows, in particular, that any two planes are congruent to one another,[†] and that in Theorem 96 the phrase *at most one* can be replaced by the phrase *just one*.

Theorem 97 can be generalized in the following way:

THEOREM 99. *Given two lines L_1, L_2 and two figures $F_1 \subset L_1$, $F_2 \subset L_2$, if $F_1 \equiv F_2$, then for every congruence function f mapping F_1 onto F_2, there exists a congruence function g mapping L_1 onto L_2 and such that $g(p) = f(p)$ for every point $p \in F_1$.*

PROOF. Let us assume that $F_1 \equiv F_2$ and let the congruence function f map figure F_1 onto figure F_2. By Theorem 97, it is sufficient to prove the theorem for the case in which figure F_1 consists of at least three points. We take any two points $a, b \in F_1$. By Theorem 97, there exists a congruence function g mapping line L_1 onto line L_2 and satisfying the condition

$$(13) \qquad\qquad g(a) = f(a) \text{ and } g(b) = f(b).$$

Let $p \in F_1 - \{a,b\}$. Then, simultaneously,

$$(14) \qquad\qquad pa \equiv f(p)f(a) \quad \text{and} \quad pb \equiv f(p)f(b),$$

and

$$(15) \qquad\qquad pa \equiv g(p)g(a) \quad \text{and} \quad pb \equiv g(p)g(b).$$

By Theorem 95, it follows from formulas (13)–(15) that $g(p) = f(p)$.

Let us now assume that lines L_1 and L_2 are identical and let $L_1 = L_2 = L$. Then Theorem 99 expresses a property of line L known as the *perfect homogeneity*.

We now consider a similar generalization of Theorem 98 (the first of the following two theorems has a preliminary character):

THEOREM 100. *For every congruence function f mapping a line L_1 of a plane P_1 onto a line L_2 of a plane P_2 there exists a congruence function g mapping P_1 onto P_2 and such that $g(p) = f(p)$ for every point $p \in L_1$.*

PROOF. Let us take any congruence function f mapping line L_1 onto line L_2. Let $P_1 = W_1 \cup L_1 \cup W_1^*$ and $P_2 = W_2 \cup L_2 \cup W_2^*$. We define a function g on plane P_1 in the following way: If $p \in L_1$, then $g(p) = f(p)$. Let us next assume that $p \sim \in L_1$. We take the perpendicular projection p_1 of point p upon line L_1. Through point $f(p_1)$ we produce in plane P_2 a

† Cf. discussion at the end of Section 19 of Chapter III.

line K_p perpendicular to line L_2. Let $A_2 = K_p \cap W_2$, $A_2^* = K_p \cap W_2^*$. If $p \in W_1$ then $g(p) \in A_2$, if $p \in W_1^*$ then $g(p) = A_2^*$, and $g(p)f(p_1) \equiv pp_1$. The reader may readily show that the function g is a congruence function mapping plane P_1 onto plane P_2.

THEOREM 101. *Given two planes P_1, P_2 and two figures $F_1 \subset P_1$, $F_2 \subset P_2$, if $F_1 \equiv F_2$, then for every congruence function f mapping F_1 onto F_2 there exists a congruence function g mapping P_1 onto P_2 such that $g(p) = f(p)$ for every point $p \in F_1$.*

PROOF. If the entire figure F_1 lies on some line $L_1 \subset P_1$, then, by Theorem 53, the entire figure F_2 lies on some line $L_2 \subset P_2$. Thus in this case the theorem follows at once from Theorems 99 and 100. If figure F_1 contains three non-collinear points, then the proof is similar to the proof of Theorem 99 (the proof being based on Theorems 98 and 96).

In particular (for $P_1 = P_2$), Theorem 101 says that every plane is *perfectly homogeneous*.

The following theorem is the space analogue to Theorem 100:

THEOREM 102. *Given two planes P_1 and P_2. For every congruence function f mapping plane P_1 onto plane P_2 there exists a congruence function g of the entire space S onto itself such that $g(p) = f(p)$ for every point $p \in P_1$.*

PROOF. The construction of the congruence function g is fully similar to the construction of congruence function g in the proof of Theorem 100. Space S plays the role of each of the two planes P_1 and P_2 (in Theorem 100), and planes P_1 and P_2 play the role of lines L_1 and L_2; the sets W_1, W_1^*, W_2, W_2^* are now half-spaces.

This theorem can also be generalized to a theorem on the *perfect homogeneity* of space S.

In conclusion, we employ Theorem 101 for the proof of a theorem on congruent triangles.

THEOREM 103. *If $\langle a_1b_1c_1 \rangle \equiv \langle a_2b_2c_2 \rangle$, then $a_1b_1c_1 \equiv a_2b_2c_2$.*

PROOF. Let us assume that $\langle a_1b_1c_1 \rangle \equiv \langle a_2b_2c_2 \rangle$ and let the function f realize the congruence of triangles $\langle a_1b_1c_1 \rangle$ and $\langle a_2b_2c_2 \rangle$. We put $P_1 = \mathbf{P}(a_1b_1c_1)$ and $P_2 = \mathbf{P}(a_2b_2c_2)$. By Theorem 101 there exists a congruence function g mapping plane P_1 onto plane P_2 and such that

$$(16) \qquad\qquad g(p) = f(p) \quad \text{for} \quad p \in \langle a_1b_1c_1 \rangle.$$

Because of the symmetry of our assumptions it is sufficient to show that $f(a_1) \in a_2b_2c_2$. As readily noted (cf. Theorems 78 and 81 of Chapter I) the vertices a, b, c of any triangle $\langle abc \rangle$ are unlike all other points of the triangle in that a line $L \subset P(abc)$ can be produced through each of them not having any more points in common with triangle $\langle abc \rangle$. We produce through point a_1 a line $L_1 \subset P_1$ such that $L_1 \cap \langle a_1 b_1 c_1 \rangle = a_1$. Then the set of points $g(L_1) \subset P_2$ is a line which intersects triangle $\langle a_2 b_2 c_2 \rangle$ only in point $f(a_1)$ (see formula (16)). Hence point $f(a_1)$ is a vertex of triangle $\langle a_2 b_2 c_2 \rangle$.

35. Parallel Half-Lines

We say that *a half-line A with origin a is parallel to a half-line B with origin b* —in symbols, $A||B$ —if one of the following two cases occurs:

Case 1. Half-lines A and B lie on the same line and have the same orientation.

Case 2. Points a and b are distinct, half-lines A and B lie in the same half-plane W with boundary $\mathbf{L}(ab)$, and in the half-pencil $\mathfrak{H}$ with vertex a, consisting of half-lines of half-plane W and ordered from $\mathbf{H}(ab)$ to $\mathbf{H}^*(ab)$, half-line A is the first half-line which does not intersect half-line B. In other words, $A \cap B = 0$, while for every half-line $D \in \mathfrak{H}$ satisfying the condition $\mathbf{B}(\mathbf{H}(ab),D,A)$ we have $D \cap B \neq 0$ (Fig. 123).

From the definition of parallel half-lines we have at once

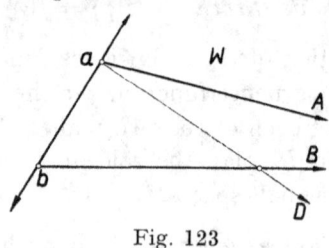

Fig. 123

THEOREM 104. *Given any half-lines A and B, if $A||B$, then $A \sim || B^*$·*

From the properties of the relation $\mathfrak{R}$ investigated in Section 15 of Chapter I we obtain immediately the properties of the parallel relation for half-lines lying on the same line (Case 1). These are formulated in the following two theorems:

THEOREM 105. *For any point a and for any half-line B on a given line K there exists just one half-line $A \subset K$ with origin a and parallel to B.*

THEOREM 106. *For any half-lines A,B,C on a given line K we have:*
(I) $A||A$;
(II)   *if $A||B$, then $B||A$;*
(III) *if $A||B$ and $B||C$, then $A||C$.*

We now pass on to Case 2.

THEOREM 107. *Given two distinct lines K and L, a half-line A ⊂ K with origin a and a half-line B ⊂ L with origin b, if A∥B, then:*

(I)　*angle* $\mathsf{H}(ab)A$ *is not greater than angle* $\mathsf{H}(ba)B^*$;

(II)　*lines K and L lie in one plane and are disjoint.*

PROOF. Let us assume that $A\|B$. Since lines K and L are distinct, then we have to do with Case 2. Let us adopt the notation introduced there.

If angle $\mathsf{H}(ab)A$ were to be greater than angle $\mathsf{H}(ba)B^*$ (Fig. 124), then by taking in half-pencil $\mathfrak{H}$ half-line D such that $\mathsf{B}(\mathsf{H}(ab),D,A)$ and $\mathsf{H}(ab)D \equiv \mathsf{H}(ba)B^*$, we would have $D \cap B = 0$ (see Theorem 35), which would contradict $A\|B$.

Obviously, lines K and L lie in one plane. Then, since $a \neq b$ and

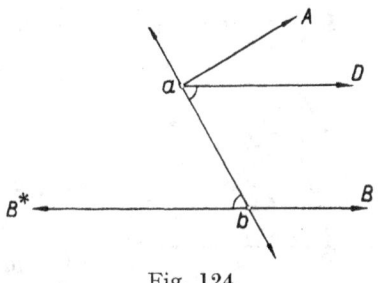

Fig. 124

$A \cap B = 0$, it would follow from $K \cap L \neq 0$ (Fig. 125) that half-line A^* intersects half-line B^* in some point q. But then angle $\mathsf{H}(ab)A$ would be greater than angle $\mathsf{H}(ba)B^*$; this would contradict part (I) of the theorem, which has already been established.

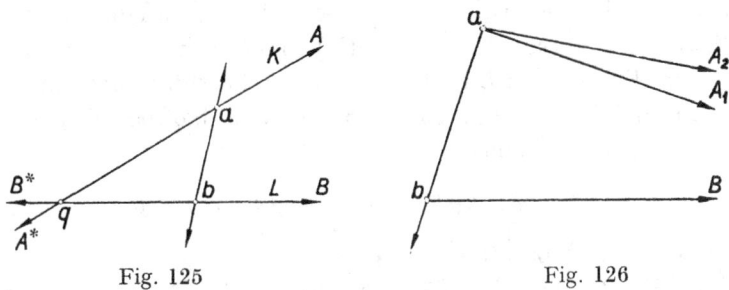

Fig. 125　　　　　　　　　　Fig. 126

THEOREM 108. *Given a line K and a half-line B ⊂ K, if a∼ ∈ K, then there exists at most one half-line A which has the origin a and is parallel to B.*

PROOF. Suppose there exist two distinct half-lines A_1 and A_2 with origin a and parallel to half-line B (Fig. 126). Then

(1)　　　　　　　　　　$A_1\|B$ and $A_2\|B$.

As is readily seen, half-lines A_1 and A_2 belong to one and the same half-pencil $\mathfrak{H}$ with end half-line $\mathbf{H}(ab)$. Thus

$$\mathbf{B}(\mathbf{H}(ab),A_1,A_2) \quad \text{or} \quad \mathbf{B}(\mathbf{H}(ab),A_2,A_1),$$

which, because of (1), implies

$$A_1 \cap B \neq 0 \quad \text{or} \quad A_2 \cap B \neq 0.$$

But this contradicts (1). Therefore, from point a at most one half-line A parallel to half-line B can be produced.

In Chapter III, by using the Axiom of Continuity, we shall prove that such a half-line A always exists.

From Theorem 106(I) it follows that the parallel relation for half-lines is reflexive in all space $\mathbf{S}$. We shall now show that it is also symmetric and transitive in all space $\mathbf{S}$. We shall begin with the transitivity in two special cases:

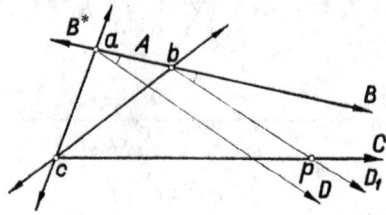

Fig. 127

THEOREM 109. *Given two distinct lines K and L and three distinct half-lines $A,B \subset K$ and $C \subset L$, if $A||B$ and $B||C$, then $A||C$.*

PROOF. We denote by a, b, c, respectively, the origins of half-lines A, B, C. We assume that $A||B$ and $B||C$. From $B||C$ it follows, by Theorem 107(II), that lines K and L lie in one plane and are disjoint; hence $a \neq c$, $A \cap C = 0$ and lines ca and cb belong to some half-pencil $\mathfrak{H}$ with end half-lines C and C^*. Hence

(2) $\mathbf{B}(C,\mathbf{H}(ca), \mathbf{H}(cb))$ or $\mathbf{B}(C,\mathbf{H}(cb), \mathbf{H}(ca))$.

We now examine separately two cases:

Case 1. $A \supset B$ (Fig. 127). Then $b \in A$, $a \in B^*$. If the first condition in (2) were fulfilled, then, by Theorem 57 of Chapter I, half-lines C and ca, and therefore also half-lines C and B^*, would lie on the same side of line cb, that is, half-lines C and B would lie on opposite sides of line cb, which is not possible, since $B||C$. Hence the second of conditions (2) is fulfilled. As a result, half-lines C and cb, and consequently also half-lines C and A, lie on the same side of line ca.

We now produce from point a any half-line D such that

(3) $\mathbf{B}(\mathbf{H}(ac),D,A)$.

In order to prove that $A\|C$, it remains to show that half-line D intersects half-line C. Since $\mathbf{H}(bc)B > \mathbf{H}(ac)A$, then there is a half-line D_1 such that $\mathbf{B}(\mathbf{H}(bc),D_1,B)$ and $AD \equiv BD_1$. From $B\|C$ it follows that half-line D_1 intersects half-line C in some point p. Since, by (3), half-line D intersects side (bc) of triangle (bcp) and it does not intersect side (bp), then half-line D intersects side $(cp) \subset C$.

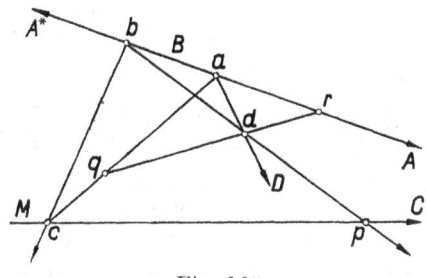

Fig. 128

Case 2. $B \supset A$ (Fig. 128). Then $a \in B$, $b \in A^*$. If the second condition in (2) were to be fulfilled, then, by Theorem 57 of Chapter I, half-lines C and ca, and consequently also half-lines C and B, would lie on opposite sides of line cb, contrary to the fact that $B\|C$. Hence the first condition in (2) holds, and therefore half-lines C and cb, and consequently also half-lines C and A^*, lie on opposite sides of line ca, that is, half-lines C and A lie on the same side of line ca. We now assume that half-line D satisfies condition (3). We choose any points $q \in (ac)$ and $r \in A$. By (3), half-line D intersects segment (qr) in some point d. It can readily be shown that point d lies in the inner domain of angle cba. Therefore $\mathbf{B}(\mathbf{H}(bc),\mathbf{H}(bd),B)$ and, since $B\|C$, half-line bd intersects half-line C in some point p. Since, as may readily be shown, points b and d lie on the same side of line $M \supset C$, then $\mathbf{B}(b,d,p)$. Further, it is readily noted that half-line D and segment (bc) lie on opposite sides of line ac, from which it follows that $D \cap (bc) = 0$. Since half-line D intersects side (bp) of triangle (bcp), but not side (bc), it must intersect side $(cp) \subset C$.

THEOREM 110. *Given two distinct lines K and L and three distinct half-lines $A \subset K$ and $B,C \subset L$, if $A\|B$ and $B\|C$, then $A\|C$.*

PROOF. We denote by a, b, c, respectively, the origins of half-lines A, B, C. We assume that $A\|B$ and $B\|C$. From $A\|B$ it follows, by Theorem 107(II), that lines K and L lie in one plane and are disjoint; therefore

$a \neq c$, $A \cap C = 0$, and half-lines ab and ac belong to some half-pencil $\mathfrak{H}$ with end half-lines A and A^*. Hence

(4) $\mathbf{B}(A,\mathbf{H}(ab),\mathbf{H}(ac))$ or $\mathbf{B}(A,\mathbf{H}(ac),\mathbf{H}(ab))$.

We shall examine separately two cases:

Case 1. $B \subset C$ (Fig. 129). Then $b \in C$ and $c \in B^*$. From $A \| B$ it readily follows that the first condition in (4) holds. Hence half-lines A and C lie

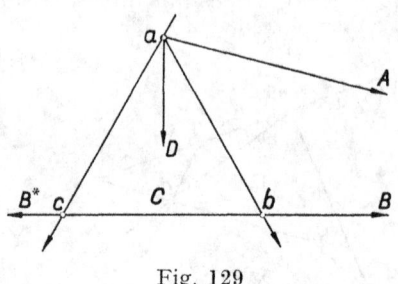

Fig. 129

on the same side of line ac. We produce from point a a half-line D satisfying condition (3). Then we have only one of the following three possibilities:

(i) $D = \mathbf{H}(ab)$ and consequently $D \cap C = b$;

(ii) $\mathbf{B}(\mathbf{H}(ac),D,\mathbf{H}(ab))$ and consequently $D \cap (cb) \neq 0$ and therefore $D \cap C \neq 0$;

(iii) $\mathbf{B}(\mathbf{H}(ab),D,A)$, from which it follows, since $A \| B$, that $D \cap B \neq 0$ and therefore $D \cap C \neq 0$.

Case 2. $C \subset B$ (Fig. 130). Thus $c \in B$ and $b \in C^*$. Now, from $A \| B$ it readily follows that the second condition in formula (4) holds. Hence

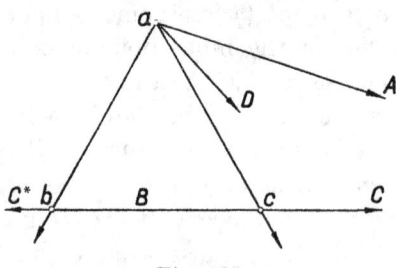

Fig. 130

half-lines A and C again lie on the same side of line ac. Let us now assume that half-line D satisfies condition (3). Then $\mathbf{B}(\mathbf{H}(ab),D,A)$, from which it follows, since $A \| B$, that

(5) $D \cap B \neq 0$.

Furthermore, $\sim B(\mathbf{HL}(ab),D,\mathbf{HL}(ac))$, that is, $D \cap (bc) = 0$. By (5), this gives $D \cap C \neq 0$.

We next prove that the parallel relation for half-lines is symmetric.

THEOREM 111. *Given any half-lines A and B, if $A\|B$, then $B\|A$.*

PROOF. By Theorem 106(II) we may assume that $\mathbf{L}(A) \neq \mathbf{L}(B)$. We denote by a and b, respectively, the origins of half-lines A and B. Let $A\|B$. Then $a \neq b$ and half-lines A and B lie in the same half-plane W with boundary $\mathbf{L}(ab)$.

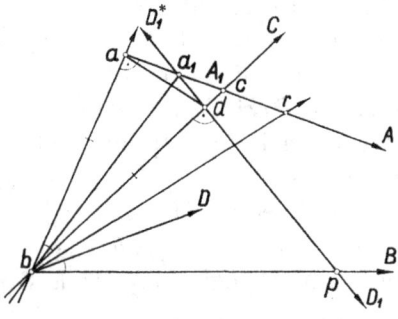

Fig. 131

We shall consider first the special case in which $\mathbf{H}(ab)A$ is a right angle (Fig. 131). Take any half-line D with origin b and such that $B(\mathbf{HL}(ba),D,B)$. In order to show that $B\|A$, it suffices to show that $D \cap A \neq 0$. Choose any point $r \in A$. We may at once restrict ourselves to the case

(6) $BD < \angle abr$.

For, if some half-line D lying between half-lines ba and B intersects half-line A, then every half-line D' lying between half-lines ba and B

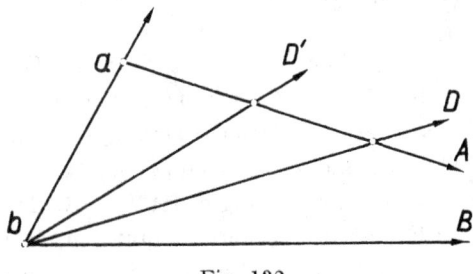

Fig. 132

and creating with half-line B an angle greater than angle BD (Fig. 132) will likewise intersect half-line A. By (6), there exists a half-line C such that $\mathbf{B}(\mathbf{H}(ba),C,\mathbf{H}(br))$ and

(7) $$\mathbf{H}(ba)C \equiv BD.$$

Half-line C intersects segment (ar) in some point c, and $bc > ba$. Consider on segment (bc) a point d such that

(8) $$bd \equiv ba.$$

From point d, on the same side of line bd as half-line B, we produce a half-line D_1 perpendicular to half-line db.

We shall show that $D_1 \cap B \neq 0$. To do so we note that angle adb is acute, since it is a base angle of isosceles triangle abd; therefore, angle adc is obtuse. Further, angle $D_1^*\mathbf{H}(dc)$ is a right angle, since it is congruent to angle $D_1\mathbf{H}(db)$. Hence, half-line D_1^* intersects segment (ac) in some point a_1. We set $A_1 = A - (aa_1\rangle$, $D_2 = (a_1d\rangle \cup D_1$. Then $A_1||A$, and, since $A||B$, it follows from Theorem 109 that $A_1||B$. Further, since angle ca_1d is acute, it is smaller than right angle baa_1 and therefore it is smaller than angle ca_1b external to triangle baa_1. Thus $\mathbf{B}(\mathbf{H}(a_1b),D_2,A_1)$, from

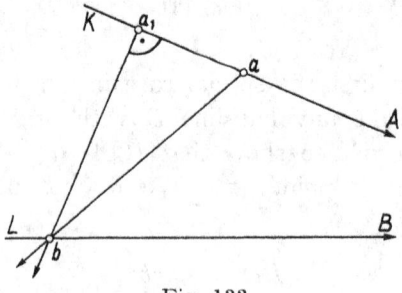

Fig. 133

which, since $A_1||B$, it follows that half-line D_2 intersects half-line B in some point p. It is readily seen that this point belongs to half-line D_1. Therefore $D_1 \cap B \neq 0$.

Let us return to half-line D. From formula (7) it follows that $\mathbf{H}(ba)D \equiv \mathbf{H}(bd)B$. Furthermore, $ab \equiv bd$ and both angles $\mathbf{H}(ab)A$ and $\mathbf{H}(db)D_1$ are right angles. Hence, by Theorem 36, it follows from $D_1 \cap B \neq 0$ that $D \cap A \neq 0$. This concludes the proof in case $\mathbf{H}(ab)A$ is a right angle.

If angle $\mathbf{H}(ab)A$ is not a right angle, let $K = \mathbf{L}(A)$, $L = \mathbf{L}(B)$, and let a_1 be the perpendicular projection of point b upon line K (Fig. 133). We denote by A_1 the half-line of line K with origin a_1 and the same orientation as half-line A. Then $A_1||A$, and, since $A||B$, it follows from Theorem 109 that $A_1||B$. Since $\mathbf{H}(a_1b)A_1$ is a right angle, it then follows from $A_1||B$ that $B||A_1$, which together with $A_1||A$, by Theorem 110, gives $B||A$.

In this way we have proved that the parallel relation for half-lines is symmetric.

We now return to the problem of transitivity. We shall examine separately several cases.

THEOREM 112. *Given two distinct lines K and L and three half-lines $A,C \subset K$ and $B \subset L$, if $A||B$ and $B||C$, then $A||C$.*

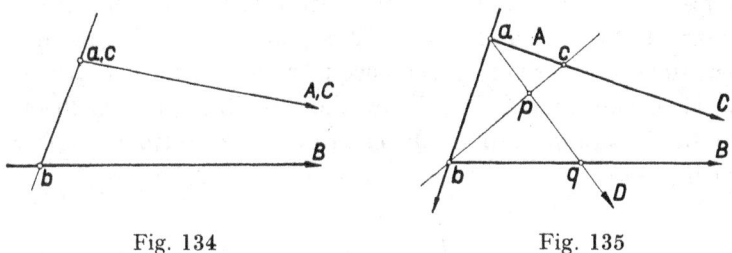

Fig. 134   Fig. 135

PROOF. Let the origins of half-lines A, B, C be points a, b, c, respectively. Assume that $A||B$ and $B||C$. Then $a \neq b$ and lines K and L lie in one plane and are disjoint. We shall examine three possible cases:

Case 1. $c = a$ (Fig. 134). From $A||B$ and $B||C$ it then follows that half-lines A, B, and C lie on the same side of line ab. Therefore $A = C$ and thus, by Theorem 106(I), $A||C$.

Case 2. $c \in A$ (Fig. 135). Let $\mathbf{B}(\mathbf{H}(ab),D,A)$. Half-line D intersects segment (bc) in some point p; since $A||B$, it also intersects half-line B, in some point q. It is readily seen that $\mathbf{B}(a,p,q)$. Therefore point a and half-line B lie on opposite sides of line bc, from which it follows, since $B||C$, that $a \sim \in C$. From $c \in A$ and $a \sim \in C$ it follows, by Theorem 36 of Chapter I, that $C \subset A$ and hence $A||C$.

Case 3. $c \in A^*$ (Fig. 136). From $A||B$ it follows that half-lines A and B

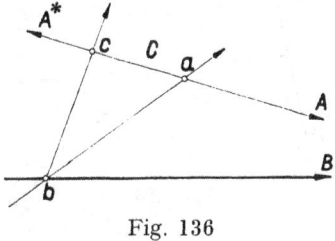

Fig. 136

lie on the same side of line ab, and therefore point c and half-line B lie on opposite sides of line ab. Thus $\mathbf{B}(\mathbf{H}(bc),\mathbf{H}(ba),B)$, from which it follows, since $B||C$, that $a \in C$, which, together with $c \sim \in A$, gives $A \subset C$; therefore $A||C$.

THEOREM 113. *Given three distinct lines* K, L, M *and three half-lines* $A \subset K$, $B \subset L$, $C \subset M$ *in a plane* P, *if* $A||B$ *and* $B||C$, *then* $A||C$.

PROOF. Let the origins of half-lines A, B, C be points a, b, c, respectively. We assume that $A||B$ and $B||C$. Then $a \neq b$ and $b \neq c$. Lines K and M do not intersect, since, if they did, two distinct half-lines parallel to half-line B would, by Theorems 109, 110, and 111, pass through their point of intersection; this would be in contradiction to Theorem 108. In particular, $A \cap C = 0$ and $a \neq c$. Thus points a, b, c are distinct.

Let us first consider the special case in which points a, b, c lie on some line N. Then half-lines A, B, C lie on the same half-plane with boundary N. We shall examine separately each of three possible cases: $\mathbf{B}(a,b,c)$, $\mathbf{B}(b,c,a)$, $\mathbf{B}(c,a,b)$.

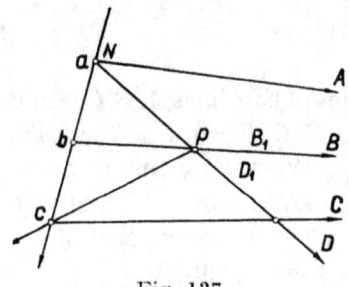

Fig. 137

Case 1. $\mathbf{B}(a,b,c)$ (Fig. 137). Let us take any half-line D such that $\mathbf{B}(\mathbf{H}(ac),D,A)$. Since $A||B$, half-line D intersects half-line B in some point p. We set $B_1 = B - \langle bp \rangle$, $D_1 = D - \langle ap \rangle$. Then $B_1||C$ and $\mathbf{B}(\mathbf{H}(pc),D_1,B_1)$. Hence half-line $D_1 \subset D$ intersects half-line C, from which it follows that $A||C$.

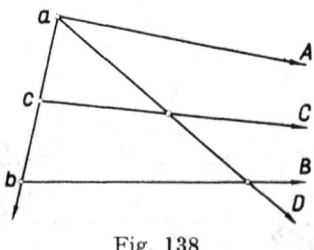

Fig. 138

Case 2. $\mathbf{B}(b,c,a)$ (Fig. 138). Let us take any half-line D such that $\mathbf{B}(\mathbf{H}(ac),D,A)$. Since $A||B$, half-line D intersects half-line B, and therefore it must intersect half-line C. We thus conclude that $A||C$.

Case 3. **B**(c,a,b). This case reduces to the previous one because of the symmetry of the parallel relation for half-lines.

Let us next suppose that points a, b, c are not collinear. We shall show that now, too, there exists a line N intersecting all three lines K, L, and M. Indeed, if lines K and M lie on opposite sides of line L, then line ac is such a line. If, however, lines K and M lie on the same side of line L

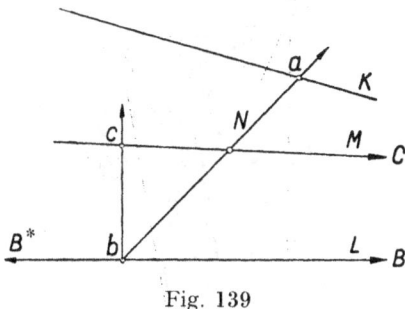

Fig. 139

(Fig. 139), then half-lines ba and bc are distinct and belong to the same half-pencil with end half-lines B and B^*. As a result

$$\text{either } \mathbf{B}(B,\mathbf{H}(ba),\mathbf{H}(bc)) \quad \text{or} \quad \mathbf{B}(B,\mathbf{H}(bc),\mathbf{H}(ba)).$$

In the first case, since $B||C$, half-line ba intersects half-line $C \subset M$ and, as a result, we may take line ab as the line N; in the second case, since $B||A$, half-line bc intersects half-line $A \subset K$ and we may take line bc as the line N.

Let us now take any line N which intersects lines K, L, M, respectively, in points a_1, b_1, c_1. Let $A_1 \subset K, B_1 \subset L, C_1 \subset M$ be half-lines with origins a_1, b_1, c_1 and the same orientations as half-lines A, B, C, respectively. Then $A_1||A, B_1||B$, and $C_1||C$. Because of Theorems 109 and 110, from $A_1||A, A||B_1$, and $B||B_1$ it follows that $A_1||B_1$, and from $B_1||B, B||C$, and $C||C_1$ it follows that $B_1||C_1$. Furthermore, since the origins of half-lines A_1, B_1, and C_1 are collinear, then from $A_1||B_1$ and $B_1||C_1$ it follows that $A_1||C_1$. Finally, from $A||A_1, A_1||C_1$, and $C_1||C$ we obtain $A||C$. This concludes the proof.

In all the cases examined thus far, half-lines A, B, C lie in one plane. We shall now consider the spatial case:

THEOREM 114. *Given three distinct lines K, L, M not lying in one plane and three half-lines $A \subset K, B \subset L$, and $C \subset M$, if $A||B$ and $B||C$, then $A||C$.*

PROOF. Let points a, b, c, respectively, be the origins of half-lines A, B, C (Fig. 140). Further, let $P_1 = \mathbf{P}(KL)$. Then $c \sim \in P_1$. Finally, let

$P_2 = \mathbf{P}(Kc)$ and $P_3 = \mathbf{P}(Lc)$. Then P_1, P_2, and P_3 are three distinct planes, and planes P_2 and P_3 intersect along some line M' passing through point c. It may easily be seen that points a, b, c are non-collinear. Let $Q = \mathbf{P}(abc)$. Half-lines A and B lie in plane P_1 on one side of line ab, and therefore, by Theorem 52 of Chapter I, they also lie on one side of plane Q; we denote by C' the half-line determined on line M' by point c and lying

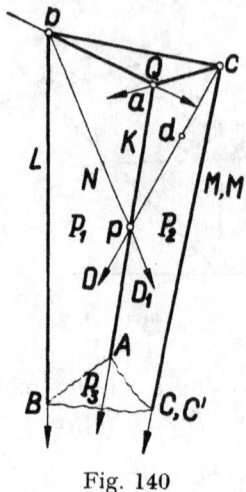

Fig. 140

on the same side of plane Q. Then, by Theorem 52 of Chapter I, half-lines A and C' lie on plane P_2 on the same side of line ac. We shall show that $C' \| A$.

Lines M' and K lie in plane P_2. We note further that lines M' and K are disjoint. Otherwise, their point of intersection would lie on both planes P_1 and P_3, and therefore, on line L, which would contradict the fact that lines K and L are disjoint. Thus, $C' \cap A = 0$ and we now have only to show that every half-line D lying between half-lines ca and C' intersects half-line A. To do this we take any point $d \in D$. Plane bcd intersects plane P_1 along some line N passing through point b. We denote by D_1 the half-line determined on line N by point b which lies in plane bcd on the same side of line bc as half-line D. Then, by Theorem 52 of Chapter I, half-lines D and D_1 lie on the same side of both plane Q and plane P_3. It thus follows that half-line D_1 lies on plane P_1 on the same side of line ab as half-line B, and on the same side of line L as half-line ba. In other words, $\mathbf{B}(\mathbf{H}(ba), D_1, B)$. Since half-lines B and A are parallel, it follows that half-line D_1 intersects half-line A in some point p. Point p lies on both planes bcd and P_2 on the same side of plane Q as point d. Therefore $p \in D$. Hence we have shown that half-line D intersects half-line A, and consequently

(9) $C'\|A.$

In the same way, it can be shown that

(10) $C'\|B.$

From $C\|B$ and from formula (10) it follows, by Theorem 108, that $C'=C$, and thus, by formula (9), we obtain $C\|A$, and consequently $A\|C$, which was to be proved.

From Theorems 106(III), 109, 110, 112, 113, and 114 the following general theorem results:

THEOREM 115. *Given any half-lines A, B, C, if $A\|B$ and $B\|C$, then $A\|C$.*

36. Parallel Axes

We shall say that *axis $\Re$ is parallel to axis $\mathfrak{L}$*—in symbols, $\Re\|\mathfrak{L}$—if some half-line A of $\Re$ is parallel to some half-line B of $\mathfrak{L}$. Thus, if $\Re\|\mathfrak{L}$, then either $\Re=\mathfrak{L}$ or axes $\Re$ and $\mathfrak{L}$ lie in one plane and have no point in common. From the transitivity of the parallel relation for half-lines we immediately derive:

THEOREM 116. *If $\Re\|\mathfrak{L}$, then every half-line of axis $\Re$ is parallel to every half-line of axis $\mathfrak{L}$.*

The properties of the parallel relation for half-lines proved in the preceding section apply to the parallel relation for axes. Indeed, from Theorem 104 we obtain:

THEOREM 117. *If $\Re\|\mathfrak{L}$, then $\Re\sim\|\mathfrak{L}^*.$*

From Theorem 108, with the help of Theorem 116, we obtain

THEOREM 118. *There is at most one axis $\Re$ passing through a given point a and parallel to a given axis $\mathfrak{L}$.*

It is seen at once that if $a \in L$, then there is just one such axis $\Re$, in fact, axis $\mathfrak{L}$ itself.

Finally, from Theorems 106(I), 111, and 115, with the help of Theorem 116, we obtain

THEOREM 119. *The parallel relation for axes is reflexive, symmetric, and transitive.*

We now divide the family of all axes into equivalence classes with respect to the parallel relation. These classes will be called the *directions of the axes*. Thus, the direction of an axis $\mathfrak{L}$ is the class of all axes parallel to $\mathfrak{L}$. By *directions of a line L* we shall understand the directions of its two axes $\mathfrak{L}$ and $\mathfrak{L}^*$.

We now fix any plane P. The axes lying in plane P can also be divided into equivalence classes with respect to the parallel relation. We shall call these classes the *directions of the axes in plane P*. Here again, by the *directions of a line $L \subset P$ in plane P* we understand the directions of both its axes $\mathfrak{L}$ and $\mathfrak{L}^*$ in plane P.

37. Parallel Lines

We say that *line K is parallel to line L*—symbolically $K \| L$—if there exists an axis $\mathfrak{K}$ of K and an axis $\mathfrak{L}$ of L such that $\mathfrak{K} \| \mathfrak{L}$. Thus, if $K \| L$, then either $K = L$ or K and L lie in the same plane and are disjoint.

Since each line may be given two orientations, then from Theorem 118 we obtain at once:

THEOREM 120. *There are at most two lines passing through a given point a and parallel to a given line L.*

It is seen at once that if $a \in L$, then there is just one such line, in fact, line L itself.

As a result of the reflexivity and symmetry of the parallel relation for axes there at once follows:

THEOREM 121. *The parallel relation for lines is reflexive and symmetric.*

We can thus speak of pairs of parallel lines. A pair of parallel lines has at least one common direction.

If lines K and L lying in one plane P neither intersect nor are parallel, then we say that they are *hyperparallel* (or *divergent*).

III

Axiom of Continuity

1. Axiom of Continuity

The fourth and last group of the axioms of absolute geometry consists of one axiom called the *Axiom of Continuity*.

AXIOM Co. *Given two arbitrary non-empty point sets X and Y, if there exists a point a such that*

(1) $\qquad\qquad p \in X \quad and \quad q \in Y \qquad implies \quad \mathbf{B}(a,p,q),$

then there exists a point b such that

(2) $\qquad\qquad p \in X - b \quad and \quad q \in Y - b \ implies \quad \mathbf{B}(p,b,q).$

The Axiom of Continuity is the only axiom of absolute geometry which involves the general notion of a set of points.

THEOREM 1. *For any axis $\mathfrak{L}$, every half-line of axis $\mathfrak{L}$, every open segment of axis $\mathfrak{L}$, and the line L of axis $\mathfrak{L}$ have the Dedekind property.*

PROOF. Consider a half-line A with origin a. Let (X,Y) be a Dedekind cut of A. Then (1) holds and, by Axiom Co, there exists a point b satisfying condition (2). It is readily seen that point b is either the last point in the class X or the first point in the class Y. Since half-line A is densely ordered, then there cannot be simultaneously the last element in class X and the first element in class Y. Thus half-line A has the Dedekind property.

In quite a similar way we prove the theorem for an open segment. As point a we now take the first (on axis $\mathfrak{L}$) of the two end points of the segment. Denote by c the other end-point. Then, for an arbitrary Dedekind cut (X,Y) of segment (ac), if b is the last point in the class X, then $X = (ab\rangle$ and $Y = (bc)$, and if b is the first point in the class Y, then $X = (ab)$ and $Y = \langle bc)$.

Consider line L. Let (X,Y) be a Dedekind cut of line L and let a be an arbitrary point of the class X. We take any point a' preceding point a. Let X' be the subclass of X consisting of all points in X which follow point a'. Since $a \in X'$, then $X' \neq 0$, and

$$p \in X' \text{ and } q \in Y \text{ implies } \mathbf{B}(a',p,q).$$

Hence, by Axiom Co, there exists a point b such that

$$p \in X' - b \text{ and } q \in Y - b \text{ implies } \mathbf{B}(p,b,q),$$

from which it follows, by the definition of class X', that

$$p \in X - b \text{ and } q \in Y - b \text{ implies } \mathbf{B}(p,b,q).$$

The rest of the proof is exactly the same as in the case of the half-line. Denote by B the half-line determined by point b on L. It is readily seen that if $b \in X$, then $X = B^* \cup b$ and $Y = B$, and if $b \in Y$, then $X = B^*$ and $Y = b \cup B$.

On the basis of Theorem 1, we shall prove an important property of the segment.

THEOREM 2. *Every open (or closed) segment is a connected set.*

PROOF. Consider an open segment (ac). Let

$$(ac) = X_1 \cup X_2, \text{ where } X_1, X_2 \neq 0 \text{ and } X_1 \cap X_2 = 0.$$

In order to prove that (ac) is connected it is sufficient to show that either in the set X_1 there is an accumulation point of the set X_2 or in the set X_2 there is an accumulation point of the set X_1.

On segment (ac) let us fix the order from point a to point c. Let

$$q_1 \in X_1 \text{ and } q_2 \in X_2.$$

Without diminishing the generality of the proof we can assume that $q_1 \prec q_2$. Let us now define sets $Y_1, Y_2 \subset (ac)$ in the following way: Point p belongs to Y_1 if and only if either $p \in X_1 \cap (aq_2)$ or there exists a point $p' \in X_1 \cap (aq_2)$ such that $p \prec p'$. Hence

$$(3) \qquad\qquad Y_1 \subset (aq_2).$$

The remaining points of segment (ac) belong to Y_2. Hence $(ac) = Y_1 \cup Y_2$. Since $q_1 \in Y_1$ and $q_2 \in Y_2$, then Y_1 and Y_2 are non-empty sets. Further, by the definition of set Y_1, if $p \in Y_1$ and $p' \prec p$, then $p' \in Y_1$, from which it follows at once that $p_1 \prec p_2$ for any points $p_1 \in Y_1$ and $p_2 \in Y_2$. Therefore the division of segment (ac) into sets Y_1 and Y_2 is a Dedekind cut. Let $b \in (ac)$ be the point determined by this cut. From formula (3) it follows at once that $b \in (aq_2\rangle$. Two cases are now possible:

Case 1. $b \in X_1$. Therefore $b \neq q_2$ and $b \prec q_2$. Taking now any neighborhood (b_1b_2) of point b such that $b \prec b_2 \prec q_2$, we have $(bb_2) \subset Y_2 \cap (aq_2)$ and consequently $(bb_2) \subset X_2$. Therefore in any neighborhood of point b there are points of set X_2 and hence point $b \in X_1$ is an accumulation point of set X_2.

Case 2. $b \in X_2$. Taking any neighborhood $(b_1 b_2)$ of point b such that $b_1 \prec b$, we have $(b_1 b) \subset Y_1$. If now point p of segment $(b_1 b)$ does not belong to set X_1, then from the definition of set Y_1 it follows that there exists a point $p' \in X_1 \cap (ab)$ such that $p' \succ p$. It is readily shown that $p' \in (b_1 b) \subset (b_1 b_2)$. Hence in any neighborhood of point b there is a point of set X_1 and therefore point $b \in X_2$ is an accumulation point of set X_1. In this way we have proved that the segment (ac) is connected.

Since the segment $\langle ac \rangle$ is the closure of the segment (ac), it is also connected.

In a similar way we can prove

THEOREM 3. *Every line is a connected set.*

The Dedekind property for segments can, in a simple way, be carried over to angles.

THEOREM 4. *Every angle (AB) ordered from its end half-line A to its end half-line B has the Dedekind property.*

PROOF. Let a pair of classes $(\mathfrak{X}_1, \mathfrak{X}_2)$ be a Dedekind cut of an angle (AB). Take any points $a \in A$ and $b \in B$. Let p be the vertex of angle AB. By Theorem 68 of Chapter I the only half-lines with origin p which belong to angle (AB) are those and only those which intersect segment (ab). We denote by X_i (for $i = 1,2$) the set of intersection points of the half-lines belonging to class $\mathfrak{X}_i$ with segment (ab). On the basis of the discussion in Section 22 of Chapter I, we can easily show that the pair of sets (X_1, X_2) is a Dedekind cut of segment (ab) ordered from end point a to end point b, and that for any point $c \in (ab)$ half-line pc is the last in class $\mathfrak{X}_1$ (or the first in class $\mathfrak{X}_2$) if and only if point c is the last in set X_1 (or the first in set X_2). Since segment (ab) has the Dedekind property, the proof is thus complete.

Finally, we show that the whole half-pencil possesses the Dedekind property.

THEOREM 5. *Every half-pencil $\mathfrak{H}$ ordered from its end half-line A to its end half-line A^* has the Dedekind property.*

PROOF. Take any Dedekind cut $(\mathfrak{X}_1, \mathfrak{X}_2)$ of a half-pencil $\mathfrak{H}$. Choose in half-pencil $\mathfrak{H}$ two half-lines $B_1 \in \mathfrak{X}_1$, $B_2 \in \mathfrak{X}_2$, and two other half-lines $C_1 \prec B_1$, $C_2 \succ B_2$. Let $\mathfrak{Y}_1 = \mathfrak{X}_1 \cap (C_1 C_2)$, $\mathfrak{Y}_2 = \mathfrak{X}_2 \cap (C_1 C_2)$. The pair of classes $(\mathfrak{Y}_1, \mathfrak{Y}_2)$ is a Dedekind cut of angle $(C_1 C_2)$ ordered from C_1 to C_2. It is readily seen that a half-line $C \in \mathfrak{H}$ is the last in class $\mathfrak{Y}_1$ (or the first in class $\mathfrak{Y}_2$) if and only if it is the last in class $\mathfrak{X}_1$ (or the first in class $\mathfrak{X}_2$). By Theorem 4 the proof is thus complete.

NOTE. Axioms of congruence were not used in the proofs of Theorems 1–5.

2. The Archimedean Postulate

We shall now prove a theorem known as the *Archimedean Postulate*.

THEOREM 6. *Consider a half-line A with origin a_0, a segment cd, and a point $b \in A$ such that $a_0 b > cd$. There then exists a natural number n such that the first n points of the points $a_1, a_2, \ldots, a_n, a_{n+1} \in A$ defined by the conditions*

(1) $\qquad a_1 \prec a_2 \prec \ldots \prec a_n \prec a_{n+1}, \quad a_{k-1} a_k \equiv cd \text{ for } k = 1, 2, \ldots n + 1,$

lie in segment $\langle ab \rangle$ and point a_{n+1} follows point b.

PROOF. Suppose that there is no natural number n such that only the first n points of the $n + 1$ points defined by conditions (1) belong to segment $\langle ab \rangle$. Then all points of the infinite sequence

$$a_1, a_2, \ldots, a_n, \ldots$$

defined on half-line A by the conditions

(2) $\qquad a_k \prec a_{k+1} \quad \text{and} \quad a_{k-1} a_k \equiv cd \quad \text{for} \quad k = 1, 2, \ldots$

belong to segment $\langle a_0 b \rangle$. Hence $b \neq a_k$ for $k = 1, 2, \ldots$, and therefore all points a_k belong to segment $\langle a_0 b \rangle$.

Let X_1 be the set of those points of segment $\langle a_0 b \rangle$ which precede any of the points a_k, and let X_2 be the set of the remaining points of segment $\langle a_0 b \rangle$. We shall show that the pair of sets (X_1, X_2) is a Dedekind cut of segment $\langle a_0 b \rangle$. From $a_k \prec a_{k+1}$ it follows that all points a_k belong to set X_1. Therefore set X_1 is non-empty. Set X_2 is also non empty. Indeed, from $a_0 b > cd$ it follows that there exists a point b_1 such that

(3) $\qquad\qquad a_0 \prec b_1 \prec b \quad \text{and} \quad bb_1 \equiv cd.$

If, for some natural number k, the point b_1 were to precede the point a_k, then we would have

$$b_1 \prec a_k \prec a_{k+1} \prec b,$$

from which it would follow that $a_k a_{k+1} < bb_1$, and thus, because of (3), we would also have $a_k a_{k+1} < cd$. This would be in contradiction to (2); hence point b_1 belongs to set X_2.

Furthermore, if $p_1 \in X_1$ and $p_2 \in X_2$, then for some natural k we have $p_1 \prec a_k \prec p_2$, from which it follows that $p_1 \prec p_2$.

Therefore the pair of sets (X_1, X_2) is a Dedekind cut of segment

(ab). Let c be the point determined by this cut. Then for any point $p \in (a_0 b)$

(4)　　　　　　if $p \prec c$, then $p \prec a_k$ for some k,

(5)　　　　　　if $p \succ c$, then $a_k \prec p$ for every k.

Since $a_k \in X_1$ for every natural k, then there is no last point in set X_1. Therefore $c \in X_2$ and $X_1 = (a_0 c)$. As a result, all points a_k belong to segment $(a_0 c)$. We can then repeat, for segment $(a_0 c)$, the same argument used for segment $(a_0 b)$. In this way we find a point $d \in (a_0 c)$ such that, for any point $p \in (a_0 c)$,

(6)　　　　　　if $p \succ d$, then $a_k \prec p$ for every k.

This, however, contradicts the condition (4), since there exists on segment $(a_0 c) \subset (a_0 b)$ a point p such that $d \prec p \prec c$.

From the Archimedean Postulate we have the following conclusions for free segments:

THEOREM 7. *If a free segment $\mathfrak{a}$ is smaller than a free segment $\mathfrak{b}$, then there exists a natural number n such that*

$$n \cdot \mathfrak{a} \leqslant \mathfrak{b}, \quad but \quad (n + 1) \cdot \mathfrak{a} > \mathfrak{b}.$$

THEOREM 8. *For any free segments $\mathfrak{a}$ and $\mathfrak{b}$, there exists an integer $k \geqslant 0$ such that*

(7)　　　　　　　　　　$$\frac{1}{2^k} \cdot \mathfrak{a} < \mathfrak{b}.$$

PROOF. If $\mathfrak{a} < \mathfrak{b}$ we take $k = 0$. We now assume that $\mathfrak{b} \leqslant \mathfrak{a}$. By Theorem 7, there exists a natural number n such that

(8)　　　　　　　　　　$$n \cdot \mathfrak{b} > \mathfrak{a}$$

(if $\mathfrak{b} = \mathfrak{a}$ it suffices to take $n = 2$). We shall show that every integer k such that $2^k \geqslant n$ satisfies inequality (7). Indeed, if

$$\frac{1}{2^k} \cdot \mathfrak{a} \geqslant \mathfrak{b},$$

then by Theorems 48(III) and 49(IV), (V) of Chapter II it would follow from $2^k \geqslant n$ that

$$\mathfrak{a} \geqslant 2^k \cdot \mathfrak{b} \geqslant n \cdot \mathfrak{b},$$

which contradicts inequality (8).

THEOREM 9. *If a free segment $\mathfrak{a}$ is less than a free segment $\mathfrak{b}$, then for each free segment $\mathfrak{c}$ there exists a dyadic number w such that*

$$\mathfrak{a} < w \cdot \mathfrak{c} < \mathfrak{b}.$$

PROOF. Assume that $a < b$. It follows from Theorem 8 that for some integer $k \geqslant 0$

(9)
$$\frac{1}{2^k} \cdot c < b - a.$$

If $\frac{1}{2^k} \cdot c > a$, then from formula (9) we obtain, with the help of Theorem 46(I) of Chapter II,

$$a < \frac{1}{2^k} \cdot c < b - a < b,$$

and the number we are seeking is $w = \frac{1}{2^k}$.

If, however, $\frac{1}{2^k} \cdot c \leqslant a$, then, by Theorem 7, there exists a natural number n such that

(10) $n \cdot \left(\frac{1}{2^k} \cdot c \right) \leqslant a$ but $(n + 1) \cdot \left(\frac{1}{2^k} \cdot c \right) \geqslant a$

and from formulas (9) and (10) we obtain, with the help of Theorem 47(II), 44(I)(IV), and 46(I) of Chapter II,

$$a \leqslant (n + 1) \cdot \left(\frac{1}{2^k} \cdot c \right) = \frac{1}{2^k} \cdot c + n \cdot \left(\frac{1}{2^k} \cdot c \right) < (b - a) + a = b.$$

The number we are seeking is thus $w = \dfrac{n + 1}{2^k}$.

In case of angles the following theorem corresponds to the Archimedean Postulate:

THEOREM 10. *Consider a half-pencil* $\mathfrak{H}$ *ordered from the end half-line* A_0 *to the end half-line* A_0^*, *an angle* CD, *and a half-line* $B \in \mathfrak{H}$ *such that* $CD < A_0B$. *Then there exists a natural number* n *such that there are* n *half-lines* $A_1, A_2, \ldots, A_n$ *in angle* (A_0B) *satisfying the conditions*

$$A_1 \prec A_2 \prec \ldots \prec A_n \text{ and } A_{k-1}A_k \equiv CD \text{ for } k = 1, 2, \ldots, n,$$

while if there is a half-line $A_{n+1} \in \mathfrak{H}$ *such that*

$$A_n \prec A_{n+1} \text{ and } A_n A_{n+1} \equiv CD,$$

then $A_{n+1} \succ B$.

The proof is similar to the proof of Theorem 6.

From Theorem 10 we have the following conclusions for free angles:

THEOREM 11. *If a free angle* $\mathfrak{A}$ *is less than a free angle* $\mathfrak{B}$, *then for some natural number* n *there exists the product* $n \cdot \mathfrak{A}$ *and* $n \cdot \mathfrak{A} \leqslant \mathfrak{B}$, *and if there exists the product* $(n+1) \cdot \mathfrak{A}$, *then* $(n+1) \cdot \mathfrak{A} > \mathfrak{B}$.

THEOREM 12. *For any free angles* $\mathfrak{A}$ *and* $\mathfrak{B}$ *there exists an integer* $k \geqslant 0$ *such that*

$$\frac{1}{2^k} \cdot \mathfrak{A} < \mathfrak{B}.$$

THEOREM 13. *If a free angle* $\mathfrak{A}$ *is less than a free angle* $\mathfrak{B}$, *then for any free angle* $\mathfrak{C}$ *there exists a dyadic number* w *such that the product* $w \cdot \mathfrak{C}$ *exists and* $\mathfrak{A} < w \cdot \mathfrak{C} < \mathfrak{B}$.

The proofs of Theorems 11–13 are analogous to the proofs of the corresponding theorems for free segments (Theorems 7–9).

3. The Saccheri Quadrangle (Conclusion)

By means of Archimedean Postulate we shall prove the following property of a Saccheri quadrangle:

THEOREM 14. *In a Saccheri quadrangle the upper base is not less than the lower base.*

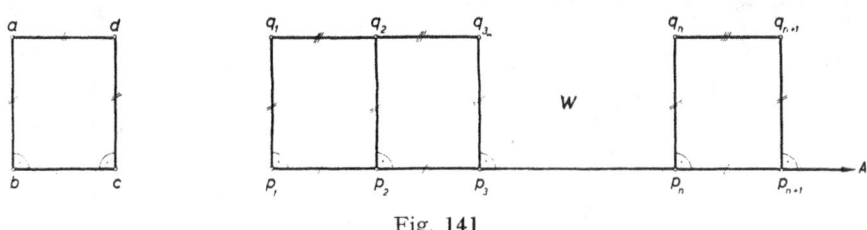

Fig. 141

PROOF. Suppose there exists a Saccheri quadrangle (a,b,c,d) with the lower base bc such that $bc > ad$ (Fig. 141). We take any half-line A and any half-plane W with the boundary $\mathbf{L}(A)$. We construct two infinite sequences of points. The first sequence

$$(p_1, p_2, \ldots, p_n, \ldots)$$

of points of half-line A is defined by the conditions: p_1 is the origin of half-line A and

(1) $p_n \prec p_{n+1}$ and $p_n p_{n+1} \equiv bc$ for $n = 1, 2, \ldots$

The second sequence

$$(q_1, q_2, \ldots, q_n, \ldots)$$

of points of half-plane W is defined by the conditions:

(2) $\qquad \mathbf{L}(q_n p_n) \perp \mathbf{L}(A)$ and $q_n p_n \equiv ab \qquad$ for $n = 1,2,\ldots$

Thus the polygonal line $(q_n, p_n, p_{n+1}, q_{n+1})$ is a Saccheri quadrangle and, by Theorem 84 of Chapter II,

(3) $\qquad q_n q_{n+1} \equiv ad \qquad$ for $n = 1,2,\ldots$

Since, by our supposition, $bc > ad$, there then exists a free segment $\mathfrak{x}$ such that

(4) $\qquad [bc] = [ad] + \mathfrak{x}.$

From formulas (1)–(4), with the help of Theorem 51 of Chapter II, we obtain for $n = 1,2\ldots$

$$n \cdot [ad] + n \cdot \mathfrak{x} = n \cdot [bc] \leqslant [ab] + n \cdot [ad] + [ab],$$

that is

$$n \cdot \mathfrak{x} \leqslant 2 \cdot [ab] \qquad \text{for } n = 1,2,\ldots,$$

which contradicts Theorem 7. This concludes the proof.

THEOREM 15. *In any Saccheri quadrangle the angles at the upper base are both right or both acute angles.*

Fig. 142

PROOF. Consider a Saccheri quadrangle (a,b,c,d) with the lower base bc (Fig. 142). By Theorem 14 there is a point $e \in (ad\rangle$ such that $ae \equiv bc$. If $e = d$, then $\sphericalangle a \equiv \sphericalangle c$ and thus the angle a is a right angle. If $e \in (ad)$, then $be < bd$ and hence $\sphericalangle a < \sphericalangle c$, from which it follows that angle a is acute. To conclude the proof let us note that, by Theorem 82 of Chapter II, the angle d is congruent to angle a.

4. The Saccheri-Legendre Theorem

We begin with an auxiliary theorem.

THEOREM 16. *The sum of two acute free angles of a right triangle is not greater than the right free angle.*

PROOF. Consider a right triangle abc with the right angle b (Fig. 143). We take a point d such that the polygonal line (a,b,c,d) is a Saccheri

Fig. 143

quadrangle with the lower base bc. By Theorem 14 either $bc \equiv ad$ or $bc < ad$, hence either $\angle bac \equiv \angle acd$ or $\angle bac < \angle acd$. Consequently $[\angle bac] + [\angle acb] \leqslant [\angle bcd]$ which completes the proof.

We shall now prove one of the most important theorems of absolute geometry, the *Saccheri-Legendre Theorem* (cf. Section 22 of Chapter II).

THEOREM 17. *The sum of any two free angles of a triangle is not greater than the angle adjacent to the third free angle of the triangle.*

PROOF.† Consider a triangle abc with acute angles a and c (Fig. 144). Let d be the perpendicular projection of the vertex b upon line ac. We take half-lines A with origin a and B with origin b such that $A \subset \mathbf{HP}(\mathbf{L}(ac)b)$, $B \subset \mathbf{HP}(\mathbf{L}(bc)a)$, and angles $A\mathbf{HL}(ac)$ and $B\mathbf{HL}(bd)$ are right angles. Now, by Theorem 16,

$$[\mathbf{H}(ab)\mathbf{HP}^*(ac)] = [\mathbf{H}(ab)A] + [A\mathbf{H}^*(ac)] \geqslant$$

$$[\angle abd] + [\angle dbc] + [\angle c] = [\angle b] + [\angle c].$$

Analogously

$$\mathbf{H}(cb)[\mathbf{H}^*(ca)] \geqslant [\angle a] + [\angle b].$$

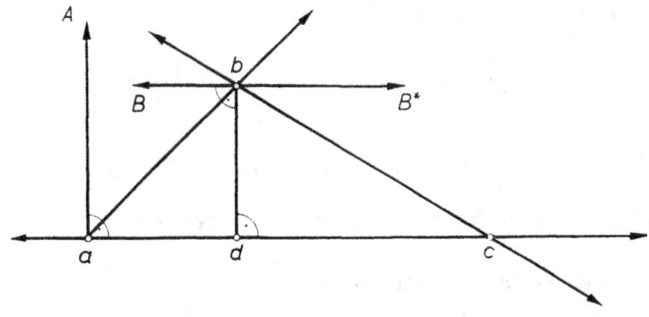

Fig. 144

† The proof of Theorem 14 outlined above and the resulting simple proof of Saccheri-Legendre Theorem were communicated to the authors by Alfred Tarski.

Then, by Theorem 63 of Chapter II and by Theorem 16,

$$[\mathbf{H^*}(ba)\mathbf{H}(bc)] = [\mathbf{H}(ba)B] + [B^*\mathbf{H}(bc)] \geqslant [\sphericalangle a] + [\sphericalangle c],$$

which concludes the proof.

As an example of a direct consequence of the above theorem, we prove

THEOREM 18. *Let A be a half-line with origin a_0 and let $b \sim \in \mathbf{L}(A)$. For any angle CD there exists a point $c \in A$ such that $\sphericalangle bca_0 < CD$.*

PROOF. We define an infinite sequence

$$(a_1, a_2, \ldots a_n, \ldots)$$

of points of half-line A in the following manner (Fig. 145):

(7) $a_n \prec a_{n+1}$ and $a_{n-1}a_n \equiv ba_{n-1}$ for $n = 1,2,\ldots$

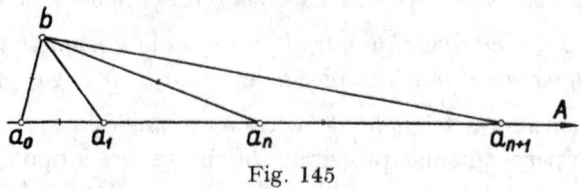

Fig. 145

Let $\mathfrak{A} = [\sphericalangle ba_1 a_0]$. We shall show that for $n = 1,2,\ldots$

(8) $$[\sphericalangle ba_n a_0] \leqslant \frac{1}{2^{n-1}} \cdot \mathfrak{A}.$$

Indeed, inequality (8) is obvious for $n = 1$. Further, by Theorem 17 and because of formula (7),

$$2 \cdot [\sphericalangle ba_{n+1}a_0] \leqslant [\sphericalangle ba_n a_0],$$

and therefore, assuming (8) holds, we have

$$[\sphericalangle ba_{n+1}a_1] \leqslant \frac{1}{2} \cdot [\sphericalangle ba_n a_0] \leqslant \frac{1}{2^n} \cdot \mathfrak{A}.$$

Hence $[\sphericalangle ba_n a_0] \leqslant \frac{1}{2^{n-1}} \cdot \mathfrak{A}$ for every natural n. By taking n sufficiently large, we obtain the point a_n which we were seeking.

5. Parallel Half-Lines (Continued)

In Section 35 of Chapter II (Theorem 108) we showed that at most one half-line A parallel to a given half-line B can be produced from a given point a. Using the Axiom of Continuity, we shall now prove that such a half-line always exists.

THEOREM 19. *There exists just one half-line A with a given origin a and parallel to a given half-line B.*

PROOF. By Theorem 105 of Chapter II, we can assume that $a \sim \in L(B)$. Let b be the origin of half-line B (Fig. 146) and let W be the half-plane having the boundary $L(ab)$ and containing half-line B. Let us consider a half-pencil $\mathfrak{H}$ with vertex a, consisting of half-lines $D \subset W$, and ordered

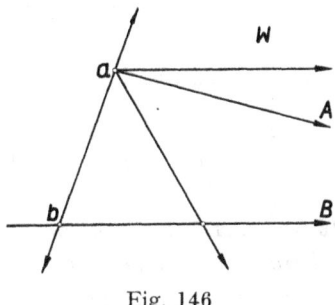

Fig. 146

from its end half-line $\mathbf{H}(ab)$ to its end half-line $\mathbf{H^*}(ab)$. We devide half-pencil $\mathfrak{H}$ into two classes $\mathfrak{X}_1$ and $\mathfrak{X}_2$ in the following manner: For any half-line $D \in \mathfrak{H}$,

$$D \in \mathfrak{X}_1 \quad \text{if} \quad D \cap B \neq 0; \quad D \in \mathfrak{X}_2 \quad \text{if} \quad D \cap B = 0.$$

As may easily be shown, the pair of classes $(\mathfrak{X}_1, \mathfrak{X}_2)$ is a Dedekind cut of half-pencil $\mathfrak{H}$, and there is no last half-line in class $\mathfrak{X}_1$. Hence, by Theorem 5, class $\mathfrak{X}_2$ has the firs thalf-line, which we denote by A. Half-line A is parallel to half-line B.

6. Parallel Axes (Continued)

With the help of Theorem 19, we can strengthen Theorem 118 of Chapter II to give us the following existence theorem:

THEOREM 20. *There is just one axis $\mathfrak{K}$ passing through a given point a and parallel to a given axis $\mathfrak{L}$.*

PROOF. Take any point $b \in L$ and let B be the half-line determined on axis $\mathfrak{L}$ by b. We produce from point a a half-line A parallel to B. By giving line $L(A)$ the orientation of half-line A, we obtain an axis $\mathfrak{K}$ parallel to $\mathfrak{L}$.

7. Angle of Parallelism

We say that *a half-line A is parallel to an axis $\mathfrak{L}$* —in symbols, $A \| \mathfrak{L}$ —if half-line A is parallel to some half-line B on axis $\mathfrak{L}$.

From Theorem 110 of Chapter II and Theorem 19 it at once follows

(i) that from a given point just one half-line A may be produced parallel to a given axis $\mathfrak{L}$, and (ii) that half-line A is parallel to every half-line on axis $\mathfrak{L}$.

We shall now consider any axis $\mathfrak{L}$ and a point $a \sim \in \mathfrak{L}$ (Fig. 147). Let point b be the perpendicular projection of point a upon axis $\mathfrak{L}$. From

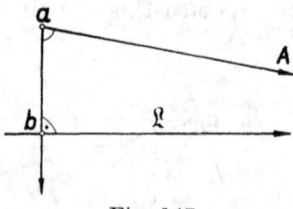

Fig. 147

point a we produce a half-line A parallel to axis $\mathfrak{L}$. We refer to the angle $\mathsf{H}(ab)A$ as the *angle of parallelism for point a with respect to axis $\mathfrak{L}$*. Let B be the half-line determined an axis $\mathfrak{L}$ by point b. Then $A\|B$ and, by Theorem 107(I) of Chapter II, angle $\mathsf{H}(ab)A$ is not greater than a right angle. We thus have:

THEOREM 21. *The angle of parallelism for a point a with respect to an axis $\mathfrak{L}$ is always an acute or right angle.*

A second fundamental theorem on the angle of parallelism is the following:

THEOREM 22. *Given two axes $\mathfrak{L}_1$ and $\mathfrak{L}_2$ and points $a_1 \sim \in \mathfrak{L}_1$ and $a_2 \sim \in \mathfrak{L}_2$. let b_1 and b_2 be the perpendicular projections of points a_1 and a_2 upon axes $\mathfrak{L}_1$ and $\mathfrak{L}_2$. Then we have:*

(I) If $a_1 b_1 \equiv a_2 b_2$, then the angle of parallelism for point a_1 with respect to axis $\mathfrak{L}_1$ is congruent to the angle of parallelism for point a_2 with respect to axis $\mathfrak{L}_2$.

(II) If $a_1 b_1 < a_2 b_2$, then the angle of parallelism for point a_1 with respect to axis $\mathfrak{L}_1$ is not smaller than the angle of parallelism for point a_2 with respect to axis $\mathfrak{L}_2$.

PROOF. We denote by A_i $(i = 1,2)$ the half-line produced from point a_i parallel to axis $\mathfrak{L}_i$; the half-line determined on axis $\mathfrak{L}_i$ by point b_i will be denoted by B_i and the half-plane having the boundary $\mathsf{L}(a_i b_i)$ and containing half-line B_i will be denoted by W_i.

(I) Assume that $a_1 b_1 \equiv a_2 b_2$ and suppose that for some special case $\mathsf{H}(a_1 b_1)A_1 > \mathsf{H}(a_2 b_2)A_2$ (Fig. 148). There then exists a half-line D_1 such that $\mathsf{B}(\mathsf{H}(a_1 b_1),D_1,A_1)$ and $\mathsf{H}(a_1 b_1)D_1 \equiv \mathsf{H}(a_2 b_2)A_2$. Since $A_1\|B_1$, then $D_1 \cap B_1 \neq 0$, from which it follows, by Theorem 36 from Chapter II, that $A_2 \cap B_2 \neq 0$, which contradicts $A_2\|B_2$. Therefore we always have

$\boldsymbol{H}(a_1b_1)A_1 \equiv \boldsymbol{H}(a_2b_2)A_2$ or $\boldsymbol{H}(a_1b_1)A_1 < \boldsymbol{H}(a_2b_2)A_2$, from which it follows, by the symmetry of the assumptions, that we always have

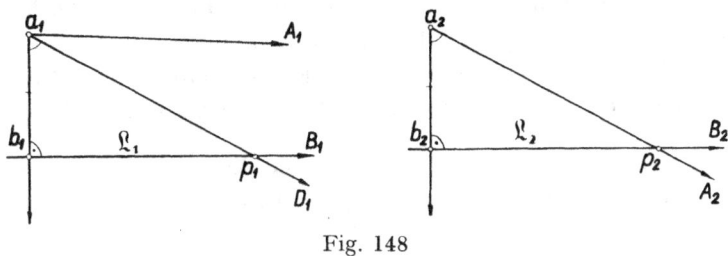

Fig. 148

$$\boldsymbol{H}(a_2b_2)A_2 \equiv \boldsymbol{H}(a_1b_1)A_1 \text{ or } \boldsymbol{H}(a_2b_2)A_2 < \boldsymbol{H}(a_1b_1)A_1.$$

Hence $\boldsymbol{H}(a_1b_1)A_1 \equiv \boldsymbol{H}(a_2b_2)A_2$.

(II) Assume that $a_1b_1 < a_2b_2$ (Fig. 149). On segment (a_2b_2) we take a point a such that $ab_2 \equiv a_1b_1$ and from point a we produce a half-line A

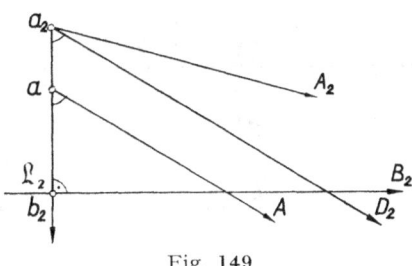

Fig. 149

parallel to axis $\mathfrak{L}_2$. If we now use part (I) of the theorem it is sufficient to show that

(1) $\boldsymbol{H}(ab_2)A \equiv \boldsymbol{H}(a_2b_2)A_2$ or $\boldsymbol{H}(ab_2)A > \boldsymbol{H}(a_2b_2)A_2$.

Suppose that in some special case $\boldsymbol{H}(ab_2)A < \boldsymbol{H}(a_2b_2)A_2$. We produce from point a_2 a half-line $D_2 \subset W_2$ such that $\boldsymbol{H}(a_2b_2)D_2 \equiv \boldsymbol{H}(ab_2)A$. Then $D_2 \cap B_2 \neq 0$, from which it follows readily that $A \cap B_2 \neq 0$. This is in contradiction to $A \| B_2$. Therefore (1) must always hold.

In connection with Theorem 21 the following question arises. Consider two axes, $\mathfrak{L}_1$ and $\mathfrak{L}_2$, and two points, $a_1 \sim \epsilon \mathfrak{L}_1$ and $a_2 \sim \epsilon \mathfrak{L}_2$. Can the angle of parallelism for point a_1 with respect to axis $\mathfrak{L}_1$ be a right angle and at the same time the angle of parallelism for point a_2 with respect to axis $\mathfrak{L}_2$ be acute? The following theorem gives the answer to this question.

THEOREM 23. *If there exists an axis $\mathfrak{R}$ and a point $c \sim \epsilon \mathfrak{R}$ such that the angle of parallelism for c with respect to $\mathfrak{R}$ is a right angle, then for each* *axis $\mathfrak{L}$ and for each point $a \sim \in \mathfrak{L}$, the angle of parallelism for a with respect to $\mathfrak{L}$ is a right angle.*

PROOF. Assume that the angle of parallelism for point c with respect to axis $\mathfrak{N}$ is a right angle, and take any axis $\mathfrak{L}$ and any point $a \sim \in \mathfrak{L}$. Let point d be the perpendicular projection of point c upon axis $\mathfrak{N}$, and point b_0 the perpendicular projection of point a upon axis $\mathfrak{L}$ (Fig. 150).

From Theorems 22 and 21 it follows at once that if $ab_0 \equiv cd$ or $ab_0 < cd$, then the angle of parallelism for point a with respect to axis $\mathfrak{L}$ is a right angle.

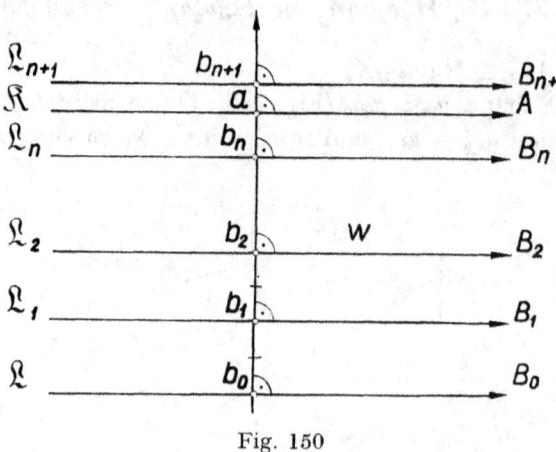

Fig. 150

It remains the case $ab_0 > cd$. We denote by B_0 the half-line determined on axis $\mathfrak{L}$ by point b_0, and by W the half-plane having the boundary $\mathbf{L}(ab_0)$ and containing half-line B_0. It follows from Theorem 6 that there exists a natural number n such that the first n points of the points $b_1, b_2, \ldots, b_n, b_{n+1} \in \mathbf{H}(b_0 a)$ defined by the conditions

$$(2) \qquad b_{k-1} \prec b_k, \text{ and } b_{k-1}b_k \equiv cd \text{ for } k = 1, 2, \ldots, n+1$$

belong to segment $(b_0 a)$, while $b_{n+1} \succeq a$. We produce from each point b_k $(k = 1, 2, \ldots, n)$ a half-line $B_k \subset W$ perpendicular to line ab_0 and denote by $\mathfrak{L}_k$ the axis determined by half-line B_k. From point a, too, we produce a half-line $A \subset W$ perpendicular to line ab_0 and denote by $\mathfrak{N}$ the axis determined by half-line A. By (2), for $k = 1, 2, \ldots, n$, the angle of parallelism for point b_{k-1} with respect to axis $\mathfrak{L}_k$ is a right angle, and therefore

$$(3) \qquad\qquad B_{k-1} \| B_k \quad \text{for } k = 1, 2, \ldots, n.$$

Since $b_n a \equiv cd$ or $b_n a < cd$, then the angle of parallelism for point b_n with respect to axis $\mathfrak{R}$ is also a right angle, and hence

(4) $$B_n \| A.$$

Because of the transitivity of the parallel relation for half-lines, we conclude from formulas (3) and (4) that $B_0 \| A$, i.e. $A \| B_0$, from which it follows that $A \| \mathfrak{L}$. Therefore the angle of parallelism for point a with respect to axis $\mathfrak{L}$ is a right angle.

8. Parallel Lines (Continued)

With the help of Theorem 20 we can strengthen Theorem 119 of Chapter II to the following existence theorem:

THEOREM 24. *There is at least one line, and there are at most two lines, passing through a given point a and parallel to a given line L.*

We shall now examine more closely the case $a \sim \in L$. Let point b be the perpendicular projection of point a upon line L (Fig. 151). From point

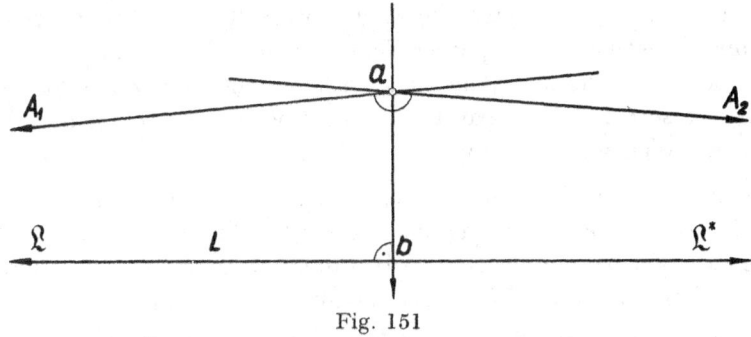

Fig. 151

a we produce half-lines A_1 and A_2 respectively parallel to axes $\mathfrak{L}$ and $\mathfrak{L}^*$ of line L. We shall say that *half-lines A_1 and A_2 are parallel to line L.* They lie in plane La on opposite sides of line ab. Angle $\mathbf{H}(ab)A_1$ is the angle of parallelism for point a with respect to axis $\mathfrak{L}$; angle $\mathbf{H}(ab)A_2$ is the angle of parallelism for point a with respect to axis $\mathfrak{L}^*$. We shall call angles $\mathbf{H}(ab)A_1$ and $\mathbf{H}(ab)A_2$ *the angles of parallelism for point a with respect to line L.* These angles are obviously distinct, but by Theorem 22(I) they are congruent. $\mathbf{L}(A_1)$ and $\mathbf{L}(A_2)$ are the two lines produced from point a parallel to line L. Obviously, $\mathbf{L}(A_1) = \mathbf{L}(A_2)$ if and only if half-lines A_1 and A_2 are complementary. We therefore have

THEOREM 25. *Let $a \sim \in L$. If the angles of parallelism for point a with respect to line L are right angles, then there is just one line K passing through a and parallel to L, and if the angles of parallelism for point a with respect* *to line L are acute, then there are just two lines* $\mathbf{L}(A_1)$ *and* $\mathbf{L}(A_2)$ *passing through a and parallel to L.*

From Theorems 23 and 25 there follows at once

THEOREM 26. *If there are a line* L_0 *and a point* $a_0 \sim \in L_0$ *such that there is only one line* K_0 *passing through* a_0 *and parallel to* L_0, *then for every line L and for every point* $a \sim \in L$ *there is only one line K passing through a and parallel to L.*

This unique line K lies in plane $P = \mathbf{P}(La)$. Let us now see how many lines on plane P, in this case, pass through point a and do not intersect line L:

THEOREM 27. *Given a line L and a point* $a \sim \in L$ *on a plane P, if there is only one line K passing through a and parallel to L, then K is the only line on P which passes through a and does not intersect L.*†

PROOF. Let K be the only line passing through point a and parallel to line L. We denote by b the perpendicular projection of point a upon line L. By Theorem 25, we have $K \perp \mathbf{L}(ab)$. If in plane P a line K' distinct from line K and not intersecting line L were to pass through point a, then line K' would not be perpendicular to line ab. It thus follows that the angle of parallelism for point a with respect to L would be acute, and, as a result, two lines parallel to line L would pass through point a. This contradicts our hypothesis.

Finally, we shall show that the existence of a rectangle is a sufficient condition that always only one line parallel to a line L is passing through a point a. By Theorem 26, given a rectangle, it suffices to show that one line L and one point $a \sim \in L$ satisfy the above condition.

THEOREM 28. *If quadrangle* (a,b,c,d) *is a rectangle, then there is only one line passing through point a and parallel to line bc (in fact, line ad).*††

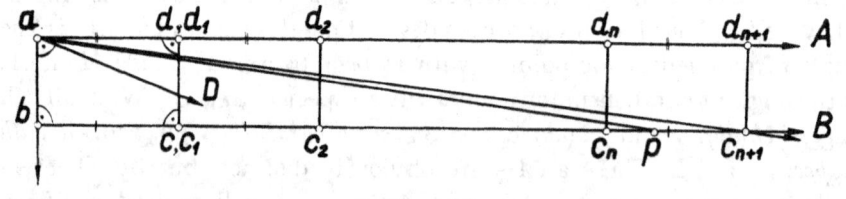

Fig. 152

PROOF. We assume that quadrangle (a,b,c,d) is a rectangle (Fig. 152) Let $A = \mathbf{H}(ad)$ and $B = \mathbf{H}(bc)$. It suffices to show that each half-line

† See footnote on page 113.
†† See footnote on page 113.

with origin a and lying between half-lines ab and A intersects half-line B. Let D be such a half-line. Thus

(5) $$\mathbf{B}\,(\mathbf{H}\,(ab),D,A).$$

By Theorem 18, there is on half-line B a point p such that

(6) $$\sphericalangle apb < DA.$$

We may assume that $p \succ c$ (on B). Let us take a natural number n such that the first n points of the points $c_1,c_2,\ldots,c_n,c_{n+1} \in B$ defined by the conditions

$$c = c_1 \prec c_2 \prec \ldots \prec c_n \prec c_{n+1} \text{ and } c_k c_{k+1} \equiv bc \text{ for } k = 1,2,\ldots,n$$

belong to segment (bp), and either $c_{n+1} = p$ or $c_{n+1} \succ p$. Then

(7) $$\sphericalangle ac_{n+1}b \leqslant \sphericalangle apb.$$

Let us take points $d_1,d_2,\ldots,d_n,d_{n+1} \in A$ defined by the conditions

$$d = d_1 \prec d_2 \prec \ldots \prec d_n \prec d_{n+1} \text{ and } d_k d_{k+1} \equiv ad \text{ for } k = 1,2,\ldots,n.$$

From Theorem 86 of Chapter II it readily follows, by induction, that quadrangle (a,b,c_{n+1},d_{n+1}) is a rectangle. Therefore

(8) $$\sphericalangle ac_{n+1}b \equiv \sphericalangle c_{n+1}ad_{n+1}.$$

From formulas (6), (7), and (8) it follows that $\mathbf{H}(ac_{n+1})A < DA$; hence $\mathbf{B}(D,\mathbf{H}(ac_{n+1}),A)$, which, together with (5), gives $\mathbf{B}(\mathbf{H}(ab),D,\mathbf{H}(ac_{n+1}))$. Thus half-line D intersects segment (bc_{n+1}) and therefore intersects half-line B. This concludes the proof.

9. Measure of Segments

By a *measure of bound segments* we understand any function φ which correlates with every segment ac a real number $\varphi(ac) > 0$ and satisfies the following two conditions:

(1) $\qquad$ if $a_1c_1 \equiv a_2c_2$, then $\varphi(a_1c_1) = \varphi(a_2c_2)$;

(2) $\qquad$ if $\mathbf{B}(a,b,c)$, then $\varphi(ac) = \varphi(ab) + \varphi(bc)$.

By a *measure of free segments* we mean any function ψ which correlates with every free segment $\mathfrak{a}$ a real number $\psi(\mathfrak{a}) > 0$ and which is additive, i.e. satisfies the following condition:

$$\psi(\mathfrak{a} + \mathfrak{b}) = \psi(\mathfrak{a}) + \psi(\mathfrak{b})$$

for every two free segments $\mathfrak{a}$ and $\mathfrak{b}$.

The simple relation between the measure of bound segments and the measure of free segments follows at once from this definition.

THEOREM 29. *If φ is a measure of bound segments, then the function ψ defined for free segments by the condition: if $\mathfrak{a} = [ab]$, then $\psi(\mathfrak{a}) = \varphi(ab)$, is a measure of free segments.*

If ψ is a measure of free segments, then the function φ defined for bound segments by the condition: if $[ab] = \mathfrak{a}$, then $\varphi(ab) = \psi(\mathfrak{a})$, is a measure of bound segments.

In further course, we shall use only the measure of bound segments. It is, however, convenient to prove the basic properties of measure for the measure of free segments, and then to apply them to the measure of bound segments, since this enables us to use the calculus of free segments.

THEOREM 30. *If ψ is a measure of free segments, then for any free segments $\mathfrak{a}$ and $\mathfrak{b}$ and for any dyadic number w,*

(I) $\mathfrak{a} < \mathfrak{b}$ *if and only if* $\psi(\mathfrak{a}) < \psi(\mathfrak{b})$;

(II) $\psi(\mathfrak{a} - \mathfrak{b}) = \psi(\mathfrak{a}) - \psi(\mathfrak{b})$ *for* $\mathfrak{a} > \mathfrak{b}$;

(III) $\psi(w \cdot \mathfrak{a}) = w \cdot \psi(\mathfrak{a})$.

The proof of this theorem is obvious.

From the definition of measure we at once have

THEOREM 31. *If function ψ is a measure of free segments, then every function of the form $\lambda \cdot \psi$, where λ is any positive number, is also a measure of free segments.*

We shall now prove the converse theorem.

THEOREM 32. *If ψ_0 is any measure of free segments, then for every measure ψ of free segments there exists a number $\lambda > 0$ such that $\psi = \lambda \cdot \psi_0$.*

PROOF. Consider any free segment $\mathfrak{a}_0$. Let

(3) $\psi_0(\mathfrak{a}_0) = x_0, \quad \psi(\mathfrak{a}_0) = x, \quad \lambda = x/x_0.$

We state that $\psi = \lambda \cdot \psi_0$, that is, that for each free segment $\mathfrak{a}$ we have $\psi(\mathfrak{a}) = \lambda \cdot \psi_0(\mathfrak{a})$. To prove this, let us suppose that for some free segment $\mathfrak{a}$ we have $\psi(\mathfrak{a}) < \lambda \cdot \psi_0(\mathfrak{a})$. Hence there exists a natural number k such that

(4) $\psi(\mathfrak{a}) + \dfrac{x}{2^k} < \lambda \cdot \psi_0(\mathfrak{a}).$

Let

(5) $\mathfrak{b} = \mathfrak{a} + \dfrac{\ulcorner 1}{2^k} \cdot \mathfrak{a}_0 > \mathfrak{a}.$

By Theorem 9, there exists a dyadic number w such that

(6) $$\mathfrak{a} < w \cdot \mathfrak{a}_0 < \mathfrak{b}$$

holds. From formulas (3) to (6), from the definition of measure, and from Theorem 30, we obtain

(7) $$w \cdot x = \psi(w \cdot \mathfrak{a}_0) < \psi(\mathfrak{b}) = \psi(\mathfrak{a}) + \frac{x}{2^k} < \lambda \cdot \psi_0(\mathfrak{a})$$

and $w \cdot x_0 = \psi_0(w \cdot \mathfrak{a}_0) > \psi_0(\mathfrak{a})$, from which it follows, by (3), that

(8) $$w \cdot x > \lambda \cdot \psi_0(\mathfrak{a}).$$

But formula (8) contradicts formula (7), and therefore the inequality $\psi(\mathfrak{a}) < \lambda \cdot \psi_0(\mathfrak{a})$ cannot hold.

In a similar manner, we prove that the inequality $\psi(\mathfrak{a}) > \lambda \cdot \psi_0(\mathfrak{a})$ cannot hold. Thus $\psi(\mathfrak{a}) = \lambda \cdot \psi_0(\mathfrak{a})$ for every free segment $\mathfrak{a}$, that is $\psi = \lambda \cdot \psi_0$, which was to be proved.

THEOREM 33. *Let $\mathfrak{a}_0$ be any free segment, and let x_0 be any real positive number. There exists just one measure ψ of free segments such that $\psi(\mathfrak{a}_0) = x_0$.*

PROOF. From Theorem 32 it follows immediately that there is at most one measure ψ correlating the number x_0 with segment $\mathfrak{a}_0$. It therefore remains to construct such a measure ψ.

We shall first construct the measure ψ correlating 1 with segment $\mathfrak{a}_0$.

Let us denote by Θ the set of positive dyadic numbers. Let $\mathfrak{a}$ be any free segment. Consider the division of set Θ into two sets,

(9) $$\Theta = \Theta_1^{(\mathfrak{a})} \cup \Theta_2^{(\mathfrak{a})},$$

defined by the conditions

$$w \in \Theta_1^{(\mathfrak{a})} \text{ if } w \cdot \mathfrak{a}_0 < \mathfrak{a}, \ w \in \Theta_2^{(\mathfrak{a})} \text{ if } w \cdot \mathfrak{a}_0 \geqslant \mathfrak{a}.$$

Obviously, sets $\Theta_1^{(\mathfrak{a})}$ and $\Theta_2^{(\mathfrak{a})}$ are disjoint, and each of them is non-empty. Furthermore, from Theorems 42(II) and 49(VII) of Chapter II it follows that every number of set $\Theta_1^{(\mathfrak{a})}$ is smaller than every number of set $\Theta_2^{(\mathfrak{a})}$. Thus division (9) is a Dedekind cut of set Θ and therefore, as known from the theory of real numbers, there exists just one positive number $x_{\mathfrak{a}}$ such that $w_1 < x_{\mathfrak{a}} < w_2$ implies $w_1 \in \Theta_1^{(\mathfrak{a})}$ and $w_2 \in \Theta_2^{(\mathfrak{a})}$, for any dyadic numbers w_1 and w_2.

We shall now show that the function ψ defined by the formula $\psi(\mathfrak{a}) = x_{\mathfrak{a}}$ is a measure of free segments, and that it correlates 1 with segment $\mathfrak{a}_0$.

First, we show that the function ψ is additive. Consider any two free segments $\mathfrak{a}$ and $\mathfrak{b}$. We have $\psi(\mathfrak{a}) = x_{\mathfrak{a}}$ and $\psi(\mathfrak{b}) = x_{\mathfrak{b}}$. We have to prove that $\psi(\mathfrak{a} + \mathfrak{b}) = x_\mathfrak{a} + x_\mathfrak{b}$, that is, that for any positive dyadic numbers w_1 and w_2 the inequality $w_1 < x_\mathfrak{a} + x_\mathfrak{b} < w_2$ implies

$$w_1 \in \Theta_1^{(\mathfrak{a} + \mathfrak{b})} \text{ and } w_2 \in \Theta_2^{(\mathfrak{a} + \mathfrak{b})}.$$

Let $w_1 < x_\mathfrak{a} + x_\mathfrak{b}$. From the arithmetic of real numbers it is known that the number w_1 can be represented as the sum of two positive dyadic numbers, $w_1 = w_1' + w_1''$, such that $w_1' < x_\mathfrak{a}$ and $w_1'' < x_\mathfrak{b}$. From the definition of the numbers $x_\mathfrak{a}$ and $x_\mathfrak{b}$ it follows that $w_1' \in \Theta_1^{(\mathfrak{a})}$ and $w_1'' \in \Theta_1^{(\mathfrak{b})}$, that is

$$w_1' \cdot \mathfrak{a}_0 < \mathfrak{a} \text{ and } w_1'' \cdot \mathfrak{a}_0 < \mathfrak{b},$$

from which, by Theorems 44(IV) and 49(I) of Chapter II, we obtain $w_1 \cdot \mathfrak{a}_0 < \mathfrak{a} + \mathfrak{b}$, that is $w_1 \in \Theta_1^{(\mathfrak{a} + \mathfrak{b})}$.

In a similar manner it follows from $w_2 > x_\mathfrak{a} + x_\mathfrak{b}$ that $w_2 \in \Theta_2^{(\mathfrak{a} + \mathfrak{b})}$. Hence the function ψ is a measure of free segments. From the definition of function ψ it follows, by Theorem 49(VII) of Chapter II, that $\psi(\mathfrak{a}_0) = 1$. As a result (see Theorem 31) function $\psi' = x_0 \cdot \psi$ is also a measure of free segments and $\psi(\mathfrak{a}_0) = x_0 \cdot 1 = x_0$. This concludes the proof.

From Theorems 32 and 33 we infer, with the help of Theorem 29, analogous theorems for the measure of bound segments.

THEOREM 34. *If φ_0 is any measure of bound segments, then for every measure φ of bound segments there exists a real number $\lambda > 0$ such that $\varphi = \lambda \cdot \varphi_0$.*

THEOREM 35. *Let $a_0 b_0$ be any segment and x_0 any real positive number. There exists just one measure φ of bound segments such that $\varphi(a_0 b_0) = x_0$.*

We shall now show that every measure runs over all positive real numbers. To show this, we shall prove

THEOREM 36. *Let φ be a measure of bound segments, and let A be a half-line with origin a. For any real number $x > 0$ there is a point $b \in A$ for which $\varphi(ab) = x$ (and, in fact, there is just one such point).*

PROOF. Consider any point $b_0 \in A$. Let $\varphi(ab_0) = x_0$. It is readily noted that

(10)   for any dyadic number w there is a point
            $p \in A$ such that $\varphi(ap) = w \cdot x_0$.

Further, from Theorem 30(I) it follows that

(11) $\varphi(ap_1) < \varphi(ap_2)$ if and only if $p_1 \prec p_2$,
            for any points $p_1, p_2 \in A$.

Take any positive number x. We divide half-line A into two sets X_1 and X_2 by including in set X_1 those points $p \in A$ for which $\varphi(ap) < x$ and in set X_2 those points $p \in A$ for which $\varphi(ap) \geqslant x$. We shall show that the pair of sets (X_1, X_2) is a Dedekind cut of half-line A. Let us take natural numbers k and n such that

(12)
$$\frac{1}{2^k} < \frac{x}{x_0} \leqslant n.$$

By (10), there exist on half-line A points q_1 and q_2 such that

$$\varphi(aq_1) = \frac{1}{2^k} \cdot x_0 \text{ and } \varphi(aq_2) = n \cdot x_0.$$

Obviously, $q_1 \in X_1$ and $q_2 \in X_2$. Thus, sets X_1 and X_2 are non-empty. Further, for any points $p_1 \in X_1$ and $p_2 \in X_2$ we have $\varphi(ap_1) < x \leqslant \varphi(ap_2)$, from which it follows, by (11), that $p_1 \prec p_2$.

Let b be the point determined on half-line A by the cut (X_1, X_2). We shall show that $\varphi(ab) = x$. If $\varphi(ab)$ were to be less than x, then by taking a dyadic number w such that $\varphi(ab) < w \cdot x_0 < x$, and by choosing on half-line A a point q such that $\varphi(aq) = w \cdot x_0$, we would have $\varphi(aq) < x$ and $b \prec q$, from which it would follow that $q \in X_1$ and, at the same time, $q \in X_2$. This would be in contradiction to the fact that X_1 and X_2 are disjoint. Therefore $\varphi(ab)$ cannot be greater than x. In a similar manner, we may prove that $\varphi(ab)$ cannot be less than x. Hence $\varphi(ab) = x$.

We shall refer to the value $\varphi(ab)$ of a measure φ on a segment pq also as the *length* of segment ab with respect to measure φ and denote it by $|ab|_\varphi$. The segment $[ab]$ determined by the condition

$$|ab|_\varphi = 1$$

is called the *unit of length* of measure φ.

NOTE: In the proof of the Thales Theorem (see Chapter V, Theorem 13) it will be convenient to use a measure defined not for all bound segments (contained in space S), but only for segments lying on some line L. More precisely, we shall understand the measure of segments on line L to be any function φ^L which correlates to every segment $ab \subset L$ some real number $\varphi^L(ab) > 0$ and which satisfies, for segments of line L, conditions (1) and (2) on page 167. It is readily seen that every measure φ^L on line L can be extended to a measure φ in all of space S. Indeed, let us take any half-line $A \subset L$. For any segment pq we choose on half-line A a point b such that $ab = pq$. Setting

$$\varphi(pq) = \varphi^L(ab)$$

we obtain the measure φ for segments in all of space **S**, which, on segments of line L, takes the same values as the measure φ^L.

By Theorem 34, we obtain

THEOREM 37. *If φ_0 is any measure of segments on a given line L, then for every measure φ of segments on line L there exists a real number $\lambda > 0$ such that $\varphi = \lambda \cdot \varphi_0$.*

10. Measure of Angles

By a *measure of bound angles* we understand any function φ which correlates with every angle AC a real number $\varphi(AC) > 0$ and satisfies the following two conditions:

(1)   if $A_1C_1 \equiv A_2C_2$, then $\varphi(A_1C_1) = \varphi(A_2C_2)$;

(2)   if **B**(A,B,C), then $\varphi(AC) = \varphi(AB) + \varphi(BC)$.

By a *measure of free angles* we understand any function ψ which correlates with every free angle $\mathfrak{A}$ a real number $\psi(\mathfrak{A})$ and which is additive, that is, satisfies the condition

$$\psi(\mathfrak{A} + \mathfrak{B}) = \psi(\mathfrak{A}) + \psi(\mathfrak{B})$$

for every pair of free angles $\mathfrak{A}$ and $\mathfrak{B}$ for which the sum $\mathfrak{A} + \mathfrak{B}$ exists.

The relation between the measure of bound angles and the measure of free angles is established by the following obvious theorem:

THEOREM 38. *If φ is a measure of bound angles, then the function ψ defined for free angles by the condition: if $\mathfrak{A} = [AB]$, then $\psi(\mathfrak{A}) = \varphi(AB)$, is a measure of free angles.*

If ψ is a measure of free angles, then the function φ defined for bound angles by the condition: if $[AB] = \mathfrak{A}$, then $\varphi(AB) = \psi(\mathfrak{A})$, is a measure of bound angles.

We shall develop the theory of measure for angles in a manner similar to that used in the theory of measure for segments. There will be, however, some essential difference.

THEOREM 39. *Let ψ be a measure of free angles. For any free angles $\mathfrak{A}$ and $\mathfrak{B}$ we have:*
(I) *$\mathfrak{A} < \mathfrak{B}$ if and only if $\psi(\mathfrak{A}) < \psi(\mathfrak{B})$;*
(II) *$\psi(\mathfrak{A} - \mathfrak{B}) = \psi(\mathfrak{A}) - \psi(\mathfrak{B})$ for $\mathfrak{A} > \mathfrak{B}$.*
For any free angle $\mathfrak{A}$ and for any dyadic number w, for which the product $w \cdot \mathfrak{A}$ exists, we have
(III) *$\psi(w \cdot \mathfrak{A}) = w \cdot \psi(\mathfrak{A})$.*

The proof is analogous to the proof of Theorem 30.

The measure of segments runs over all positive real numbers, but the measure of angles is bounded from above. In fact, from Theorem 72 of Chapter II and Theorem 39(I) we derive the following

THEOREM 40. *If a measure ψ takes the value x_0 for the free right angle, then for any free angle $\mathfrak{A}$ we have $\psi(\mathfrak{A}) < 2x_0$.*

From the definition of measure we at once have

THEOREM 41. *If function ψ is a measure of free angles, then every function of the form $\lambda \cdot \psi$, where λ is any positive number, is also a measure of free angles.*

We turn to the converse theorem.

THEOREM 42. *If ψ_0 is any measure of free angles, then for every measure ψ of free angles there exists a number $\lambda > 0$ such that $\psi = \lambda \cdot \psi_0$.*

PROOF. The proof of this theorem is similar to the proof of Theorem 32. It differs only in that the natural number k must now satisfy, in addition to the former condition $\psi(\mathfrak{A}) + x/2^k < \lambda \cdot \psi_0(\mathfrak{A})$, one more condition, in fact, that the sum $\mathfrak{A} + 1/2^k \cdot \mathfrak{A}_0$ exists.
By Theorem 55 of Chapter II and Theorem 12, such a number k indeed exists.

Theorem 33, too, can be extended to the measure of free angles by means of a proof analogous to the proof of Theorem 33. Passing on to the measure of bound angles we thus obtain, by means of Theorem 38,

THEOREM 43. *Let A_0B_0 be any angle and x_0 any real positive number. There then exists just one measure φ of bound angles such that $\varphi(A_0B_0) = x_0$.*

We shall show that if a measure φ assigns the number x_0 to a right angle, then the range of φ coincides with the interval $(0, 2x_0)$.

THEOREM 44. *Let φ be a measure of bound angles assigning the number x_0 to a right angle A_0B_0, and let $\mathfrak{H}$ be a half-pencil ordered from its end half-line A to its end half-line A^*. By these assumptions, for any real positive number $x < 2x_0$ there is a half-line $B \in \mathfrak{H}$ for which $\varphi(AB) = x$ (and, in fact, there is just one such half-line).*

PROOF. The proof of this theorem is similar to the proof of Theorem 36. The difference lies in the fact that condition (10) is now satisfied only for dyadic numbers $w < 2$ (see Theorem 40), and therefore instead of a natural number $n \geqslant x/x_0$, it is necessary to take a dyadic number w such that $2 > w \geqslant x/x_0$.

From Theorem 43 it follows, in particular, that there exists just one measure φ_0 assigning the number $\pi/2$ to right angles. Thus, the fact that from among all free angles we have singled out the right free angle implies that from among all measures of angles we can single out the measure φ_0 which assigns $\pi/2$ to all right angles. We shall call this measure the *natural* measure of angles, and henceforth we shall use only this measure. Instead of $\varphi_0(AB)$ we shall write $|AB|$; we shall refer to the real number $|AB|$ as the *(natural) size* of angle AB.

From Theorem 40 it immediately follows that for any angle AB we have $0 < |AB| < \pi$. With the help of Theorem 71 of Chapter II, we obtain

THEOREM 45. *For any angle AB,*

$$|AB| + |A^*B| = \pi.$$

NOTE: The choice of the number $\pi/2$ as the measure of right angles is non-essential; any other positive real number as well could have been taken for the measure of right angles.

11. The Saccheri-Legendre Theorem Formulated in Terms of Measure

By making use of Theorem 45, we can give the Saccheri-Legendre Theorem (Theorem 17) the following simple form:

THEOREM 46. *In every triangle, the sum of the sizes of the interior angles is no greater than π.*

In connection with Theorem 46 there arises the question of whether it is possible that in one triangle the sum of the sizes of the angles equals π, and, at the same time, in another triangle this sum is less than π. To answer this questions we shall first prove

THEOREM 47. *If there exists a triangle for which the sum of the sizes of the angles is equal to π, then there exists a rectangle.*

PROOF. Assume that the angles of a triangle abc satisfy the condition

(1) $$|\sphericalangle a| + |\sphericalangle b| + |\sphericalangle c| = \pi.$$

At least two of the three angles, say a and b, are acute (Fig. 153). Then the projection d of vertex c on line ab lies bewteen points a and b. Let ω_1 denote the sum of the sizes of the angles of triangle acd, and ω_2 the sum of the sizes of the angles of triangle bcd. By (1), we have $\omega_1 + \omega_2 = 2\pi$; on the other hand, by Theorem 46, we have $\omega_1 \leqslant \pi$ and $\omega_2 \leqslant \pi$. Therefore $\omega_1 = \pi$. We now take in plane abc a point e such that $\mathbf{B}(e,\mathbf{L}(ac),d)$

and $ea \equiv cd$ and $ec \equiv ad$. It may easily be shown that quadrangle a,e,c,d) is a rectangle.

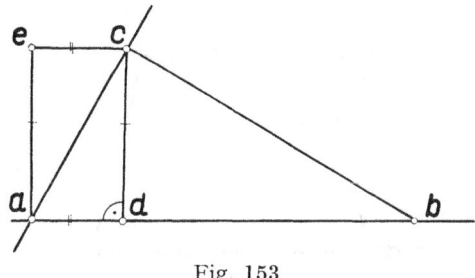

Fig. 153

Considering, in turn, Theorems 47, 28, 26, 27 and Theorem 64 of Chapter II, we arrive at the answer to the question following Theorem 46.

THEOREM 48. *If there is a triangle* $a_0 b_0 c_0$ *such that*

$$|\angle a_0| + |\angle b_0| + |\angle c_0| = \pi,$$

then for every triangle abc we have

$$|\angle a| + |\angle b| + |\angle c| = \pi.$$

This theorem could also be proved without using the theorems on parallel lines, but entirely on the basis of two theorems on rectangles (see Theorems 85 and 86 of Chapter II). The main element of such a proof is the construction of a rectangle with sufficiently large sides under the assumption that some rectangle exists (cf. the proof of Theorem 28).

On the basis of Theorem 48, the sum of the sizes of the three angles of a triangle is either always equal to π or always less than π. Which of these two possibilities actually occurs cannot be settled on the basis of absolute geometry. This will be shown in Chapter IV.

The following conclusion may be drawn from Theorem 46:

THEOREM 49. *In every convex quadrangle the sum of the sizes of the angles is not greater than* 2π.

PROOF. Consider a convex quadrangle (a,b,c,d) (Fig. 154). Applying the Saccheri-Legendre Theorem to triangles abc and acd, we obtain

(2) $$|\angle bac| + |\angle b| + |\angle bca| \leqslant \pi$$
and
(3) $$|\angle cad| + |\angle acd| + |\angle d| \leqslant \pi.$$

By Theorem 93(II) of Chapter I, we have

$$|\angle bac| + |\angle cad| = |\angle a| \quad \text{and} \quad |\angle bca| + |\angle acd| = |\angle c|.$$

From the above inequalities we obtain

(4)
$$|\sphericalangle a| + |\sphericalangle b| + |\sphericalangle c| + |\sphericalangle d| \leqslant 2\pi.$$

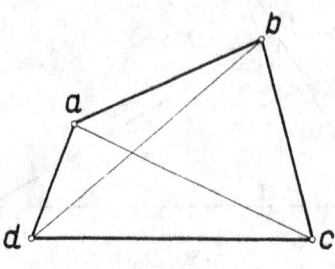

Fig. 154

12. Distance Between Two Points. Space *S* as a Metric Space

Take an arbitrary measure φ of segments and put

$$\varrho_\varphi(a,b) = \begin{cases} 0 & \text{if } a = b, \\ |ab|_\varphi & \text{if } a \neq b \end{cases}$$

for any two points a and b. Then, obviously,

(1) $\varrho_\varphi(a,b) = 0$ if and only if $a = b$,

(2) $\varrho_\varphi(a,b) = \varrho_\varphi(b,a)$,

and from Theorem 50 of Chapter II, from the definition of measure, and from Theorem 30(I) it is seen at once that

(3) $\varrho_\varphi(a,b) + \varrho_\varphi(b,c) \geqslant \varrho_\varphi(a,c)$

for any points a,b,c. Thus function ϱ_φ is a metric in space *S* (see Introduction, Section 9); we refer to it as the *metric induced by measure* φ. A metric ϱ_φ induced by a measure φ will be called a *proper* metric. Henceforth by *metrics* we shall understand proper metrics only.

In what follows we shall use any arbitrarily chosen, but always the same, measure φ. Therefore, when designating the length of a segment or the distance between two points, we may omit the symbol φ. In fact, we shall write $|ab|$ instead of $|ab|_\varphi$, and $\varrho(a,b)$ instead of $\varrho_\varphi(a,b)$.

13. Distance of a Point From a Figure

By the *distance $\varrho(a,F)$ of a point a from a figure F* we undertstand the lower bound of the distances $\varrho(a,p)$ for $p \in F$. It is readily seen that the distance $\varrho(a,F)$ always exists and is a non-negative number.

In particular, we can speak about the distance $\varrho(a,L)$ of a point a from a line L and about the distance $\varrho(a,P)$ of a point a from a plane P. From Theorems 80 and 94 of Chapter II we at once have

THEOREM 50. *The distance of a point a from a line L (or plane P) coincides with the distance of point a from its perpendicular projection a_1 upon line L (or plane P).*

Thus, the distance of point a from line L (or plane P) is equal to zero only if point a lies on line L (or plane P).

14. A Characterization of Betweenness and Equidistance Relations in Terms of Distance

From the definition of the distance between two points and from Theorem 50 of Chapter II we obtain

THEOREM 51. *Given any three points a,b,c, we have* **B** (a,b,c) *if and only if points a,b,c, are distinct and* $\varrho(a,b) + \varrho(b,c) = \varrho(a,c)$.

From the definition of distance, from Axiom C1 and from Theorem 2 of Chapter II there follows

THEOREM 52. *Given any four points a,b,c,d, we have* **E** $(a,b;c,d)$ *if and only if* $\varrho(a,b) = \varrho(c,d)$.

15. Similitudes

The transformation f of a figure F_1 onto a figure F_2 is called a *similitude* if there exists a positive constant λ such that

$$(1) \qquad \varrho(f(p),f(q)) = \lambda \cdot \varrho(p,q)$$

for every two points $p,q \in F_1$. The constant λ is called the *coefficient* of similitude f. If there exists a similitude f mapping figure F_1 onto figure F_2, then we say that *figure F_1 is similar to figure F_2.* It is easily seen that the similitude f correlates distinct points of figure F_2 with distinct points of figure F_1. Indeed, for any points $p,q \in F_1$, if $p \neq q$, then $\varrho(p,q) \neq 0$, from which it follows, by (1), that $\varrho(f(p),f(q)) \neq 0$, that is, $f(p) \neq f(q)$. We therefore have

THEOREM 53. *Similitudes are one to one transformations.*

Taking $F_1 = F_2 = S$, we obtain similitudes of the whole space S onto itself. Henceforth in this section, we shall understand by similitudes exclusively these similitudes.

It is readily shown that if a transformation f is a similitude with the coefficient λ, then the inverse transformation f^{-1} is a similitude with the coefficient $1/\lambda$; and if a transformation f_1 is a similitude with the coefficient λ_1 and a transformation f_2 is a similitude with the coefficient λ_2, then the superposition $f_2 f_1$ of these two transformations is a similitude with the coefficient $\lambda_1 \cdot \lambda_2$. It thus follows:

THEOREM 54. *The similitudes of space* **S** *onto itself form a group of transformations.*

We shall now show that all the primitive notions of absolute geometry are invariants of the group of similitudes.

THEOREM 55. *Given any similitude f of space* **S** *onto itself, we have:*
(I) *if* **B**(a,b,c), *then* **B**$(f(a),f(b),f(c))$;
(II) *if* **E**$(a,b;c,d)$, *then* **E**$(f(a),f(b);f(c),f(d))$;
(III) *if L is a line, then f(L) is also a line;*
(IV) *if P is a plane, then f(P) is also a plane.*

PROOF. (I) results immediately from Theorems 51 and 53; (II) results immediately from Theorem 52. Finally, (III) and (IV) follow, by (I) and Theorem 53, from the fact that the notions of a line and of a plane are expressible in terms of the relation **B** and the identity relation $=$ (see Theorems 71 and 72 of Chapter I). We shall illustrate this by a rigorous proof for case (III).

Consider any line L and any two distinct points a and b on it. Then $L = \mathbf{L}(ab)$ and, by Theorem 71 of Chapter I, a point p lies on line L if and only if

$$(2) \qquad p = a \text{ or } p = b \text{ or } \mathbf{B}(p,a,b) \text{ or } \mathbf{B}(a,p,b) \text{ or } \mathbf{B}(a,b,p).$$

By Theorems 53 and 55 (I), condition (2) is equivalent to the following:

$$(3) \qquad \begin{aligned} & f(p) = f(a) \text{ or } f(p) = f(b) \text{ or } \mathbf{B}(f(p),f(a),f(b)), \\ & \text{or } \mathbf{B}(f(a),f(p),f(b)) \text{ or } \mathbf{B}(f(a),f(b),f(p)). \end{aligned}$$

Therefore, set $f(L)$ consists of points $f(p)$ defined by condition (3), and, as a result, set $f(L)$ is the line determined by points $f(a)$ and $f(b)$.

The result obtained in Theorem 55 can be extended step by step to all the defined geometrical notions discussed in this work; i.e., each of these notions proves to be an invariant of the group of similitudes. In particular, we have the following two theorems:

THEOREM 56. *Given any similitude f of space* **S** *onto itself, we have:*
(I) *If A is a half-line of line L with origin a, then f(A) is a half-line of line f(L) with origin f(a).*
(II) $f(A^*) = (f(A))^*$ *for any half-line A.*
(III) *If half-lines A, B, C belong to one pencil, are co-half-pencilar, and* **B**(A,B,C), *then half-lines f(A), f(B), f(C) also belong to one pencil, are co-half-pencilar, and* **B**$(f(A),f(B),f(C))$.
(IV) *For any segments ab and cd, if ab $\equiv$ cd, then f(a)f(b) $\equiv$ f(c)f(d).*
(V) *For any angles AB and CD, if AB $\equiv$ CD, then f(A)f(B) $\equiv$ f(C)f(D).*
(VI) *If AB is a right angle, then f(A)f(B) is a right angle.*

The simple proof of the above theorem is left to the reader. (The proof of part III of Theorem 55 may serve as an example of proofs of this type.)

THEOREM 57. *Given any similitude f of space **S** onto itself, and any measure φ of angles, let*

$$\varphi^{(f)}(AB) = \varphi(f(A)f(B)) \tag{4}$$

for any angle AB. Then we have:

(I) *Function $\varphi^{(f)}$ is a measure of angles.*

(II) *If φ is the natural measure of angles, then $\varphi^{(f)}$ is also the natural measure of angles.*

PROOF. From Theorem 56(III),(V) it follows immediately that function $\varphi^{(f)}$ is a measure of angles.

Assume now that φ is the natural measure of angles, that is, $\varphi(AB) = \pi/2$ for every right angle AB. If AB is a right angle, then, by Theorem 56(VI), angle $f(A)f(B)$ is also a right angle, and, by (4), we have $\varphi^{(f)}(AB) = \pi/2$, that is, $\varphi^{(f)}$ is the natural measure of angles.

There exists only one natural measure φ_0. From Theorem 57(II) it thus follows that for every similitude f

$$\varphi_0^{(f)} = \varphi_0.$$

Consequently, for every similitude f and for every angle AB we have, by (4),

$$\varphi_0(f(A)f(B)) = \varphi_0(AB),$$

that is,

$$f(A)f(B) \equiv AB.$$

Thus we have:

THEOREM 58. *For every similitude f of space **S** onto itself, and for every angle AB,*

$$f(A)f(B) \equiv AB.$$

As we have seen, the fact that, in absolute geometry, we may single out the free right angle from among all the free angles plays an essential role in the proof of this theorem.

16. Topology in Space Induced by a Metric

Having established a metric for space **S**, we have thus set up a topology in space **S**. In this topology, the family of ε-neighborhoods (see Introduction, Section 9) constitutes the base of neighborhoods. The topology of the entire space **S** induces a topology on every plane and on every line. We shall now show that, on every plane P, the topology defined by the metric of space **S** coincides with the topology (introduced in

Section 27 of Chapter I on the basis of the axioms of incidence and order) in which the family of open triangles constitutes the base of neighborhoods.

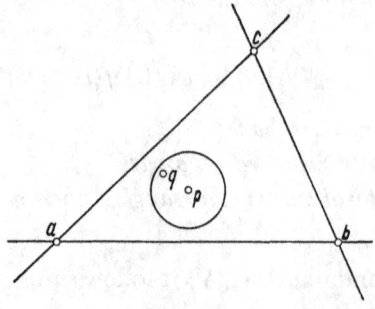

Fig. 155

Consider any open triangle $(abc) \subset P$ and a point $p \in (abc)$ (Fig. 155). Let ε be any positive number not greater than any of the numbers $\varrho(p,\mathbf{L}(ab))$, $\varrho(p,\mathbf{L}(ac))$, $\varrho(p,\mathbf{L}(bc))$. We shall show that the ε-neighborhood of point p is included in triangle (abc). Indeed, if $\varrho(p,q) < \varepsilon$, then, by Theorem 50, the segment $\langle pq \rangle$ does not contain points of any of the lines $\mathbf{L}(ab)$, $\mathbf{L}(bc)$, $\mathbf{L}(ac)$, and therefore $q \in (abc)$.

We next consider an ε-neighborhood of a point $p \in P$. We produce in plane P from point p three not co-half-pencilar half-lines A,B,C (Fig. 156) such that no two of them lie on one line (by Theorem 67(I) of Chapter I,

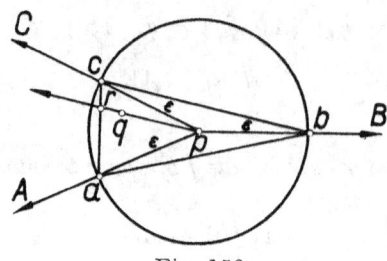

Fig. 156

such half-lines A,B,C exist), and we choose points $a \in A$, $b \in B$, and $c \in C$ such that

(1) $$\varrho(p,a) = \varrho(p,b) = \varrho(p,c) = \varepsilon.$$

From the construction it follows that $p \in (abc)$. We shall show that triangle (abc) is contained in the ε-neighborhood of point p. Indeed, if $q \in (abc)$ and $q \neq p$, then, by Theorem 80 of Chapter I, half-line pq intersects the boundary of triangle (abc) in some point r, and, by formula (1), it is readily noted that $\varrho(p,r) < \varepsilon$. Since $\varrho(p,q) < \varrho(p,r)$, then $\varrho(p,q) < \varepsilon$ and q belongs to the ε-neighborhood of point p.

Since on each plane P the topology defined by the metric coincides with the topology defined by the base of open triangles, it follows that on every line L, as well, the topology defined by the metric coincides with the topology in which the family of open segments constitutes the base of neighborhoods (see Chapter I, Section 11).

17. Distance As a Continuous Function of Two Points

First of all, we shall show how it follows from the triangle inequality that the distance ϱ is a continuous function of two variable points p and q (see Introduction, Section 9).

Using the triangle inequality four times, we obtain, for any four points p,p',q,q', the inequalities

$$\varrho(p,p') + \varrho(p',q') + \varrho(q',q) \geqslant \varrho(p,q)$$

and

$$\varrho(p',p) + \varrho(p,q) + \varrho(q,q') \geqslant \varrho(p',q').$$

Therefore, if

$$\varrho(p,p') < \frac{\varepsilon}{2} \quad \text{and} \quad \varrho(q,q') < \frac{\varepsilon}{2},$$

then

$$|\varrho(p,q) - \varrho(p',q')| < \varepsilon,$$

which proves the continuity (and even the uniform continuity) of the function ϱ.

We now consider the distance of a point from a line.

THEOREM 59. *The distance of a point p from a fixed line L is a continuous function of point p.*

PROOF. We denote by q and q', respectively, the perpendicular projections of points p and p' upon line L. By Theorem 50, we have

$$\varrho(p,L) + \varrho(p',p) \geqslant \varrho(p',q) \geqslant \varrho(p',L)$$

and

$$\varrho(p',L) + \varrho(p',p) \geqslant \varrho(p,q') \geqslant \varrho(p,L),$$

from which we obtain

$$|\varrho(p',L) - \varrho(p,L)| \leqslant \varrho(p',p).$$

This establishes the continuity (and even the uniform continuity) of the distance of a point p from line L.

We shall now prove a theorem which we shall use several times in the remaining part of this book. The reader will find the definition of an increasing function of a point in Section 16 of Chapter I.

THEOREM 60. *Given on a plane P an obtuse angle AB with vertex a, let line $K \subset P$ intersect half-line B in some point a' at a right angle. Then the function φ defined for points $p \in a \cup A$ by the formula $\varphi(p) = \varrho(p, K)$ is continuous, increasing, and unbounded.*

PROOF. We denote by p_x the point on the closed half-line $a \cup A$ such that $\varrho(a, p_x) = x$, and we consider a non-negative function $\psi(x) = \varrho(p_x, K)$. The distance $\varrho(a, p)$ of the fixed point a from a point $p \in a \cup A$ is, as may readily be shown, a continuous, increasing, and unbounded function of the variable p. Thus, instead of proving that the function φ of a point is continuous, increasing, and unbounded, it is sufficient to show that these properties are also possessed by the function ψ of a real variable.

For any $x \geqslant 0$ we denote by p'_x the projection of point p_x upon line K (Fig. 157). We add to x an increment $h > 0$ and examine the convex

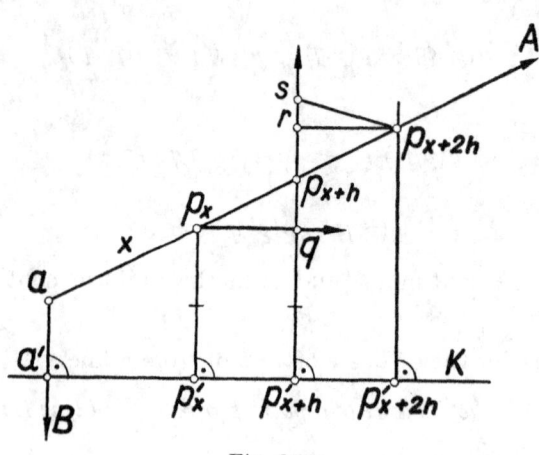

Fig. 157

quadrangles (a, a', p'_x, p_x) and $(p_x, p'_x, p'_{x+h}, p_{x+h})$. In the first of these, two angles are right angles and one is obtuse, and therefore, by Theorem 49, the fourth angle, i.e. angle $ap_xp'_x$ is acute. It thus follows that, in the second quadrangle, angle $p'_xp_xp_{x+h}$ is obtuse. Laying off on half-line $p'_{x+h}p_{x+h}$ segment $p'_{x+h}q$ congruent to p'_xp_x, we obtain quadrangle (p_x, p'_x, p'_{x+h}, q) which, as may easily be shown, is a Saccheri quadrangle with the lower base $p'_xp'_{x+h}$. By Theorem 15, $\sphericalangle p'_xp_xq$ is an acute or right angle. It thus follows that half-line p_xq lies in the inner domain of angle $p'_xp_xp_{x+h}$. As may readily be shown, point p'_{x+h} also lies in the inner domain of angle $p'_xp_xp_{x+h}$. As a result point q lies between points p_{x+h} and p'_{x+h}. Hence we obtain

$$\psi(x) < \psi(x + h) \quad \text{for } h > 0,$$

and therefore ψ is an increasing function. Moreover, angle $p_x q p_{x+h}$ is either a right or obtuse angle, and therefore $q p_{x+h} < p_x p_{x+h}$, that is,

$$\psi(x+h) - \psi(x) < h,$$

from which it follows that ψ is a continuous (and even a uniformly continuous) function in the entire set of non-negative numbers.

It remains to be shown that the function ψ increases without bound. To do this, we increase the variable x by the number $2h$ and consider on half-line $p'_{x+h} p_{x+h}$ two points r and s defined by the conditions:

$$\mathbf{B}(q, p_{x+h}, r), \quad q p_{x+h} \equiv p_{x+h} r, \quad p'_{x+h} s \equiv p'_{x+2h} p_{x+2h}.$$

Since ψ is an increasing function, we have

$$\mathbf{B}(q, p_{x+h}, s).$$

The triangles $p_{x+h} q p_x$ and $p_{x+h} r p_{x+2h}$ are congruent; specifically, the angle $p_{x+h} r p_{x+2h}$ is congruent to angle $p_{x+h} q p_x$, and therefore it is a right or obtuse angle. The angle $p_{x+h} s p_{x+2h}$ is a right or acute angle since it is the angle at the upper base of the Saccheri quadrangle $(s, p'_{x+h} p'_{x+2h} p_{x+2h})$. Thus $\mathbf{B}(p_{x+h}, r, s)$ or $r = s$, from which it follows that

$$\varrho(p'_{x+2h}, p_{x+2h}) \geqslant \varrho(p'_{x+h}, r) = \varrho(p'_{x+h}, p_{x+h}) + \varrho(p_{x+h}, r) =$$

$$\varrho(p'_{x+h}, p_{x+h}) + \varrho(p'_{x+h}, p_{x+h}) - \varrho(p'_x p_x),$$

that is,

$$\psi(x+2h) \geqslant \psi(x+h) + \psi(x+h) - \psi(x).$$

The last inequality can be put into the form

$$\psi(x+2h) - \psi(x+h) \geqslant \psi(x+h) - \psi(x),$$

from which we obtain, by induction, the inequality

$$\psi(x+kh) - \psi(x+(k-1)h) \geqslant \psi(x+h) - \psi(x) \text{ for } k = 1, 2, \ldots$$

Summing these inequalities for $k = 1, 2, \ldots, n$, we obtain

$$\psi(x+nh) \geqslant \psi(x) + n \cdot [\psi(x+h) - \psi(x)],$$

from which, since $\psi(x+h) - \psi(x) > 0$, it follows that the function ψ increases without bound. This is what we had to prove.

NOTE 1. If in Theorem 60 we assume that angle AB is a right angle (and not obtuse), then, by proceeding in a similar manner, we arrive at the conclusion that ψ is a continuous, non-decreasing function of the variable x.

We thus have:

THEOREM 61. *If a_1 and b_1 are the perpendicular projections of points a and b upon a line L (on a plane P), then $\varrho(a_1,b_1) \leqslant \varrho(a,b)$.*

PROOF. Letting $A = \mathbf{H}(a_1a)$, $B = \mathbf{H}(a_1b_1)$, and $K = \mathbf{L}(b_1b)$ (Fig. 158), we have

$$\varrho(a_1,b_1) = \varrho(a_1,K) \leqslant \varrho(a,K) \leqslant \varrho(a,b).$$

NOTE 2. Theorem 60 remains valid if the line K, perpendicular to line $\mathbf{L}(B)$, passes through the origin a of half-line B instead of intersecting

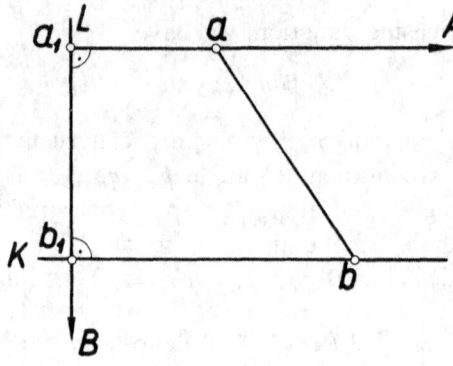

Fig. 158

it. In this case, the proof differs from the proof of Theorem 60 only in that instead of quadrangle (a,a',p_x',p_x) we consider the triangle $ap_x'p_x$, in which the angle ap_xp' is acute.

We therefore have:

THEOREM 62. *Let AC be an angle with vertex a. The distance $\varrho(p,K)$ of a point $p \in a \cup A$ from line $K \supset C$ is a continuous, increasing, and unbounded function of p.*

18. Coordinates on a Line. Metric Type of Lines

Given a line L. We fix on it an arbitrary point a_0, and from the two half-lines determined by point a_0 on line L we choose one, say A. Let p be any point on line L. We define a real number x^p in the following manner:

$$x^p = \begin{cases} |a_0p| & \text{if } p \in A, \\ 0 & \text{if } p = a_0, \\ -|a_0p| & \text{if } p \in A^*. \end{cases}$$

We call the function Φ defined for any point $p \in L$ by the condition

$$\Phi(p) = x^p$$

the *coordinate system on line L*, and the number x^p the *coordinate* of the point p in the system Φ.

The coordinate system Φ on line L is uniquely determined (since the measure of segments is fixed) by point a_0 and half-line A; we shall call point a the *origin* and half-line A the *positive half-line*, of the system Φ.

Because of Theorem 36, the function Φ maps line L onto one-dimensional Cartesian space $\mathbf{C}_1$. Furthermore, it is readily noted that if $\Phi(p) = x^p$ and $\Phi(q) = x^q$, then the distance between points p and q is expressed by the formula

$$\varrho(p,q) = |x^p - x^q|.$$

The same number $|x^p - x^q|$ is the distance between the Cartesian points $\Phi(p)$ and $\Phi(q)$. We therefore have

THEOREM 63. *Any coordinate system Φ on line L isometrically maps line L onto Cartesian space $\mathbf{C}_1$.*

In this manner we have shown that every line is isometric with space $\mathbf{C}_1$, and therefore does not differ as regards its metric properties from Cartesian space $\mathbf{C}_1$. We may therefore say that the axioms of absolute geometry determine the metric type of lines.

We take on line L any two distinct points p and q. Let Φ be any coordinate system on line L. It is readily shown that if $\Phi(p) = x^p$ and $\Phi(q) = x^q$, then segment $\langle pq \rangle$ is mapped by Φ onto the closed interval of real numbers $\langle x^p, x^q \rangle$. Since function Φ, as an isometry, is a homeomorphism, then segment $\langle pq \rangle$ turns out to be an arc. The length of this arc coincides, by Theorem 51, with the length of segment pq. This remark makes it clear why a metric induced by a measure is called a proper metric.

19. Absolute Coordinates on a Plane. Topological Type of Planes

Consider a plane P. We fix on plane P any line K_0, and we choose one of the two half-planes, say W, determined on plane P by line K_0. On line K_0 we establish any coordinate system Φ_0. Also, on any line $L \subset P$ perpendicular to line K_0 (Fig. 159) we establish a coordinate system Φ_L in the following manner: the point $a_L = L \cap K_0$ is the origin, and the half-line $A_L = L \cap W$ is the positive half-line, of the system Φ_L.

Now let p be any point on plane P. We define two real numbers x_1^p and x_2^p in the following manner: Let L be the line produced from point p perpendicular to line K_0; then

$$x_1^p = \Phi_0(a_L), \quad x_2^p = \Phi_L(p).$$

Therefore

$$x_2^p = \begin{cases} |a_L p| & \text{if } p \in W, \\ 0 & \text{if } p \in K_0, \\ -|a_L p| & \text{if } p \in W^*. \end{cases}$$

The function Φ defined for any point $p \in P$ by the condition

$$\Phi(p) = (x_1^p, x_2^p)$$

will be called the *absolute coordinate system on plane P*, and the numbers x_1^p and x_2^p, the *first* and *second absolute coordinates* of point p in system Φ.

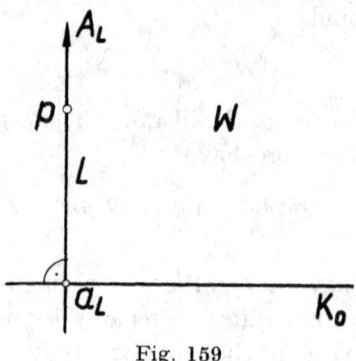

Fig. 159

The coordinate system Φ on plane P is uniquely determined (since the measure of segments is fixed) by line K_0, half-plane W, and coordinate system Φ_0 on line K_0; we shall call line K_0 and half-plane W the *origin line* and the *positive half-plane* of system Φ, respectively.

It may readily be shown that function Φ maps plane P onto two-dimensional Cartesian space $\mathbf{C}_2$ in a one-to-one way.

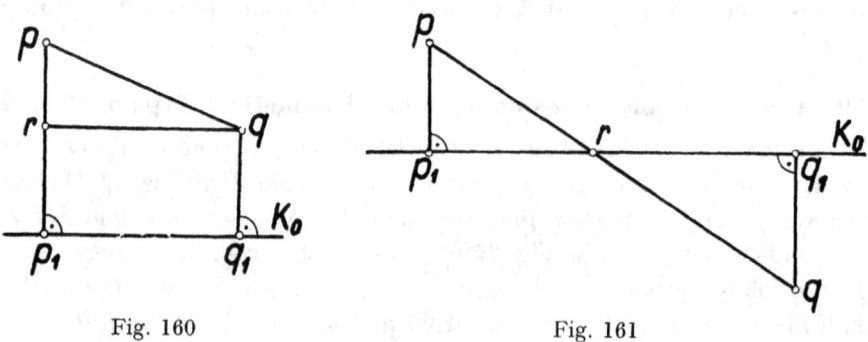

Fig. 160   Fig. 161

We shall now show that transformation Φ is bicontinuous. To do this we first show that for every two points $p, q \in P$

(1)
$$|x_1^p - x_1^q| \leqslant \varrho(p,q)$$

and

(2)
$$|x_2^p - x_2^q| \leqslant \varrho(p,q).$$

Inequality (1) is a direct consequence of Theorems 61 and 63. Inequality (2) is obvious when $x_1^p = x_1^q$ or $x_2^p = x_2^q$. We therefore assume that $x_1^p \neq x_1^q$ and $x_2^p \neq x_2^q$. We denote by p_1 and q_1 the perpendicular projections of points p and q upon line K_0. We distinguish two cases:

Case 1. $x_2^p \cdot x_2^q \geqslant 0$ (Fig. 160). Let $r = \Phi^{-1}(x_1^p, x_2^q)$. Then angle prq is either a right or obtuse angle, and therefore

$$|x_2^p - x_2^q| = \varrho(p,r) < \varrho(p,q).$$

Case 2. $x_2^p \cdot x_2^q < 0$ (Fig. 161). Then $\mathbf{B}(p, K_0, q)$, and hence there exists on line K_0 a point r such that $\mathbf{B}(p,r,q)$ and

$$x_2^p - x_2^q| = |x_2^p| + |x_2^q| = \varrho(p,p_1) + \varrho(q,q_1) \leqslant \varrho(p,r) + \varrho(r,q) = \varrho(p,q).$$

Hence inequality (2) is always satisfied.

It follows immediately from inequalities (1) and (2) that if for a constant p we take a point q such that

$$\varrho(p,q) < \frac{\varepsilon\sqrt{2}}{2},$$

we obtain in $\mathbf{C}_2$

$$\varrho(\Phi(p),\Phi(q)) < \varepsilon.$$

This shows that Φ is a continuous (and even uniformly continuous) function on the entire plane P.

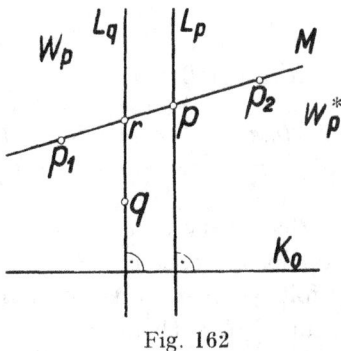

Fig. 162

In order to show that the function Φ^{-1} is also continuous, we take any point $p \in P$ and any number $\varepsilon > 0$ (Fig. 162). The line $L_p \subset P$ passing through point p and perpendicular to line K_0 determines in plane P two half-planes: W_p consisting of points whose first absolute coordinate is less than x_1^p, and W_p^* consisting of points whose first absolute coordinate is greater than x_1^p. We produce a line $M \neq L_p$ through point p and pick two points $p_1 \in M \cap W_p$ and $p_2 \in M \cap W_p^*$ such that

$$(3) \qquad \varrho(p,p_1) < \frac{\varepsilon}{3} \quad \text{and} \quad \varrho(p,p_2) < \frac{\varepsilon}{3}.$$

We thus have $x_1^{p_1} < x_1^p < x_1^{p_2}$. The continuity of the function Φ^{-1} will be established when we show that the inequalities

$$(4) \qquad x_1^{p_1} < x_1^q < x_1^{p_2}$$

and

$$(5) \qquad x_2^p - \frac{\varepsilon}{3} < x_2^q < x_2^p + \frac{\varepsilon}{3}$$

imply the inequality $\varrho(p,q) < \varepsilon$. Hence, we assume that point q satisfies inequalities (4) and (5); let $L_q \subset P$ denote the line produced through point q and perpendicular to line K_0. By (4), points p_1 and p_2 lie on opposite sides of line L_q, and consequently, there exists a point $r \in L_q$ such that $\mathbf{B}(p_1,r,p_2)$. Because of (3), we have $\varrho(p,r) < \varepsilon/3$, from which, by formula (2), we obtain $|x_2^p - x_2^r| < \varepsilon/3$.
The latter inequality and inequality (5) give us

$$\varrho(q,r) = |x_2^q - x_2^r| \leqslant |x_2^q - x_2^p| + |x_2^p - x_2^r| < \frac{\varepsilon}{3} + \frac{\varepsilon}{3} = \frac{2}{3}\varepsilon.$$

Finally, we have

$$\varrho(p,q) \leqslant \varrho(p,r) + \varrho(r,q) < \frac{1}{3}\varepsilon + \frac{2}{3}\varepsilon = \varepsilon,$$

which is what we wished to prove.

We may therefore state the following

THEOREM 64. *Any absolute coordinate system Φ on a plane P maps plane P homeomorphically onto Cartesian space $\mathbf{C}_2$.*

Thus, we have shown that every plane is homeomorphic to space $\mathbf{C}_2$, and consequently the topological type of planes is uniquely determined.

From Theorem 64 it follows, in particular, that every two planes (of space $\mathbf{S}$) are homeomorphic. From Theorem 98 of Chapter II it furthermore follows that every two planes (of space $\mathbf{S}$) are isometric. However, unlike the case of the line (see Theorem 63), the metric type of planes cannot be established on the basis of absolute geometry. In fact, in Chapter IV we shall discuss two spaces (namely Cartesian space $\mathbf{C}_3$ and

Klein space K_3) such that in each of them all the axioms of absolute geometry are satisfied; nevertheless the planes of one of the spaces are not isometric with the planes of the other.

20. Absolute Coordinates in Space. Topological Type of Space

We fix a plane P_0 in space S and choose one of the two half-spaces, say W, into which plane P_0 divides space S. We fix on plane P_0 any arbitrary absolute coordinate system Φ_0. Also, on any line L perpendicular to plane P_0 (Fig. 163) we establish a coordinate system Φ_L in the following manner: the point $a_L = L \cap P_0$ is the origin, and the half-line $A_L = L \cap W$ is the positive half-line of the system Φ_L. For any point p we now define three real numbers x_1^p, x_2^p, x_3^p as follows: Let L be the line produced through point p perpendicular to plane P_0; then

$$(x_1^p, x_2^p) = \Phi_0(a_L), \quad x_3^p = \Phi_L(p).$$

Hence

$$x_3^p = \begin{cases} |a_L p| & \text{if } p \in W, \\ 0 & \text{if } p \in P_0, \\ -|a_L p| & \text{if } p \in W^*. \end{cases}$$

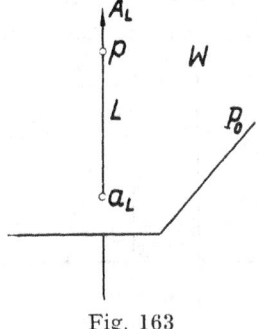

Fig. 163

The function Φ defined for any point p by the condition

$$\Phi(p) = (x_1^p, x_2^p, x_3^p)$$

will be called the *absolute coordinate system in space* S, and the numbers x_1^p, x_2^p, x_3^p, the *first*, *second*, and *third absolute coordinates* of point p in the system Φ.

The function Φ maps every point $p \in S$ on some point $x^p = \Phi(p)$ of three dimensional Cartesian space C_3. Just as in the case of the plane, we may state that Φ is a one-to-one transformation. We thus have

THEOREM 65. *The space* S *has the power of the continuum.*

We shall now show that the transformation Φ is bicontinuous. We take any two points p and q; denote by p_1 and q_1 the perpendicular projections of p and q upon plane P_0. By Theorem 61, we have $\varrho(p_1,q_1) \leqslant \varrho(p,q)$, from which, because of inequalities (1) and (2) of the preceding section, we obtain

$$|x_1^p - x_1^q| \leqslant \varrho(p,q) \quad \text{and} \quad |x_2^p - x_2^q| \leqslant \varrho(p,q).$$

In addition, we have

$$|x_3^p - x_3^q| \leqslant \varrho(p,q),$$

the proof of which, by means of Theorem 92 of Chapter II, is entirely similar to the proof of inequality (2) of the preceding section. From the last two formulas it follows that the function Φ is continuous (and even uniformly continuous). To prove that the function Φ^{-1} is also continuous, we shall need the following:

LEMMA. *Let (a,b,p,q) be a Saccheri quadrangle with a fixed side ab and a variable side pq in space S. Then for each $\varepsilon > 0$ we may choose an $\eta > 0$ such that $\varrho(b,p) < \eta$ implies $\varrho(a,q) < \varepsilon$.*

PROOF. We take any $\varepsilon > 0$. We pick any plane P containing line ab and we assume at first that the Saccheri quadrangle (a,b,p,q) lies in

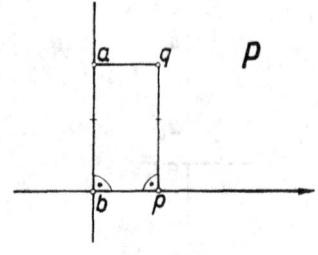

Fig. 164

plane P. We now establish on plane P the absolute coordinate system Φ_P by selecting line bp as the origin line and the half-plane $L(bp)a$ as the positive half-plane (Fig. 164); on the origin line bp we fix the coordinate system with the origin b and with the positive half-line bp. Then

(1) $\Phi_P(a) = (0, |ab|), \quad \Phi_P(q) = (|bp|, |ab|).$

By Theorem 64, there exists an $\eta > 0$ such that if in space C_2

$$\varrho(\Phi_P(a),\Phi_P(q)) < \eta,$$

then on plane P

$$\varrho(a,q) < \varepsilon.$$

But, by (1), we have $\varrho(\Phi_P(a), \Phi_P(q)) = \varrho(b,p)$, and hence

(2)   if $\varrho(b,p) < \eta$, then $\varrho(a,q) < \varepsilon$ for $p,q \in P$.

Consider now any Saccheri quadrangle (a,b,p,q) such that $\varrho(b,p) < \eta$. As may readily be noted, there exist in plane P points p_1, q_1 such that quadrangle (a,b,p_1,q_1) is a Saccheri quadrangle with the lower base bp_1 and $bp_1 \equiv bp$. Then, by Theorem 84 of Chapter II, we have $aq_1 \equiv aq$. Since $\varrho(b,p_1) < \eta$, then, by (2), we have $\varrho(a,q_1) < \varepsilon$. Hence, also, $\varrho(a,q) < \varepsilon$. This concludes the proof.

Let us fix a point p and consider a variable point q. We denote by p_1 and q_1 the perpendicular projections of points p and q upon plane P_0 (Fig. 165), and we assume that

(3) $p \neq p_1 \neq q_1$.

Let $r = \Phi^{-1}(x_1^q, x_2^q, x_3^p)$. It is readily noted that (p,p_1,q_1,r) is a Saccheri quadrangle with a fixed side pp_1. It follows from the lemma that there exists a number $\eta' > 0$ such that the inequality

(4) $\varrho(p_1,q_1) < \eta'$

implies the inequality

(5) $\varrho(p,r) < \dfrac{\varepsilon}{2}.$

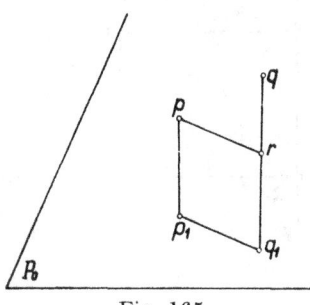

Fig. 165

Furthermore, as a result of Theorem 64, there exists a positive number $\eta < \dfrac{\varepsilon}{2}$ such that (4) and therefore also (5) follow from

(6) $|x_1^p - x_1^q| < \eta$ and $|x_2^p - x_2^q| < \eta$.

If, in addition,

(7) $|x_3^p - x_3^q| < \eta,$

then

$$\varrho(r,q) = |x_3^p - x_3^q| < \eta < \frac{\varepsilon}{2},$$

and hence, formulas (6) and (7) imply together that

$$\varrho(p,q) \leqslant \varrho(p,r) + \varrho(r,q) < \varepsilon.$$

The discarding of condition (3) does not cause any new difficulty. We have thus proved that the function Φ^{-1} is continuous. Therefore we may state the following

THEOREM 66. *Any absolute coordinate system Φ of space $\mathbf{S}$ maps space $\mathbf{S}$ homeomorphically onto Cartesian space $\mathbf{C}_3$.*

21. Rectangular Coordinates

On a plane P, we fix two perpendicular lines K_1 and K_2 intersecting in a point a_0 (Fig. 166). We fix on line K_i ($i = 1,2$) a coordinate system Φ_i with origin a_0 and with the positive half-line A_i. Half-line A_i determines on line K_i the axis $\mathfrak{R}_i$.

Let p be any point on plane P. We denote by p_i the perpendicular projection of point p upon line K_i and we put

$$\Phi_i(p_i) = \xi_i^p \quad \text{for} \quad i = 1,2.$$

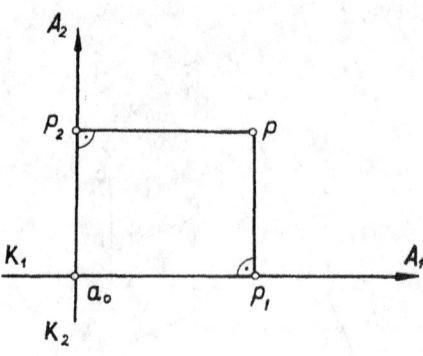

Fig. 166

The function Ψ defined for any point $p \in P$ by the formula

$$\Psi(p) = (\xi_1^p, \xi_2^p)$$

will be called the *rectangular coordinate system on plane P*, and the numbers ξ_1^p and ξ_2^p will be called the *first* and *second rectangular coordinates* of point p in the system Ψ. Obviously, the pair of axes $(\mathfrak{R}_1, \mathfrak{R}_2)$ uniquely determines the coordinate system Ψ. The number ξ_1^p is also called the *abcissa*, and the number ξ_2^p the *ordinate*, of point p.

The function Ψ correlates with every point p of plane P a point $\Psi(p)$ of Cartesian space $\mathbf{C}_2$. It is readily noted that if $p \neq q$, then $\Psi(p) \neq \Psi(q)$.

On the basis of absolute geometry, however, it cannot be proved that every point $(\xi_1, \xi_2) \in \mathbf{C}_2$ is the image of some point $p \in P$, since, as we shall see in Section 29 of Chapter VI, it cannot be shown that in plane P the line perpendicular to K_1 and passing through point $\Phi_1^{-1}(\xi_1) \in K_1$ always intersects the line perpendicular to K_2 and passing through point $\Phi_2^{-1}(\xi_2) \in K_2$.

In a similar manner, an ordered triple of axes $\mathfrak{K}_1, \mathfrak{K}_2, \mathfrak{K}_3$ passing through some point a_0 and such that any two of them are perpendicular determines a *rectangular coordinate system* Ψ *in space* $\mathbf{S}$. The function Ψ transforms space $\mathbf{S}$ into Cartesian space $\mathbf{C}_3$, but again, on the basis of absolute geometry it cannot be shown that this is the transformation onto all of space $\mathbf{C}_3$.

This is the reason why, instead of rectangular coordinates, we have introduced absolute coordinates both on the plane and in the space.

IV

Models of Absolute Geometry

1. Problems of Consistency, Independence, and Categoricity of an Axiom System of Geometry. Interpretation. Model

The most important problem in connection with an axiom system of geometry is the problem of *consistency*. Axioms are *consistent* if no antinomy can be derived from them, i.e. if no two contradictory statements are both geometrical theorems.

A second problem in connection with a system of axioms of geometry is the problem of their *mutual independence*. We say that *a sentence* S *is independent of axioms* $A_1, A_2, \ldots, A_n$ if it cannot be derived from them. Axioms are *mutually independent* if each of them is independent of the remaining ones. The mutual independence of axioms is a problem of much less importance than their consistency. As regards geometry, however, it was precisely the investigation of the independence of axioms that led to the discovery of new geometries, for example, Bolyai-Lobachevskian geometry.

The proof of the consistency, as well as the proof of the independence, of an axiom system of geometry is carried out by the interpretation method. To prove the consistency of an axiom system we give some interpretation to the primitive notions of the system and we show that in this interpretation all the axioms are satisfied. In other words, we find sets and relations which satisfy all the axioms. In order to prove that a certain sentence S is independent of axioms $A_1, A_2, \ldots, A_n$, we find sets and relations satisfying axioms $A_1, A_2, \ldots, A_n$, but not sentence S. We now define such sets and relations precisely (for the example of the system of primitive notions of space geometry) and explain what is meant by saying that they satisfy certain sentences of geometry.

As the primitive notions of geometry (see page 19) we have taken: a set S, two classes $\mathfrak{S}\mathfrak{L}$ and $\mathfrak{P}\mathfrak{L}$ of subsets of S, and two relations, a three-termed relation B and a four-termed relation E, among elements of S. Let us now assume that in some axiomatic theory T we have defined a certain set X, two classes $\mathfrak{X}_1$ and $\mathfrak{X}_2$ of subsets of X, and two relations, a three-termed relation R_1 and a four-termed relation R_2, among elements of X The system of notions

$$X, \mathfrak{X}_1, \mathfrak{X}_2, R_1, R_2$$

will be called an *interpretation* of the system of primitive notions of geometry in the theory T. Since our system of geometry is based upon set theory and the arithmetic of real numbers we assume that T is based upon these two theories as well. In practice we shall always take for T the algebra of real numbers (provided with a set theoretical basis).

Once some interpretation (I) is given, we shall refer to the elements of set X as I-points, to the sets of class $\mathfrak{X}_1$ as I-*lines*, and to sets of class $\mathfrak{X}_2$ as I-*planes*; the relations R_1 and R_2 will be denoted by $\boldsymbol{B}_\mathrm{I}$ and $\boldsymbol{E}_\mathrm{I}$. Henceforth we shall write the interpretation (I) in the following way:

(I)   I-points, I-lines, I-planes, $\boldsymbol{B}_\mathrm{I}$, $\boldsymbol{E}_\mathrm{I}$.

We say that *interpretation* (I) *satisfies sentence* S (of geometry) *in the theory* T if, after replacing in sentence S the terms

(1)   *points, lines, planes,* **B**, **E**

by the corresponding terms of system (I), we obtain a theorem of theory T. Interpretation (I) will then be called a *model for sentence* S. If the interpretation (I) is a model for each of the sentences $S_1, S_2, \ldots, S_n$, then we say that it is a *model for the system of sentences†* $S_1, S_2, \ldots, S_n$.

To prove that a given axiom system of geometry is consistent we find, in some consistent theory T, an interpretation (I) that satisfies all the axioms of the system. To prove that an axiom A is independent of the remaining axioms of the system we find, in some consistent theory T, an interpretation (I) which does not satisfy axiom A, but which satisfies all the remaining axioms of the system.

We say that the interpretation

(I_1) I_1-points, I_1-lines, I_1-planes, $\boldsymbol{B}_{I_1}$, $\boldsymbol{E}_{I_1}$

in theory T_1 is *isomorphic* with the interpretation

(I_2) I_2-points, I_2-lines, I_2-planes, $\boldsymbol{B}_{I_2}$, $\boldsymbol{E}_{I_2}$

in theory T_2, if there exists a one-to-one transformation f of the set of I_1-points onto the set of I_2-points such that the classes of lines and planes and the relations **B** and **E** are preserved, i.e., if the following four conditions are satisfied:

---

† In general the expressions *interpretation* and *model* are used as synonyms. This is not the case in this book, where the notion of interpretation is broader than the notion of model. Among all possible interpretations we shall seek models for individual sentences or for systems of sentences of geometry. It may be said that models are good interpretations.

(I) L is an I_1-line if and only if $f(L)$ is an I_2-line.

(II) P is an I_1-plane if and only if $f(P)$ is an I_2-plane.

(III) $\boldsymbol{B}_{I_1}(a,b,c)$ if and only if $\boldsymbol{B}_{I_2}(f(a),f(b),f(c))$, for any three I_1-points a, b, c.

(IV) $\boldsymbol{E}_{I_1}(a,b;c,d)$ if and only if $\boldsymbol{E}_{I_2}(f(a),f(b);f(c),f(d))$, for any four I_1-points a, b, c, d.

Thus, if two interpretations (I_1) and (I_2) are isomorphic, then there is no geometrical sentence S such that (I_1) satisfies S while (I_2) satisfies the negation of S.

Next in importance after consistency as regards the axioms of geometry is the question of *categoricity*. An axiom system is called *categorical* if any two of its models are isomorphic. A categorical axiom system is strong enough to determine its model uniquely up to isomorphism.

In the further sections of this chapter we shall consider the problems of consistency, mutual independence, and categoricity as regards the following two theories: space absolute geometry based on the system of primitive notions (1) and on the system of axioms

(GA_3) I1—I9, O1—O9, C1—C7, Co,

and plane absolute geometry based on the system of primitive notions

points, lines, $\boldsymbol{B}$, $\boldsymbol{E}$

and on the system of axioms

(GA_2) I1—I4, O1—O9, C1—C7, Co.

Axiom I4 takes the following form in the plane case:

AXIOM I4. *There exist three non-collinear points a, b, c.*

In section 3 we shall prove that the system (GA_3) is consistent by constructing the *Cartesian model* for it. In section 4, by a modification of the Cartesian model, we shall show that the axiom of continuity is independent of the remaining axioms of system (GA_3). In Section 7 we shall build the *Klein model* for the axiom system (GA_3). This model turns out not to be isomorphic with the Cartesian model. In this connection, there arises the question of supplementing the axiom system (GA_3) to a categorical axiom system. In Chapters V and VI we shall show that for this purpose it suffices to add to (GA_3) one of the following two sentences:

E. *For any plane P, any line L ⊂ P, and any point a ∈ P−L there exists at most one line K ⊂ P passing through point a and not intersecting line L.*

BL. *For some plane P_0, some line $L_0 ⊂ P_0$, and some point $a_0 ∈ P_0−L_0$ there exist at least two distinct lines $K_1 ⊂ P_0$ and $K_2 ⊂ P_0$ passing through point a_0 and not intersecting line L_0.*

It is clear that sentence BL is the negation of sentence E.

In this way, from space absolute geometry we shall pass over either to space *Euclidean geometry* or to space *Bolyai-Lobachevskian geometry*.

A similar investigation of plane absolute geometry in the same sections and chapters leads to the conclusion that the axiom system (GA$_2$) is not categorical, but becomes categorical by adding to it one of the sentences E or BL, which in the plane case have the form:

E. *For any line L and any point $a ∼ ∈ L$ there exists at most one line K passing through point a and not intersecting line L.*

BL. *For some line L_0 and some point $a_0 ∼ ∈ L_0$ there exist at least two distinct lines K_1 and K_2 passing through point a and not intersecting line L_0.*

In this way, from plane absolute geometry we shall pass over either to plane Euclidean geometry or to plane Bolyai-Lobachevskian geometry.

The Cartesian and Klein models are built up on the basis of the arithmetic of real numbers. The Cartesian models consist of notions of analytic geometry of Cartesian spaces C_2 and C_3; the Klein models consist of notions of analytic geometry of projective spaces P_2 and P_3 (see Introduction, Section 10). We precede the construction of the Cartesian models by a short introduction to the analytic geometry of Cartesian spaces (Section 2). The construction of the Klein models is preceded by the theory of the Cartesian circle (Section 5) and by a short introduction to the analytic geometry of projective spaces (Section 6). At the same time, in addition to the notions necessary for the construction of the Klein model, we shall also discuss those notions of analytic geometry which will be employed in building up the models for projective geometry (Part II, Chapter IX).

2. The Cartesian Space C_n

As we have already mentioned in the Introduction (Section 10), the n-dimensional Cartesian space C_n is a metric space whose points are all the sequences $x = (x_1, x_2, \ldots, x_n)$ of real numbers and in which the distance ρ of point $x = (x_1, x_2, \ldots, x_n)$ from point $y = (y_1, y_2, \ldots, y_n)$ is given by the formula

(1) $$\rho(x,y) = \sqrt{\sum_{i=1}^{n}(x_i - y_i)^2}.$$

The numbers $x_1, x_2, \ldots, x_n$ are called the *coordinates* of the point $x = (x_1, x_2, \ldots, x_n)$. A point $x = (x_1)$ of space $\mathbf{C}_1$ will be identified with its coordinate x_1. Thus the space $\mathbf{C}_1$ will be regarded as the set of real numbers in which the distance ρ between the numbers x and y is defined by the formula

$$\rho(x,y) = \sqrt{(x-y)^2} = |x-y|.$$

We shall introduce several operations for points of the space $\mathbf{C}_n$. Let $x = (x_1, x_2, \ldots, x_n)$, $y = (y_1, y_2, \ldots, y_n)$ and let λ be any real number. Then

(2) $$x + y = (x_1 + y_1, x_2 + y_2, \ldots, x_n + y_n),$$

(3) $$-x = (-x_1, -x_2, \ldots, -x_n),$$

(4) $$x - y = x + (-y),$$

(5) $$\lambda \cdot x = x \cdot \lambda = (\lambda x_1, \lambda x_2, \ldots, \lambda x_n),$$

(6) $$x \cdot y = \sum_{i=1}^{n} x_i y_i,$$

(7) $$x^2 = x \cdot x,$$

(8) $$|x| = \sqrt{x^2}.$$

These operations in the space $\mathbf{C}_1$ coincide with the operations with real numbers denoted in the same way. In the space $\mathbf{C}_n$, for $n > 1$, the operations of *addition* and *subtraction of points*, *multiplication of a point by a number*, the formation of an *opposite point*, and *absolute value of a point* are subject to the same laws of calculation as the corresponding operations with real numbers. On the other hand, for the *scalar multiplication* of points, some laws hold and others fail. Thus, e.g., for any three points $x, y, z \in \mathbf{C}_n$

$$x \cdot y = y \cdot x \quad \text{and} \quad x \cdot (y + z) = x \cdot y + x \cdot z,$$

but:

(i) the scalar multiplication of points is not associative;

(ii) $x \cdot y = 0$ does not imply that one of the two points x and y must coincide with the point $0 = (0, 0, \ldots, 0)$;

(iii) $(x \cdot y)^2 \leqslant x^2 \cdot y^2$, and $(x \cdot y)^2 = x^2 \cdot y^2$ only if there exist numbers λ and μ that do not vanish simultaneously and such that $\lambda \cdot x = \mu \cdot y$.

The expressions involving the symbols of the introduced operations should in general be regarded only as computational abbreviations without any geometrical content; some of them, however, have a geometrical sense, e.g., the term $|x - y|$, since

$$|x - y| = \rho(x,y),$$

and the term $\frac{1}{2}(x + y)$, which is the only solution (for z) of the system of equations

$$\rho(x,z) = \rho(y,z) = \tfrac{1}{2}\rho(x,y).$$

We shall refer to the point $\frac{1}{2}(x + y)$ as the *midpoint* of points x and y.

An ordered pair of points (x,y) will be called a *bound vector* and denoted by $\overrightarrow{xy}$. The distance $\rho(x,y)$ will be called the *length of vector* $\overrightarrow{xy}$. The coordinates of point $y - x$ will be called the *coordinates of vector* $\overrightarrow{xy}$. Two vectors $\overrightarrow{xy}$ and $\overrightarrow{x'y'}$ have the same coordinates if $y - x = y' - x'$, i.e. if $\frac{1}{2}(x' + y) = \frac{1}{2}(x + y')$, which already has geometrical sense. The class of all bound vectors having the same coordinates $\alpha_1, \alpha_2, \ldots, \alpha_n$ will be called the *free* (Cartesian) *vector* with coordinates $\alpha_1, \alpha_2, \ldots, \alpha_n$ and will be denoted by $[\alpha_1, \alpha_2, \ldots, \alpha_n]$. Obviously, every sequence $[\alpha_1, \alpha_2, \ldots, \alpha_n]$ of real numbers denotes some free vector. If a free vector $\mathfrak{a} = [\alpha_1, \alpha_2, \ldots, \alpha_n]$ is given, then by $(\mathfrak{a})$ we understand the point $(\alpha_1, \alpha_2, \ldots, \alpha_n)$; if the point $x = (x_1, x_2, \ldots, x_n)$ is given, then by $[x]$ we understand the vector $[x_1, x_2, \ldots, x_n]$. If $\overrightarrow{xy} \in \mathfrak{a}$, then we call the bound vector $\overrightarrow{xy}$ the *representative* of free vector $\mathfrak{a}$. Then $\mathfrak{a} = [y - x]$ and $y = x + (\mathfrak{a})$. Thus there exists just one representative of free vector $\mathfrak{a}$ with a given origin x.

We shall now define geometrically (i.e. invariantly with respect to isometries) some operations with free vectors.

Absolute value. Every representative $\overrightarrow{xy}$ of a vector $\mathfrak{a}$ has the same length $\rho(x,y) = |y - x| = |(\mathfrak{a})|$ which we call the *length* or *absolute value of free vector* $\mathfrak{a}$ and denote it by $|\mathfrak{a}|$. Thus

(9) $$|\mathfrak{a}| = |(\mathfrak{a})|.$$

The vector $0 = [0,0,\ldots,0]$ is the only vector of length 0. Vectors of length 1 will be called *versors*. Let

(10) $$\delta_i^j = \begin{cases} 1 & \text{if } i = j, \\ 0 & \text{if } i \neq j. \end{cases}$$

The following vectors

(11) $$\mathfrak{w}_j = [\delta_1^j, \delta_2^j, \ldots, \delta_n^j] \quad \text{for } j = 1,2,\ldots,n$$

are of course versors.

Addition of vectors. Let there be given two free vectors $\mathfrak{a}$ and $\mathfrak{b}$. We take any point x (Fig. 167). Let $\overrightarrow{xy}$ be lthe representative of vector $\mathfrak{a}$ with origin x, and $\overrightarrow{yz}$ the representative of vector $\mathfrak{b}$ with origin y. By the *sum* $\mathfrak{a} + \mathfrak{b}$ *of free vectors* $\mathfrak{a}$ *and* $\mathfrak{b}$ we understand the free vector whose representative is $\overrightarrow{xz}$. This sum is independent of the choice of point x, since $y = x + (\mathfrak{a})$, $z = (x + (\mathfrak{a})) + (\mathfrak{b}) = x + ((\mathfrak{a}) + (\mathfrak{b}))$, i.e. $z - x = (\mathfrak{a}) + (\mathfrak{b})$, from which it follows that

(12) $$\mathfrak{a} + \mathfrak{b} = [(\mathfrak{a}) + (\mathfrak{b})].$$

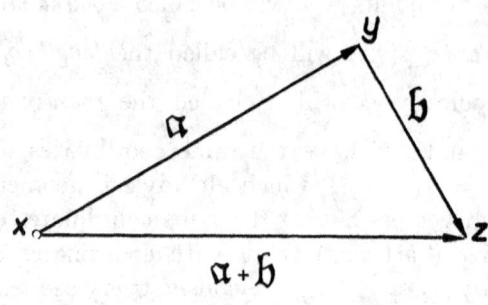

Fig. 167

Opposite vector. Given a free vector $\mathfrak{a}$ with a representative $\overrightarrow{xy}$. We define the vector $-\mathfrak{a}$, *opposite* to vector $\mathfrak{a}$, as the free vector with the representative $\overrightarrow{yx}$. Vector $-\mathfrak{a}$ is independent of the choice of point x, since $y - x = (\mathfrak{a})$, i.e. $x - y = -(\mathfrak{a})$, from which it follows that

(13) $$-\mathfrak{a} = [-(\mathfrak{a})].$$

We set

(14) $$\mathfrak{a} - \mathfrak{b} = \mathfrak{a} + (-\mathfrak{b}).$$

Multiplication of a vector by a number. We say that *free vectors* $\mathfrak{a}$ *and* $\mathfrak{b}$ *are parallel* if

(15) $$|\mathfrak{a}| + |\mathfrak{b}| = |\mathfrak{a} + \mathfrak{b}| \quad \text{or} \quad |\mathfrak{a}| + |\mathfrak{b}| = |\mathfrak{a} - \mathfrak{b}|.$$

In the first case we say that parallel vectors $\mathfrak{a}$ and $\mathfrak{b}$ have the *same sense*; in the second case, that they have the *opposite sense*. It is easy to show that formula (15) is equivalent to the identity

$$((\mathfrak{a}) \cdot (\mathfrak{b}))^2 = (\mathfrak{a})^2 \cdot (\mathfrak{b})^2,$$

while this is equivalent to the existence of two numbers λ and μ that do not vanish simultaneously and such that $\lambda \cdot (\mathfrak{a}) = \mu \cdot (\mathfrak{b})$. If $\mathfrak{a}, \mathfrak{b} \neq 0$, this last condition is equivalent to the existence of a number $\varkappa \neq 0$ such that $(\mathfrak{a}) = \varkappa \cdot (\mathfrak{b})$. It is readily seen that if $\varkappa > 0$, then the parallel vectors $\mathfrak{a}$ and $\mathfrak{b}$ have the same sense, and if $\varkappa < 0$, the opposite sense. We refer to the class of all vectors parallel to a given free vector $\mathfrak{a} \neq 0$ as the *direction* of $\mathfrak{a}$.

Let there be given a free vector $\mathfrak{a}$ and a real number λ. By the product $\lambda \cdot \mathfrak{a}$ we understand (*i*) if $\lambda \geqslant 0$, the vector of length $\lambda \cdot |\mathfrak{a}|$ parallel to vector $\mathfrak{a}$ and with the same sense as $\mathfrak{a}$; (ii) if $\lambda < 0$, the vector of length $-\lambda \cdot |\mathfrak{a}|$ parallel to vector $\mathfrak{a}$ and with the sense opposite to the sense of $\mathfrak{a}$. It is readily shown that only the vector $[\lambda \cdot (\mathfrak{a})]$ satisfies these conditions. Therefore

(16) $$\lambda \cdot \mathfrak{a} = [\lambda \cdot (\mathfrak{a})].$$

Scalar product. Let there be given two free vectors $\mathfrak{a}$ and $\mathfrak{b}$. Then

$$|\mathfrak{a} + \mathfrak{b}|^2 = |(\mathfrak{a}) + (\mathfrak{b})|^2 = (\mathfrak{a})^2 + 2 \cdot (\mathfrak{a}) \cdot (\mathfrak{b}) + (\mathfrak{b})^2 = |\mathfrak{a}|^2 + 2 \cdot (\mathfrak{a}) \cdot (\mathfrak{b}) + |\mathfrak{b}|^2,$$

from which we have

(17) $$(\mathfrak{a}) \cdot (\mathfrak{b}) = \tfrac{1}{2} (|\mathfrak{a} + \mathfrak{b}|^2 - |\mathfrak{a}|^2 - |\mathfrak{b}|^2).$$

We call the number $(\mathfrak{a}) \cdot (\mathfrak{b})$ the *scalar product of vectors* $\mathfrak{a}$ and $\mathfrak{b}$ and denote it by $\mathfrak{a} \cdot \mathfrak{b}$.
Thus

(18) $$\mathfrak{a} \cdot \mathfrak{b} = (\mathfrak{a}) \cdot (\mathfrak{b}).$$

It is seen at once from formula (17) that, for any three points $x, y, z \in \mathbf{C}_n$, the scalar product $[y - x] \cdot [z - y]$ is uniquely determined by the distances $\rho(x,y)$, $\rho(y,z)$, and $\rho(x,z)$.

We say that *vectors* $\mathfrak{a}$ *and* $\mathfrak{b}$ *are perpendicular* —in symbols, $\mathfrak{a} \perp \mathfrak{b}$ — if $\mathfrak{a} \cdot \mathfrak{b} = 0$.

Let $\mathfrak{a} = [\alpha_1, \alpha_2, \ldots, \alpha_n]$, $\mathfrak{b} = [\beta_1, \beta_2, \ldots, \beta_n]$ and let λ be any real number. Then formulas (12) and (13) take the form

(2') $$\mathfrak{a} + \mathfrak{b} = [\alpha_1 + \beta_1, \alpha_2 + \beta_2, \ldots, \alpha_n + \beta_n],$$

(3') $$-\mathfrak{a} = [-\alpha_1, -\alpha_2, \ldots, -\alpha_n].$$

We write formula (14) once again:

(4') $$\mathfrak{a} - \mathfrak{b} = \mathfrak{a} + (-\mathfrak{b}).$$

Formulas (16) and (18) take the form

(5') $$\lambda \cdot \mathfrak{a} = [\lambda \alpha_1, \lambda \alpha_2, \ldots, \lambda \alpha_n],$$

(6')
$$a \cdot b = \sum_{i=1}^{n} \alpha_i \beta_i.$$

We put

(7')
$$a^2 = a \cdot a.$$

Then formula (9) takes the form

(8')
$$|a| = \sqrt{a^2}.$$

Comparing these formulas with formulas (2)–(8), by means of which the operations on points were defined, we see that the difference involves only the change in the interpretation of the n-termed sequences, which, instead of points, now denote free vectors. It thus follows that the laws governing operations on vectors do not differ from the laws governing the corresponding operations on points of space C_n. The nature of these operations, however, undergoes a change. With respect to vectors, these operations already have the character of invariant operations.

Let us consider any isometry f mapping the point $x = (x_1, x_2, \ldots, x_n) \in C_n$ onto the point $\bar{x} = (\bar{x}_1, \bar{x}_2, \ldots, \bar{x}_n) \in C_n$. Let $f(0) = a$ and $f(w_i) = a_i$ (see formula (11)), where

(19) $a = (\alpha_{10}, \alpha_{20}, \ldots, \alpha_{n0})$, $a_i = [\alpha_{1i}, \alpha_{2i}, \ldots, \alpha_{ni}]$ for $i = 1, 2, \ldots, n$.

Then vectors $a_1, a_2, \ldots, a_n$ are versors perpendicular to one another, i.e.

(20)
$$a_i \cdot a_j = \delta_i^j.$$

For any point $x = (x_1, x_2, \ldots, x_n) \in C_n$ we have

$$[x-0] = x_1 \cdot w_1 + x_2 \cdot w_2 + \ldots + x_n \cdot w_n,$$

from which it follows that

$$[\bar{x}-a] = x \cdot a_1 + x_2 \cdot a_2 + \ldots + x_n \cdot a_n,$$

i.e.

(21) $\bar{x} = a + x_1 \cdot (a_1) + x_2 \cdot (a_2) + \ldots + x_n \cdot (a_n),$

i.e.

(22) $\bar{x}_i = \alpha_{i0} + \alpha_{i1} x_1 + \alpha_{i2} x_2 + \ldots + \alpha_{in} x_n$ for $i = 1, 2, \ldots, n.$

By formulas (19) and (20), the matrix $[\alpha_{ij}]$ $(i, j = 1, 2, \ldots, n)$ is orthogonal.
    Conversely, if $\bar{x} = f(x)$ is a transformation given by formula (22), where the matrix $[\alpha_{ij}]$ $(i, j = 1, 2, \ldots, n)$ is orthogonal, then, assuming (19) we have (20) and (21), and for any two points $x, y \in C_n$

$$\varrho(\bar{x},\bar{y})^2 = (\bar{x}-\bar{y})^2 = (\sum_{i=1}^{n}(x_i-y_i) \cdot a_i)^2 = \sum_{i=1}^{n}(x_i-y_i)^2 = \varrho(x,y)^2.$$

Hence f is an isometry.

Thus we see that isometries of space $\mathbf{C}_n$ onto itself are identical with the transformations $\bar{x} = f(x)$ given by formula (22), where the matrix $[\alpha_{i,j}]$ $(i,j = 1,2,\ldots,n)$ is orthogonal.

We shall make use below of the following statement on the existence of isometries:†

STATEMENT 1. *For any three points* $p,q,r \in \mathbf{C}_n$ $(n > 1)$ *there exists an isometry* f *of space* $\mathbf{C}_n$ *onto itself such that*

$$f(p) = (0,0,\ldots,0), \quad f(q) = (a_1,0,0,\ldots,0), \quad f(r) = (b_1,b_2,0,0,\ldots,0),$$

where $a_1 \geqslant 0$ and $b_2 \geqslant 0$.

This isometry f is obtained as the superposition of the *translation* given by the formula

$$g(x) = x-p \quad \text{for} \quad x \in \mathbf{C}_n$$

and several suitably chosen *rotations*, each $\bar{x} = h(x)$ being expressed by formulas of the form

(23)
$$\begin{aligned}
\bar{x}_k &= x_k \cos \alpha - x_l \sin \alpha, \\
\bar{x}_l &= x_k \sin \alpha + x_l \cos \alpha, \\
\bar{x}_i &= x_i \text{ for } i \neq k,l,
\end{aligned}$$

where $0 \leqslant \alpha < 2\pi$ and $1 \leqslant k < l \leqslant n$.

We refer to the subsets of space $\mathbf{C}_n$ which are isometric with space $\mathbf{C}_1$ as *Cartesian lines* or briefly *lines*. A special case of the Cartesian line is the set of points of the form $x = (x_1,0,0,\ldots,0)$. This line will be denoted by $\mathbf{C}_{n,1}$ (for $n > 1$).

We shall now establish some properties of Cartesian lines.

STATEMENT 2. *No proper part of a line is a line.*

PROOF. Since this property of a line is an *internal* property (i.e. independent of the space $\mathbf{C}_n$ in which the line lies), it then suffices to prove the theorem for the line $\mathbf{C}_1$. Hence we have to show that if f is an isometry and $f(\mathbf{C}_1) = X \subset \mathbf{C}_1$, then $X = \mathbf{C}_1$.

† Theorems of the arithmetic of real numbers will be called STATEMENTS to distinguish them from theorems of geometry, which we call THEOREMS.

Let $f(0) = a$. Then $a \in X$. We now take any real number $x \neq a$; let $|x-a| = \lambda$. Thus $x = a + \lambda$ or $x = a - \lambda$. The numbers $a + \lambda$ and $a - \lambda$ are all the points of space $\mathbf{C_1}$ at the distance λ from point a; and the numbers λ and $-\lambda$ are different points of space $\mathbf{C_1}$ at the distance λ from the point 0. Thus either $f(-\lambda) = x$ or $f(\lambda) = x$. Consequently $x \in X$. This concludes the proof.

STATEMENT 3. *For any two distinct points a and b of space $\mathbf{C_n}$ there is just one line passing through them. It is identical with the set L of points of the form*

(24) $x(t) = (1-t) \cdot a + t \cdot b,$

where the parameter t is an arbitrary real number.

PROOF. Consider the mapping of set L onto space $\mathbf{C_1}$ as given by the formula

$$f(x(t)) = \rho(a,b) \cdot t.$$

It is readily shown that the transformation f is an isometry. Indeed, we have

$$|x(t_1) - x(t_2)| = |-(t_1-t_2)\cdot a + (t_1-t_2)\cdot b| = |b-a|\cdot|t_1-t_2| = \rho(a,b)\cdot|t_1-t_2|$$

in $\mathbf{C_n}$, and we have

$$|\rho(a,b) \cdot t_1 - \rho(a,b) \cdot t_2| = \rho(a,b) \cdot |t_1-t_2|$$

in $\mathbf{C_1}$. Hence the set L is a line. This line passes through points a and b, since $a = x(0)$ and $b = x(1)$.

We now take any line K passing through points a and b. There exists an isometry g such that $g(\mathbf{C_1}) = K$. Let $g(\alpha) = a$ and $g(\beta) = b$. We may assume that $\alpha < \beta$. We take any point $p \in K$ distinct from points a and b, and let $p = g(\xi)$. Suppose first that

$$\alpha < \beta < \xi.$$

Then

$$|\beta-\alpha| + |\xi-\beta| = |\xi-\alpha|,$$

from which it follows that

$$|b-a| + |p-b| = |p-a| = |(p-b) + (b-a)|.$$

Hence vectors $[b-a]$ and $[p-b]$ are parallel, and, as a result, there exists a number $\varkappa \neq 0$ such that

$$p-b = \varkappa \cdot (b-a),$$

that is,

$$p = -\varkappa \cdot a + (\varkappa + 1) \cdot b.$$

Therefore $p = x(\varkappa + 1)$ and hence $p \in L$.

In a similar way it can be proved that $p \in L$ for the remaining cases: $\xi < \alpha < \beta$ or $\alpha < \xi < \beta$. Hence $K \subset L$, from which we conclude, by Statement 1, that $K = L$. We have thus proved the statement.

We call formula (24) the *parametric equation of line L*.

STATEMENT 4. *For every line L of space* $\mathbf{C}_n$ $(n > 1)$ *there exists an isometry f (of space* $\mathbf{C}_n$ *onto itself) such that* $f(L) = \mathbf{C}_{n,1}$.

PROOF. Take on line L two distinct points a and b. By Statement 1, there exists an isometry f such that $f(a), f(b) \in \mathbf{C}_{n,1}$. On the other hand, the points $f(a)$, $f(b)$ are distinct and lie on line $f(L)$. Thus, by Statement 3, we have $f(L) = \mathbf{C}_{n,1}$.

Because of Statement 4, to establish a property of the line in space $\mathbf{C}_n$ (*external* property, i.e. property relating the line to the entire space $\mathbf{C}_n$, as well) it is sufficient to prove it for the line $\mathbf{C}_{n,1}$.

Let L be any line. We say that a bound vector $\overrightarrow{xy}$ lies on line L if $x, y \in L$. We say that *a free vector* $\mathfrak{a}$ *is parallel to line L* (or that *line L is parallel to vector* $\mathfrak{a}$)—in symbols, $\mathfrak{a} \| L$—if some representative of vector $\mathfrak{a}$ lies on line L. Therefore vectors parallel to line $\mathbf{C}_{n,1}$ have the form $[\alpha_1, 0, 0, \ldots, 0]$. Taking for L the line $\mathbf{C}_{n,1}$, we may readily prove

STATEMENT 5. *For any line L, any vectors* $\mathfrak{a}$, $\mathfrak{b}$, *and any points a, b in space* $\mathbf{C}_n$: (I) *if* $\mathfrak{a} \| L$, $a \in L$, *and* $\overrightarrow{ab}$ *is a representative of vector* $\mathfrak{a}$, *then* $b \in L$; (II) *if* $\mathfrak{a} \| L$ *and* $\mathfrak{b} \| L$, *then* $\mathfrak{a} \| \mathfrak{b}$; (III) *if* $\mathfrak{a} \neq 0$, $\mathfrak{a} \| L$, *and* $\mathfrak{b} \| \mathfrak{a}$, *then* $\mathfrak{b} \| L$.

The direction of all vectors parallel to line L is called the *direction of line L*.

For any point a and any vector $\mathfrak{a} \neq 0$ in space $\mathbf{C}_n$, the line L determined by points a and $a + (\mathfrak{a})$ is the only line which passes through point a and is parallel to vector $\mathfrak{a}$. Line L has the parametric equation

$$x(t) = (1-t) \cdot a + t \cdot (a + (\mathfrak{a})),$$

that is

(25) $$x(t) = a + t \cdot (\mathfrak{a}).$$

This is another form of the parametric equation of a line.

We say *that vector* $\mathfrak{a}$ *is perpendicular to line L*—in symbols, $\mathfrak{a} \perp L$—if it is perpendicular to every vector parallel to line L. We shall now limit ourselves to space $\mathbf{C}_2$. Vectors perpendicular to line $\mathbf{C}_{2,1}$ have the form $[0, \alpha_2]$. Taking line $\mathbf{C}_{2,1}$ for line L, we may readily prove

STATEMENT 6. *For any line L and for any vectors $\mathfrak{a}$ and $\mathfrak{b}$ in space $\mathbf{C}_2$:*
(I) *if $\mathfrak{a} \neq 0$, $\mathfrak{a} \perp L$, and $\mathfrak{b} \perp \mathfrak{a}$, then $\mathfrak{b} \| L$;* (II) *all vectors perpendicular to line L form one direction of vectors.*

This direction we call the *direction perpendicular to line L.*

For any point a and any vector $\mathfrak{a} \neq 0$ in space $\mathbf{C}_2$ there exists just one line L passing through point a and perpendicular to vector $\mathfrak{a}$ (since for the point $a = (0,0)$ and for the vector $\mathfrak{a}=[0,\alpha_2]$ the only such line is the line $L=\mathbf{C}_{2,1}$). As may easily be shown, this line consists of all points $x = (x_1,x_2)$ satisfying the condition $\mathfrak{a} \cdot [x-a] = 0$ which is equivalent to the condition

(26) $$\alpha_0 + \alpha_1 x_1 + \alpha_2 x_2 = 0,$$

where $\alpha_0 = -(\mathfrak{a}) \cdot a$ and $[\alpha_1,\alpha_2] = \mathfrak{a}$. We call formula (26) the *linear equation* of line $L \subset \mathbf{C}_2$. Conversely, every equation of this form (for $\mathfrak{a} = [\alpha_1,\alpha_2] \neq 0$) is the equation of some line in space $\mathbf{C}_2$.

It is readily shown that the two equations

$$\alpha_0 + \alpha_1 x_1 + \alpha_2 x_2 = 0 \text{ and } \beta_0 + \beta_1 x_1 + \beta_2 x_2 = 0,$$

where

$$\mathfrak{a} = [\alpha_1,\alpha_2] \neq 0 \text{ and } \mathfrak{b} = [\beta_1,\beta_2] \neq 0,$$

represent one and the same line if and only if the triples of numbers $(\alpha_0, \alpha_1, \alpha_2)$ and $(\beta_0, \beta_1, \beta_2)$ are proportional.

Line $\mathbf{C}_{2,1}$ has the linear equation

(27) $$x_2 = 0.$$

Let us take any line L with linear equation (26). We assume that $L \neq \mathbf{C}_{2,1}$. Then at least one of the coefficients α_0,α_1 is different from zero, and therefore the system of equations (26) and (27) has a solution for $x = (x_1,x_2)$ if and only if $\alpha_1 \neq 0$, that is, if and only if the direction of line L is different from the direction of line $\mathbf{C}_{2,1}$. We therefore have a general statement as follows:

STATEMENT 7. *Two distinct lines K and L in space $\mathbf{C}_2$ have a common point if and only if they have different directions.*

We fix in space $\mathbf{C}_2$ any line L. For any point $x \in \mathbf{C}_2$ let K_x denote the line passing through point x and perpendicular to line L, and let $c_x = K_x \cap L$. By the *symmetry* with respect to line L we understand the transformation mapping point $x \in \mathbf{C}_2$ onto the point $\bar{x} \in \mathbf{C}_2$ defined by the condition

(28) $$c_x = \tfrac{1}{2} \cdot (x + \bar{x}).$$

If $x \in L$, then $c_x = x$, and hence $\bar{x} = x$. Therefore the symmetry with respect to line L coincides on line L with the identity transformation. Condition (28) is equivalent to the condition

$$\rho(x,c_x) = \rho(c_x,\bar{x}) = \tfrac{1}{2}\,\rho(x,\bar{x})$$

(see the discussion on page 199), from which it follows that

(29) $$\rho(x,c_x) + \rho(\bar{x},c_x) = \rho(x,\bar{x}).$$

The symmetry $\bar{x} = f_0(x)$ with respect to line $\mathbf{C}_{2,1}$ is expressed, as readily shown, by the formulas

$$\bar{x}_1 = x_1,$$
$$\bar{x}_2 = -x_2;$$

therefore it is an isometry. Hence the symmetry with respect to any line L is an isometry.

Subsets of space $\mathbf{C}_n$ isometric with space $\mathbf{C}_2$ will be called *Cartesian planes* or briefly *planes*. A special case of the Cartesian plane is thus the set of points of the form $x = (x_1,x_2,0,\dots,0)$. We shall denote this plane by $\mathbf{C}_{n,2}$ (for $n > 2$).

We shall now establish several properties of the Cartesian plane.

STATEMENT 8. *If points a and b are distinct and lie on a plane P, then the entire line L determined by points a and b is included in P.*

PROOF. From the definition of the plane, it follows that there exists an isometry f such that $P = f(\mathbf{C}_2)$. In particular, $a = f(a')$, $b = f(b')$, where $a',b' \in \mathbf{C}_2$. By Statement 3, a line $L' \subset \mathbf{C}_2$ passes through points a' and b'. The set $f(L') \subset P$ is the line passing through points a and b.

STATEMENT 9. *On every plane there are three non-collinear points.*

PROOF. By Statement 8, this is an internal property of the plane and, consequently, it is sufficient to carry out the proof for the plane $\mathbf{C}_2$. We take in $\mathbf{C}_2$ the points $a = (0,0)$, $b = (1,0)$, $c = (0,1)$. Points a and b determine the line $\mathbf{C}_{2,1}$, which does not contain point c. Therefore points a, b, c, are non-collinear.

STATEMENT 10. *If on a plane P three non-collinear points a, b, c are given, then every line $K \subset P$ intersects at least one of the lines L_{ab} or L_{ac} determined by points a and b or a and c, respectively.*

PROOF. We may assume that $P = \mathbf{C}_2$ and that $a = (0,0)$, $b = (b_1,0)$. Then line L_{ab} has the equation

$$x_2 = 0,$$

and line L_{ac} has an equation of the form

$$x_1 = \beta x_2.$$

If line K does not intersect line L_{ab} then it has an equation of the form

$$x_2 = \gamma$$

and, consequently, it intersects line L_{ac} in the point $(\beta\gamma,\gamma)$.

STATEMENT 11. *For every three points a, b, c of space $\mathbf{C}_n$ (where $n \geqslant 2$) there is a plane passing through them; if points a, b, c are non-collinear there is only one such plane and it coincides with the set P of all points of the form*

$$x(\lambda,\mu,\nu) = \lambda \cdot a + \mu \cdot b + \nu \cdot c \quad \text{for } \lambda + \mu + \nu = 1.$$

PROOF. By Statement 1, there exists an isometry f (of space $\mathbf{C}_n$ onto itself) such that $f(a), f(b), f(c) \in \mathbf{C}_{3,2}$. Then the plane $f^{-1}(\mathbf{C}_{3,2})$ contains the points a, b, c.

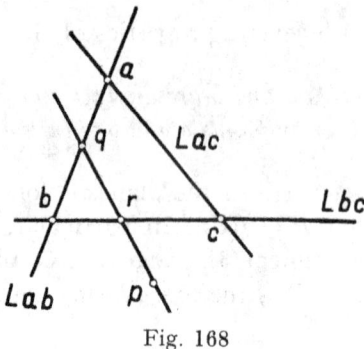

Fig. 168

We assume now that points a, b, c are non-collinear. Let Q be any plane passing through points a, b, c; and, for arbitrary distinct points $p,q \in Q$, let L_{pq} be the line determined by points p and q.

Let us take any point $x(\lambda,\mu,\nu)$ of set P. At least one of the numbers λ,μ,ν, say the number ν, is different from 1. We may then write

$$x(\lambda,\mu,\nu) = (1-\nu)\left(\frac{\lambda}{1-\nu} \cdot a + \frac{\mu}{1-\nu} \cdot b\right) + \nu \cdot c,$$

and, by Statement 8,

$$q = \frac{\lambda}{1-\nu} \cdot a + \frac{\mu}{1-\nu} \cdot b \in L_{ab} \subset Q$$

and

$$x(\lambda,\mu,\nu) = (1-\nu) \cdot q + \nu \cdot c \in L_{qc} \subset Q.$$

We have thus shown that $P \subset Q$.

Let us now take any point $p \in Q$. Point p does not lie on at least one of the lines L_{ab}, L_{bc}, L_{ac}, say $p \sim \in L_{ab}$ (Fig. 168). We pick on line L_{ab} any point $q \neq a,b$. By Statement 10, the line L_{pq} intersects at least one of the two lines L_{bc} and L_{ac}, say line L_{bc}, in some point r. Obviously $r \neq q$. Thus, for some real numbers t, u, v,

$$q = (1-t) \cdot a + t \cdot b,$$
$$r = (1-u) \cdot b + u \cdot c,$$

$$p = (1-v) \cdot q + v \cdot r = (1-v)(1-t) \cdot a + ((1-v)t + v(1-u)) \cdot b + vu \cdot c.$$

Since

$$(1-v)(1-t) + (1-v)t + v(1-u) + vu = 1,$$

then $p \in P$. We have thus shown that $Q \subset P$, which, together with $P \subset Q$, gives $P = Q$. This completes the proof.

STATEMENT 12. *For every plane P of space $\mathbf{C}_n$ ($n > 2$) there exists an isometry f (of space $\mathbf{C}_n$ onto itself) such that $f(P) = \mathbf{C}_{n,2}$.*

PROOF. By Statement 9, three non-collinear points a, b, c may be found on plane P. By Statement 1, there exists an isometry f such that $f(a), f(b), f(c) \in \mathbf{C}_{n,2}$. Points $f(a), f(b), f(c)$ are non-collinear and lie in the plane $f(P)$. Hence $f(P) = \mathbf{C}_{n,2}$.

By Statement 12, to establish a property of the plane in $\mathbf{C}_n$ it suffices to prove it for the plane $\mathbf{C}_{n,2}$.

Let P be any plane. In exactly the same way as in the case of the line, we introduce the notion of a *free vector* $\mathfrak{a}$ *parallel* or *perpendicular to plane* P (in symbols, $\mathfrak{a}||P$ or $\mathfrak{a} \perp P$). Vectors parallel to plane $\mathbf{C}_{n,2}$ have the form $[\alpha_1, \alpha_2, 0, 0, \ldots, 0]$. Taking for plane P the plane $\mathbf{C}_{n,2}$, we may readily prove, for any plane $P \subset \mathbf{C}_n$, statements analogous to statements 5(I) and 5(III), which have been proved for a line $L \subset \mathbf{C}_n$. Hence, the directions of vectors parallel to plane P consist of vectors parallel to plane P; we shall call them the *directions parallel to plane P*. We now limit ourselves to the space $\mathbf{C}_3$. Vectors perpendicular to plane $\mathbf{C}_{3,2}$ have the form $[0, 0, \alpha_3]$. Taking for plane P the plane $\mathbf{C}_{3,2}$ we readily prove, for any plane $P \subset \mathbf{C}_3$, statements analogous to statements 6(I) and 6(II), which have been proved for a line $L \subset \mathbf{C}_2$. Hence all vectors perpendicular to plane P form one direction of vectors, which we shall call the *direction perpen-*

dicular to plane P, and the directions perpendicular to this direction coincide with the directions parallel to plane P.

Also, in exactly the same way as in the case of a line $L \subset \mathbf{C}_2$, we deduce the linear equation of a plane $P \subset \mathbf{C}_3$ passing through a point a and perpendicular to a vector $\mathfrak{a} = [\alpha_1, \alpha_2, \alpha_3] \neq 0$:

(30) $\alpha_0 + \alpha_1 x_1 + \alpha_2 x_2 + \alpha_3 x_3 = 0,$

where $\alpha_0 = -a \cdot (\mathfrak{a})$. Conversely, every equation of this form (for $\mathfrak{a} = [\alpha_1, \alpha_2, \alpha_3] \neq 0$) is the equation of some plane of space $\mathbf{C}_3$. It is readily shown that the two equations

$$\alpha_0 + \alpha_1 x_1 + \alpha_2 x_2 + \alpha_3 x_3 = 0 \quad \text{and} \quad \beta_0 + \beta_1 x_1 + \beta_2 x_2 + \beta_3 x_3 = 0,$$

where $\mathfrak{a} = [\alpha_1, \alpha_2, \alpha_3] \neq 0$ and $\mathfrak{b} = [\beta_1, \beta_2, \beta_3] \neq 0$, represent one and the same plane if and only if the quadruples of numbers $(\alpha_0, \alpha_1, \alpha_2, \alpha_3)$ and $(\beta_0, \beta_1, \beta_2, \beta_3)$ are proportional.

Consider four points: $a = (a_1, a_2, a_3), b = (b_1, b_2, b_3), c = (c_1, c_2, c_3), d = (d_1, d_2, d_3)$ in space $\mathbf{C}_3$. These points lie in one plane if and only if there exist real numbers $\alpha_0, \alpha_1, \alpha_2, \alpha_3$ such that at least one of the numbers $\alpha_1, \alpha_2, \alpha_3$ is distinct from 0 and equation (30) is satisfied by each of the points a, b, c, d. This is equivalent to the condition

(31)
$$\begin{vmatrix} 1 & a_1 & a_2 & a_3 \\ 1 & b_1 & b_2 & b_3 \\ 1 & c_1 & c_2 & c_3 \\ 1 & d_1 & d_2 & d_3 \end{vmatrix} = 0.$$

From the above we immediately obtain the following equation of the plane determined by three non-collinear points a, b, c:

(32)
$$\begin{vmatrix} 1 & x_1 & x_2 & x_3 \\ 1 & a_1 & a_2 & a_3 \\ 1 & b_1 & b_2 & b_3 \\ 1 & c_1 & c_2 & c_3 \end{vmatrix} = 0.$$

In conclusion we shall investigate the intersections of two lines and of two planes.

STATEMENT 13. *Two distinct lines K and L on a plane P in space $\mathbf{C}_3$ have a common point if and only if they have different directions.*

PROOF. It is sufficient to examine the case $P = \mathbf{C}_{3,2}$ and $K = \mathbf{C}_{3,1}$. The proof is then similar to the proof of Statement 7.

STATEMENT 14. *Two distinct planes P and Q of space $\mathbf{C}_3$ have a common point if and only if the directions perpendicular to them are different.*

PROOF. It is sufficient to examine the case $P = \mathbf{C}_{3,2}$. The proof is then similar to the proof of Statement 7.

STATEMENT 15. *If two distinct planes P and Q of space $\mathbf{C}_3$ have a point in common, then their intersection is a line L; and every line L of space $\mathbf{C}_3$ can be represented as the intersection of two distinct planes $P,Q \subset \mathbf{C}_3$.*

PROOF. We may assume that $P = \mathbf{C}_{3,2}$. Then plane P has the equation

$$x_3 = 0,$$

and plane Q has an equation of the form

$$\alpha_0 + \alpha_1 x_1 + \alpha_2 x_2 + \alpha_3 x_3 = 0,$$

where at least one of the two numbers α_1 and α_2 does not vanish. The intersection $P \cap Q$ is the set of points $(x_1,x_2,x_3) \in \mathbf{C}_3$ defined by the condition

$$\alpha_0 + \alpha_1 x_1 + \alpha_2 x_2 = 0 \quad \text{and} \quad x_3 = 0,$$

which is isometric with the set of points $(x_1,x_2) \in \mathbf{C}_2$ defined by the condition

$$\alpha_0 + \alpha_1 x_1 + \alpha_2 x_2 = 0.$$

This is a line in $\mathbf{C}_2$. Therefore $L = P \cap Q$ is a line in $\mathbf{C}_3$.

In order to prove the second part of the theorem it is sufficient to note that line $L = \mathbf{C}_{3,1}$ is the intersection of planes P and Q respectively defined by the equations $x_2 = 0$ and $x_3 = 0$.

3. The Cartesian Model. Consistency of Absolute Geometry

As we have already mentioned, the Cartesian model for the axiom system (GA$_3$) is constructed from the notions of analytical geometry of three-dimensional Cartesian space $\mathbf{C}_3$. From the name CARTESIUS (latinized form of DESCARTES), we shall call it the *model* (C). By ρ we shall denote, as in Section 2, the distance between two points of space $\mathbf{C}_3$. We define C-*points*, C-*lines*, and C-*planes*, respectively, as points, lines, and planes of space $\mathbf{C}_3$.

For any arbitrary C-points a, b, c, we put

$$\mathbf{B}_\mathrm{C}(a,b,c)$$

if and only if

(1)   C-points a, b, c, are distinct and $\rho(a,b) + \rho(b,c) = \rho(a,c)$.

For any arbitrary C-points a, b, c, d, we put

(2) $\mathbf{E}_\mathrm{C}(a,b;c,d)$ if and only if $\rho(a,b) = \rho(c,d)$.

We shall now prove†

PROPOSITION 1. *The interpretation*

C)   C-*points*, C-*lines*, C-*planes*, $\mathbf{B}_C$, $\mathbf{E}_C$

is a model for the axiom system (GA₃) *and satisfies the sentence* E.

PROOF. From the discussion of the previous section, it readily follows that the interpretation (C) satisfies all the axioms of incidence (page 21). Indeed, from the definition of the Cartesian line it follows at once that interpretation (C) satisfies Axiom I1, and from Statements 3, 9, 11, 8, 15 it follows, in turn, that this interpretation also satisfies Axioms I2–I8. In order to verify Axiom I9 it is sufficient to note that the C-points $a = (0,0,0)$, $b = (1,0,0)$, $c = (0,1,0)$, $d = (0,0,1)$ are not C-coplanar, since they do not satisfy condition (31) on page 210.

We now pass on to the linear axioms of order (page 26). We shall begin with Axiom O1. We assume that $\mathbf{B}_C(a,b,c)$. Then (1) holds; therefore

$$|b-a| + |c-b| = |(c-b) + (b-a)|,$$

i.e. $[c-b]\|[b-a]$. Let us denote by L the C-line determined by C-points a and b. We have $[b-a]\|L$, and, by Statement 5(III), we also have $[c-b]\|L$, from which it follows, by Statement 5(I), that $c \in L$. Hence C-points a, b, c are C-collinear. We have thus shown that interpretation (C) satisfies Axiom O1.

Let L be any C-line. Take any of its parametric equations

(3) $x(t) = a + t \cdot (\mathfrak{a})$.

Then

(4) $\rho[x(t_1),x(t_2)] = |t_1-t_2| \cdot |\mathfrak{a}|$.

Further, let $p = x(t_p)$, $q = x(t_q)$, and $r = x(t_r)$ be three distinct C-points of C-line L. By (4), we then have $\mathbf{B}_C(p,q,r)$ if and only if

$$|t_p-t_q| + |t_q-t_r| = |t_p-t_r|,$$

which, as the calculations readily show, is equivalent to t_q being greater than one of the two numbers t_p and t_r, and smaller than the other. Thus, for any C-points $p,q,r \in L$, we have

(5) $\mathbf{B}_C(p,q,r)$ if and only if $t_p < t_q < t_r$ or $t_p > t_q > t_r$.

---

† Theorems on geometry (belonging to metageometry) will be called PRO-POSITIONS.

Formula (5) allows us to verify Axioms O2–O8 in a simple way. Indeed, taking as L the C-line determined by the C-points a and b and putting in formula (3) the vector $[b-a]$ for $\mathfrak{a}$, we have $t_a = 0$, $t_b = 1$, and hence $t_a < t_b$. Thus, that interpretation (C) satisfies Axioms O2–O8 follows, in turn, from the following elementary properties of the relation *less than* for real numbers:

(I)   If $t_a < t_b < t_c$, then $t_c > t_b > t_a$.

(II) If $t_a < t_b < t_c$, then neither $t_b < t_c < t_a$ nor $t_c < t_a < t_b$.

(III) If $t_a < t_b$, then for any number $t_c \neq t_a, t_b$ it follows that $t_c < t_a < t_b$ or $t_a < t_c < t_b$ or $t_a < t_b < t_c$.

(IV) If $t_a < t_b$, then there exists a number t_c such that $t_a < t_b < t_c$.

(V)   If $t_a < t_b$, then there exists a number t_c such that $t_a < t_c < t_b$.

(VI) If $t_a < t_b < t_c$ and $t_b < t_c < t_d$, then $t_a < t_b < t_d$.

(VII) If $t_a < t_b < t_d$ and $t_b < t_c < t_d$, then $t_a < t_b < t_c$.

The further axioms of absolute geometry involve the following defined notions: half-lines, the betweenness relation **B** for a line and two points (see page 41), and half-planes. We shall now consider the interpretation of these notions in turn. Obviously, this interpretation is uniquely determined by the interpretation (C) of the primitive notions.

We return to a C-line L given by parametric equation (3). Taking in formula (5) for q the C-point a, we see that, since $t_a = 0$, for any C-points $p, r \in L - a$, we have

$$\mathbf{B}_C(p, a, r) \quad \text{if and only if} \quad t_p t_r < 0.$$

It thus follows at once that C-half-lines A and A^* determined on C-line L by C-point a have the parametric equations

(6) $$x(t) = a + t \cdot (\mathfrak{a}), \quad \text{where } t > 0,$$

and

$$x(t) = a + t \cdot (\mathfrak{a}), \quad \text{where } t < 0.$$

In particular, let us take for L the C-line $\mathbf{C}_{3,1}$. Putting $a = (0,0,0)$ and $\mathfrak{a} = [1,0,0]$ in formula (6), we see that the set of C-points

$$(t, 0, 0) \quad \text{for} \quad t > 0$$

is one of the C-half-lines of C-line $\mathbf{C}_{3,1}$ with C-origin $(0,0,0)$. We shall denote this C-half-line by $\mathbf{C}_{3,1}^+$.

We shall now consider the formula $\mathbf{B}_C(a, K, c)$. We shall need the corresponding analytical formula only when C-points a and c and C-line K lie in C-plane $\mathbf{C}_{3,2}$ and moreover C-line K coincides with C-line $\mathbf{C}_{3,1}$. Thus, let $K = \mathbf{C}_{3,1}$. Take two distinct C-points $a = (a_1,a_2,0)$ and $c = (c_1,c_2,0)$ belonging to the set $\mathbf{C}_{3,2}$—$\mathbf{C}_{3,1}$. Putting, in formula (3), $\mathfrak{a} = [c{-}a]$, we obtain a parametric equation of the C-line L determined by C-points a and c. Then $a = x(0)$, $c = x(1)$. Let $x(t) = (x_1(t),x_2(t),x_3(t))$. Then $x_3(t) = 0$ for every t and

$$x_2(0) = a_2 \neq 0 \quad \text{and} \quad x_2(1) = c_2 \neq 0.$$

The term $x_2(t)$ is a polynomial of at most the first degree with respect to t. If, therefore, $x_2(0) \cdot x_2(1) > 0$, then for every number t of the interval $(0,1)$ we have $x_2(t) \neq 0$, from which it follows, by formula (5), that no C-point of C-segment (ac) lies on C-line $\mathbf{C}_{3,1}$. If, however, $x_2(0) \cdot x_2(1) < 0$, then, for some number t_0 of the interval $(0,1)$, it must be that $x_2(t_0) = 0$. Thus C-point $x(t_0)$ lies on C-line $\mathbf{C}_{3,1}$ and at the same time, because of formula (5), it belongs to C-segment (ac). Hence, we have shown that for any C-points $a,c \in \mathbf{C}_{3,2}$ we have

(7) $\mathbf{B}_\mathrm{C}(a,\mathbf{C}_{3,1},c)$ if and only if $a_2 \cdot c_2 < 0$.

It thus follows immediately that the set of C-points

$$(x_1,x_2,0) \quad \text{for} \quad x_2 > 0$$

is one of the C-half-planes of C-plane $\mathbf{C}_{3,2}$ with C-boundary $\mathbf{C}_{3,2}$. We shall denote this C-half-plane by $\mathbf{C}_{3,2}^+$.

We now return to the verification of the axioms. Axiom O9 is the only axiom of order (page 42) that still remains to be verified. All notions of interpretation (C) were defined invariantly with respect to the group of isometries (of space $\mathbf{C}_3$ onto itself). We can therefore assume that $P = \mathbf{C}_{3,2}$, $L = \mathbf{C}_{3,1}$, $a = (a_1,a_2,0)$, $b = (b_1,b_2,0)$, $c = (c_1,c_2,0)$ and $c_2 > 0$ (see Statement 1). Let us assume that $\mathbf{B}_\mathrm{C}(a,\mathbf{C}_{3,1},b)$. Then, by equivalence (7), we have $a_2 \cdot b_2 < 0$. Thus

$$a_2 \cdot c_2 < 0 \quad \text{or} \quad b_2 \cdot c_2 < 0,$$

from which it follows, by equivalence (7), that

$$\mathbf{B}_\mathrm{C}(a,\mathbf{C}_{3,1},c) \quad \text{or} \quad \mathbf{B}_\mathrm{C}(b,\mathbf{C}_{3,1},c).$$

We have thus shown that interpretation (C) satisfies the plane axiom of order.

We now go on to the axioms of congruence (pages 80–82). The fact that Axioms C1 and C2 are satisfied by interpretation (C) results immediately from the following basic properties of the distance ρ:

$$\rho(x,y) = 0 \text{ if and only if } x = y$$

and

$$\rho(x,y) = \rho(y,x).$$

Axiom C3 is verified at once, since if $\rho(a,b) = \rho(p,q)$ and $\rho(a,b) = \rho(r,s)$, then $\rho(p,q) = \rho(r,s)$.

In the remaining axioms of congruence the defined relation $\equiv$ appears. From the interpretation of the primitive relation $\mathbf{E}$ it follows that for any C-segments ab and cd we have

$$ab \equiv_c cd \quad \text{if and only if} \quad \rho(a,b) = \rho(c,d).$$

Interpretation (C) satisfies Axiom C4, since if $\rho(a_1,b_1) + \rho(b_1,c_1) = \rho(a_1,c_1)$, $\rho(a_2,b_2) + \rho(b_2,c_2) = \rho(a_2,c_2)$, $\rho(a_1,b_1) = \rho(a_2,b_2)$, and $\rho(b_1,c_1) = \rho(b_2,c_2)$, then $\rho(a_1,c_1) = \rho(a_2,c_2)$.

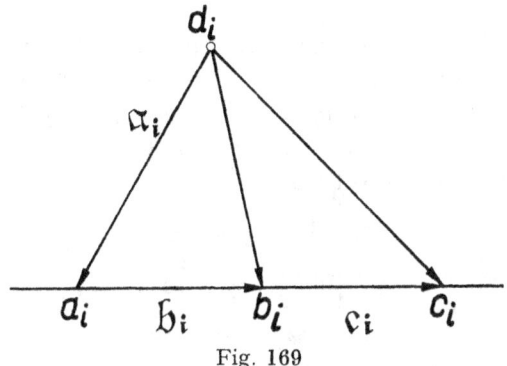

Fig. 169

Axiom C5 may also be checked without difficulty. Indeed, we may assume that $a = (0,0,0)$ and $A = \mathbf{C}_{3,1}^{+}$. Then C-point $b = (\rho(p,q),0,0)$ is the only C-point of C-half-line $\mathbf{C}_{3,1}^{+}$ such that $ab \equiv pq$.

We now come to Axiom C6. For $i = 1,2$ we put (Fig. 169)

$$\mathfrak{a}_i = [a_i - d_i], \quad \mathfrak{b}_i = [b_i - a_i], \quad \mathfrak{c}_i = [c_i - b_i].$$

Then

$$\mathfrak{a}_i + \mathfrak{b}_i = [b_i - d_i], \quad \mathfrak{b}_i + \mathfrak{c}_i = [c_i - a_i], \quad \mathfrak{a}_i + \mathfrak{b}_i + \mathfrak{c}_i = [c_i - d_i],$$

and by our hypothesis

$$|\mathfrak{a}_1| = |\mathfrak{a}_2|, \ |\mathfrak{b}_1| = |\mathfrak{b}_2|, \ |\mathfrak{c}_1| = |\mathfrak{c}_2|, \ |\mathfrak{a}_1 + \mathfrak{b}_1| = |\mathfrak{a}_2 + \mathfrak{b}_2|,$$

$$|\mathfrak{b}_1| + |\mathfrak{c}_1| = |\mathfrak{b}_1 + \mathfrak{c}_1|, \quad |\mathfrak{b}_2| + |\mathfrak{c}_2| = |\mathfrak{b}_2 + \mathfrak{c}_2|,$$

and

$$\mathfrak{b}_1 + \mathfrak{c}_1 = \lambda_1 \cdot \mathfrak{b}_1, \quad \mathfrak{b}_2 + \mathfrak{c}_2 = \lambda_2 \cdot \mathfrak{b}_2 \quad \text{for some } \lambda_1, \lambda_2 > 0.$$

Therefore

$$|\mathfrak{b}_1 + \mathfrak{c}_1| = |\mathfrak{b}_2 + \mathfrak{c}_2| \quad \text{and} \quad \lambda_1 = \lambda_2.$$

Consequently (see formulas (17) and (18) on page 201),

$$\mathfrak{a}_1 \cdot \mathfrak{b}_1 = \mathfrak{a}_2 \cdot \mathfrak{b}_2,$$

and since

$$a_1 \cdot (b_1 + c_1) = a_1 \cdot (\lambda_1 \cdot b_1) = \lambda_1 \cdot (a_1 \cdot b_1),$$

$$a_2 \cdot (b_2 + c_2) = a_2 \cdot (\lambda_2 \cdot b_2) = \lambda_2 \cdot (a_2 \cdot b_2),$$

then also

$$a_1 \cdot (b_1 + c_1) = a_2 \cdot (b_2 + c_2).$$

From

$$a_1 \cdot (b_1 + c_1) = a_2 \cdot (b_2 + c_2), \quad |a_1| = |a_2|, \quad \text{and} \quad |b_1 + c_1| = |b_2 + c_2|$$

it follows at once that

$$|a_1 + b_1 + c_1| = |a_2 + b_2 + c_2|,$$

i.e. $d_1 c_1 \equiv d_2 c_2$. Therefore interpretation (C) satisfies Axiom C6.

We now come to Axiom C7. It may be assumed that $W = \mathbf{C}_{3,2}^+$, $a = (0,0,0)$, $b = (b_1,0,0)$, and $b_1 > 0$. Let $\mathfrak{a} = [p\text{—}r]$, $\mathfrak{b} = [q\text{—}p]$. Then $\mathfrak{a} + \mathfrak{b} = [q\text{—}r]$ and $b_1 = |\mathfrak{b}|$. We seek a C-point $c = (x_1, x_2, 0)$ whose coordinates satisfy the conditions

(8)
$$x_1^2 + x_2^2 = \mathfrak{a}^2,$$

(9)
$$(x_1 - |\mathfrak{b}|)^2 + x_2^2 = (\mathfrak{a} + \mathfrak{b})^2,$$

(10)
$$x_2 > 0.$$

Subtracting both sides of equation (8) from equation (9), we obtain just one solution for x_1:

(11)
$$x_1 = - \frac{(\mathfrak{a} + \mathfrak{b})^2 - \mathfrak{a}^2 - \mathfrak{b}^2}{2|\mathfrak{b}|} = - \frac{\mathfrak{a} \cdot \mathfrak{b}}{|\mathfrak{b}|}.$$

From (8) and (11) it follows that

(12)
$$x_2^2 = \mathfrak{a}^2 - x_1^2 = \frac{\mathfrak{a}^2 \cdot \mathfrak{b}^2 - (\mathfrak{a} \cdot \mathfrak{b})^2}{\mathfrak{b}^2} > 0,$$

since $\mathfrak{a} \sim \|\mathfrak{b}$ (see (iii) on page 198). Because of the inequality (10), formula (12) gives just one solution for x_2. We have thus shown that interpretation (C) also satisfies the last axiom of congruence.

We now consider the Axiom of Continuity (page 151). Let X_1 and X_2 be two arbitrary non-empty sets of C-points, and let a C-point a satisfy the following condition:

$$p_1 \in X_1 \quad \text{and} \quad p_2 \in X_2 \quad \text{implies} \quad \mathbf{B}_\mathrm{C}(a, p_1, p_2).$$

Then sets X_1 and X_2 are included in a C-half-line A with C-origin a. We take for C-half-line A the parametric equation (6) and we consider two sets of real numbers Θ_1 and Θ_2 defined by the condition

$$t \in \Theta_i \quad \text{if and only if} \quad x(t) \in X_i, \quad \text{for} \quad i = 1,2.$$

By equivalence (5)

$$t_1 \in \Theta_1 \quad \text{and} \quad t_2 \in \Theta_2 \quad \text{implies} \quad 0 < t_1 < t_2,$$

and from the continuity of the set of real numbers it follows that for some real number t_b

$$t_1 \in \Theta_1 \quad \text{and} \quad t_2 \in \Theta_2 \quad \text{implies} \quad t_1 < t_b < t_2.$$

Putting $b = x(t_b)$ we conclude, by equivalence (5), that

$$p_1 \in X_1 \quad \text{and} \quad p_2 \in X_2 \quad \text{implies} \quad \mathbf{B}_C(p_1,b,p_2).$$

We have thus shown that interpretation (C) is a model for the axiom system (GA₃).

We still have to show that interpretation (C) satisfies the sentence E (page 197). We take in C-plane P a C-line L and a C-point $a \sim \in L$. If in C-plane P two distinct C-lines K_1 and K_2 such that $K_i \cap L = 0$ $(i = 1,2)$ were to pass through C-point a, then, by choosing on C-line K_i a C-point $b_i \neq a$, we would obtain three C-non-collinear C-points a, b_1, b_2, and, by Statement 10, C-line L would intersect at least one of the two C-lines K_1 and K_2, in contradiction to our assumption.

This completes the proof of the proposition.

Since the analytical geometry of Cartesian spaces enters into the realm of the arithmetic of real numbers, we obtain, as direct conclusion from Proposition 1:

PROPOSITION 2. *If the arithmetic of real numbers is consistent, then the axiom system* (GA₃) *is consistent.*

NOTE. From Proposition 1 it follows that if to the system of axioms (GA₃) we add sentence E as a new axiom, then the system of axioms of geometry thus extended will also be consistent (under the assumption that the arithmetic of real numbers is consistent).

Thus far we have been considering space geometry. In conclusion we say a few words about plane geometry.

The axiom system (GA₂) of plane absolute geometry (see page 197) is consistent. The proof of the consistency of the system (GA₂) is carried out in the same way as the proof of the consistency of the axiom system (GA₃). Thus the *plane Cartesian interpretation*

(C)   C-*points*, C-*lines*, $\mathbf{B}_C$, $\mathbf{E}_C$,

is constructed, where C-points and C-lines are points and lines of two-dimensional Cartesian space $\mathbf{C}_2$ and for any C-points a, b, c, d the formulas $\mathbf{B}_C(a,b,c)$ and $\mathbf{E}_C(a,b;c,d)$ are defined, respectively, by conditions (1) and

(2) on page 211. Proceeding precisely in the same way as in the proof of Proposition 1, we may prove that plane interpretation (C) is a model for the axiom system (GA_2).

Again, by proceeding in the same way as in the proof of Proposition 1, we may prove that plane interpretation (C) satisfies sentence E (as formulated on page 197, line 14). Hence the axiom system (GA_2) strengthened by sentence E remains a consistent axiom system.

It is readily seen from the considerations on plane model (C) that in the analytical geometry of space C_2 we may employ the theorems of plane geometry which are consequences of the axiom system (GA_2) and sentence E as well as the purely arithmetical theorems. We shall apply this mixed method in Section 5 where we investigate the properties of the circle.

Henceforth, in denoting the notions of the models (C) for (GA_2) and (GA_3) we shall omit the prefix (or index) "C". Thus, instead of C-*line* we shall write simply *line*; instead of "B_C", we shall write simply "B", and so on.

4. Independence of the Axiom of Continuity

In order to show that the Axiom of Continuity is independent, we shall construct an arithmetical interpretation of the system of primitive notions of geometry that will be a model for the axioms of incidence, order, and congruence, but not for the Axiom of Continuity. We obtain this interpretation from interpretation (C) by narrowing space C_3 to some denumerable space $C_3^* \subset C_3$ in such a way that for points of space C_3^* we take only those points of space C_3 whose coordinates belong to some denumerable set X of real numbers. At first glance, it may appear that it suffices to take for X the set of rational numbers. It turns out, however, that when verifying the axioms (e.g. the axiom of congruence C7) we encounter problems leading to equations of the second degree. Owing to this, we take for the set X a set Λ which we define in the following way:

Let us denote by Λ the common part of all sets of real numbers which contain 1 and which are closed with respect to the operations of addition,† subtraction, multiplication, division by a number different from zero, and taking the square root (of non-negative numbers).

The set Λ thus satisfies the following four conditions:

(I) $1 \in \Lambda$;

(II) if $x,y \in \Lambda$, then $x + y$, $x-y$, $x \cdot y \in \Lambda$;

(III) if $x,y \in \Lambda$ and $y \neq 0$, then $x/y \in \Lambda$;

(IV) if $x \in \Lambda$ and $x \geqslant 0$, then $\sqrt{x} \in \Lambda$.

---

† We say that a set X of real numbers is *closed* with respect e.g. to the operation of addition if the sum of two numbers of set X always belongs to set X.

We shall show that the set $\varLambda$ is denumerable. To do this we define a sequence of its subsets

$$(\varLambda_1, \varLambda_2, \ldots, \varLambda_n, \ldots)$$

in the following manner:

$\varLambda_1$ is the set of the rational numbers. Assuming that the set $\varLambda_n$ has already been defined, we define $\varLambda_{n+1}$ as the set composed of all numbers of the form

$$x + y, \quad x-y, \quad x/y \text{ (for } y \neq 0), \text{ and } \sqrt{x} \text{ (for } x \geqslant 0),$$

where $x,y \in \varLambda_n$. By means of induction we readily prove that for every natural n the set $\varLambda_n$ is a denumerable subset of $\varLambda$, and that $1 \in \varLambda_n$, from which, since $x \cdot 1 = x$, we conclude that for $n = 1,2,\ldots$

$$(1) \qquad\qquad\qquad \varLambda_n \subset \varLambda_{n+1}.$$

We shall show that $\varLambda = \overset{\infty}{\underset{n=1}{\cup}} \varLambda_n$. Since $\varLambda_n \subset \varLambda$, it follows that $\overset{\infty}{\underset{n=1}{\cup}} \varLambda_n \subset \varLambda$; it thus remains to be shown that $\varLambda \subset \overset{\infty}{\underset{n=1}{\cup}} \varLambda_n$. We have $1 \in \overset{\infty}{\underset{n=1}{\cup}} \varLambda_n$. Further, if $x,y \in \overset{\infty}{\underset{n=1}{\cup}} \varLambda_n$, then there exist two natural numbers k and l such that $x \in \varLambda_k$ and $y \in \varLambda_l$. Taking as m the larger of the two numbers k and l, we conclude, by formula (1), that $x,y \in \varLambda_m$; hence the numbers

$$x + y, \quad x \cdot y, \; x/y \text{ (for } y \neq 0), \text{ and } \sqrt{x} \text{ (for } x \geqslant 0)$$

belong to the set $\varLambda_{m+1}$ and therefore to the sum $\overset{\infty}{\underset{n=1}{\cup}} \varLambda_n$. Hence the set $\overset{\infty}{\underset{n=1}{\cup}} \varLambda_n$ contains 1 and is closed with respect to the operations of addition subtraction, multiplication, division, and taking the square root. As a result, the set $\varLambda$, as the common part of all sets with these properties, is included in the set $\overset{\infty}{\underset{n=1}{\cup}} \varLambda_n$. Thus, finally, $\varLambda = \overset{\infty}{\underset{n=1}{\cup}} \varLambda_n$, and, since the sets $\varLambda_n$ are denumerable, we have

STATEMENT 16. *The set $\varLambda$ is denumerable.*

We now pass on to Cartesian space $\mathbf{C}_3$. The points $x = (x_1,x_2,x_3) \in \mathbf{C}_3$, where the numbers x_1, x_2, x_3 belong to the set $\varLambda$, will be called $\varLambda$-*points.* By $\varLambda$-*lines* we shall understand those lines of space $\mathbf{C}_3$ which are determined by two distinct $\varLambda$-points; similarly, by $\varLambda$-*planes* we shall understand those planes of space $\mathbf{C}_3$ which are determined by three non-collinear $\varLambda$-points. We shall call segments whose ends are $\varLambda$-points $\varLambda$-*segments*; similarly, triangles whose vertices are $\varLambda$-points will be called

Λ-*triangles.* Finally, by Λ-*vector* we shall understand every free vector whose coordinates belong to the set Λ.

We next discuss the properties of the notions introduced here.

It is readily noted that the set of Λ-points is not a plane set, that is:

STATEMENT 17. *For every plane P, there exists a Λ-point p not lying on P.*

Further, we have

STATEMENT 18. *For any two distinct Λ-points a and b there exists a Λ-point c such that* $\mathbf{B}(a,b,c)$ *and a Λ-point d such that* $\mathbf{B}(a,d,b)$.

PROOF. The line ab has the parametric equation

$$x(t) = a + t \cdot (b-a).$$

Points $c = x(2)$ and $d = x(\frac{1}{2})$ are obviously Λ-points, where $\mathbf{B}(a,b,c)$ and $\mathbf{B}(a,d,b)$.

We now consider Λ-lines and Λ-planes. We shall begin with their analytical characterization:

STATEMENT 19. *A line $L \subset \mathbf{C}_3$ is a Λ-line if and only if L has a parametric equation of the form*

(2) $$x(t) = a + t \cdot (\mathfrak{a}),$$

where a is a Λ-point and $\mathfrak{a}$ is a Λ-vector.

PROOF. If L is a Λ-line, two distinct Λ-points a and b lie on it. Then vector $\mathfrak{a} = [a-b]$ is a Λ-vector and line L has the parametric equation $x(t) = a + t(\mathfrak{a})$. On the other hand, if (2) is a parametric equation of line L, then there are two distinct Λ-points on L, in fact $x(0) = a$ and $x(1) = a + (\mathfrak{a})$. Therefore L is a Λ-line.

STATEMENT 20. *A plane $P \subset \mathbf{C}_3$ is a Λ-plane if and only if P has an equation of the form*

(3) $$\alpha_0 + \alpha_1 x_1 + \alpha_2 x_2 + \alpha_3 x_3 = 0, \quad \text{where } \alpha_0, \alpha_1, \alpha_2, \alpha_3 \in \Lambda.$$

PROOF. If P is a Λ-plane, then there are three non-collinear Λ-points, $a = (a_1, a_2, a_3)$, $b = (b_1, b_2, b_3)$, $c = (c_1, c_2, c_3)$ on P. Then

$$\begin{vmatrix} 1 & x_1 & x_2 & x_3 \\ 1 & a_1 & a_2 & a_3 \\ 1 & b_1 & b_2 & b_3 \\ 1 & c_1 & c_2 & c_3 \end{vmatrix} = 0$$

is the equation of plane P (see formula (32) on page 210). It is readily noted that all coefficients of this equation belong to set Λ.

We next assume that formula (3) is the equation of a plane P. Then at least one of the numbers α_1, α_2, α_3, say α_1, is different from 0. It is easily shown that the points

$$a = \left(-\frac{\alpha_0}{\alpha_1}, 0, 0\right), \quad b = \left(-\frac{\alpha_0 + \alpha_2}{\alpha_1}, 1, 0\right), \quad c = \left(-\frac{\alpha_0 + \alpha_3}{\alpha_1}, 0, 1\right)$$

are $\varLambda$-points, that they all lie on plane P, and that they are non-collinear (since $a, b \in C_{3,2}$ and $c \sim \in C_{3,2}$). Hence plane P is a $\varLambda$-plane, which was to be proved.

We next investigate the intersection of two $\varLambda$-lines and two $\varLambda$-planes.

STATEMENT 21. *The intersection point of two distinct $\varLambda$-lines is a $\varLambda$-point.*

PROOF. Consider any two $\varLambda$-lines K and L. By Statement 19, $\varLambda$-lines K and L have the respective parametric equations

$$x(t) = a + t \cdot (\mathfrak{a}) \quad \text{and} \quad y(u) = b + u \cdot (\mathfrak{b}),$$

where a and b are $\varLambda$-points, $\mathfrak{a}$ and $\mathfrak{b}$ are $\varLambda$-vectors. If K and L intersect in some point c, then the equation

$$x(t) = y(u)$$

has exactly one pair of roots t_c and u_c, where, as readily noted, the numbers t_c and u_c belong to the set $\varLambda$. The point $c = x(t_c) = y(u_c)$ is therefore a $\varLambda$-point, which was to be proved.

STATEMENT 22. *The intersection line of two distinct $\varLambda$-planes is a $\varLambda$-line.*

PROOF. We consider any two $\varLambda$-planes P_1 and P_2. By Statement 20 $\varLambda$-planes P_1 and P_2 have the respective linear equations

(4)
$$\alpha_0 + \alpha_1 x_1 + \alpha_2 x_2 + \alpha_3 x_3 = 0, \quad \text{where} \quad \alpha_0, \alpha_1, \alpha_2, \alpha_3 \in \varLambda,$$
$$\beta_0 + \beta_1 x_1 + \beta_2 x_2 + \beta_3 x_3 = 0, \quad \text{where} \quad \beta_0, \beta_1, \beta_2, \beta_3 \in \varLambda.$$

If P_1 and P_2 intersect along some line L, then the matrix

$$\left\| \begin{matrix} \alpha_1 & \alpha_2 & \alpha_3 \\ \beta_1 & \beta_2 & \beta_3 \end{matrix} \right\|$$

is of the rank two. Let, for example,

$$\left| \begin{matrix} \alpha_1 & \alpha_2 \\ \beta_1 & \beta_2 \end{matrix} \right| \neq 0.$$

Setting $x_3 = 0$ in (4) we find a $\varLambda$-point $a = (a_1, a_2, 0)$ belonging to line L. Similarly, setting $x_3 = 1$ in (4) we find a $\varLambda$-point $b = (b_1, b_2, 1)$ belonging to line L. Therefore L is a $\varLambda$-line.

By the *sphere* S (in space $\mathbf{C}_3$) with *center* a and *radius* λ we understand the set of all points $x \in \mathbf{C}_3$ at the distance λ from point a. If $a = (0,0,0)$, then the sphere S has the equation

$$x_1^2 + x_2^2 + x_3^2 = \lambda^2.$$

We shall now prove

STATEMENT 23. *If* $\lambda \in \Lambda$, *then the intersection points of any* Λ-*line* L *and the sphere* S *given by the equation* $x_1^2 + x_2^2 + x_3^2 = \lambda^2$ *are* Λ-*points.*

PROOF. Assume that $\lambda \in \Lambda$. Λ-line L has an equation $x(t) = a + t \cdot (\mathfrak{a})$, where $a = (a_1, a_2, a_3)$ is a Λ-point, and $\mathfrak{a} = [\alpha_1, \alpha_2, \alpha_3]$ is a Λ-vector. The value of the parameter t for which $x(t) \in S$ is found from the equation

$$(a_1 + t\alpha_1)^2 + (a_2 + t\alpha_2)^2 + (a_3 + t\alpha_3)^2 = \lambda^2,$$

which, after rearranging, takes the form

$$\mathfrak{a}^2 \cdot t^2 \cdot + 2 \left((\mathfrak{a}) \cdot a \right) t + (a^2 - \lambda^2) = 0.$$

If $L \cap S \neq 0$, then this equation has the real roots t_1 and t_2 which, as may readily be shown, belong to the set Λ. Therefore the points $x(t_1)$ and $x(t_2)$ of the intersection of Λ-line L with sphere S are Λ-points.

Finally, we consider Λ-segments and Λ-triangles.

STATEMENT 24. *Let* pq *be a* Λ-*segment. If a segment* ab *lies on a* Λ-*line* L, *and* a *is a* Λ-*point, and* $ab \equiv pq$, *then* b *is also a* Λ-*point.*

PROOF. Λ-line L has a parametric equation

$$x(t) = a + t(\mathfrak{a}),$$

where $\mathfrak{a}$ is a Λ-vector. The value of the parameter t for which $x(t) = b$ is one of the roots of the equation

$$\mathfrak{a}^2 t^2 = (q - p)^2.$$

The roots of this equation belong to the set Λ and therefore point b is a Λ-point.

STATEMENT 25. *Let* pqr *be a* Λ-*triangle. If a triangle* abc *lies on a* Λ-*plane* P, *and* ab *is a* Λ-*segment, and* $ab \equiv pq$, $ac \equiv pr$, $bc \equiv qr$, *then* c *is a* Λ-*point·*

PROOF. We have $c = a + (c - a)$; it thus suffices to show that vector $[c - a]$ is a Λ-vector. We set $[c - a] = [x_1, x_2, x_3]$, $[a - b] = [\beta_1, \beta_2, \beta_3]$ and denote by σ the scalar product of the vectors $[r-p]$ and $[p-q]$. Then $\beta_1, \beta_2, \beta_3,\ \sigma \in \Lambda$. Further, let

$$\alpha_0 + \alpha_1 x_1 + \alpha_2 x_2 + \alpha_3 x_3 = 0, \text{ where } \alpha_0, \alpha_1, \alpha_2, \alpha_3 \in \Lambda,$$

be an equation of Λ-plane P. We then have the following equations:

(5) $$\alpha_1 x_1 + \alpha_2 x_2 + \alpha_3 x_3 = 0,$$

(6) $$\beta_1 x_1 + \beta_2 x_2 + \beta_3 x_3 = \sigma,$$

(7) $$x_1^2 + x_2^2 + x_3^2 = (r-p)^2.$$

By Statement (22), the intersection of the Λ-planes given by equations (5) and (6) is some Λ-line L; and the intersection points of L and the sphere given by equation (7) are, by Statement 23, Λ-points. Therefore the vector $[c-a] = [x_1, x_2, x_3]$ is a Λ-vector. This is what we had to prove.

We now come to the main purpose of our considerations, that is, the construction of a model for the axioms of incidence, order, and congruence which does not satisfy the Axiom of Continuity.

For any figure $F \subset \mathbf{C}_3$ we denote by F^* the set of all Λ-points belonging to figure F. We define: *-*points* as points of the set $\mathbf{C}_3^*$, that is, simply as Λ-points; *-*lines* as the sets L^*, where L is any Λ-line; *-*planes* as the sets P^*, where P is any Λ-plane.

The *-*betweenness* relation $\mathbf{B}^*$ is defined as the betweenness relation $\mathbf{B}$ restricted to *-points; hence, for any *-points a, b, c, we have

(8) $$\mathbf{B}^*(a,b,c) \text{ if and only if } \mathbf{B}(a,b,c).$$

Similarly, we define the *-*equidistance* relation $\mathbf{E}^*$ as the equidistance relation $\mathbf{E}$ restricted to *-points; hence, for any *-points a, b, c, d, we have

(9) $$\mathbf{E}^*(a,b,c,d) \text{ if and only if } \mathbf{E}(a,b,c,d).$$

The interpretation

(C*) $$\text{*-points, *-lines, *-planes, } \mathbf{B}^*, \mathbf{E}^*$$

of the primitive notions determines the interpretation of the defined notions. At this moment we are interested in the defined notions which enter into the axioms. We shall examine the relations which hold between their interpretations in (C) and (C*).

With the help of Statement 17, we may readily prove

STATEMENT 26. *Any *-points a, b, c are *-collinear if and only if they are collinear, and any *-points a, b, c, d are *-coplanar if and only if they are coplanar.*

Further, by means of Statement 21 and equivalence (8), we obtain

STATEMENT 27. *For any Λ-line K and for any Λ-points a and c,*

$$(10) \qquad\qquad \mathbf{B}^*(a,K,c) \text{ if and only if } \mathbf{B}(a,K,c).$$

We shall call every half-line determined on a Λ-line by a Λ-point, a Λ-*half-line*, and every half-plane determined on a Λ-plane by a Λ-line a Λ-*half-plane*. On the basis of equivalences (8) and (10) and the definitions of the half-line and half-plane (see Sections 12 and 19, Chapter I), we may readily prove (the second part, by means of Statement 21)

STATEMENT 28. *The $*$-half-lines are identical with the sets A^*, where A is any Λ-half-line, the $*$-origin of $*$-half-line A^* coinciding with the origin of half-line A; the $*$-half-planes are identical with the sets W^*, where W is any Λ-plane, and if line K is the boundary of half-plane W, then $*$-line K^* is the $*$-boundary of $*$-half-plane W^*.*

Further, from equivalence (9), we obtain

STATEMENT 29. *Given two $*$-segments ab and cd, then $ab \equiv^* cd$ if and only if $ab \equiv cd$.*

Finally, we obviously have

STATEMENT 30. *Given any Λ-line L and any Λ-plane P, then $L^* \subset P^*$ if and only if $L \subset P$.*

It may readily be shown that interpretation (C*) is a model for all the axioms of incidence, order and congruence. This follows at once from the fact that interpretation (C) is a model for these axioms, by Statements 17, 22, 26 (required for verifying the axioms of incidence), Statements 18, 26, 27, 30 (required for verifying the axioms of order), and Statements 24, 25, 26, 28, 29 (required for verifying the axioms of congruence).

Interpretation (C*) does not, however, satisfy the Axiom of Continuity. Otherwise, interpretation (C*) would be a model for the entire axiom system (GA$_3$) and, as a result, $*$-space $\mathbf{C}^*$ would, by Theorem 65 of Chapter III, have the power of the continuum, despite the fact that $\mathbf{C}_3^*$ is, by Statement 16, a denumerable set.

We have therefore proved the following:

PROPOSITION 3. *If the arithmetic of real numbers is consistent, then the Axiom of Continuity is independent of the remaining axioms of the system* (GA$_3$).

NOTE. From the fact that interpretation (C) satisfies sentence E (see Proposition 1) it follows, by Statement 21, that interpretation (C*) also satisfies sentence E. Therefore, if to the system of axioms (GA$_3$) we add sentence E as a new axiom, then the Axiom of Continuity thereafter remains independent of the remaining axioms.

5. The Cartesian Circle

We call the spheres of Cartesian space $\mathbf{C}_2$ *circles*. Therefore a circle S with center a and radius λ is the set of points $x \in \mathbf{C}_2$ defined by the condition

$$\rho(x,a) = \lambda,$$

which is equivalent, for $x = (x_1, x_2)$ and $a = (a_1, a_2)$, to the condition

$$(x_1 - a_1)^2 + (x_2 - a_2)^2 = \lambda^2.$$

The following theorem establishes the number of the intersection points of a circle and a line.

STATEMENT 31. *Consider a line L and a circle S with center a and radius λ. If $\rho(a,L) > \lambda$, then line L does not intersect circle S; if $\rho(a,L) = \lambda$, then line L intersects circle S in one point; if $\rho(a,L) < \lambda$, then line L intersects circle S in two points.*

PROOF. We may assume that $L = \mathbf{C}_{2,1}$. We obtain the points $x = (x_1, x_2)$ of the set $S \cap L$ from the system of equations

$$(x_1 - a_1)^2 + (x_2 - a_2)^2 = \lambda^2,$$
$$x_2 = 0,$$

which is equivalent to the system of equations

(1) $$x_1^2 - 2\, a_1 x_1 + a_1^2 + a_2^2 - \lambda^2 = 0,$$
$$x_2 = 0.$$

Since $\rho(a,L) = |a_2|$, then

$$a_1^2 - (a_1^2 + a_2^2 - \lambda^2) = \lambda^2 - a_2^2 = \lambda^2 - \rho(a,L)^2.$$

Thus the number of roots of equation (1) is equal to zero when $\rho(a,L) > \lambda$, equal to 1 when $\rho(a,L) = \lambda$, and equal to 2 when $\rho(a,L) < \lambda$.

From Statement 31 it follows, in particular, that any three points on a circle are non-collinear.

Let S be a circle with center a and radius λ. The set of points $x \in \mathbf{C}_2$ satisfying the condition

$$\rho(x,a) < \lambda \quad (\text{or } \rho(x,a) > \lambda)$$

will be called the *inner (or outer) domain* of circle S.

Statement 31 allows the points belonging to the inner domain G of circle S to be characterized in the following manner:

STATEMENT 32. *A point p belongs to the inner domain G of a circle S if and only if every line L passing through point p intersects circle S in two points.*

PROOF. Let λ be the radius of circle S. If $p \in G$, then, for every line L passing through point p, we have

$$\rho(a,L) \leqslant \rho(a,p) < \lambda$$

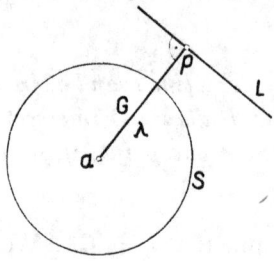

Fig. 170

and, consequently, line L intersects circle S in two points. If $p\sim \in G$ (Fig. 170), then, by producing through point p a line L perpendicular to the line ap, we obtain

$$\rho(a,L) = \rho(a,p) \geqslant \lambda,$$

and, as a result, line L intersects circle S in at most one point.

We shall now establish several properties of the inner domain of a circle.

STATEMENT 33. *Let G denote the inner domain of a circle S. If points p and q are distinct and $p,q \in G \cup S$, then $(pq) \subset G$.*

PROOF. We denote the center of circle S by a and its radius by λ (Fig. 171). As may easily be shown, for any point $r \in (pq)$ we have

$$\rho(a,r) < \rho(a,p) \quad \text{or} \quad \rho(a,r) < \rho(a,q),$$

and therefore $\rho(a,r) < \lambda$, that is, $r \in G$.

From Statement 33 it follows, in particular, that the inner domain G of a circle S is a convex figure.

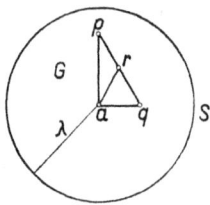 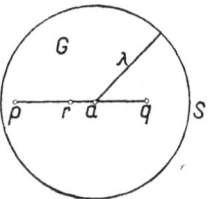

Fig. 171

STATEMENT 34. *Let G denote the inner domain of a circle S. If a line L intersects circle S in two points p and q, then $L \cap G = (pq)$.*

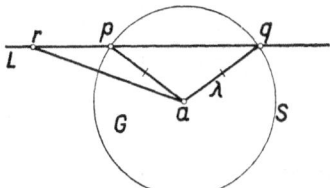 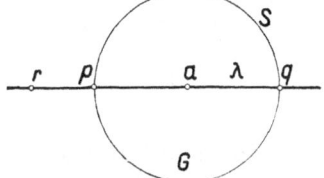

Fig. 172

PROOF. We denote the center of circle S by a and its radius by λ. From Statement 33 it follows that $(pq) \subset L \cap G$. If $r \in L - (pq)$ (Fig. 172), then, as readily seen,

$$\rho(a,r) \geqslant \rho(a,p) \quad \text{or} \quad \rho(a,r) \geqslant \rho(a,q),$$

from which it follows that $\rho(a,p) \geqslant \lambda$ and, consequently, $r \sim \in G$. Therefore $L \cap G = (pq)$.

STATEMENT 35. *If a point b belongs to the inner domain G a of circle S, then every half-line B with origin b intersects circle S in one point.*

PROOF. By Statement 32, line $L \supset B$ intersects circle S in two points p and q (Fig. 173). By Statement 34 we have $b \in (pq)$. Therefore either $p \in B$ or $q \in B$.

Consider a circle S with center $a = (a_1, a_2)$ and radius λ. Circle S has the equation

(2) $$(x_1 - a_1)^2 + (x_2 - a_2)^2 = \lambda^2.$$

Let

(3) $$x_1(\varphi) = a_1 + \lambda \cos \varphi,$$

(4) $$x_2(\varphi) = a_2 + \lambda \sin \varphi.$$

Then the solution $x = (x_1, x_2)$ of equation (2) may be represented para-metrically in the form

$$x(\varphi) = (x_1(\varphi),\ x_2(\varphi)),$$

where the parameter φ runs over any arbitrary interval of the form $< \alpha,\ 2\pi + \alpha)$.

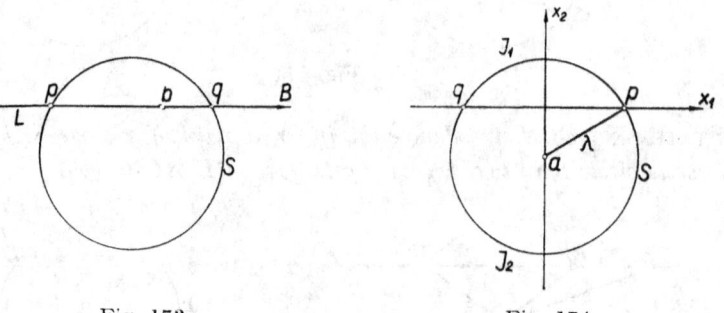

Fig. 173   Fig. 174

We take on circle S two distinct points $p = x(\alpha)$ and $q = x(\beta)$, where $0 \leqslant \alpha < \beta < \pi$. The set $S - \{p,q\}$ may be represented as the sum of two sets: J_1 consisting of all points $x(\varphi)$ for $\alpha < \varphi < \beta$ and J_2 consisting of all points $x(\varphi)$ for $\beta < \varphi < 2\pi + \alpha$. Since x as a function of parameter φ is a homeomorphism in both intervals (α,β) and $(\beta, 2\pi + \alpha)$, then the sets J_1 and J_2 are open arcs with the end points p and q.

We shall investigate the position of arcs J_1 and J_2 with respect to line pq. In these considerations it is convenient to assume that $p \in C_{2,1}^{+}$, $q = -p$, and $a = (0,a_2)$ where $a_2 \leqslant 0$ (Fig. 174). Then

(5) $$x_1(\alpha) = \lambda \cos \alpha > 0,$$

(6) $$x_2(\alpha) = a_2 + \lambda \sin \alpha = 0.$$

From (6) and (5) we obtain

(7) $$\sin \alpha = \frac{-a_2}{\lambda} \geqslant 0$$

and $\cos \alpha > 0$; therefore $0 \leqslant \alpha < \dfrac{\pi}{2}$ and

(8)
$$\sin \varphi > \sin \alpha \quad \text{if} \quad \alpha < \varphi < \pi - \alpha,$$
$$\sin \varphi < \sin \alpha \quad \text{if} \quad \pi - \alpha < \varphi < 2\pi + \alpha.$$

Furthermore, we have

(9) $$x_1(\beta) = -\lambda \cos \alpha = \lambda \cos (\pi - \alpha) = x_1 (\pi - \alpha),$$

(10) $$x_2(\beta) = 0 = a_2 + \lambda \sin (\pi - \alpha) = x_2 (\pi - \alpha).$$

Therefore $\beta = \pi - \alpha$ and by formulas (4), (7), and (8) we conclude

$$x_2(\varphi) > 0 \quad \text{if} \quad \alpha < \varphi < \beta,$$

$$x_2(\varphi) < 0 \quad \text{if} \quad \beta < \varphi < 2\pi + \alpha.$$

Thus arcs J_1 and J_2 lie on opposite sides of line $\mathbf{C}_{2,1}$.

We therefore have

STATEMENT 36. *Given on a circle S two distinct points p and q, let W and W* be the half-planes determined by the line pq on plane* $\mathbf{C}_2$*. Then the figure S*$-${*p,q*} *is the sum of two disjoint open arcs* $J_1 = S \cap W$ *and* $J_2 = S \cap W^*$ *with the common end points p and q.*

We shall now prove

STATEMENT 37. *Given a circle S and two distinct points p,q* $\in S$*, if r* $\in (pq)$ *then any line L* $\neq \mathbf{L}(pq)$ *passing through point r intersects each of the arcs* J_1 *and* J_2 *of circle S with end points p and q.*

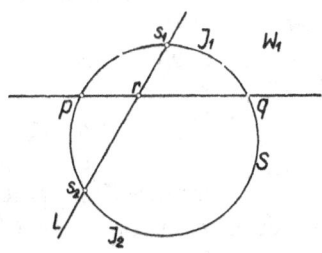

Fig. 175

PROOF. Since point r belongs to the inner domain of circle S, then (see Statement 32) line L intersects circle S in two points, say s_1 and s_2 (Fig. 175). By Statement 34, we have $r \in (s_1 s_2)$. If both points s_1 and s_2 were to lie on the same arc, e.g. arc J_1, then, by Statement 37, they would lie on the same half-plane W with boundary $\mathbf{L}(pq)$. Then the entire segment $(s_1 s_2)$ and, in particular, point r would lie in half-plane W, in contradiction to the fact that r $\in \mathbf{L}(pq)$.

Finally, we have two theorems concerning angles.

STATEMENT 38. *Given a circle S and points $b_1, b_2, b_3, p \in S$, if $p \neq b_1, b_2, b_3$, then half-lines pb_1, pb_2, pb_3 are co-half-pencilar.*

PROOF. Let a and λ denote the center and the radius of circle S, respectively (Fig. 176). We produce through point p a line L perpendicular to line ap; let $W = \mathbf{HP}(La)$. If half-line ab_i $(i = 1,2,3)$ does not

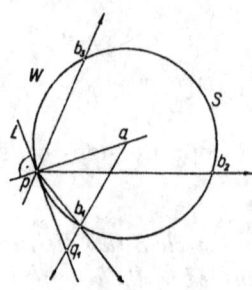

Fig. 176

intersect line L, then, obviously, $b_i \in W$, and if half-line ab_i intersects line L in some point q_i, then $\rho(a, q_i) > \rho(a, p) = \lambda$ and, therefore $\mathbf{B}(a, b_i, q_i)$, from which it follows that $b_i \in W$. Thus, points b_1, b_2, b_3 lie in half-plane W. Consequently, half-lines pb_1, pb_2, pb_3 lie in half-plane W and are therefore co-half-pencilar.

STATEMENT 39. *Let S be a circle with center a, and let b and c be two distinct points on S not collinear with a. For every point $p \in S$ lying on the same side of line bc as point a, we then have*

(10) $$|\sphericalangle bpc| = \tfrac{1}{2} \cdot |\sphericalangle bac| < \frac{\pi}{2}.$$

PROOF. By hypothesis p is an intersection point of circle S and line ap; let q be the second intersection point of S and $\mathbf{L}(ap)$. By Statement 35, we have $\mathbf{B}(p,a,q)$. We put $A = \mathbf{H}(pa) = \mathbf{H}(pq)$, $B = \mathbf{H}(pb)$, $C = \mathbf{H}(pc)$. Then $\sphericalangle bpc = BC$ and, by Statement 38,

$A = B$ or $A = C$ or $\mathbf{B}(A,B,C)$ or $\mathbf{B}(B,C,A)$ or $\mathbf{B}(C,A,B)$.

Because of the symmetry of our hypothesis with respect to the variables b and c it is sufficient to investigate the following three cases:

Case 1. $A = B$ (Fig. 177), Then $q = b$ and angle bac is adjacent to angle pac. Hence, by Theorem 64 of Chapter II, we have

$$2|\sphericalangle bpc| = 2|\sphericalangle apc| = |\sphericalangle bac|,$$

which implies formula (10).

Case 2. $\mathbf{B}(A,B,C)$ (Fig. 178). Then half-line B intersects segment (ac) in some point d. Since point d lies in the inner domain of circle S, then $\mathbf{B}(b,d,p)$ and, consequently, $\mathbf{B}(\mathbf{H}(ab), \mathbf{H}(ac), \mathbf{H}(ap))$, from which it follows that

(11) $\mathbf{B}(\mathbf{H}(aq), \mathbf{H}(ab), \mathbf{H}(ac))$.

From $\mathbf{B}(A,B,C)$ and (11) we conclude (see Case 1) that

$$|BC| = |AC| - |AB| = \tfrac{1}{2}|\sphericalangle\, qac| - \tfrac{1}{2}|\sphericalangle\, qab| = \tfrac{1}{2}|\sphericalangle\, bac|.$$

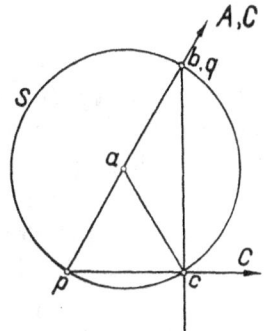

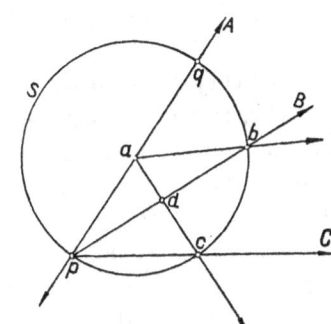

Fig. 177   Fig. 178

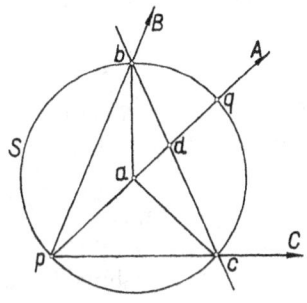

Fig. 179

Case 3. $\mathbf{B}(C,A,B)$ (Fig. 179). Then half-line A intersects segment (bc) in some point d. Since $p \in \mathbf{HP}(\mathbf{L}(bc)a)$, then $d \in \mathbf{H}(aq)$; and since point d lies in the inner domain of circle S, then $\mathbf{B}(a,d,q)$. Thus, similarly to Case 2, we obtain

$$|BC| = |AC| + |AB| = \tfrac{1}{2}|\sphericalangle\, qac| + \tfrac{1}{2}|\sphericalangle\, qab| = \tfrac{1}{2}|\sphericalangle\, bac|,$$

which concludes the proof.

6. Projective Space P_n

As already mentioned in the Introduction (Section 10), the points of n-dimensional projective space P_n are the sets $[x_0, x_1, \ldots, x_n]$ composed of all sequences $(\lambda x_0, \lambda x_1, \ldots, \lambda x_n)$, where not all the numbers $x_0, x_1, \ldots, x_n$ vanish simultaneously and $\lambda \neq 0$. Each of the sequences $(\lambda x_0, \lambda x_1, \ldots, \lambda x_n)$ will be called the *(projective) coordinate system of the point* $[x_0, x_1, \ldots, x_n]$. Obviously, $[x_0, x_1, \ldots, x_n] = [y_0, y_1, \ldots, y_n]$ if and only if sequences $(x_0, x_1, \ldots, x_n)$ and $(y_0, y_1, \ldots, y_n)$ are proportional, i.e. if there exist numbers λ and μ not vanishing simultaneously and such that

$$\lambda x_i + \mu y_i = 0 \quad \text{for } i = 0, 1, \ldots, n.$$

We call every point of the form $[1, x_1, x_2, \ldots, x_n] \in P_n$ a *proper point* and identify it with the point $(x_1, x_2, \ldots, x_n) \in C_n$. In this way Cartesian space C_n becomes a part of projective space P_n. We call every point of the form $[0, x_1, x_2, \ldots, x_n] \in P_n$ an *improper point* and identify it with the direction of the Cartesian vector $[x_1, x_2, \ldots, x_n]$. It can therefore be said that projective space P_n consists of points and directions of Cartesian space C_n.

As we already said in the Introduction (Section 10), the analytic geometry of projective space P_n investigates invariants of the group of linear transformations of space P_n onto itself, that is, the *projective notions* in space P_n. We now introduce the most fundamental of these notions.

We take any two distinct points $a = [a_0, a_1, \ldots, a_n]$ and $b = [b_0, b_1, \ldots, b_n]$ of space P_n. Let L denote the set of points

$$(1) \qquad x(\lambda, \mu) = [\lambda a_0 + \mu b_0, \lambda a_1 + \mu b_1, \ldots, \lambda a_n + \mu b_n]$$

of space P_n for all λ and μ not vanishing simultaneously. It is readily shown that set L does not depend on the choice of the coordinate systems $(a_0, a_1, \ldots, a_n)$ and $(b_0, b_1, \ldots, b_n)$ of points a and b, but only on the points a and b themselves, and that

$$(2) \qquad x(\lambda_1, \mu_1) = x(\lambda_2, \mu_2) \quad \text{if and only if} \quad \begin{vmatrix} \lambda_1 & \mu_1 \\ \lambda_2 & \mu_2 \end{vmatrix} = 0.$$

Set L will be called the *(projective) line determined by points a and b* and formula (1) will be referred to as the *parametric equation of L*. By *(projective) line in space P_n* we shall understand the line determined by any two distinct points a and b of space P_n.

The line determined by points $[1,0,0,\ldots,0]$ and $[0,1,0,\ldots,0]$ consists, by formula (1), of all points of the form $[\lambda,\mu,0,\ldots,0]$; we denote it by $\boldsymbol{P}_{n,1}$.

We now subject space $\boldsymbol{P}_n$ to linear transformation f mapping the point $x = [x_0,x_1,\ldots,x_n] \in \boldsymbol{P}_n$ onto the point $\bar{x} = [\bar{x}_0,\bar{x}_1,\ldots,\bar{x}_n]$ by means of the formula

$$(3) \qquad \bar{x}_i = \sum_{j=0}^{n} \alpha_{ij} x_j \quad \text{for } i = 0,1,\ldots,n,$$

where the determinant $|\alpha_{ij}|$ $(i,j = 0,1,\ldots,n)$ is different from zero. Linear transformation f maps any point x (λ,μ) of line L onto the point $f(x(\lambda,\mu)) = [\bar{x}_0,\bar{x}_1,\ldots,\bar{x}_n]$, where, for $i = 0,1,\ldots,n$,

$$(4) \qquad \bar{x}_i = \sum_{j=0}^{n} \alpha_{ij}(\lambda a_j + \mu b_j) = \lambda \cdot \sum_{j=0}^{n} \alpha_{ij} a_j + \mu \cdot \sum_{j=0}^{n} \alpha_{ij} b_j = \lambda \bar{a}_i + \mu \bar{b}_i.$$

Let $\bar{L}$ be the line determined by points $\bar{a}$ and $\bar{b}$. Let us consider its parametric equation

$$(5) \qquad \bar{x}(\lambda,\mu) = [\lambda \bar{a}_0 + \mu \bar{b}_0, \lambda \bar{a}_1 + \mu \bar{b}_1,\ldots,\lambda \bar{a}_n + \mu \bar{b}_n].$$

From formula (4) it follows that

$$(6) \qquad f(x(\lambda,\mu)) = \bar{x}(\lambda,\mu).$$

Hence the point set $f(L)$ is identical with the line $\bar{L}$.

In this way we have shown that the notion of the line is an invariant of linear transformations; in other words, linear transformations are *collineations*.

We shall now investigate the relation between the projective line and the Cartesian line. Let us take a projective line L determined by points $a,b \in \boldsymbol{P}_n$. We shall examine three cases separately:

Case 1. $a,b \in \boldsymbol{C}_n$. Then we can assume $a_0 = b_0 = 1$, and the parametric equation (1) of line L takes the form

$$x(\lambda,\mu) = [\lambda + \mu, \lambda a_1 + \mu b_1,\ldots,\lambda a_n + \mu b_n].$$

If $\lambda + \mu \neq 0$, then

$$x(\lambda,\mu) = \left[1, \frac{\lambda}{\lambda + \mu} a_1 + \frac{\mu}{\lambda + \mu} b_1,\ldots, \frac{\lambda}{\lambda + \mu} a_n + \frac{\mu}{\lambda + \mu} b_n\right],$$

and consequently $x(\lambda,\mu)$ runs over all points of Cartesian line K determined by points a and b. If $\lambda + \mu = 0$, then $x(\lambda,\mu)$ is the direction of Cartesian vector $[a_1-b_1, a_2-b_2, \ldots, a_n-b_n]$, that is, the direction of Cartesian line K.

Case 2. $a \in \mathbf{C}_n$, b is an improper point. Then $b_0 = 0$, and we may take $a_0 = 1$; thus the parametric equation (1) of line L takes the form

$$x(\lambda,\mu) = [\lambda, \lambda a_1 + \mu b_1, \ldots, \lambda a_n + \mu b_n].$$

If $\lambda \neq 0$, then

$$x(\lambda,\mu) = [1, a_1 + \frac{\mu}{\lambda}b_1, \ldots, a_n + \frac{\mu}{\lambda}b_n],$$

and therefore $x(\lambda,\mu)$ runs over all points of Cartesian line L passing through point a and having the direction b. If $\lambda = 0$, then $x(\lambda,\mu)$ is the direction b of line K.

Case 3. Both points a and b are improper. Since in space $\mathbf{P}_1$ there is just one improper point $[0,1]$, this case can occur only in the spaces $\mathbf{P}_n$ for $n > 1$. Since $a_0 = b_0 = 0$, then the parametric equation (1) for line L takes the form

$$x(\lambda,\mu) = [0, \lambda a_1 + \mu b_1, \ldots, \lambda a_n + \mu b_n].$$

Hence all points $x(\lambda,\mu)$ of line L are improper, and the points

$$y(\lambda,\mu) = [\lambda a_1 + \mu b_1, \lambda a_2 + \mu b_2, \ldots, \lambda a_n + \mu b_n] \in \mathbf{P}_{n-1}$$

form a line of space $\mathbf{P}_{n-1}$.

From this discussion it follows that there exist two kinds of projective lines in space $\mathbf{P}_n$, in fact (i) *proper lines* each of which is obtained by adjoining to a Cartesian line of space $\mathbf{C}_n$ the direction of this line as a new point, and (ii) *improper lines* (only for $n > 1$) each of which consists of improper points $[0, x_1, x_2, \ldots, x_n]$ such that points $[x_1, x_2, \ldots, x_n]$ form a (proper or improper) line in space $\mathbf{P}_{n-1}$.

From this, on the basis of the discussion in Section 2, we obtain the following statements:

STATEMENT 40. *At least three distinct points lie on every line $L \subset \mathbf{P}_n$.*

In the case where L is an improper line the proof is carried out by means of induction with respect to n.

STATEMENT 41. *There is just one line $L \subset \mathbf{P}_n$ passing through any two distinct points $a, b \in \mathbf{P}_n$.*

In case both points a and b are improper, the proof is carried out by means of induction with respect to n.

STATEMENT 42. *Consider in space $\boldsymbol{P}_n$ three proper points a, b, c (that is, a,b,c $\in \boldsymbol{C}_n$). Then points a, b, c are collinear in space $\boldsymbol{P}_n$ if and only if they are collinear in space $\boldsymbol{C}_n$.*

Space $\boldsymbol{P}_1$ consists of Cartesian points $[1,x_1]$, where x_1 is any real number, which form the Cartesian line $\boldsymbol{C}_1$, and of improper point $[0,1]$, which is the direction of line $\boldsymbol{C}_1$. Consequently, space $\boldsymbol{P}_1$ is a projective line. Since, in space $\boldsymbol{C}_1$, except for $\boldsymbol{C}_1$ itself, there are no other Cartesian lines, then in space $\boldsymbol{P}_1$, except for $\boldsymbol{P}_1$ itself, there are no other projective lines.

We now consider space $\boldsymbol{P}_2$. A proper line $L \subset \boldsymbol{P}_2$ including a Cartesian line $K \subset \boldsymbol{C}_2$ with the linear equation

(7) $$\alpha_0 + \alpha_1 x_1 + \alpha_2 x_2 = 0$$

has the linear equation

(8) $$\alpha_0 x_0 + \alpha_1 x_1 + \alpha_2 x_2 = 0.$$

Indeed, a point $[1,x_1,x_2] \in \boldsymbol{P}_2$ satisfies equation (8) if and only if the point $(x_1,x_2) \in \boldsymbol{C}_2$, identified with it, satisfies equation (7); and a point $[0,x_1,x_2] \in \boldsymbol{P}_2$ satisfies equation (8) if and only if $\alpha_1 x_1 + \alpha_2 x_2 = 0$, that is, if Cartesian vector $[x_1,x_2]$ is perpendicular to Cartesian vector $[\alpha_1,\alpha_2]$, that is, if point $[0,x_1,x_2]$ is the direction of line K.

There exists just one improper line L in space $\boldsymbol{P}_2$. It consists of all the points of the form $[0,x_1,x_2]$, where $(x_1,x_2) \in \boldsymbol{C}_2$, that is, of all the improper points of space $\boldsymbol{P}_2$. Line L therefore has the equation

$$x_0 = 0.$$

It follows from this discussion that every projective line of space $\boldsymbol{P}_2$ has an equation of the form (8), where not all the coefficients α_0, α_1, α_2 vanish simultaneously, and that every equation of the form (8), where not all the coefficients α_0, α_1, α_2 vanish simultaneously, is an equation of a projective line in space $\boldsymbol{P}_2$.

From Statement 7 on lines of space $\boldsymbol{C}_2$ we obtain at once the following statement on lines of space $\boldsymbol{P}_2$:

STATEMENT 43. *In space $\boldsymbol{P}_2$ each two lines have a common point.*

Let us take in space $\boldsymbol{P}_n$ any line L and on it any two distinct points $a = [a_0,a_1,\ldots,a_n]$ and $b = [b_0,b_1,\ldots,b_n]$. Then (1) is a parametric equation of line L. Let us take on line L any four points p, q, r, s such that $p \neq s$ and $q \neq r$. Let $p = x\,(\lambda_p,\mu_p)$, $q = x\,(\lambda_q,\mu_q)$, $r = x\,(\lambda_r,\mu_r)$, $s = x\,(\lambda_s,\mu_s)$. By the *cross ratio* of the ordered quadruple of points p, q, r, s we shall understand the number

(9)
$$(p,q;r,s) = \frac{\begin{vmatrix} \lambda_p & \mu_p \\ \lambda_r & \mu_r \end{vmatrix} \cdot \begin{vmatrix} \lambda_q & \mu_q \\ \lambda_s & \mu_s \end{vmatrix}}{\begin{vmatrix} \lambda_p & \mu_p \\ \lambda_s & \mu_s \end{vmatrix} \cdot \begin{vmatrix} \lambda_q & \mu_q \\ \lambda_r & \mu_r \end{vmatrix}}.$$

By (2), the denominator of this fraction is different from zero.

From formula (2) it is seen at once that the number $(p,q;r,s)$ does not depend on the choice of parameters λ_p and μ_p, λ_q and μ_q, λ_r and μ_r, λ_s and μ_s of points p, q, r, s. The number $(p,q;r,s)$ also does not depend on the choice of points a and b by means of which we defined line L. Indeed, let us take two other distinct points $c = [c_0, c_1, \ldots, c_n]$ and $d = [d_0, d_1, \ldots, d_n]$ on line L; then line L has the parametric equation

(10) $y(\lambda',\mu') = [\lambda'c_0 + \mu'd_0, \lambda'c_1 + \mu'd_1, \ldots, \lambda'c_n + \mu'd_n].$

Let $a = y(\lambda'_a, \mu'_a)$, $b = y(\lambda'_b, \mu'_b)$. Then for $i = 0, 1, \ldots, n$

$$\lambda a_i + \mu b_i = \lambda(\lambda'_a c_i + \mu'_a d_i) + \mu(\lambda'_b c_i + \mu'_b d_i) = (\lambda\lambda'_a + \mu\lambda'_b)c_i + (\lambda\mu'_a + \mu\mu'_b)d_i,$$

from which it follows that

$$x(\lambda,\mu) = y(\lambda\lambda'_a + \mu\lambda'_b, \lambda\mu'_a + \mu\mu'_b).$$

If, therefore, we calculate the cross ratio of the quadruple of points p, q, r, s by means of parametric equation (10), then each of the four determinants occurring in the fraction defining this ratio (see formula (9)) should be multiplied by the same determinant

$$\begin{vmatrix} \lambda'_a & \lambda'_b \\ \mu'_a & \mu'_b \end{vmatrix} \neq 0.$$

This, obviously, does not change the value of the fraction. Hence the number $(p,q;r,s)$ is a function of the points p, q, r, s only.

Let us assume that $p \neq q$. Taking, in particular, $a = p$, $b = q$, we may assume that $\lambda_p = 1$, $\mu_p = 0$, $\lambda_q = 0$, $\mu_q = 1$ and we obtain, for the cross ratio of the quadruple of points p, q, r, s, the formula

(11) $$(p,q;r,s) = \frac{\mu_r \lambda_s}{\lambda_r \mu_s}.$$

If we subject space $\boldsymbol{P}_n$ to a linear transformation f mapping every point $x = [x_0, x_1, \ldots, x_n] \in \boldsymbol{P}_n$ onto point $f(x) = \bar{x} = [\bar{x}_0, \bar{x}_1, \ldots, \bar{x}_n] \in \boldsymbol{P}_n$ by means of formula (3), then points a and b are mapped onto points $\bar{a}$ and $\bar{b}$, respectively, line L onto line $\bar{L}$ with parametric equation (5) and points $p, q, r, s \in L$ onto points $\bar{p}, \bar{q}, \bar{r}, \bar{s} \in \bar{L}$ given by the equalities

$$\bar{p} = \bar{x}\,(\lambda_p,\mu_p),\ \bar{q} = \bar{x}\,(\lambda_q,\mu_q),\ \bar{r} = \bar{x}\,(\lambda_r,\mu_r),\ \bar{s} = \bar{x}\,(\lambda_s,\mu_s)$$

(see formula (6)). From this it follows at once that

$$(p,q;r,s) = (\bar{p},\bar{q};\bar{r},\bar{s}).$$

Therefore the cross ratio is an invariant of linear transformations. We shall now establish the basic properties of the cross ratio. From formulas (9) and (2) it follows at once that

(12) $(p,q;r,s) = 0$ if and only if $p = r$ or $q = s$.

Simple calculation leads from formula (9) to the following

STATEMENT 44. *Given on a line L four distinct points p, q, r, s, let* $(p,q;r,s) = \alpha$. *Then*

$$(r,s;p,q) = \alpha,\ (q,p;r,s) = \frac{1}{\alpha},\ (p,r;q,s) = 1-\alpha.$$

By means of formula (11) we derive

STATEMENT 45. *Let p,q,r be three distinct points on a given line L. For any arbitrary number* α *there exists just one point* $s \in L$ *such that* $(p,q;r,s) = \alpha$.

PROOF. Let $p = [p_0,p_1,\ldots,p_n]$, $q = [q_0,q_1,\ldots,q_n]$. The equality

$$\frac{\mu_r \lambda_s}{\lambda_r \mu_s} = \alpha$$

is equivalent to the equality

$$\frac{\lambda_s}{\mu_s} = \frac{\alpha \lambda_r}{\mu_r},$$

which uniquely defines the point

$$s = [(\alpha\lambda_r)p_0 + \mu_r q_0,\ (\alpha\lambda_r)p_1 + \mu_r q_1,\ldots,\ (\alpha\lambda_r)p_n + \mu_r q_n].$$

We now assume that line L is a proper line and that points p, q, r, s are proper (that is, $p,q,r,s \in \mathbf{C}_n$) and distinct. We may take $p_0 = q_0 = 1$. Then $p = (p_1,\ldots,p_n)$, $q = (q_1,\ldots,q_n)$, line L has the parametric equation

$$x(\lambda,\mu) = [\lambda + \mu, \lambda p_1 + \mu q_1,\ldots,\lambda p_n + \mu q_n]$$

and $\lambda_r + \mu_r,\ \lambda_s + \mu_s \neq 0$. Put

$$\lambda_r' = \frac{\lambda_r}{\lambda_r + \mu_r},\ \mu_r' = \frac{\mu_r}{\lambda_r + \mu_r},\ \lambda_s' = \frac{\lambda_s}{\lambda_s + \mu_s},\ \mu_s' = \frac{\mu_s}{\lambda_s + \mu_s}.$$

Then $\lambda_r' + \mu_r' = 1$, $\lambda_s' + \mu_s' = 1$ and

$$r = \lambda_r' p + \mu_r' q, \qquad s = \lambda_s' p + \mu_s' q,$$

from which it follows that

$$r - p = \mu_r'(q - p), \qquad s - p = \mu_s'(q - p),$$
$$r - q = -\lambda_r'(q - p), \qquad s - q = -\lambda_s'(q - p).$$

Thus

(11) $$(p,q;r,s) = \frac{\mu_r'\lambda_s'}{\lambda_r'\mu_s'} = \pm \frac{\rho(p,r) \cdot \rho(q,s)}{\rho(p,s) \cdot \rho(q,r)}$$

(ρ is the Cartesian distance in space $\mathbf{C}_n$), where we take the minus sign only if one of the Cartesian vectors $[r-p]$, $[r-q]$, $[s-p]$, $[s-q]$ has the sense opposite to the remaining ones.

By a (*projective*) *plane* in space $\mathbf{P}_3$ we understand every point set $P \subset \mathbf{P}_3$ defined by the condition

$$[x_0,x_1,x_2,x_3] \in P \text{ if and only if } \alpha_0 x_0 + \alpha_1 x_1 + \alpha_2 x_2 + \alpha_3 x_3 = 0,$$

where $\alpha_0, \alpha_1, \alpha_2, \alpha_3$ are any real constants not all vanishing simultaneously. We call the expression

(12) $$\alpha_0 x_0 + \alpha_1 x_1 + \alpha_2 x_2 + \alpha_3 x_3 = 0,$$

a *linear equation of plane P*.†

If at least one of the coefficients α_1, α_2, α_3 does not vanish, then we call plane P a *proper plane*. The proper points $[1,x_1,x_2,x_3]$ of plane P then form a Cartesian plane $Q \subset \mathbf{C}_3$ described by the equation

$$\alpha_0 + \alpha_1 x_1 + \alpha_2 x_2 + \alpha_3 x_3 = 0,$$

and the improper points $[0,x_1,x_2,x_3]$ of plane P satisfy the equation

$$\alpha_1 x_1 + \alpha_2 x_2 + \alpha_3 x_3 = 0,$$

i.e. coincide with the directions parallel to plane Q and form an improper line of space $\mathbf{P}_3$.

If $\alpha_1 = \alpha_2 = \alpha_3 = 0$, then $\alpha_0 \neq 0$ and equation (12) takes the form

$$x_0 = 0.$$

We call the plane P with this equation an *improper plane*; it consists of all improper points of space $\mathbf{P}_3$. Thus

$$[0,x_1,x_2,x_3] \in P \text{ if and only if } [x_1,x_2,x_3] \in \mathbf{P}_2.$$

---

† The plane could have been introduced in another way, and at once in an arbitrary space $\mathbf{P}_n$ for $n \geqslant 2$, by means of the parametric equation, similarly to the scae of the line.

The proper plane with the equation $x_3 = 0$ consists of all points of the form $[x_0,x_1,x_2,0]$; we denote it by $\boldsymbol{P}_{3,2}$.

From the above discussion concerning lines and planes of space $\boldsymbol{P}_3$ we have, on the basis the considerations of Section 2, the following statements:

STATEMENT 46. *There are three non-collinear points a, b, c on every plane* $P \subset \boldsymbol{P}_3$.

STATEMENT 47. *For every three points a, b, c of space $\boldsymbol{P}_3$, there is a plane passing through them, and only one such plane if points a, b, c are non-collinear.*

STATEMENT 48. *Consider in space $\boldsymbol{P}_3$ four proper points a, b, c, d (that is, a,b,c,d $\in \boldsymbol{C}_3$). Then points a, b, c, d are coplanar in space $\boldsymbol{P}_3$ if and only if they are coplanar in space $\boldsymbol{C}_3$.*

STATEMENT 49. *If points a,b $\in \boldsymbol{P}_3$ are distinct and lie in a plane $P \subset \boldsymbol{P}_3$, then the entire line L determined by points a and b is included in plane P.*

STATEMENT 50. *Every two lines lying in the same plane $P \subset \boldsymbol{P}_3$ have a common point.*

STATEMENT 51. *Every two distinct planes P and Q of space $\boldsymbol{P}_3$ intersect along some line $L \subset \boldsymbol{P}_3$; and every line L of space $\boldsymbol{P}_3$ can be represented as the intersection of two distinct planes $P,Q \subset \boldsymbol{P}_3$.*

We now pass to linear transformations. In order to formulate the basic statement here (on the unique determination of a linear transformation) so as to apply simultaneously to spaces $\boldsymbol{P}_1, \boldsymbol{P}_2, \boldsymbol{P}_3$, it is convenient to use the following definition: We shall say that points $a^0, a^1, \ldots, a^n$ of space $\boldsymbol{P}_n$ ($n = 1,2,3$) are *linearly independent* if one of the following three cases holds:

Case 1. $n = 1$ and points a^0 and a^1 are distinct;
Case 2. $n = 2$ and points a^0, a^1, a^2 are not collinear;
Case 3. $n = 3$ and points a^0, a^1, a^2, a^3 are not coplanar.

From the definition of the point of projective space $\boldsymbol{P}_1$ and from the form of the linear equations of the line in space $\boldsymbol{P}_2$ and of the plane in space $\boldsymbol{P}_3$ it follows at once (cf. derivation of formula (32) on page 210) that points

$$a^j = [a_0^j, a_1^j, \ldots, a_n^j] \quad \text{for} \quad n = 1,2,3 \quad \text{and} \quad j = 0,1,\ldots,n$$

are linearly independent if and only if the determinant $|a_i^j|$ $(i,j = 0,1,\ldots,n)$ is different from zero.

STATEMENT 52. *Consider in space $\boldsymbol{P}_n$ ($n = 1,2,3$) points $a^0, a^1, \ldots, a^{n+1}$ each $n + 1$ point of which are linearly independent. Let* † $p^j = [\delta_0^j, \delta_1^j, \ldots, \delta_n^j]$

† The definition of the function δ may be found on page 199.

for $j = 0,1,\ldots,n$, and let $p^{n+1} = [1,1,\ldots,1]$. There then exists just one linear transformation f of space $\mathbf{P}_n$ onto itself such that

$$(13) \qquad\qquad f(p^j) = a^j \quad \text{for } j = 0,1,\ldots,n+1.$$

PROOF. Let $a^j = [a_0^j, a_1^j, \ldots, a_n^j]$. Then linear transformation f mapping the point $x = [x_0, x_1, \ldots, x_n] \in \mathbf{P}_n$ onto point $\bar{x} = [\bar{x}_0, \bar{x}_1, \ldots, \bar{x}_n] \in \mathbf{P}_n$ by formulas

$$\bar{x}_i = \alpha_{i0}x_0 + \alpha_{i1}x_1 + \ldots + \alpha_{in}\alpha_n, \quad \text{for } i = 0,1,\ldots,n$$

satisfies condition (13) if and only if there exist numbers $\lambda_0, \lambda_1, \ldots, \lambda_{n+1} \neq 0$ such that simultaneously

$$(14) \qquad\qquad \lambda_j a_i^j = \alpha_{ij} \text{ for } i,j = 0,1,\ldots,n,$$

$$\lambda_{n+1} a_i^{n+1} = \alpha_{i0} + \alpha_{i1} + \ldots + \alpha_{in} \text{ for } i = 0,1,\ldots,n.$$

The system of equations (14) is equivalent to the system of equations

$$\lambda_0 a_i^0 + \lambda_1 a_i^1 + \ldots + \lambda_n a^n - \lambda_{n+1} a_i^{n+1} = 0 \text{ for } i = 0,1,\ldots,n,$$

which, up to a constant factor, has just one solution $\lambda_0, \lambda_1, \ldots, \lambda_{n+1} \neq 0$ (see the analytical condition for the linear independence of points given above). This completes the proof of the theorem.

As a direct conclusion from Statement 52, we obtain at once

STATEMENT 53. *Consider in space $\mathbf{P}_n$ ($n = 1,2,3$) points $a^0, a^1, \ldots, a^{n+1}$ of which each $n+1$ points are linearly independent, and points $b^0, b^1, \ldots, b^{n+1}$ each $n+1$ points of which are also linearly independent. There exists just one linear transformation f of space $\mathbf{P}_n$ onto itself such that $f(a^j) = b^j$ for $j = 0,1,\ldots,n+1$.*

Linear transformations f of space $\mathbf{P}_n$ onto itself satisfying the condition $f(\mathbf{C}_n) = \mathbf{C}_n$ are called *affine transformations*. It is readily shown that affine transformations are only those linear transformations which can be defined by formulas of the form

$$\bar{x}_0 = x_0,$$

$$\bar{x}_i = \alpha_{i0}x_0 + \alpha_{i1}x_1 + \ldots + \alpha_{in}x_n \text{ for } i = 1,2,\ldots,n,$$

the determinant $|\alpha_{ij}|$ ($i,j = 1,2,\ldots,n$) of the matrix $[\alpha_{ij}]$ ($i,j = 1,2,\ldots,n$) being different from zero. Affine transformations f for which this matrix is orthogonal coincide in space $\mathbf{C}_n$ with isometries; we call them *isometries of space $\mathbf{P}_n$*. In particular, we can speak of the symmetries of space $\mathbf{P}_2$.

Consider two lines K and L in space $\mathbf{P}_n$. A transformation f of K onto L will be called a *linear transformation* if there exists a linear transformation g of space $\mathbf{P}_n$ onto itself such that

$$f(p) = g(p) \quad \text{for } p \in K.$$

Henceforth we shall be interested primarily in linear transformations of lines in space $\mathbf{P}_2$.

STATEMENT 54. *Consider two lines K and L, three distinct points $a^0, a^1, a^2 \in K$ and three distinct points $b^0, b^1, b^2 \in L$ in space $\mathbf{P}_2$. There exists just one linear transformation f of K onto L such that*

$$(15) \qquad\qquad f(a^j) = b^j \text{ for } j = 0,1,2.$$

PROOF. By Statement 52 (for $n=2$), there exist linear transformations g_1 and g_2 of space $\mathbf{P}_2$ onto itself such that $g_1(K) = \mathbf{P}_{2,1}$ and $g_2(L) = \mathbf{P}_{2,1}$. Let

$$g_1(a^i) = [a_{i0}, a_{i1}, 0], \; g_2(b^i) = [b_{i0}, b_{i1}, 0] \quad \text{for } i = 0,1,2.$$

We take in space $\mathbf{P}_1$ points

$$g_1'(a^i) = [a_{i0}, a_{i1}], \quad g_2'(b^i) = [b_{i0}, b_{i1}] \qquad \text{for } i = 0,1,2.$$

Every linear transformation $\bar{x} = h(x)$ of line $\mathbf{P}_{2,1} \subset \mathbf{P}_2$ onto itself can be defined, as may readily be shown, by formulas of the form

$$(16) \qquad
\begin{aligned}
\bar{x}_0 &= \alpha_{00} x_0 + \alpha_{01} x_1, \\
\bar{x}_1 &= \alpha_{10} x_0 + \alpha_{11} x_1, \\
\bar{x}_2 &= x_2
\end{aligned}
\quad \text{where} \quad
\begin{vmatrix} \alpha_{00} & \alpha_{01} \\ \alpha_{10} & \alpha_{11} \end{vmatrix} \neq 0,$$

and, conversely, every linear transformation $\bar{x} = h(x)$ of space $\mathbf{P}_2$ onto itself defined by formulas (16) transforms line $\mathbf{P}_{2,1}$ onto itself.

Let us consider the transformation $\bar{x} = h'(x)$ of space $\mathbf{P}_1$ onto itself as defined by the formulas

$$\begin{aligned}
\bar{x}_0 &= \alpha_{00} x_0 + \alpha_{01} x_1, \\
\bar{x}_1 &= \alpha_{10} x_0 + \alpha_{11} x_1.
\end{aligned}$$

It is readily noted that the condition

$$(17) \qquad\qquad h(g_1(a^i)) = g_2(b^i) \quad \text{for } i = 0,1,2$$

is equivalent to the condition

$$(18) \qquad\qquad h'(g_1'(a^i)) = g_2'(b^i) \quad \text{for } i = 0,1,2.$$

By Statement 53 (for $n = 1$), there exists just one linear transformation h' of space $\mathbf{P}_1$ onto itself satisfying condition (18). Consequently, there exists just one linear transformation h of line $\mathbf{P}_{2,1} \subset \mathbf{P}_2$ onto itself satisfying condition (17). Hence the only linear transformation f of line K onto line L satisfying condition (15) is the transformation $g_2^{-1} h g_1$.

The following statement has a more special character:

STATEMENT 55. *If a linear transformation f of a line $L \subset \mathbf{P}_2$ onto itself has the property that, for some two distinct points $a,b \in L$,*

(19) $$f(a) = b \ \text{and} \ f(b) = a,$$

then $ff(p) = p$, for every point $p \in L$ (that is, f is an involution).

PROOF. Let $a^0 = [1,0,0]$, $b^0 = [0,1,0]$. Let us assume at first that $L = \mathbf{P}_{2,1}$, $a = a^0$, $b = b^0$. Every linear transformation $\bar{x} = f(x)$ of line $\mathbf{P}_{2,1}$ onto itself can be defined by formulas (16). If $f(a^0) = b^0$ and $f(b^0) = a^0$, then $\alpha_{00}=0$ and $\alpha_{11}=0$. Thus $\alpha_{01},\alpha_{10} \neq 0$ and $f([x_0,x_1,0])=[\alpha_{01}x_1,\alpha_{10}x_0,0]$, from which it follows that

$$ff([x_0,x_1,0]) = [\alpha_{01}\alpha_{10}x_0, \alpha_{10}\alpha_{01}x_1, 0] = [x_0,x_1,0]$$

that is $f(p) = p$ for every point $p \in \mathbf{P}_{2,1}$.

Now let L be any line of space $\mathbf{P}_2$, a and b any two distinct points on line L, and let a linear transformation f of line L onto itself satisfy condition (19). By Statement 54, there exists a linear transformation g of line $\mathbf{P}_{2,1}$ onto line L such that $g(a^0) = a$ and $g(b^0) = b$. Then

$$fg(a^0) = g(b^0) \ \text{and} \ fg(b^0) = g(a^0),$$

which implies

$$g^{-1}fg(a^0) = b^0 \ \text{and} \ g^{-1}fg(b^0) = a^0.$$

By what has been proved above, we therefore have for any point $q \in \mathbf{P}_{2,1}$

$$g^{-1}fgg^{-1}fg(q) = q, \ \text{i.e.} \ g^{-1}fg(q) = q,$$

that is, $ffg(q)=g(q)$ for every point $q \in \mathbf{P}_{2,1}$, that is, $ff(p)=p$ for every point $p \in L$.

This completes the proof.

In space $\mathbf{P}_2$ we fix any line L and any point $a \sim \in L$. Let p be any point of the set $\mathbf{P}_2-a$. The point p_1 in which the line M_p determined by points a and p intersects line L (see Statement 43) is called the *central projection of point p upon line L from center a.* Let us take any line $K \subset \mathbf{P}_2-a$. The function f mapping every point $p \in K$ onto its central projection p_1 upon line L from center a is called the *perspective transformation* of line K onto line L from *center a*.

STATEMENT 56. *For any lines K and L and for any point $a \sim \in K \cup L$ in space P_2, the perspective transformation f of K onto L from centre a is a linear transformation.*

PROOF. By Statement 52 (for $n = 2$), we may assume that $L = P_{2,1}$ and that $a = [0,0,1]$. Then lines K and L have the equations

$$K: \alpha_0 x_0 + \alpha_1 x_1 + x_2 = 0,$$
$$L: \qquad\qquad x_2 = 0,$$

and any line M passing through point a has the equation

$$M: \beta_0 x_0 + \beta_1 x_1 = 0,$$

from which it follows that

$$M \cap K = [-\beta_1, \beta_0, \alpha_0\beta_1 - \alpha_1\beta_0], \; M \cap L = [-\beta_1, \beta_0, 0].$$

It is now easy to show that the linear transformation $\bar{x} = g(x)$ given by the formulas

$$\bar{x}_0 = x_0,$$
$$\bar{x}_1 = x_1,$$
$$\bar{x}_2 = \alpha_0 x_0 + \alpha_1 x_1 + x_2$$

maps line K onto line L in such a way that $g(M \cap K) = M \cap L$. Thus on line K transformation g is identical with the perspective transformation f.

Let us take in space P_3 any plane P and let us fix on P any line L and any point $a \sim \in L$. In exactly the same way as in space P_2 we define the *central projection* of any point $p \in P$ upon line L from *center a* (see Statement 50) and the *perspective transformation* of any line $K \subset P - a$ onto line L from *center a*. Corresponding to Statement 56 we now have

STATEMENT 57. *Let there be given in space P_3 a plane P, two lines $K, L \subset P$ and a point $a \in P - (K \cup L)$. The perspective transformation f of line K onto line L from center a is a linear tranformation.*

PROOF. By Statement 52 (for $n = 3$) we may assume that $P = P_{3,2}$, $L = P_{3,1}$, $a = [0,0,1,0]$. Under this assumption the proof is similar to the proof of Statement 56.

We call the point set $F \subset P_n$ the *algebraic variety* in space P_n if there exists a homogeneous polynomial $\varphi(x_0, x_1, \ldots, x_n)$ such that for any point $[x_0, x_1, \ldots, x_n] \in P_n$

(20) $\qquad [x_0, x_1, \ldots, x_n] \in F$ if and only if $\varphi(x_0, x_1, \ldots, x_n) = 0.$

We say that an algebraic variety $F \subset P_n$ is *of degree* k if among the homogeneous polynomials $\varphi(x_0, x_1, \ldots, x_n)$ satisfying condition (20) there exists a polynominal of degree k, but no polynomial of degree $< k$. It is readily shown that the two notions, algebraic variety and its degree, are invariants of linear transformations.

In space P_2 algebraic varieties of degree 1 coincide with lines in P_2. Algebraic varieties of degree 2 in space P_2 are called *algebraic curves* of degree 2. The general equation of an algebraic curve F of degree 2 has the form

(21) $\qquad \alpha_{00}x_0^2 + \alpha_{11}x_1^2 + \alpha_{22}x_2^2 + 2\alpha_{01}x_0x_1 + 2\alpha_{02}x_0x_2 + 2\alpha_{12}x_1x_2 = 0.$

Let us consider the parametric equation (1) of a line L. Examining the system of equations (1) and (21), we readily conclude that line L is either included in curve F or has zero, one, or two points in common with curve F. If $L \subset F$ or if L has one point in common with F, then we say that *line* L *is tangent to curve* F or *curve* F *is tangent to line* L. Since linear transformations are collineations and transform space P_2 onto itself in a one-to-one way, then the notion of a tangent to an algebraic curve of degree 2 is an invariant of linear transformations.

To algebraic curves of degree 2 belong, among others, Cartesian circles, in particular, the circle S with center $[1,0,0]$ and radius 1, which in space P_2 has the equation

(22) $\qquad\qquad\qquad\qquad x_1^2 + x_2^2 - x_0^2 = 0.$

We say that a curve $F \subset P_2$ is a *conic* if there exists a linear transformation of space P_2 onto itself, which maps curve F onto circle S. Since circle S does not include any line (see Statement 31), then no conic includes a line. Hence a line is tangent to a conic if and only if it has just one point in common with it.

Finally, we shall prove one more special statement concerning conics.

STATEMENT 58. *Consider in space* P_2 *three non-collinear points* a, b, c *and two lines* K *and* L, *the first of which passes through point* a, *but does not contain points* b *and* c, *while the second passes through point* b, *but does not contain points* a *and* c. *There then exists just one conic passing through point* c *and tangent to line* K *at point* a *and to line* L *at point* b.

PROOF. Let d denote the intersection point of lines K and L. It is plain that among the four points a, b, c, d no three are collinear. By

Statement 52 (for $n = 2$), there thus exists a linear transformation of space $\boldsymbol{P_2}$ onto itself whereby points a, b, c, d are mapped onto points $[1,0,0]$, $[0,1,0]$, $[0,0,1]$, $[1,1,1]$, respectively. Because of the projective character of the theorem, we may assume that $a = [1,0,0]$, $b = [0,1,0]$ $c = [0,0,1]$, $d = [1,1,1]$.

Any conic has an equation of the form (21). If points a, b, c satisfy this equation, then $\alpha_{00} = \alpha_{11} = \alpha_{22} = 0$; thus the equation of the conic we are seeking must have the form

$$(23) \qquad \alpha_{01} x_0 x_1 + \alpha_{02} x_0 x_2 + \alpha_{12} x_1 x_2 = 0.$$

Line K has the equation $x_1 - x_2 = 0$; if it is tangent at point a to the curve with equation (23), then $\alpha_{01} = -\alpha_{02}$. Similarly, line L has the equation $x_0 - x_2 = 0$; if it is tangent at point b to the curve with equation (23), then $\alpha_{01} = -\alpha_{12}$. Therefore the conic we are seeking may be only the curve given by the equation

$$x_0 x_1 - x_0 x_2 - x_1 x_2 = 0.$$

This curve is a conic, since the linear transformation f mapping the point $x = [x_0, x_1, x_2] \in \boldsymbol{P_2}$ onto the point $\bar{x} = [\bar{x}_0, \bar{x}_1, \bar{x}_2]$ defined by the formulas

$$x_0 = \bar{x}_1,$$
$$x_1 = \bar{x}_0 - \bar{x}_1 - \bar{x}_2,$$
$$x_2 = -\bar{x}_0 + \bar{x}_1 - \bar{x}_2$$

maps this curve onto the circle S given by equation (22). It is readily shown that this conic satisfies the required conditions.

7. The Klein-Beltrami Model

The *Klein-Beltrami model* was constructed by F. KLEIN on the basis of the earlier ideas of E. BELTRAMI. We shall describe this model in detail only for the axiom system (GA_2) of plane absolute geometry (see page 197).

We regard Cartesian space $\boldsymbol{C_2}$ as a sub-space of projective space $\boldsymbol{P_2}$ and we take in $\boldsymbol{C_2}$ a circle S with center $(0,0)$ and radius 1. We shall call circle S the *absolute*. We denote the inner domain of circle S by $\boldsymbol{K_2}$ (the index 2 indicates the dimension of the domain).

We shall say that a linear transformation f of space $\boldsymbol{P_2}$ onto itself is a K-*isometry* if $f(S) = S$. It is seen at once that we have

STATEMENT 59. *The K-isometries form a group of transformations.*

A point p of space $\boldsymbol{P_2}$ belongs to domain $\boldsymbol{K_2}$ if and only if every projective line passing through point p intersects circle S in two points (see Statement 32). Since, as may readily be seen, this property is an invariant of K-isometries, then

STATEMENT 60. *If f is a K-isometry, then $f(\mathbf{K}_2) = \mathbf{K}_2$.*

Since the K-isometries, as linear transformations, are continuous, then

STATEMENT 61. *Every K-isometry f maps an arc of circle S with end points $p,q \in S$ onto an arc of circle S with end points $f(p),f(q) \in S$.*

Since K-isometries, as linear transformations, are collineations, then, by Statement 60, we also have

STATEMENT 62. *Every K-isometry maps the open segment of domain $\mathbf{K}_2$ with end points $p,q \in \mathbf{K}_2 \cup S$ onto the open segment of domain $\mathbf{K}_2$ with end points $f(p),f(q) \in \mathbf{K}_2 \cup S$.*

As an immediate conclusion from the above we obtain

STATEMENT 63. *If f is a K-isometry, then for any three points $p,q,r \in \mathbf{K}_2 \cup S$ it follows from $\mathbf{B}(p,q,r)$ that $\mathbf{B}(f(p),f(q),f(r))$.*

Obviously, every isometry of space $\mathbf{P}_2$ which leaves the point $(0,0)$ invariant is a K-isometry. It could be proved that, conversely, every K-isometry which leaves the point $(0,0)$ invariant is an isometry. For our purposes a weaker statement suffices, in fact

STATEMENT 64. *Let f be a K-isometry and let $a = (0,0)$. If $f(a) = a$, then $\rho(a,b) = \rho(a,f(b))$ for any point $b \in \mathbf{K}_2$.*

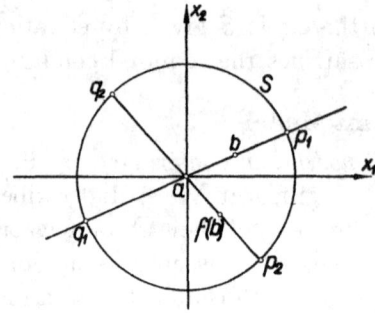

Fig. 180

PROOF. We may assume that $a \neq b$ (Fig. 180). Line ab intersects circle S in two distinct points p_1 and q_1. Let

$$f(p_1) = p_2 \text{ and } f(q_1) = q_2.$$

Since $p_2,q_2 \in S$ and $p_2 \neq q_2$, then

$$\rho(a,p_1) = \rho(a,p_2), \; \rho(a,q_1) = \rho(a,q_2), \; \rho(p_1,q_1) = \rho(p_2,q_2).$$

Thus the *K*-isometry f is an isometry for three points a,p_1,q_1 of line ab, and, consequently, it is an isometry for the entire line ab (see Theorems 99 and 100 of Chapter II which are satisfied by plane model (C), and Statement 54). In particular, we have $\rho(a,b) = \rho(a,f(b))$.

We shall now give a sufficient condition that a linear transformation f of space $\mathbf{P_2}$ onto itself be a K-isometry.

STATEMENT 65. *In order that a linear transformation f of space $\mathbf{P_2}$ onto itself be a K-isometry, it is sufficient that there exist three distinct points $p,q,s \in S$ such that*

(I) $f(p)$, $f(q)$, $f(s) \in S$;

(II) *if r is the intersection point of the tangents to circle S at points p and q, then $f(r)$ is the intersection point of the tangents to circle S at points $f(p)$ and $f(q)$.*

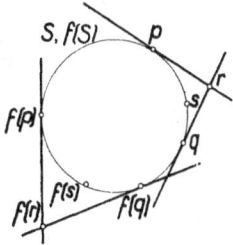

Fig. 181

PROOF. Let the linear transformation f satisfy conditions (I) and (II). Then $f(p)$, $f(q)$, $f(r)$ lie on both conics S and $f(S)$ (Fig. 181), and, since transformation f maps projective line rp tangent to circle S at point p onto projective line $f(r)f(p)$ tangent to conic $f(S)$ at point $f(p)$ and projective line rq tangent to circle S at point q onto projective line $f(r)f(q)$ tangent to conic $f(S)$ at point $f(q)$, hence projective line $f(r)f(p)$ is tangent at point $f(p)$ and projective line $f(r)f(q)$ is tangent at point $f(q)$ to both conics S and $f(S)$. Thus, by Statement 58, we at once obtain $f(S) = S$, that is, f is a K-isometry.

For constructing the Klein-Beltrami model we shall need special K-isometries whose existence is guaranteed by the following statements:

STATEMENT 66. *Given points $a_i \in \mathbf{K_2}$ and $p_i \in S$, where $i = 1,2$, let J_i be one of the open arcs into which line $a_i p_i$ divides circle S. Then there exists just one K-isometry f such that*

(1) $f(a_1) = a_2,\ f(p_1) = p_2,\ f(J_1) = J_2.$

PROOF. By hypothesis p_i is an intersection point of circle S and line $a_i p_i$; let q_i be the second intersection point of S and $\mathbf{L}(a_i p_i)$ (Fig. 182).

Further, we denote by r_i the intersection point of the tangents to circle S at points p_i and q_i; let s_i be the intersection point of line $r_i a_i$ and arc J_i (see Theorem 37). If a linear transformation f is a K-isometry and satisfies condition (1), then

(2) $$f(p_1) = p_2, \ f(q_1) = q_2, \ f(r_1) = r_2, \ f(s_1) = s_2.$$

Since, for $i = 1,2$, every three of the points p_i, q_i, r_i, s_i are non-collinear, then, by Statement 53 for $n = 2$, the system of equalities (2) defines just

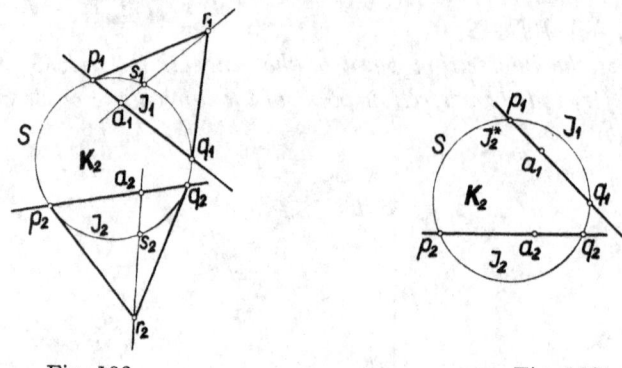

Fig. 182   Fig. 183

one linear transformation f. By Statement 65, transformation f is a K-isometry, and, on the basis of Statement 61, we may directly show that it satisfies condition (1). This is what we had to prove.

STATEMENT 67. *For any points $a_1, a_2 \in K_2$ and $p_1, p_2 \in S$ there exist just two K-isometries f such that*

(3) $$f(a_1) = a_2 \ and \ f(p_1) = p_2,$$

and both are identical on line $a_1 p_1$.

PROOF. By hypothesis p_i $(i = 1,2)$ is an intersection point of circle S and line $a_i p_i$; let q_i be the second intersection point of S and $\mathbf{L}(a_i p_i)$ (Fig. 183). Further, we denote by J_1 either of the arcs of circle S with end points p_1 and q_1, and by J_2 and J_2^* the arcs of circle S with end points p_2 and q_2. If a K-isometry f satisfies condition (3), then $f(q_1) = q_2$, and, because of Statement 61,

either $f(J_1) = J_2$ or $f(J_1) = J_2^*$.

By Statement 66, each of the systems of equalities

$$f(a_1) = a_2, \ f(p_1) = p_2, \ f(J_1) = J_2$$

and

$$f(a_1) = a_2, \ f(p_1) = p_2, \ f(J_1) = J_2^*$$

determines just one K-isometry f. Therefore there exists just two K-isometries f' and f'' satisfying condition (3). We have

$$f'(a_1) = f''(a_1), \quad f'(p_1) = f''(p_1), \quad f'(q_1) = f''(q_1),$$

and, by Statement 54, the K-isometries f' and f'' are identical on line a_1p_1. This concludes the proof.

STATEMENT 68. *For any two distinct points $a_1, a_2 \in \mathbf{K}_2$ there exists a K-isometry f such that*

$$f(a_1) = a_2 \quad and \quad f(a_2) = a_1.$$

PROOF. Let p_1 and p_2 be the intersection points of line a_1a_2 and the absolute S (Fig. 184). By Statement 67, there exists a K-isometry f such that

$$f(a_1) = a_2 \quad and \quad f(p_1) = p_2.$$

It is readily noted that we then have $f(p_2) = p_1$, and thus, by Statement 55, we conclude that the K-isometry f is an involution on projective line p_1p_2, and, in particular,

$$f(a_2) = f(a_1) = a_1.$$

The transformation f is thus the K-isometry we are seeking.

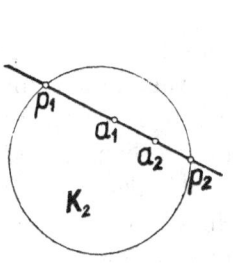

Fig. 184

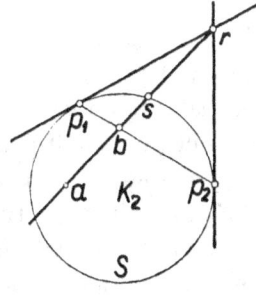

Fig. 185

We now pass on to the construction of the Klein-Beltrami model. We shall call this model simply *model* (K). We define K-*points* as points of the domain $\mathbf{K}_2$, K-*lines* as the sets $L \cap \mathbf{K}_2 \neq 0$, where L is any (Cartesian) line, that is, as the segments $(pq) \subset \mathbf{C}_2$, where p and q are distinct points of the absolute S (Fig. 185). The points p and q are not K-points; we shall call them *improper points of* K-*line* (pq). The domain $\mathbf{K}_2$ will also be called the K-*plane* or the *two-dimensional Klein space*.

We define the K-*betweenness relation* $\mathbf{B}_{\mathrm{K}}$ as the betweenness relation $\mathbf{B}$ for points of plane $\mathbf{C}_2$ restricted to K-points. Hence for any three K-points a, b, c we have

(4) $\mathbf{B}_{\mathrm{K}}(a,b,c)$ if and only if $\mathbf{B}(a,b,c)$.

We define the K-*congruence relation* $\mathbf{E}_{\mathrm{K}}$ in the following way: For any four K-points a,b,c,d we have $\mathbf{E}_{\mathrm{K}}(a,b;c,d)$ if and only if there exists a K-isometry f such that $f(a) = c$ and $f(b) = d$.

We shall now show that the interpretation

(K)   K-points, K-lines, $\mathbf{B}_{\mathrm{K}}$, $\mathbf{E}_{\mathrm{K}}$

satisfies all the axioms of absolute geometry.

From the fact that the plane interpretation (C) is a model for the axiom system (GA_2) it follows in a straightforward way that the interpretation (K) satisfies the axioms of incidence and order. In verifying Axiom O9 we make use of the convexity of the domain $\mathbf{K}_2$.

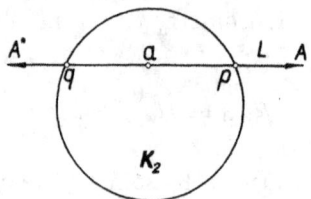

Fig. 186

Interpretation (K) of the primitive notions determines the interpretation of the defined notions, in particular, the notions of the open segment, half-line, and half-plane. From equivalence (4) it follows at once that K-*open* K-*segments* coincide with open segments (ab) with end points $a,b \in \mathbf{K}_2$. Further, on the basis of equivalence (4) it is readily seen that if a point $a \in \mathbf{K}_2$ determines half-lines A and A^* on line L (Fig. 186), then K-point a determines K-*half-lines* $A \cap \mathbf{K}_2$ and $A^* \cap \mathbf{K}_2$ on K-line $L \cap \mathbf{K}_2$. In other words, if line L intersects circle S in points p and q, then K-half-lines of K-line (pq) with K-origin a coincide with Cartesian segments (ap) and (aq). We shall call point p the *improper point* of K-*half-line* (ap). Also, from formula (4) and from the convexity of domain $\mathbf{K}_2$ it follows directly that for any K-line $L \cap \mathbf{K}_2$ and for any K-points a and c we have

$$\mathbf{B}_{\mathrm{K}}(a,L \cap \mathbf{K}_2,c) \text{ if and only if } \mathbf{B}(a,L,c).$$

From the above we conclude at once that if $L \cap \mathbf{K}_2 \neq 0$ and line L determines half-planes W and W^* in space $\mathbf{C}_2$, then K-line $L \cap \mathbf{K}_2$ determines K-*half-planes* $W \cap \mathbf{K}_2$ and $W^* \cap \mathbf{K}_2$ in space $\mathbf{K}_2$.

We turn to the axioms of congruence.

Axiom C1 is verified immediately, since if $\mathbf{E}_K (a,a;p,q)$ for some K-points a, p, q, then, by the definition of relation $\mathbf{E}_K$, there exists a K-isometry f such that $f(a) = p$ and $f(a) = q$. Hence $p = q$.

From Statement 68 and from the fact that the identity transformation is a K-isometry it at once follows that interpretation (K) satisfies Axiom C2.

We now pass on to Axiom C3. If $\mathbf{E}_K(a,b;p,q)$ and $\mathbf{E}_K(a,b;r,s)$, then there exist K-isometries f and g such that

$$f(a) = p, \ f(b) = q, \ g(a) = r, \ g(b) = s.$$

Then K-isometry $h = gf^{-1}$ (see Statement 59) transforms K-points p and q onto K-points r and s, respectively, and therefore $\mathbf{E}_K(p,q;r,s)$. Hence interpretation (K) satisfies Axiom C3.

Since interpretation (K) satisfies Axioms C1–C3, then it also satisfies Theorems 1–6 of Chapter II. Thus, for any K-segments ab and cd, we have

(5) $\qquad\qquad ab \equiv_K cd$ if and only if $\mathbf{E}_K(a,b;c,d)$,

and the relation $\equiv_K$ is reflexive, symmetric and transitive.

We now take two K-congruent K-triangles $a_1b_1c_1$ and $a_2b_2c_2$ in which

$$a_1b_1 \equiv_K a_2b_2, \ b_1c_1 \equiv_K b_2c_2, \ a_1c_1 \equiv_K a_2c_2.$$

From formula (5) it follows that there exist three K-isometries f', f'', f''' such that

$$f'(a_1) = a_2 \text{ and } f'(b_1) = b_2,$$
$$f''(b_1) = b_2 \text{ and } f''(c_1) = c_2,$$
$$f'''(a_1) = a_2 \text{ and } f'''(c_1) = c_2.$$

These three K-isometries do not necessarily have to coincide. We have however the following:

STATEMENT 69. *Given two* K-*triangles* $a_1b_1c_1$ *and* $a_2b_2c_2$, *if*

(6) $\qquad\qquad a_1b_1 \equiv_K a_2b_2, \ b_1c_1 \equiv_K b_2c_2, \ a_1c_1 \equiv_K a_2c_2,$

then there exists a K-*isometry* f *such that*

$$f(a_1) = a_2, \ f(b_1) = b_2, \ f(c_1) = c_2.$$

PROOF. The K-isometry f we are seeking turns out to be a superposition of three K-isometries which we shall now determine:

K-*isometry* g. We denote by a the center of circle S. By Statement 67 there exists a K-isometry g such that

(7) $$g(a_2) = a.$$

Let

(8) $$g(b_2) = b, \quad g(c_2) = c.$$

Then

(9) $$a_2 b_2 \equiv_K ab, \quad b_2 c_2 \equiv_K bc, \quad a_2 c_2 \equiv_K ac.$$

Since the hypothesis is symmetric with respect to the variables b_2 and c_2 we may assume that

(10) $$ab \leqslant ac$$

(in the Cartesian sense). We denote by W the half-plane (of plane $\mathbf{C}_2$) determined by line ab and containing K-point c.

K-*isometry* g'. From $a_1 b_1 \equiv_K a_2 b_2$ and $a_2 b_2 \equiv_K ab$ (see formulas (6) and (9)) it follows that $a_1 b_1 \equiv_K ab$. There thus exists a K-isometry g' such that

(11) $$g'(a_1) = a, \quad g'(b_1) = b.$$

We put

(12) $$g'(c_1) = c'.$$

Then

(13) $$a_1 c_1 \equiv_K ac', \quad b_1 c_1 \equiv_K bc'.$$

K-*isometry* g''. If $c' \in W$, then g'' is the indentity transformation in space $\mathbf{P}_2$. If $c' \in W^*$, then g'' is the symmetry with respect to line ab in space $\mathbf{P}_2$. We thus have

(14) $$g''(a) = a, \quad g''(b) = b.$$

We put

(15) $$g''(c') = d.$$

Then $d \in W$ (by formula (29) on page 207 it follows that in case $c' \in W^*$ points c' and d lie on opposite sides of line ab) and

(16) $$ac' \equiv_K ad, \quad bc' \equiv_K bd.$$

K-*isometry* f. Let $f = g^{-1} g'' g'$. From formulas (7), (8), (11), (12), (14), and (15) it follows that

$$f(a_1) = a_2, \quad f(b_1) = b_2, \quad f(c_1) = g^{-1}(d).$$

The theorem will therefore be proved if we show that $g^{-1}(d) = c_2$, that is (see formula (8)), if we show that $d = c$.

Let us suppose that $d \neq c$ (Fig. 187). By formulas (9), (6), (13), and (16) we have

(17) $$ac \equiv_K ad \text{ and } bc \equiv_K bd.$$

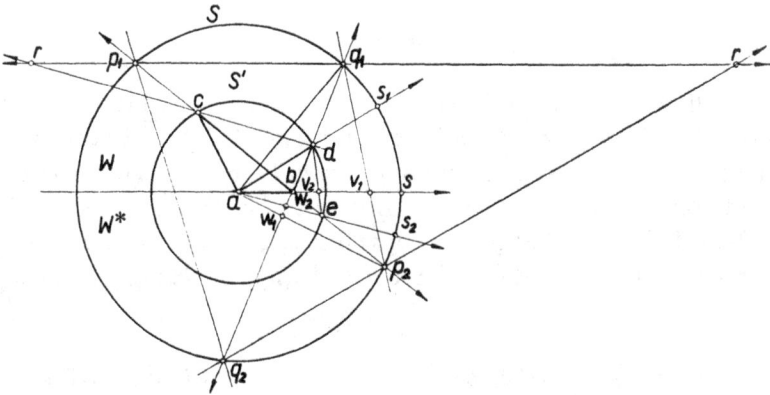

Fig. 187

From $ac \equiv_K ad$ it follows that there exists a K-isometry h such that $h(a) = a$ and $h(c) = d$. By Statement 64, we have $ac \equiv ad$ in the Cartesian sense, and, therefore, points c and d both lie on some circle S' with center a.

From $bc \equiv_K bd$ (see formula (17)) it follows that there exists a K-isometry h' such that

(18) $$h'(b) = b, \ h'(c) = d.$$

We put

$$S \cap \boldsymbol{H}(bc) = p_1, \ S \cap \boldsymbol{H}^*(bc) = p_2, \ S \cap \boldsymbol{H}(bd) = q_1, \ S \cap \boldsymbol{H}^*(bd) = q_2.$$

From $c \neq d$ it follows that $\boldsymbol{H}(bc) \neq \boldsymbol{H}(bd)$. From $c, d \in W$ it follows that $\boldsymbol{H}(bc), \boldsymbol{H}(bd) \subset W$. Therefore $p_1 \neq q_1$ and $p_1, q_1 \in W$; also $p_2 \neq q_2$ and $p_2, q_2 \in W^*$. Let r be the intersection point of projective lines $p_1 q_1$ and $p_2 q_2$. By formula (18), the K-isometry h' maps projective line bc onto projective line bd, where (see Statement 63)

$$h'(p_1) = q_1, \quad h'(p_2) = q_2.$$

Hence at three distinct points b, p_1, p_2 the K-isometry h' takes the same values as the perspective transformation from center r of projective line bc onto projective line bd. Therefore, K-isometry h' coincides with the perspective transformation from center r on the entire projective line bc (see Statements 57 and 54), and, in particular, line cd passes through point r.

Since $p_1, q_1 \in W$, then half-lines $\boldsymbol{H}(ba)$, $\boldsymbol{H}(bp_1)$, $\boldsymbol{H}(bq_1)$ are co-half-pencilar. Without deminishing the generality of the proof we may assume that

(19) $\boldsymbol{B}(\boldsymbol{H}(ba),\ \boldsymbol{H}(bp_1),\ \boldsymbol{H}(bq_1))$.

Then

(20) $\boldsymbol{B}(\boldsymbol{H}(ba),\ \boldsymbol{H}(bq_2),\ \boldsymbol{H}(bp_2))$.

We shall now determine the position of point r in two different ways:

Point r as the intersection point of lines $p_1 q_1$ and $p_2 q_2$. From formulas (19) and (20) it follows that points a and p_1 lie on the same side of line $q_1 q_2$ and points a and q_2 lie on the same side of line $p_1 p_2$. Thus, by Statement 39, it follows that angles $q_2 p_1 q_1$ and $p_1 q_2 p_2$ are both acute. Therefore r is a Cartesian point, and we have $r = \boldsymbol{H}(p_1 q_1) \cap \boldsymbol{H}(q_2 p_2)$. Hence

(21) $r \in \boldsymbol{H}(p_1 q_1)$.

Point r as the intersection point of lines $p_1 q_1$ and cd. We shall examine first the case $ab < ac$ (see formula (10)). Then point b belongs to the inner domain of circle S' and, by Statement 35, half-line bp_2 intersects circle S' in some point e. We put

$$\boldsymbol{H}^*(ba) \cap S = s, \quad \boldsymbol{H}(ad) \cap S = s_1, \quad \boldsymbol{H}(ae) \cap S = s_2.$$

From formula (19) it follows that $\boldsymbol{B}(\boldsymbol{H}(bq_1), \boldsymbol{H}(bs), \boldsymbol{H}(bp_2))$. As a result, half-line bs intersects segment $(q_1 p_2)$ in some point v_1 and segment (de) in some point v_2. Since $\boldsymbol{B}(v_1, b, a)$ and $\boldsymbol{B}(p_2, b, p_1)$, then points a and p_1 lie on the same side of line $q_1 p_2$ and, by Statement 39,

(22) $|\measuredangle q_1 p_1 p_2| = \tfrac{1}{2}|\measuredangle q_1 a p_2|$.

Since $\boldsymbol{B}(v_2, b, a)$ and $\boldsymbol{B}(e, b, c)$, then points a and c lie on the same side of line de, and, by Statement 39,

(23) $|\measuredangle dce| = \tfrac{1}{2}|\measuredangle s_1 a s_2|$.

From formula (20) it follows that half-line bq_2 intersects segment (ap_2) in some point w_1 and segment (ae) in some point w_2. From $\boldsymbol{B}(b, e, p_2)$ it follows that $\boldsymbol{B}(b, w_2, w_1)$. Therefore, on half-line $q_1 q_2$ the points are ordered in the following way:

$$q_1 \prec d \prec b \prec w_2 \prec w_1.$$

It thus follows at once that $\measuredangle s_1 a s_2 < \measuredangle q_1 a p_2$, which, together with formulas (22) and (23), implies that

(24) $\measuredangle dcb = \measuredangle dce < \measuredangle q_1 p_1 p_2$.

We arrive at the same conclusion in case $ab \equiv ad$ (Fig. 188). The argument is analogous, but simpler, since points e, s_2, v_2, w_2 do not enter into it.

From formula (24) it follows at once that $r = \mathbf{H^*}(p_1q_1) \cap \mathbf{H^*}(cd)$. Therefore

$$(25) \qquad\qquad r \in \mathbf{H^*}(p_1q_1).$$

Comparing formulas (21) and (25), we arrive at a contradiction. Hence points c and d are identical. This concludes the proof.

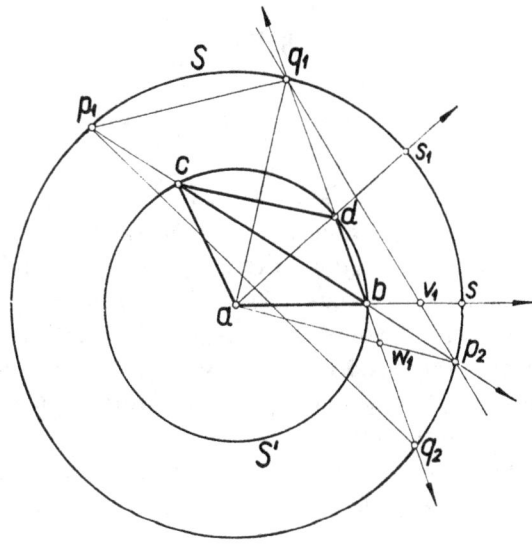

Fig. 188

We return to the verification of the axioms of congruence. For the time being, we omit Axiom C4 and take up Axiom C5.

Let a_1b_1 be any K-segment and let A be any K-half-line with the K-origin $a \in \mathbf{K_2}$ and the improper point $p \in S$ (Fig. 189). K-half line A is therefore identical with segment $(ap) \subset \mathbf{C_2}$. We have to show that there is just one K-point b on K-half-line A such that K-segment a_1b_1 is K-congruent to K-segment ab. We denote by p_1 the improper point of K-half-line a_1b_1; K-half-line a_1b_1 is therefore identical with segment $(a_1p_1) \subset \mathbf{C_2}$. By Statement 67, there exists a K-isometry f such that

$$f(a_1) = a \text{ and } f(p_1) = p.$$

By Statement 62, the K-isometry f maps open segment (a_1p_1) onto open segment (ap), and, in particular, K-point b_1 onto some point $b \in (ap)$. Hence $b \in A$ and $a_1b_1 \equiv_K ab$.

Now let b' be any point on K-half-line A such that $a_1b_1\equiv_K ab'$. There then exists a K-isometry f' such that $f'(a_1) = a$ and $f'(b_1) = b'$. Then, by Statement 63, we have $f'(p_1) = p$ and, by Statement 67, K-isometries f and f' are identical on line a_1p_1, from which it follows that $b' = b$. Therefore point b is the only K-point on K-half-line A such that $a_1b_1\equiv_K ab$, and consequently interpretation (K) satisfies Axiom C5.

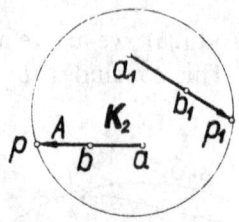

Fig. 189

We return to Axiom C4. Take any K-points a_1, b_1, c_1 and a_2, b_2, c_2 and assume that $\mathbf{B}_K(a_1,b_1,c_1)$, $\mathbf{B}_K(a_2,b_2,c_2)$, $a_1b_1\equiv_K a_2b_2$, and $b_1c_1\equiv_K b_2c_2$. We have to show that $a_1c_1\equiv_K a_2c_2$. From $a_1b_1\equiv_K a_2b_2$ it follows that there exists a K-isometry f such that $f(a_1) = a_2$ and $f(b_1) = b_2$. The K-isometry f maps K-point c_1 on some K-point c_2' of K-half-line b_2c_2 and thus $b_1c_1\equiv_K b_2c_2'$. Since Axiom C5 has already been verified, we therefore have $c_2' = c_2$. Hence $f(a_1) = a_2$ and $f(c_1) = c_2$, from which it follows that $a_1c_1\equiv_K a_2c_2$, which was to be proved.

We now come to Axiom C6. Given two K-lines L_1 and L_2 and K-points $a_1,b_1,c_1 \in L_1$, $d_1 \sim \in L_1$, and $a_2,b_2,c_2 \in L_2$, $d_2\sim \in L_2$, let us assume that

$$(26) \qquad \mathbf{B}_K(a_1,b_1,c_1), \quad \mathbf{B}_K(a_2,b_2,c_2), \quad b_1c_1\equiv_K b_2c_2,$$

and

$$(27) \qquad a_1b_1\equiv_K a_2b_2, \quad d_1a_1\equiv_K d_2a_2, \quad d_1b_1\equiv_K d_2b_2.$$

We have to show that $d_1c_1\equiv_K d_2c_2$. Indeed, from (27) it follows, by Statement 69, that there exists a K-isometry f such that

$$f(a_1) = a_2, \; f(b_1) = b_2, \; f(d_1) = d_2.$$

Because of Statement 63 and Axiom C5 (already verified) it follows from (26) that $f(c_1) = c_2$. Therefore $d_1c_1\equiv_K d_2c_2$.

We consider the last axiom of congruence, C7. Given a K-half-plane $W \cap \mathbf{K}_2$ with K-boundary $K \cap \mathbf{K}_2$ (W is a Cartesian half-plane with boundary K), a K-segment $ab \subset K$, and any K-triangle $a_1b_1c_1$, let us assume that

$$(28) \qquad a_1b_1\equiv_K ab.$$

We have to prove that on half-plane W there is just one K-point c such that $a_1c_1\equiv_K ac$ and $b_1c_1\equiv_K bc$. We put

$$(29) \qquad \mathbf{H}(ab) \cap S = p, \qquad \mathbf{H}(a_1b_1) \cap S = p_1,$$

and $\mathbf{HP}(\mathbf{L}(a_1p_1)c_1) = W_1$. Let us consider the arcs $J = S \cap W$ and $J_1 = S \cap W_1$ of circle S (see Statement 36). By Statement 66, there exists a K-isometry f such that

$$(30) \qquad f(a_1) = a,$$

$$(31) \qquad f(p_1) = p,$$

$$(32) \qquad f(J_1) = J.$$

From formulas (28)–(31) and from Axiom C5 (already verified) it follows that

$$(33) \qquad f(b_1) = b.$$

We put

$$(34) \qquad f(c_1) = c.$$

Then $c \in \mathbf{K}_2$, and from formulas (30), (33), and (34) it follows that $a_1c_1\equiv_K ac$ and $b_1c_1\equiv_K bc$. We still have to show that $c \in W$. We put $r = \mathbf{H}(ac) \cap S$, $r_1 = \mathbf{H}(a_1c_1) \cap S$. Obviously

$$(35) \qquad r_1 \in J_1.$$

Further, from formulas (30) and (34) it follows that $f(r_1) = r$, which implies, by formulas (32) and (35), that $r \in J$; hence $c \in W$.

Now let c' be any K-point of half-plane W such that $a_1c_1\equiv_K ac'$ and $b_1c_1\equiv_K bc'$. Since, in addition, formula (28) holds, then, by Statement 69, there exists a K-isometry f' such that

$$(36) \qquad f'(a_1) = a, \ f'(b_1) = b, \ f'(c_1) = c'.$$

Then, as may readily be noted,

$$f'(p_1) = p \text{ and } f'(J_1) = J,$$

and therefore, by Statement 66, K-isometry f' is identical with K-isometry f. Hence, by formulas (34) and (36), K-point c' coincides with K-point c.

We turn to the Axiom of Continuity. Since domain $\mathbf{K}_2$ is convex, and the

Axiom of Continuity is satisfied by interpretation (C), then it is also satisfied by interpretation (K).

In this manner we have shown that interpretation (K) satisfies all the axioms of absolute geometry.

At last, we shall consider sentence BL (as formulated on page 197, line 16). Take any K-line L and any K-point $a \sim \epsilon L$ (Fig. 190). Let p and q be the improper points of K-line L. Then K-lines $K_1 = \mathbf{L}(ap) \cap \mathbf{K_2}$ and $K_2 = \mathbf{L}(aq) \cap \mathbf{K_2}$ are distinct; they both pass through K-point a and do not intersect K-line L. Therefore interpretation (K) satisfies sentence BL.

Thus, we have

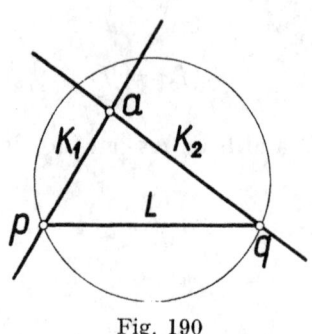

Fig. 190

PROPOSITION 4. *The interpretation*

(K)   K-*points*, K-*lines*, $\mathbf{B}_K$, $\mathbf{E}_K$

is a model for the axiom system (GA_2) *and satisfies the sentence* BL.

The Klein-Beltrami model for the axiom system (GA_3) of space absolute geometry consists of the notions of three-dimensional projective space $\mathbf{P_3}$. In its construction the same ideas are used as in the plane case. In order to obtain from the definitions of the notions of the Klein-Beltrami model for the axiom system (GA_2) definitions of the notions of the Klein-Beltrami model for the axiom system (GA_3), it is sufficient:

1. To regard the sphere with center $(0,0,0)$ and radius equal to 1 as the absolute S; to replace the domain $\mathbf{K_2}$ by the inner domain $\mathbf{K_3}$ of sphere S; to replace the space $\mathbf{P_2}$ everywhere by the space $\mathbf{P_3}$, and the space $\mathbf{C_2}$ everywhere by the space $\mathbf{C_3}$.

2. To define additionally K-planes as non-empty sets of the form $P \cap \mathbf{K_3}$, where P is a plane of space $\mathbf{C_3}$.

No new difficulties are involved in checking that the interpretation thus obtained is actually a model for the axiom system (GA_3).

Domain $\mathbf{K_2}$ may be treated as one of the K-planes of three-dimensional Klein space $\mathbf{K_3}$ (Fig. 191), and K-points and K-lines of the plane model may be simultaneously treated as K-points and K-lines of space K_3. Thus, from Proposition 4 we obtain immediately

PROPOSITION 5. *The Klein-Beltrami model for the axiom system* (GA_3) *satisfies sentence* BL.

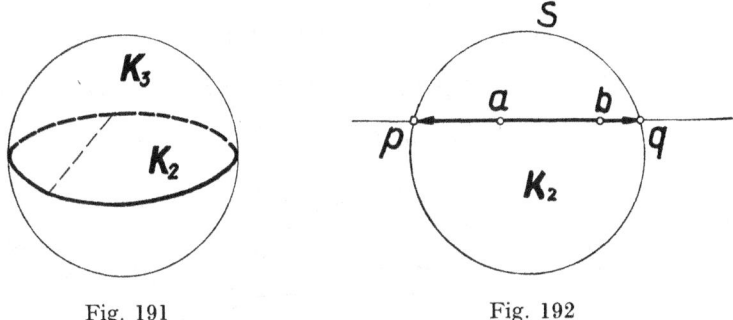

Fig. 191   Fig. 192

8. Formula for Distance in Klein Space K_2

Given two distinct points $a,b \in K_2$. We denote by p and q the inter-section points of line ab and the absolute S (Fig. 192). Since the sense of vectors $[p-a]$ and $[p-b]$ is opposite to the sense of vectors $[q-a]$ and $[q-b]$, then (see formula (11) on page 238) the cross ratio of the four points a, b, p, q is expressed by the formula

$$(a,b;p,q) = \frac{\rho(a,p) \cdot \rho(b,q)}{\rho(a,q) \cdot \rho(b,p)}.$$

Let ln denote the natural logarithm. Then, by Statement 44,

(1) $|\ln(a,b;p,q)| = |\ln(b,a;p,q)| = |\ln(a,b;q,p)| = |\ln(b,a;q,p)|,$

and therefore the number $|\ln(a,b;p,q)|$ may be regarded as a function of K-segment ab:

$$\varphi_0(ab) = |\ln(a,b;p,q)|.$$

We shall show that the function φ_0 is a K-measure of K-segments. To do this we should check that φ_0 satisfies the following two conditions:

(I) $a_1c_1 \equiv_K a_2c_2$ implies $\varphi_0(a_1c_1) = \varphi_0(a_2c_2)$, for any K-segments a_1c_1 and a_2c_2.

(II) $B_K(a,b,c)$ implies $\varphi_0(ab) + \varphi_0(bc) = \varphi_0(ac)$ for any K-points a, b, c.

That condition (I) is satisfied results at once from the fact that the cross ratio is an invariant of the linear transformations and, in particular, of the K-isometries.

We pass on to condition (II). Assume that $B_K(a,b,c)$. We denote by

p and q respectively the improper points of K-half-lines bc and ba (Fig. 193). We order the points of the segment $< pq >$ from point p to point q; thus $p \prec c \prec b \prec a \prec q$. Then $ap > bp$ and $bq > aq$ and hence $(a,b;p,q) > 1$; similarly, $bp > cp$ and $cq > bq$, and hence $(b,c;p,q) > 1$. It thus follows that

$$\varphi_0(ab) + \varphi_0(bc) = |\ln(a,b;p,q)| + |\ln(b,c;p,q)| =$$

$$|\ln(a,b;p,q) + \ln(b,c;p,q)| = |\ln((a,b;p,q) \cdot (b,c;p,q))| =$$

$$|\ln(a,c;p,q)| = \varphi_0(ac).$$

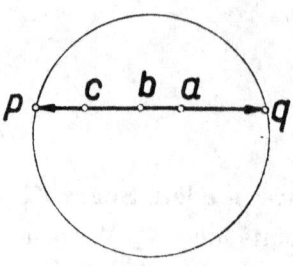

Fig. 193

Therefore condition (II) is also satisfied and the function φ_0 is a K-measure of K-segments.

Thus an arbitrary K-measure of K-segments has the form $\lambda \cdot \varphi_0$, where λ is a positive constant. For our purposes it is convenient to fix in $\mathbf{K}_2$ the K-measure $\frac{1}{2}\varphi_0$; the formula for the K-distance ρ_K induced by this K-measure looks as follows

$$\rho_K(a,b) = \frac{1}{2} |\ln(a,b;p,q)| = \frac{1}{2}\left|\ln \frac{\rho(a,p) \cdot \rho(b,q)}{\rho(a,q) \cdot \rho(b,p)}\right|.$$

By this formula the K-distance $\rho_K(a,b)$ of K-points a and b has been expressed in terms of the Cartesian distance ρ.

We shall now find a formula determining $\rho_K(a,b)$ as a function of the Cartesian coordinates of K-points a and b. Circle S has the equation

(2) $x_1^2 + x_2^2 = 1,$

the domain $\mathbf{K}_2$ consists of points (x_1, x_2) satisfying the inequality

(3) $x_1^2 + x_2^2 < 1.$

We assume that $a \neq b$; let $a = (a_1, a_2)$, $b = (b_1, b_2)$.

The intersection points p and q of line ab and circle S (Fig. 192) may be represented either in the form

$$p = a + t_1 \cdot (a-b), \quad q = a + t_2 \cdot (a-b),$$

where t_1 and t_2 are the roots of the equation

$$(a-b)^2 t^2 + 2(a \cdot (a-b))t + a^2 - 1 = 0$$

(see equation (2)), or in the form

$$q = b + u_1 \cdot (b-a), \quad p = b + u_2 \cdot (b-a),$$

where u_1 and u_2 are the roots of the equation

$$(b-a)^2 u^2 + 2(b \cdot (b-a))u + b^2 - 1 = 0.$$

We put

(4) $\quad \alpha = (a-b)^2 = (b-a)^2, \quad \beta = a \cdot (a-b), \quad \beta_1 = b \cdot (b-a),$

(5) $\quad \gamma = 1-a^2, \quad \gamma_1 = 1-b^2, \quad \delta = \beta^2 + \alpha\gamma, \quad \delta_1 = \beta_1^2 + \alpha\gamma_1.$

Then

(6) $\quad\quad\quad \beta + \gamma = 1 - a \cdot b, \quad \beta_1 + \gamma_1 = 1 - b \cdot a,$

and from inequality (3) it follows that

$$\alpha, \gamma, \gamma_1, \beta + \gamma, \beta_1 + \gamma_1 > 0.$$

From formulas (4)–(6) we obtain successively

$$\beta + \beta_1 = \alpha, \quad \beta + \gamma = \beta_1 + \gamma_1, \quad \delta = \delta_1, \quad \beta\beta_1 + \delta = \alpha(\beta + \gamma),$$
$$(\beta + \gamma)^2 - \delta = \gamma\gamma_1.$$

Let

(7)
$$\tau = \sqrt{\frac{\gamma\gamma_1}{(\beta+\gamma)(\beta_1+\gamma_1)}} = \frac{\sqrt{1-a^2}\,\sqrt{1-b^2}}{\sqrt{1-ab}\,\sqrt{1-ba}} =$$
$$= \frac{\sqrt{1-a_1^2-a_2^2}\,\sqrt{1-b_1^2-b_2^2}}{1-a_1b_1 - a_2b_2}.$$

We now have

$$\rho_K(a,b) = \left| \ln \frac{\rho(a,p) \cdot \rho(b,q)}{\rho(a,q) \cdot \rho(b,p)} \right| = \left| \ln \frac{t_1 u_1}{t_2 u_2} \right| = \left| \ln \frac{(\beta+\sqrt{\delta})(\beta_1+\sqrt{\delta_1})}{(\beta-\sqrt{\delta})(\beta_1-\sqrt{\delta_1})} \right| =$$

$$= \left| \ln \frac{(\beta+\sqrt{\delta})(\beta_1+\sqrt{\delta})}{(\beta-\sqrt{\delta})(\beta_1-\sqrt{\delta})} \right| = \left| \ln \frac{(\beta\beta_1+\delta) + (\beta+\beta_1)\sqrt{\delta}}{(\beta\beta_1+\delta) - (\beta+\beta_1)\sqrt{\delta}} \right| =$$

$$= \left| \ln \frac{\alpha(\beta+\gamma) + \alpha\sqrt{\delta}}{\alpha(\beta+\gamma) - \alpha\sqrt{\delta}} \right| = \left| \ln \frac{1 + \dfrac{\sqrt{\delta}}{\beta+\gamma}}{1 - \dfrac{\sqrt{\delta}}{\beta+\gamma}} \right| = \ln \frac{1 + \dfrac{\sqrt{\delta}}{\beta+\gamma}}{1 - \dfrac{\sqrt{\delta}}{\beta+\gamma}}.$$

Moreover we have

$$\frac{\delta}{(\beta+\gamma)^2} = 1 - \frac{(\beta+\gamma)^2-\delta}{(\beta+\gamma)^2} = 1 - \frac{\gamma\gamma_1}{(\beta+\gamma)^2} = 1 - \tau^2,$$

and therefore

(8) $$\rho_K(a,b) = \tfrac{1}{2} \ln \frac{1 + \sqrt{1-\tau^2}}{1 - \sqrt{1-\tau^2}}.$$

Substituting in formula (8) the value for τ from formula (7), we obtain, after simple rearrangement, the sought-for relation between $\rho_K(a,b)$ and the coordinates of the K-points a and b:

(9) $$\rho_K(a,b) = \tfrac{1}{2} \ln \frac{(1-a_1b_1-a_2b_2) + \sqrt{(a_1-b_1)^2+ (a_2-b_2)^2 - (a_1b_2-a_2b_1)^2}}{(1-a_1b_1-a_2b_2) - \sqrt{(a_1-b_1)^2+ (a_2-b_2)^2 - (a_1b_2-a_2b_1)^2}}.$$

It is readily shown that formula (9) remains valid when $a = b$.

Henceforth, by two-dimensional Klein space $\boldsymbol{K_2}$ we shall always understand the metric space in which the metric ρ_K is defined by formula (9).

9. Non-Categoricity of Absolute Geometry. Euclidean Geometry. Bolyai-Lobachevskian Geometry

In Sections 3 and 7 we have described two models for the axiom system (GA_3) of absolute geometry, the Cartesian model (C) and the Klein model (K). By Propositions 1 and 5, model (C) satisfies sentence E and model (K) satisfies sentence BL, which is the negation of E. Therefore models (C) and (K) are non-isomorphic. Thus the axiom system (GA_3) is not categorical, i.e. it is too weak to determine its model uniquely up to isomorphism. System (GA_3) should therefore be strengthened. To do so we shall add to (GA_3) a sentence S as a new axiom. Obviously this sentence cannot be contradictory to the remaining axioms of system (GA_3). On the other hand, if this strengthening is to have any essential significance, then sentence S must be independent of the axioms of (GA_3).

From Propositions 1 and 5 it at once follows that each of the sentences E and BL satisfies these two conditions. By adding sentence E to the axiom system (GA_3) we obtain a consistent axiom system (GE_3) of space Euclidean geometry and by adding sentence BL to the axiom system (GA_3) we obtain a consistent axiom system (GBL_3) of space Bolyai-Lobachevskian geometry. In Section 13 of Chapter V and in Section 31 of Chapter VI we shall show that (GE_3) and (GBL_3) constitute categorical axiom systems of geometry. We shall develop Euclidean geometry (in Chapter V) and Bolyai-Lobachevskian geometry (in Chapter VI) sufficiently to be able to prove in them the theorems necessary to establish the categoricity of the systems (GE_3) and (GBL_3).

When supplementing the axiom system (GA_3) to make it a categorical axiom system we could of course use, instead of sentences E and BL, any other sentences which are equivalent to them on the basis of absolute geometry. As an example, we shall show that each of the following five sentences is equivalent to sentence E:

S1. *In every triangle the sum of the sizes of the angles is equal to π.*

S2. *There exists a triangle in which the sum of the sizes of the angles is equal to π.*

S3. *There exists a rectangle.*

S4. *There exists a line L_0 and a point $a_0 \sim \in L_0$ such that there is at most one line K passing through point a_0 and parallel to line L_0.*

S4. *There exists a line L_0 and a point $a_0 \sim \in L_0$ such that there is at most one line K passing through a_0 and parallel to L_0.*

Indeed, by Theorem 64 of Chapter II and Theorem 45 of Chapter III, sentence E implies sentence S1; since there exists a triangle, then sentence S1 implies sentence S2; next, because of Theorems 47, 28, 26, and 24 of Chapter III, each of the sentences S2, S3, S4 implies the sentence following it; finally, by Theorems 27 and 24 of Chapter III, sentence S5 leads back to sentence E. Therefore, on the basis of absolute geometry, sentence E is equivalent to each of the sentences S1–S5 and hence sentence BL is equivalent to the negation of each of the sentences S1–S5.

Since both geometries, Euclidean and Bolyai-Lobachevskian, are consistent and categorical theories, they are equally satisfactory from the point of view of logic. They give, however, two different pictures of the space. Does either of these pictures conform with reality, and if so, then which? This question can be settled, if at all, only by experience. It is known that the abstract scheme of the phenomena of size and shape given by Euclidean geometry exhibits good agreement with reality. There is not, however, a sufficient basis for the opinion that this agreement is complete and that another abstract scheme, e.g. Bolyai-Lobachevskian geometry, is not in better agreement with reality. Since the time of the theory of relativity, the ability of Euclidean geometry to describe the geometrical relations of the entire universe has been seriously questioned. Although it does not seem probable that, in this regard, Bolyai-Lobachevskian geometry accurately reflects reality, its discovery, however, has contributed to the final rejection of the erroneous views that the origin of the geometrical notions and theorems is independent of experience.

V

Euclidean Geometry

1. The Axiom of Euclid

By the *Axiom of Euclid* we understand the sentence (see page 197):

AXIOM E. *For any plane P, any line $L \subset P$, and any point $a \in P-L$ there exists at most one line $K \subset P$ passing through point a and not intersecting line L.*

The axioms of absolute geometry, together with Axiom E, constitute the entire axiom system of Euclidean geometry.

Axiom E and Theorem 39 of Chapter II give us at once

THEOREM 1. *For any plane P, any line $L \subset P$ and any point $a \in P-L$ there exists just one line $K \subset P$ passing through point a and not intersecting line L.*

2. The Sum of the Angles of a Triangle and of a Quadrangle

From Theorem 1 and from Theorems 64 of Chapter II and 45 of Chapter III we obtain the following important theorem:

THEOREM 2. *The sum of the sizes of the angles in any triangle equals π.*

As a simple consequence of Theorem 2 we have

THEOREM 3. *The sum of the sizes of the angles in any convex quadrangle equals 2π.*

The proof follows at once from the proof of Theorem 49 of Chapter III (on the sum of the sizes of the angles of a convex quadrangle in absolute geometry) if, in formulas (2)–(4) on page 175, we change the sign "$\leqslant$" to the sign "$=$".

From Theorem 3 and from Theorem 82 of Chapter II we obtain immediately

THEOREM 4. *Every Saccheri quadrangle is a rectangle.*

3. Parallel Lines (Conclusion)

In Euclidean geometry we can characterize parallel lines in the following manner (cf. definition on page 150):

THEOREM 5. *If $K \neq L$, then $K \parallel L$ if and only if lines K and L lie in one plane and are disjoint.*

PROOF. We know that if lines K and L are parallel and distinct, then they lie in one plane and are disjoint (see page 150). Let us now assume that lines K and L both lie in a plane P and are disjoint (Fig. 194). Let $a \in K$. By Theorem 24 of Chapter III, there is a line K' passing through point a and parallel to L. Since lines K' and L lie in one plane, then $K' \subset P$; moreover $K' \cap L = 0$. Thus, by Axiom E, we conclude that $K' = K$. Hence $K \parallel L$.

From Theorems 1 and 5 we get at once

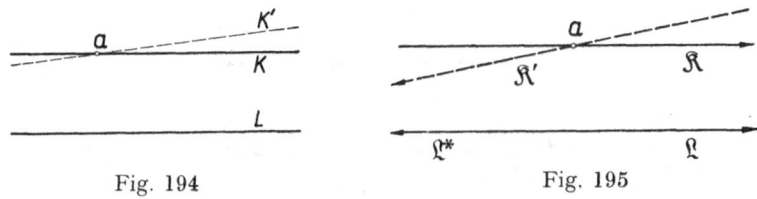

Fig. 194   Fig. 195

THEOREM 6. *For any point a and any line L there is just one line K passing through point a and parallel to L.*

We now consider two axes, $\Re$ of a line K and $\mathfrak{L}$ of a line L. We assume that $\Re \parallel \mathfrak{L}$ (Fig. 195). Then $K \parallel L$. Let $a \in K$. By Theorem 20 of Chapter III, there is an axis $\Re'$ passing through point a and parallel to axis $\mathfrak{L}^*$. Let K' be the line of axis $\Re'$. Then $K' \parallel L$, and, by Theorem 6, $K' = K$. Since, by Theorem 117 of Chapter II, we have $\Re \sim \parallel \mathfrak{L}^*$, then $\Re^* \parallel \mathfrak{L}^*$. We have thus shown that

(1) $\text{if } \Re \parallel \mathfrak{L}, \text{ then } \Re^* \parallel \mathfrak{L}^*.$

It readily follows from (1) that the parallel relation for lines is transitive:

THEOREM 7. *Given any three lines K, L, M, if $K \parallel L$ and $L \parallel M$, then $K \parallel M$.*

PROOF. If $K \parallel L$, and $L \parallel M$, then there exist an axis $\Re$ of line K and an axis $\mathfrak{L}$ of line L such that $\Re \parallel \mathfrak{L}$. Because of (1), there exists an axis $\mathfrak{M}$ of line M such that $\mathfrak{L} \parallel \mathfrak{M}$. Then $\Re \parallel \mathfrak{M}$ and therefore $K \parallel M$.

4. Parallel Half-Lines (Conclusion)

If half-lines A and B lie on one line, then they are parallel if and only if they have the same orientation (see Case 1 in the definition on page 138). Half-lines that are parallel, but which do not lie on one line, may be characterized in Euclidean geometry in the following simple way (see Case 2 in the definition on page 138):

THEOREM 8. *Given two half-lines, $A \subset K$ with origin a and $B \subset L$ with origin b, if $K \neq L$ and $a \neq b$, then $A \parallel B$ if and only if $K \parallel L$ and half-lines A and B lie on the same side of line ab.*

PROOF. Assume that $K \neq L$ and $a \neq b$. If $A \parallel B$, then, from Theorem 107(II) of Chapter II and Theorem 5, it immediately follows that $K \parallel L$, and, by the definition of parallel half-lines, half-lines A and B lie on the

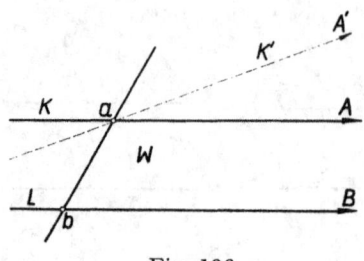

Fig. 196

same side of line ab. We next assume that $K \parallel L$ and that half-lines A and B lie on the same half-plane W with boundary $\mathbf{L}(ab)$ (Fig. 196). By Theorem 19 of Chapter III, a half-line $A' \subset K'$ parallel to half-line B may be produced from point a. Then $K' \parallel L$, which, by Theorem 6, implies that $K' = K$. Further, from $A' \parallel B$ it follows that $A' \subset W$, and consequently $A' = A$. Therefore $A \parallel B$.

From Theorem 8 it follows at once that, for any two half-lines A and B,

(1)   if $A \parallel B$,   then $A^* \parallel B^*$.

We may then readily obtain

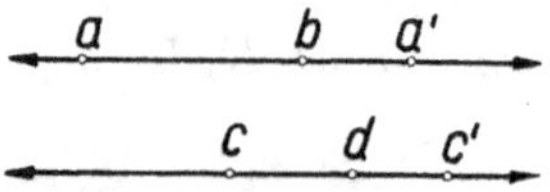

Fig. 197

THEOREM 9. *For any points a, b, c, d, if $a \neq b$ and $c \neq d$, then*

$$\mathbf{H}(ab) \parallel \mathbf{H}(cd) \quad \text{implies} \quad \mathbf{H}(ba) \parallel \mathbf{H}(dc).$$

PROOF. Take points a' and c' (Fig. 197) such that $\mathbf{B}(a,b,a')$ and $\mathbf{B}(c,d,c')$. From $\mathbf{H}(ab) \parallel \mathbf{H}(cd)$ it follows that $\mathbf{H}(ba') \parallel \mathbf{H}(dc')$, from which it follows, by formula (1), that $\mathbf{H}(ba) \parallel \mathbf{H}(dc)$.

5. Two Angles With Respectively Parallel Sides

THEOREM 10. *Given angles A_1B_1 and A_2B_2 in a plane P, if $A_1 \parallel A_2$ and $B_1 \parallel B_2$, then $A_1B_1 \equiv A_2B_2$.*

PROOF. Assume that $A_1 \parallel A_2$ and $B_1 \parallel B_2$. Let us first assume that one pair of the parallel half-lines, say A_1 and A_2, lies on one line L

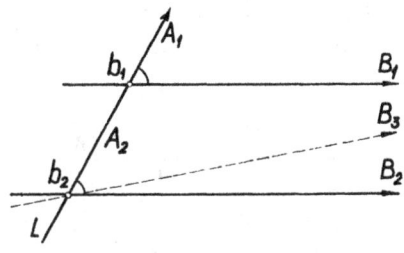

Fig. 198

(Fig. 198). We denote by b_1 and b_2 the origins of half-lines B_1 and B_2. If $b_1 = b_2$, then it is readily seen that angles A_1B_1 and A_2B_2 are identical, and, therefore, congruent. We may thus henceforth assume that b_1 and b_2 are distinct points on line L. Half-lines B_1 and B_2 lie on one side of line L. On the same side we produce from point b_2 a half-line B_3 such that

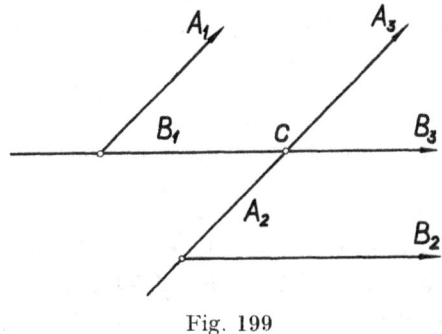

Fig. 199

$A_1B_1 \equiv A_2B_3$. From Theorem 5 it follows that $L(B_3) \parallel L(B_1)$. Since $L(B_1) \parallel L(B_2)$, then, by Theorem 7, $L(B_3) \parallel L(B_2)$, from which it follows that $L(B_3) = L(B_2)$. Hence $B_3 = B_2$, and, as a result, $A_1B_1 \equiv A_2B_2$, which was to be proved.

Let us next assume that neither of the pairs of half-lines, A_1, A_2 and B_1, B_2, lies on one line (Fig. 199). Let $c = L(A_2) \cap L(B_1)$. From point c we produce half-lines $A_3 \parallel A_1$ and $B_3 \parallel B_1$. It is readily seen that $A_3 \subset L(A_2)$ and $B_3 \subset L(B_1)$, and, by Theorem 115 of Chapter II, we have $A_3 \parallel A_2$ and $B_3 \parallel B_2$. Because of what we have already proved and by

Theorem 111 of Chapter II, we obtain $A_1B_1 \equiv A_3B_3$ and $A_3B_3 \equiv A_2B_2$; hence $A_1B_1 \equiv A_2B_2$.

Thus the theorem is proved.

6. Parallel Projection upon a Line

Let there be given a line M and a line L not parallel to M on a plane P (Fig. 200). For every point $p \in P$, the line M_p passing through point p and parallel to line M intersects, by Theorem 7, line L is some point p_1.

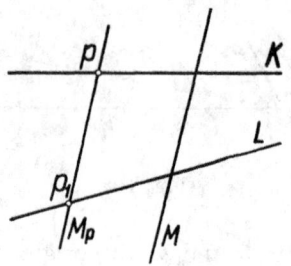

Fig. 200

The point p_1 is called the *parallel projection of point p upon line L in the direction of line M*. Take a line $K \subset P$ not parallel to line M. The function f mapping every point $p \in K$ onto the projection $f(p) = p_1 \in L$ is called the *parallel projectivity of line K upon line L in the direction of line M*.

THEOREM 11. *Every parallel projectivity f of a line K upon a line L on a plane P establishes a one-to-one correspondence between the points of K and the points of L, where $\mathbf{B}(p,q,r)$ implies $\mathbf{B}(f(p),f(q),f(r))$ for any three points $p,q,r \in K$.*

PROOF. Let f be an arbitrary parallel projection of line K upon line L. From Theorems 5 and 7 it follows that $f(K) = L$, and, by Theorem 6, the transformation f is one-to-one. The proof of the remaining part of the theorem is quite similar to the proof of Theorem 79 of Chapter II (on perpendicular projection).

7. Parallelograms

A quadrangle (a,b,c,d) whose opposite sides lie on parallel lines, that is, $\mathbf{L}(ab) \parallel \mathbf{L}(cd)$ and $\mathbf{L}(bc) \parallel \mathbf{L}(ad)$, will be called a *parallelogram*. It is readily seen that the parallelogram is a convex quadrangle.

THEOREM 12. *In order that a plane quadrangle (a,b,c,d) be a parallelogram it is necessary and sufficient that $\mathbf{H}(ab) \parallel \mathbf{H}(dc)$ and $ab \equiv cd$.*

PROOF. We assume that a plane quadrangle (a,b,c,d) is a parallelogram (Fig. 201). Because of the convexity of quadrangle (a,b,c,d) it follows, by Theorem 8, that $\mathbf{H}(ab) \parallel \mathbf{H}(dc)$ and that $\mathbf{H}(ad) \parallel \mathbf{H}(bc)$. By Theorem 10, we therefore have

$$\sphericalangle \, bac \equiv \mathbf{H}^*(cd)\,\mathbf{H}^*(ca) \equiv \sphericalangle \, dca \qquad \text{and} \qquad \sphericalangle \, dac \equiv \mathbf{H}^*(cb)\,\mathbf{H}^*(ca) \equiv \sphericalangle \, bca\,;$$

thus the triangles abc and cda are congruent and $ab \equiv cd$.

We next assume that $\mathbf{H}(ab) \parallel \mathbf{H}(dc)$ and $ab \equiv cd$ (Fig. 202). Suppose that lines bc and ad intersect in some point p. From $\mathbf{H}(ab) \parallel \mathbf{H}(dc)$ it follows that $\sim\mathbf{B}(b,p,c)$; let for example, $\mathbf{B}(b,c,p)$. Then, by Theorem 11,

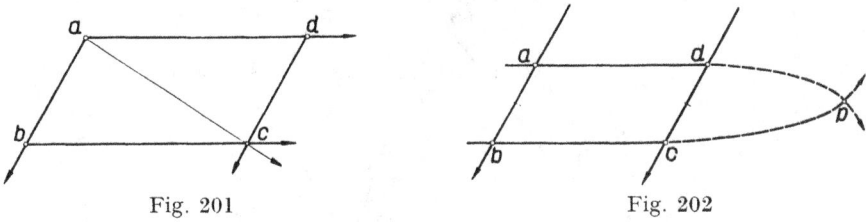

Fig. 201   Fig. 202

we also have $\mathbf{B}(a,d,p)$. Therefore $\mathbf{H}(bp) \parallel \mathbf{H}(cp)$ and $\mathbf{H}(ap) \parallel \mathbf{H}(dp)$; thus, since $\mathbf{H}(ab) \parallel \mathbf{H}(dc)$, we conclude, by Theorems 9 and 10, that $\sphericalangle \, b \equiv \sphericalangle \, dcp$ and $\sphericalangle \, a \equiv \sphericalangle \, cdp$. But this would then give us $bp \equiv cp$, which contradicts the fact that $bp > cp$. Therefore lines bc and ad do not have any point in common, which means that $\mathbf{L}(bc) \parallel \mathbf{L}(ad)$. Since also $\mathbf{L}(ab) \parallel \mathbf{L}(cd)$, then quadrangle (a,b,c,d) is a parallelogram.

8. The Theorem of Thales

We shall now prove the *Thales Theorem*, which, with the help of the notions of parallel projectivity and similitude, may be formulated in the following way:

THEOREM 13. *Any parallel projectivity f of a line K upon a line L on a plane P is a similitude.*

PROOF. Let f be any parallel projectivity of line K upon line L. If $K \parallel L$ then, because of Theorem 12, the transformation f is an isometry, and therefore a similitude. We next assume that lines K and L intersect in some point p (Fig. 203). We associate with every segment ab of line K a number

$$\varphi(ab) = |f(a)f(b)|.$$

We shall show that the function φ defined in this way satisfies the following two conditions:

(I) If a_1b_1 and a_2b_2 are two segments of line K, then from $a_1b_1 \equiv a_2b_2$ it follows that $\varphi(a_1b_1) = \varphi(a_2b_2)$.

(II) if $a,b,c \in K$ and $\mathbf{B}(a,b,c)$, then $\varphi(ac) = \varphi(ab) + \varphi(bc)$.

From Theorem 11 it follows at once that condition (II) is satisfied. We now pass to the proof of condition (I). Take two segments a_1b_1 and a_2b_2 on line K such that

$$(1) \qquad\qquad a_1b_1 \equiv a_2b_2.$$

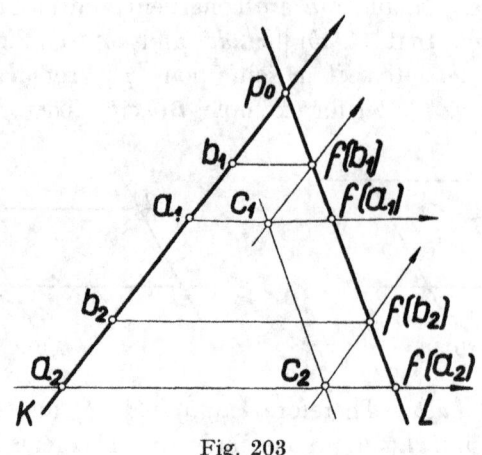

Fig. 203

We may assume that half-lines a_1b_1 and a_2b_2 have the same orientation (Fig. 203); then

$$(2) \qquad\qquad \mathbf{H}(a_1b_1) \parallel \mathbf{H}(a_2b_2).$$

Through point $f(b_i)$ $(i = 1,2)$ we produce the line parallel to line K; we denote by c_i the point of intersection of this line with the line $a_i f(a_i)$. The quadrangle $(a_i, b_i, f(b_i), c_i)$ is a parallelogram. Using Theorem 12, we obtain

$$(3) \qquad a_i b_i \equiv c_i f(b_i) \quad \text{and} \quad \mathbf{H}(a_i b_i) \parallel \mathbf{H}(c_i f(b_i)) \quad (i = 1,2).$$

From formulas (1)–(3) it follows immediately that

$$(4) \qquad c_1 f(b_1) \equiv c_2 f(b_2) \quad \text{and} \quad \mathbf{H}(c_1 f(b_1)) \parallel \mathbf{H}(c_2 f(b_2)).$$

Using Theorem 12 again, we conclude that quadrangle $(c_1, f(b_1), f(b_2), c_2)$ is a parallelogram, and therefore $\mathbf{L}(c_1 c_2) \parallel L$. Hence quadrangle

$$(c_1, f(a_1), f(a_2), c_2)$$

is also a parallelogram, from which it follows that

$$(5) \qquad c_1 f(a_1) \equiv c_2 \mathbf{H} f(a_2) \quad \text{and} \quad \mathbf{H}(c_1 f(a_1)) \parallel \mathbf{H}(c_2 f(a_2)).$$

From the second formulas in (4) and (5) we conclude, by Theorem 10, that $\sphericalangle f(a_1)c_1f(b_1) \equiv \sphericalangle f(a_2)c_2f(b_2)$, which, together with the first formulas in (4) and (5), gives us $f(a_1)f(b_1) \equiv f(a_2)f(b_2)$, that is

$$\varphi(a_1b_1) = \varphi(a_2b_2).$$

This concludes the proof of condition (I).

By conditions (I) and (II), the function φ is a measure of segments on line K. By Theorem 34 of Chapter III there thus exists a constant $\lambda > 0$ such that

$$\varphi(ab) = |f(a)f(b)| = \lambda \cdot |ab|$$

for any segment ab on line K. Thus f is a similitude, which is what we had to prove.

As a consequence of Theorem 13 we obtain

THEOREM 14. *Consider in a plane P two lines K, L intersecting in some point p_0 and two parallel lines M_1, M_2 not passing through point p_0 and intersecting line K in points a_1 and a_2, and line L in points b_1 and b_2, respectively. Then*

$$\frac{\rho(p_0,a_1)}{\rho(p_0,a_2)} = \frac{\rho(p_0,b_1)}{\rho(p_0,b_2)} = \frac{\rho(a_1,b_1)}{\rho(a_2,b_2)}.$$

PROOF. The first of these equalities is obvious by Thales' Theorem. We shall show that the second equality also holds. To do this we produce

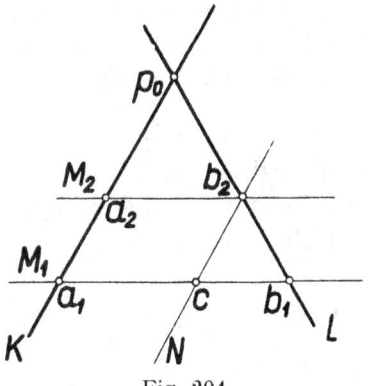

Fig. 204

through point b_2 a line $N \parallel K$ (Fig. 204). It intersects line M_1 in some point c. From Thales' Theorem (applied to the parallel projection of line L upon line M_1 in the direction of line K) we conclude that

$$\frac{\rho(p_0,b_1)}{\rho(p_0,b_2)} = \frac{\rho(a_1,b_1)}{\rho(a_1,c)},$$

since $a_1c \equiv a_2b_2$, then

$$\frac{\varrho(p_0,b_1)}{\varrho(p_0,b_2)} = \frac{\varrho(a_1,b_1)}{\varrho(a_2,b_2)}.$$

This concludes the proof.

9. Similar Triangles

In Chapter III (page 177) we introduced the general notion of two similar figures. In this section we are interested in similar triangles. The following theorem gives a sufficient condition that two triangles be similar:

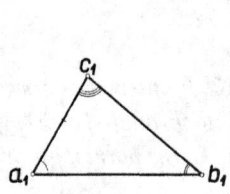 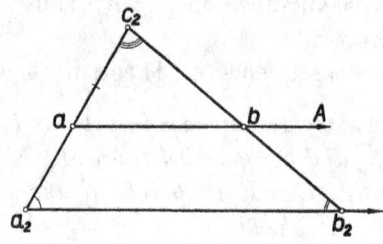

Fig. 205

THEOREM 15. *Given two triangles $a_1b_1c_1$ and $a_2b_2c_2$, if $\sphericalangle a_1 \equiv \sphericalangle a_2$, $\sphericalangle b_1 \equiv \sphericalangle b_2$, and $\sphericalangle c_1 \equiv \sphericalangle c_2$, then triangle $a_1b_1c_1$ is similar to triangle $a_2b_2c_2$, and the function f defined by the conditions $f(a_1) = a_2$, $f(b_1) = b_2$, $f(c_1) = c_2$ realizes this similitude.*

PROOF. If $a_1c_1 \equiv a_2c_2$, then the theorem follows at once from Theorem 22(II) of Chapter II. We therefore assume that, for example, $a_1c_1 < a_2c_2$; we take a point $a \in (a_2c_2)$ (Fig. 205) such that

(1) $a_1c_1 \equiv ac_2$.

We produce from point a a half-line $A \parallel \mathbf{H}(a_2b_2)$. It intersects side (b_2c_2) in some point b. We shall examine triangles $a_1b_1c_1$ and abc_2. We have $\sphericalangle a_1 \equiv \sphericalangle a_2 \equiv \sphericalangle c_2ab$, which, together with $\sphericalangle c_1 \equiv \sphericalangle c_2$ and formula (1), gives

(2) $a_1b_1 \equiv ab, \ a_1c_1 \equiv ac_2, \ b_1c_1 \equiv bc_2$.

Using now Theorem 14, we obtain

$$\frac{\varrho(a_2,c_2)}{\varrho(a,c_2)} = \frac{\varrho(b_2,c_2)}{\varrho(b,c_2)} = \frac{\varrho(a_2,b_2)}{\varrho(a,b)} = \lambda$$

from which it follows, by formula (2), that

$$\varrho(a_2,c_2) = \lambda \cdot \varrho(a_1,c_1), \quad \varrho(b_2,c_2) = \lambda \cdot \varrho(b_1,c_1), \quad \varrho(a_2,b_2) = \lambda \cdot \varrho(a_1,b_1),$$

which is what we had to prove.

10. Pythagorean Theorem

We shall now prove the *Pythagorean Theorem*.

THEOREM 16. *If in a triangle abc the angle b is a right angle, then*

$$\varrho(a,b)^2 + \varrho(b,c)^2 = \varrho(a,c)^2.$$

PROOF. Let d denote the perpendicular projection of point b upon line ac (Fig. 206). It is readily seen that

(1) $\mathbf{B}(a,d,c).$

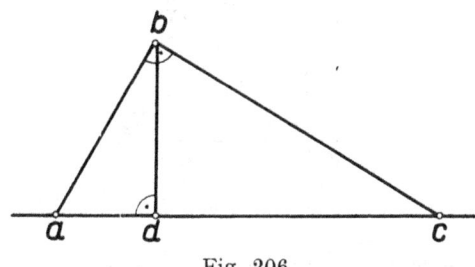

b

a   d   c

Fig. 206

We now consider triangles abc and adb. We have $\sphericalangle a \equiv \sphericalangle a$, $\sphericalangle b \equiv \sphericalangle adb$, from which it follows, by Theorem 2, that $\sphericalangle c \equiv \sphericalangle abd$. By applying Theorem 15, we thus conclude that

$$\frac{\varrho(a,b)}{\varrho(a,c)} = \frac{\varrho(a,d)}{\varrho(a,b)},$$

that is,

(2) $\varrho(a,b)^2 = \varrho(a,c) \cdot \varrho(a,d).$

Similarly, by considering triangles abc and bdc we obtain

(3) $\varrho(b,c)^2 = \varrho(a,c) \cdot \varrho(d,c).$

From formulas (1)–(3) it follows that

$$\varrho(a,b)^2 + \varrho(b,c)^2 = \varrho(a,c) \cdot (\varrho(a,d) + \varrho(d,c)) = \varrho(a,c)^2,$$

which was to be proved.

11. Metric Type of Planes

In Chapter III (Theorem 63) we have shown that every line of absolute geometry is isometric with the Cartesian space $\mathbf{C_1}$. As regards the planes of absolute geometry, we have proved only that they are homeomorphic with the Cartesian space $\mathbf{C_2}$ (Theorem 64 of Chapter III). On the basis of Euclidean geometry we are however in a position to establish the metric type of planes.

Let P be an arbitrary plane. We fix an absolute coordinate system Φ on P. Consider any two points $p, q \in P$; let $\Phi(p) = (x_1^p, x_2^p)$ and $\Phi(q) = (x_1^q, x_2^q)$. We assume initially that $x_1^p \neq x_1^q$ and $x_2^p \neq x_2^q$. Let us denote by p_1 and q_1 the perpendicular projections of points p and q upon the origin line K_0 of Φ (Fig. 207), and let $r = \Phi^{-1}(x_1^p, x_2^q)$. We have

(1) $$\varrho(p, r) = |x_2^p - x_2^q|.$$

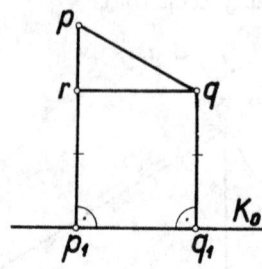

Fig. 207

By Theorem 4, the Saccheri quadrangle (r, p_1, q_1, q) is a rectangle. Hence $\sphericalangle prq$ is a right angle and

(2) $$\varrho(r, q) = \varrho(p_1, q_1) = |x_1^p - x_1^q|.$$

Using the Pythagorean Theorem for the triangle pqr, we obtain from formulas (1) and (2)

$$\varrho(p, q) = \sqrt{(x_1^p - x_1^q)^2 + (x_2^p - x_2^q)^2}.$$

Obviously, this formula also holds if $x_1^p = x_1^q$ or $x_2^p = x_2^q$.

The number $\sqrt{(x_1^p - x_1^q)^2 + (x_2^p - x_2^q)^2}$ is also the distance between the points $x^p = \Phi(p) = (x_1^p, x_2^p)$ and $x^q = \Phi(q) = (x_1^q, x_2^q)$ of Cartesian plane $\mathbf{C}_2$. Since, in addition, by Theorem 64 of Chapter III, the function Φ maps plane P onto $\mathbf{C}_2$, we have

THEOREM 17. *Any absolute coordinate system on a plane P maps plane P isometrically onto Cartesian space $\mathbf{C}_2$.*

12. Metric Type of Space

In Chapter III (Theorem 66) we have shown that the space $\mathbf{S}$ of absolute geometry is homeomorphic with the Cartesian space $\mathbf{C}_3$. We shall now show that the space $\mathbf{S}$ of Euclidean geometry is isometric with space $\mathbf{C}_3$. We proceed as in the previous section. We fix in space $\mathbf{S}$ a system Φ of absolute coordinates. Consider any two points p and q; let $\Phi(p) = (x_1^p, x_2^p, x_3^p)$ and $\Phi(q) = (x_1^q, x_2^q, x_3^q)$. We assume initially that $x_3^p \neq x_3^q$ and that at least one of the inequalities: $x_1^p \neq x_1^q$ or $x_2^p \neq x_2^q$, holds. We denote by p_1 and q_1

the perpendicular projections of points p and q upon the origin plane P_0, and by r the point $\Phi^{-1}(x_1^p, x_2^p, x_3^q)$. We shall consider the right triangle pqr (Fig. 208). We have

(1) $$\varrho(p,r) = |x_3^p - x_3^q|$$

and, by Theorem 17,

(2) $$\varrho(r,q) = \varrho(p_1,q_1) = \sqrt{(x_1^p - x_1^q)^2 + (x_2^p - x_2^q)^2}.$$

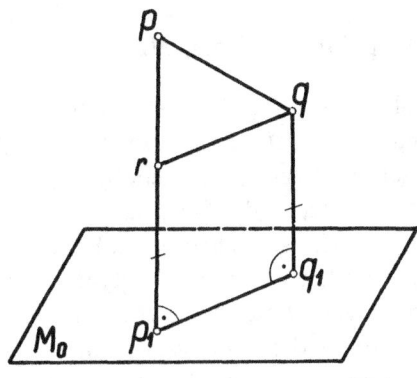

Fig. 208

Applying the Pythagorean Theorem to triangle pqr, we obtain from formulas (1) and (2)

$$\varrho(p,q) = \sqrt{(x_1^p - x_1^q)^2 + (x_2^p - x_2^q)^2 + (x_3^p - x_3^q)^2}.$$

It is readily noted that this formula also holds if $x_3^p = x_3^q$ or if both $x_1^p = x_1^q$ and $x_2^p = x_2^q$ hold simultaneously.

The number $\sqrt{(x_1^p - x_1^q)^2 + (x_2^p - x_2^q)^2 + (x_3^p - x_3^q)^2}$ is also the distance between the points $x^p = \Phi(p) = (x_1^p, x_2^p, x_3^p)$ and $x^q = \Phi(q) = (x_1^q, x_2^q, x_3^q)$ of the Cartesian space $\mathbf{C}_3$. Since, in addition, by Theorem 66 of Chapter III, the function Φ maps the space $\mathbf{S}$ onto the space $\mathbf{C}_3$, then we have

THEOREM 18. *The space $\mathbf{S}$ is isometric with the Cartesian space $\mathbf{C}_3$.*

With this we conclude our discussion of Euclidean geometry. The isometry of the space $\mathbf{S}$ of Euclidean geometry with the space $\mathbf{C}_3$ and the fact that lines, planes, and the betweenness and equidistance relations have been expressed in terms of distance (Theorems 51 and 52 of Chapter III and Theorems 71 and 72 of Chapter I) constitute a sufficient basis for the proof of categoricity.

13. Categoricity of Euclidean Geometry

In Chapter IV, Section 3 we constructed for the axiom system (GE_3) of Euclidean geometry the arithmetical model (C). At present we shall show that any two models for the axiom system (GE_3) are isomorphic, in other words, that (C) is up to isomorphism the unique model for (GE_3).

Let us consider any two particular models for the axiom system (GE_3):

$$(M_1) \quad M_1\text{-points, } M_1\text{-lines, } M_1\text{-planes, } B_{M_1}, E_{M_1}$$

in a theory T_1 and

$$(M_2) \quad M_2\text{-points, } M_2\text{-lines, } M_2\text{-planes, } B_{M_2}, E_{M_2}$$

in a theory T_2. Models (M_1) and (M_2) satisfy all the theorems of Euclidean geometry. In particular, the set of M_i-points $(i = 1,2)$ becomes a metric space by the introduction of any proper metric ϱ_{M_i} for the M_i-points. As such metric spaces, both sets, of M_1-points and of M_2-points, are isometric with the Cartesian space C_3 (see Theorem 18). Consequently, the set of M_1-points is isometric with the set of M_2-points. Let the function f establish this isometry. We shall show that the function f establishes also the isomorphism of models (M_1) and (M_2).

Indeed, from the definition of the function f we obtain immediately that the function f maps the set of M_1-points onto the set of M_2-points in a one-to-one way. We shall now show that the function f satisfies conditions (I) to (IV) on page 196.

We consider first the betweenness relation. By Theorem 51 of Chapter III, for any M_1-points a, b, c we have $B_{M_1}(a,b,c)$ if and only if M_1-points a, b, c are distinct and $\varrho_{M_1}(a,b) + \varrho_{M_1}(b,c) = \varrho_{M_1}(a,c)$. Since f is an isometry, then this condition holds if and only if M_2-points $f(a)$, $f(b)$, $f(c)$ are distinct and $\varrho_{M_2}(f(a),f(b)) + \varrho_{M_2}(f(b),f(c)) = \varrho_{M_2}(f(a),f(c))$, that is, (by Theorem 51 of Chapter III) if and only if $B_{M_2}(f(a),f(b),f(c))$. Therefore

(I′) $B_{M_1}(a,b,c)$ if and only if $B_{M_2}(f(a),f(b),f(c))$, for any M_1-points a, b, c.

In a similar manner, by employing Theorem 52 of Chapter III, we may prove

(II′) $E_{M_1}(a,b;c,d)$ if and only if $E_{M_2}(f(a),f(b);f(c),f(d))$, for any M_1-points a, b, c, d.

We now turn to lines. Consider any M_1-line, and two distinct M_1-points a and b on it. By Theorem 71 of Chapter I, an M_1-point p lies on L if and only if

(1) $p = a$ or $p = b$ or $B_{M_1}(p,a,b)$ or $B_{M_1}(a,p,b)$ or $B_{M_1}(a,b,p)$.

By (I′), condition (1) is equivalent to the following:

(2) $f(p) = f(a)$ or $f(p) = f(b)$ or $\boldsymbol{B}_{M_2}(f(p),f(a),f(b))$

   or $\boldsymbol{B}_{M_2}(f(a),f(p),f(b))$ or $\boldsymbol{B}_{M_2}(f(a),f(b),f(p))$.

Therefore the set $f(L)$ consists of M_2-points $f(p)$ determined by condition (2) and, as a result (see Theorem 71 of Chapter I), $f(L)$ is the M_2-line determined by M_2-points $f(a)$ and $f(b)$. In this way we have shown that if L is an M_1-line, then $f(L)$ is an M_2-line. In the same way, by using the fact that the transformation f^{-1} is also an isometry, we may prove that if for a set L of M_1-points the set $f(L)$ is an M_2-line, then set L itself is an M_1-line. Hence

(III′) L is an M_1-line if and only if $f(L)$ is an M_2-line.

Finally, we turn to planes. For any M_i-points a, c and for any M_i-line K ($i = 1,2$) let the formula $\boldsymbol{B}_{M_i}(a,K,c)$ denote that $a,c \sim \in K$ and that for some M_i-point $b \in K$ the relation $\boldsymbol{B}_{M_i}(a,b,c)$ holds. From (I′) and (III′) it follows at once that

(3) $\boldsymbol{B}_{M_1}(a,K,c)$ if and only if $\boldsymbol{B}_{M_2}(f(a),f(K),f(c))$.

We employ equivalence (3) and Theorem 72 of Chapter I in the argument similar to that used in the case of the line. In this way we get

(IV′) P is an M_1-plane if and only if $f(P)$ is an M_2-plane.

We have thus proved

PROPOSITION 6. *The axiom system* (GE$_3$) *of space Euclidean geometry is categorical.*

In conclusion, we consider the axiom system (GE$_2$) of plane Euclidean geometry, which is obtained from the axiom system (GA$_2$) of plane absolute geometry (see page 197) by adding Axiom E (as formulated on page 197, line 14). The proof of categoricity for the axiom system (GE$_2$) is quite similar to the proof of categoricity for the axiom system (GE$_3$). Such a proof would be based on Theorems 51 and 52 of Chapter III, on Theorem 71 of Chapter I, and on Theorem 17 (as modified for the plane case). All these theorems were derived from axioms of the system (GE$_2$) without the use of the space axioms I5–I9. Therefore we have

PROPOSITION 7. *The axiom system* (GE$_2$) *of plane Euclidean geometry is categorical.*

VI

Bolyai-Lobachevskian Geometry

1. The Axiom of Bolyai-Lobachevski

By the *Axiom of Bolyai-Lobachevski* we understand the sentence (see page 197):

BL. *For some plane P_0, some line $L_0 \subset P_0$, and some point $a_0 \in P_0 - L_0$ there exist at least two distinct lines $K_1 \subset P_0$ and $K_2 \subset P_0$ passing through point a_0 and not intersecting line L_0.*

The axioms of absolute geometry, together with Axiom BL, constitute the entire axiom system of Bolyai-Lobachevskian geometry.

The following theorem is an immediate consequence of Axiom BL (see Chapter IV, page 263).

THEOREM 1. *For any plane P, any line L, and any point $a \in P - L$ there exist at least two distinct lines $K_1 \subset P$ and $K_2 \subset P$ passing through point a and not intersecting line L.*

2. The Sum of the Angles of a Triangle and of a Quadrangle

From Axiom BL we at once have the following theorem (see Chapter IV, page 263):

THEOREM 2. *In every triangle the sum of the sizes of the angles is less than π.*

As a direct consequence of Theorem 2 we have

THEOREM 3. *In every convex quadrangle the sum of the sizes of the angles is less than 2π.*

We obtain the proof at once from the proof of Theorem 49 of Chapter III (on the sum of the sizes of the angles of a convex quadrangle in absolute geometry) by changing in formulas (2)–(4) on page 175 the sign "$\leqslant$" to the sign "$<$".

From Theorem 3 we get immediately

THEOREM 4. *No quadrangle is a rectangle.*

From Theorem 3 and from Theorem 82 of Chapter II we get immediately

THEOREM 5. *In every Saccheri quadrangle, the upper base angles are acute.*

3. Relations Between Sides and Angles of Two Triangles (Conclusion)

THEOREM 6. *Given two triangles $a_1b_1c_1$ and $a_2b_2c_2$, if $\sphericalangle\, a_1 \equiv \sphericalangle\, a_2$, $\sphericalangle\, b_1 \equiv \sphericalangle\, b_2$, and $\sphericalangle\, c_1 \equiv \sphericalangle\, c_2$, then $a_1b_1 \equiv a_2b_2$, $b_1c_1 \equiv b_2c_2$, and $a_1c_1 \equiv a_2c_2$.*

PROOF. Assume that $\sphericalangle\, a_1 \equiv \sphericalangle\, a_2$, $\sphericalangle\, b_1 \equiv \sphericalangle b_2$, $\sphericalangle\, c_1 \equiv \sphericalangle\, c_2$, and suppose that $a_1b_1 > a_2b_2$. Let us take a point $b \in (a_1b_1)$ such that $a_1b \equiv a_2b_2$, and

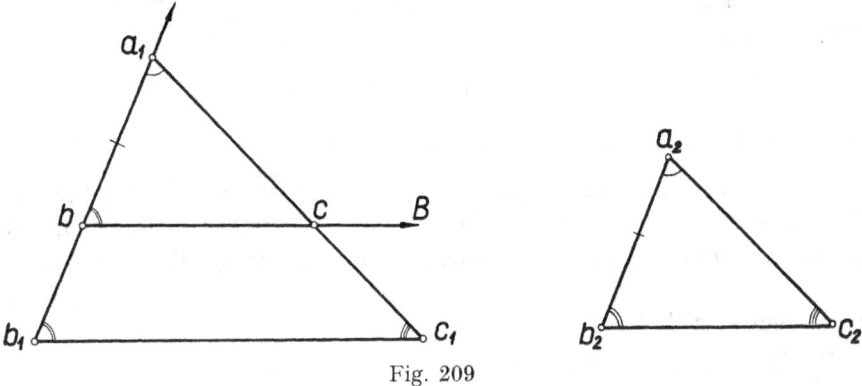

Fig. 209

from point b, in half-plane $\mathbf{L}(a_1b_1)c_1$, let us produce a half-line B such that the angle $B\mathbf{L}(ba_1)$ is congruent to angle b_1 (Fig. 209). Half-line B does not intersect segment (b_1c_1), and therefore it intersects segment (a_1c_1) in some point c. In the congruent triangles a_1bc and $a_2b_2c_2$ we have $\sphericalangle\, a_1cb \equiv \sphericalangle\, c_2 \equiv \sphericalangle\, c_1$. Now let

$$(1) \qquad \omega_1 = |\sphericalangle\, a_1| + |\sphericalangle\, b_1| + |\sphericalangle\, c_1| = |\sphericalangle\, a_1| + |\sphericalangle\, a_1bc| + |\sphericalangle\, a_1cb|$$

and let ω_2 denote the sum of the sizes of the angles of the convex quadrangle (b_1,b,c,c_1). By formula (1) we have

$$\omega_1 + \omega_2 = \omega_1 + 2\pi$$

from which it follows that $\omega_2 = 2\pi$, contrary to Theorem 3. Thus segment a_1b_1 cannot be greater than segment a_2b_2; also, because of the symmetry of our hypothesis, segment a_1b_1 cannot be smaller than segment a_2b_2 and therefore $a_1b_1 \equiv a_2b_2$. From this it follows that $b_1c_1 \equiv b_2c_2$ and $a_1c_1 \equiv a_2c_2$.

In Section 7 of Chapter II we gave two sufficient conditions for the congruence of triangles (which are valid for both Euclidean and Bolyai-Lobachevskian geometries). From Theorem 6 we obtain a third sufficient condition for the congruence of two triangles, which applies only to Lobachevskian geometry:

Two triangles are congruent if the three angles of one triangle are congruent to the three angles of the other.

4. Similitudes

THEOREM 7. *Every similitude of space* **S** *onto itself is an isometric transformation.*

PROOF. Take any similitude f and any distinct points $p,q \in$ **S**. Let r be any point non-collinear with points p and q. By Theorem 58 of Chapter III, for triangles pqr and $f(p)f(q)f(r)$, we have

$$\sphericalangle p \equiv \sphericalangle f(p), \quad \sphericalangle q \equiv \sphericalangle f(q), \quad \sphericalangle r \equiv \sphericalangle f(r),$$

from which, by Theorem 6, it follows that $pq \equiv f(p)f(q)$.

In Section 13 we shall connect this fact with the equally unexpected fact that in Bolyai-Lobachevskian geometry (as opposed to Euclidean geometry) a free segment can be singled out from among all free segments.

5. Defect of a Triangle

By Theorem 2, for every triangle abc the number

$$\pi - (|\sphericalangle a| + |\sphericalangle b| + |\sphericalangle c|)$$

is positive. For certain reasons (which will become clear at the end of this section) it is convenient to regard this number as a function not of an ordinary triangle abc, but of a closed triangle abc:

$$\Delta(\langle abc \rangle) = \pi - (|\sphericalangle a| + |\sphericalangle b| + |\sphericalangle c|).$$

This is possible, since, by Theorems 89 and 83 of Chapter I, triangle $\langle abc \rangle$ uniquely determines triangle abc. The number $\Delta(\langle abc \rangle)$ will be called the *defect of triangle* $\langle abc \rangle$.

From Theorem 103 of Chapter II we get at once

THEOREM 8. *If closed triangles* T_1 *and* T_2 *are congruent, then*

$$\Delta(T_1) = \Delta(T_2).$$

The following two theorems establish the additivity of the defect:

THEOREM 9. *Given a triangle* $\langle abc \rangle$, *if* **B**(a,d,c), *then*

$$\Delta(\langle abc \rangle) = \Delta(\langle abd \rangle) + \Delta(\langle bcd \rangle).$$

PROOF. Take any point d on side (ac) (Fig. 210). Then

$$\Delta(\langle abd \rangle) + \Delta(\langle bcd \rangle) =$$

$$\pi - (|\sphericalangle a| + |\sphericalangle abd| + |\sphericalangle adb|) + \pi - (|\sphericalangle c| + |\sphericalangle cbd| + |\sphericalangle cdb|) =$$

$$2\pi - (|\sphericalangle a| + |\sphericalangle b| + |\sphericalangle c| + \pi) = \pi - (|\sphericalangle a| + |\sphericalangle b| + |\sphericalangle c|) = \Delta(\langle abc \rangle).$$

The pair of triangles $\langle abd \rangle$ and $\langle bcd \rangle$ constitute a triangulation of triangle $\langle abc \rangle$. We shall now show that Theorem 9 can be extended to any arbitrary triangulation $\mathfrak{T}$ of triangle $\langle abc \rangle$:

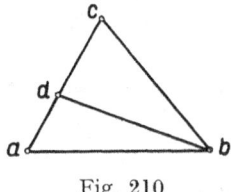

Fig. 210

THEOREM 10. *If $\mathfrak{T}$ is any triangulation of triangle $\langle abc \rangle$, then*

$$\Delta(\langle abc \rangle) = \sum_{T \in \mathfrak{T}} \Delta(T).$$

PROOF. Let k be the number of vertices of triangulation $\mathfrak{T}$ belonging to the figure $(abc) \cup (ab)$. We shall refer to the number k as the *index* of triangulation $\mathfrak{T}$ (of triangle $\langle abc \rangle$) with respect to the side (ab).

Let $k = 0$. We denote by n the number of triangles of the triangulation $\mathfrak{T}$. We employ induction with respect to n. If $n = 1$, the theorem is obvious. Let us now assume that the theorem is proved (for $k = 0$) for some $n \geqslant 1$, and that the triangulation $\mathfrak{T}$ consists of $n + 1$ triangles. Since $k = 0$, then one of the triangles $T \in \mathfrak{T}$ has as its vertices points a and b, and the third vertex d belongs to one of the two segments (ac) and (bc), say to segment (ac) (Fig. 210). Let us put $\mathfrak{T}' = \mathfrak{T} - \{\langle abd \rangle\}$. Because of our inductive assumption we then have

$$\Delta(\langle abc \rangle) = \Delta(\langle abd \rangle) + \Delta(\langle bcd \rangle) = \Delta(\langle abd \rangle) + \sum_{T \in \mathfrak{T}'} \Delta(T) = \sum_{T \in \mathfrak{T}} \Delta(T).$$

We now assume that $k > 0$ and that the theorem is valid for every triangle $\langle a'b'c' \rangle$ and for every triangulation $\mathfrak{T}'$ of triangle $\langle a'b'c' \rangle$ if the index of triangulation $\mathfrak{T}'$ with respect to side $(a'b')$ is smaller than k. Among the vertices of the triangulation $\mathfrak{T}$ of triangle $\langle abc \rangle$ belonging to the figure $(abc) \cup (ab)$ we take a vertex d such that there is no vertex of triangulation $\mathfrak{T}$ lying in the inner domain of angle acd (Fig. 211). Let $L = \mathbf{L}(cd)$ and $e = L \cap (ab)$. Further, let $T = \langle pqr \rangle$ be any triangle of triangulation $\mathfrak{T}$. We denote its boundary by F. Suppose that $(pqr) \cap L \neq 0$. Then boundary F has just two points s_1 and s_2 in common with line L. Two cases are possible:

Case 1. Points s_1 is one of the vertices of triangle $\langle pqr \rangle$, say $s_1 = p$ (Fig. 212). Then $s_2 \in (qr)$ and

(1) $$\Delta(\langle pqr \rangle) = \Delta(\langle pqs_2 \rangle) + \Delta(\langle prs_2 \rangle).$$

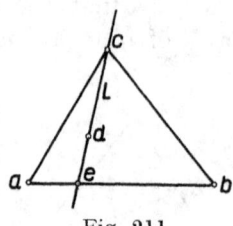

Fig. 211

Case 2. None of the points s_1 and s_2 is a vertex of triangle $\langle pqr \rangle$ (Fig. 213); assume that $s_1 \in (pr)$ and $s_2 \in (qr)$. Then the class

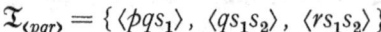

$$\mathfrak{T}_{\langle pqr \rangle} = \{ \langle pqs_1 \rangle, \ \langle qs_1s_2 \rangle, \ \langle rs_1s_2 \rangle \}$$

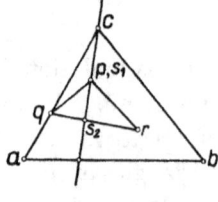
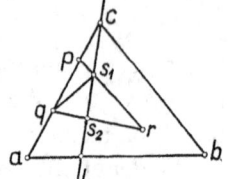

Fig. 212   Fig. 213

is a triangulation of triangle $\langle pqr \rangle$ whose index with respect to side (pq) is equal to zero, and consequently

(2) $$\Delta(\langle pqr \rangle) = \Delta(\langle pqs_1 \rangle) + \Delta(\langle qs_1s_2 \rangle) + \Delta(\langle rs,s_2 \rangle).$$

In triangulation $\mathfrak{T}$, let us replace every triangle $\langle pqr \rangle$ such that $(pqr) \cap L \neq 0$ by two triangles $\langle pqs_2 \rangle$ and $\langle prs_2 \rangle$ in Case 1 and by three triangles $\langle pqs_1 \rangle$, $\langle qs_1s_2 \rangle$, and $\langle rs_1s_2 \rangle$ in Case 2. We obtain in this way a new triangulation $\mathfrak{T}_1$ of triangle $\langle abc \rangle$; if $\mathfrak{T}_1$ has any new vertices (with respect to triangulation $\mathfrak{T}$) then they belong to segment (ce) only, while from formulas (1) and (2) it follows that

(3) $$\underset{T \in \mathfrak{T}}{\Sigma} \Delta(T) = \underset{T \in \mathfrak{T}_1}{\Sigma} \Delta(T).$$

As may readily be seen, triangulation $\mathfrak{T}_1$ can be devided into two disjoint classes $\mathfrak{T}_a$ and $\mathfrak{T}_b$, of which the first is the triangulation of triangle $\langle ace \rangle$

and the second, of triangle $\langle bce \rangle$. But the index of the triangulation $\mathfrak{T}_a$ of triangle $\langle ace \rangle$ with respect to side (ae) is equal to 0 and the index of triangulation $\mathfrak{T}_b$ of triangle $\langle bce \rangle$ with respect to side (be) is smaller than k. Therefore, because of the inductive assumption and formula (3),

$$\Delta(\langle abc \rangle) = \Delta(\langle ace \rangle) + \Delta(\langle bce \rangle) =$$
$$= \sum_{T \in \mathfrak{T}_a} \Delta(T) + \sum_{T \in \mathfrak{T}_b} \Delta(T) = \sum_{T \in \mathfrak{T}_1} \Delta(T) = \sum_{T \in \mathfrak{T}} \Delta(T),$$

which was to be proved.

The properties of the defect proved in Theorems 8 and 10 are of great importance to Bolyai-Lobachevskian geometry in connection with the theory of the area of plane figures. In particular, it can be shown that the area of a triangle is proportional to its defect and consequently has an upper bound, which actually equals $\lambda \cdot \pi$, where λ is the coefficient of proportionality. We shall not develop here the general theory of area. For our purposes it is sufficient to extend the notion of the defect to any plane figure.

6. Defect of a Polygon

We shall now extend the notion of the defect to polygons.

THEOREM 11. *If $\mathfrak{T}_1$ and $\mathfrak{T}_2$ are two triangulations of a polygon E then*

(1)
$$\sum_{T \in \mathfrak{T}_1} \Delta(T) = \sum_{T \in \mathfrak{T}_2} \Delta(T).$$

PROOF. Because of Theorem 100 of Chapter I there exists a triangulation $\mathfrak{T}$ of polygon E such that every triangle $T \in \mathfrak{T}_1 \cup \mathfrak{T}_2$ has a triangulation $\mathfrak{T}_T \subset \mathfrak{T}$. From this and from Theorem 10 we conclude that for $i = 1,2$

$$\sum_{T \in \mathfrak{T}_i} \Delta(T) = \sum_{T \in \mathfrak{T}_i} (\sum_{T' \in \mathfrak{T}_T} \Delta(T')) = \sum_{T' \in \mathfrak{T}} \Delta(T'),$$

which implies formula (1).

The theorem just proved permits the following definition to be introduced. We define the *defect* $\Delta(E)$ *of a polygon* E to be the sum of the defects of triangles constituting any triangulation of polygon E.

The following three theorems establish the basic properties of the defect of a polygon:

THEOREM 12. *If polygons E_1 and E_2 are congruent then $\Delta(E_1) = \Delta(E_2)$.*

PROOF. This is a straightforward conclusion from Theorem 8.

THEOREM 13. *If polygon E is the sum of two polygons E_1 and E_2 with disjoint interiors, then $\Delta(E) = \Delta(E_1) + \Delta(E_2)$.*

Proof. It is sufficient to note that the sum $\mathfrak{T}$ of any triangulation $\mathfrak{T}_1$ of polygon E_1 and any triangulation $\mathfrak{T}_2$ of polygon E_2 is a triangulation of polygon E.

Theorem 14. *For any two polygons E_1 and E_2, if $E_1 \subset E_2$, then $\Delta(E_1) \leqslant \Delta(E_2)$.*

Proof. By Theorem 99 of Chapter I, polygon E_2 is the sum of polygons E_1 and $\overline{E_2 - E_1}$ which, as readily seen, have disjoint interiors. By Theorem 13, it thus follows at once that $\Delta(E_1) \leqslant \Delta(E_2)$.

7. Defect of a Plane Figure

Let F be any figure lying on a plane P. We define the *defect $\Delta(F)$ of figure F* to be the least upper bound of the defects of all polygons $E \subset F$, and in case there are no such polygons we assume $\Delta(F) = 0$. It is clear that the number $\Delta(F)$ so defined is non-negative (it may also be equal to $+\infty$). From Theorem 14 it follows at once that if figure F is a polygon, then the definition of its defect is in conformity with the definition of the defect of a polygon given in the preceding section.

From Theorem 12 we obtain the following:

Theorem 15. *If figures F_1 and F_2 are congruent, then $\Delta(F_1) = \Delta(F_2)$.*

Furthermore, from the definition of the defect of a plane figure we conclude

Theorem 16. *For every two (plane) figures F_1 and F_2, if $F_1 \subset F_2$, then $\Delta(F_1) \leqslant \Delta(F_2)$.*

It is readily noted that if a figure F is the sum of two plane figures F_1 and F_2 with disjoint interiors, then

$$\Delta(F) \geqslant \Delta(F_1) + \Delta(F_2).$$

In some cases we have $\Delta(F) = \Delta(F_1) + \Delta(F_2)$. For example, from Theorem 98 of Chapter I and from Theorem 13 we derive at once

Theorem 17. *Given a plane P and a line $K \subset P$, if $P = W \cup K \cup W^*$, then for every figure $F \subset P$*

$$\Delta(F) = \Delta(F \cap \overline{W}) + \Delta(F \cap \overline{W^*}).$$

8. Parallel Lines and Hyperparallel Lines

We now come to the theory of parallel and hyperparallel lines. From Axiom BL (see discussion on sentences equivalent to Axiom BL on page 263) and from Theorems 24 and 25 of Chapter III we obtain the following:

THEOREM 18. *If $a \sim \in L$, then the angles of parallelism for point a with respect to line L are acute, and there are just two distinct lines K_1 and K_2 passing through a and parallel to L.*

Let P be the plane determined by point a and line L (Fig. 214). We denote by b the perpendicular projection of point a upon line L.

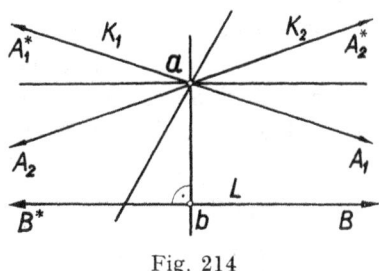

Fig. 214

Let $L = B \cup b \cup B^*$. We produce from point a half-lines $A_1 \| B$ and $A_2 \| B^*$. Lines $K_1 \supset A_1$ and $K_2 \supset A_2$ are the two lines which pass through point a and are parallel to line L. Obviously, $K_1, K_2 \subset P$. We now divide all lines of plane P passing through point a and distinct from lines K_1 and K_2 into two classes $\mathfrak{X}_1$ and $\mathfrak{X}_2$ in the following manner: A line M belongs to class $\mathfrak{X}_1$ if it contains a half-line $C \in (A_1 A_2)$, and to class $\mathfrak{X}_2$ if it contains a half-line $C \in (A_1 A_2^*)$.

Every line M of class $\mathfrak{X}_1$ intersects line L. In fact, line ab intersects line L in point b, and a line $M \neq \mathbf{L}(ab)$ either contains half-line C lying between half-lines ab and A_1, and consequently intersects half-line B, or it contains a half-line C lying between half-lines ab and A_2, and consequently intersects half-line B^*.

On the other hand, every line M of class $\mathfrak{X}_2$ does not intersect line L. In fact, its half-line $C \in (A_1 A_2^*)$ and line L lie on opposite sides of line K_1, and therefore do not intersect. Similarly, complementary half-line C^* and line L lie on opposite sides of line K_2, and therefore also do not intersect. Lines M and L are therefore disjoint, and since they are not parallel, they form a pair of hyperparallel lines (see page 150).

We therefore have:

THEOREM 19. *Given a line L and a point $a \sim \in L$, if A_1 and A_2 are half-lines with origin a and parallel to line L, then every line containing a half-line of angle $(A_1 A_2)$ intersects L, and every line containing a half-line of angle $(A_1 A_2^*)$ is hyperparallel to L.*

9. Pairs of Parallel Lines

Let us consider two distinct parallel lines K and L. Then one of the axes $\mathfrak{K}$ of line K is parallel to one of the axes $\mathfrak{L}$ of line L. Let a be any point on line K. Since the angles of parallelism for point a with respect to line L are acute (see Theorem 18), then

$$\mathfrak{K} \parallel \mathfrak{L} \quad \text{implies} \quad \mathfrak{K}^* \sim \parallel \mathfrak{L}^*.$$

In addition, by Theorem 117 of Chapter II, it follows from $\mathfrak{K} \parallel \mathfrak{L}$ that $\mathfrak{K} \sim \parallel \mathfrak{L}^*$ and $\mathfrak{K}^* \sim \parallel \mathfrak{L}$. Therefore there is only one way to orient lines K and L so that the resulting axes $\mathfrak{K}$ and $\mathfrak{L}$ be parallel. We then say that *the pair of parallel axes $\mathfrak{K}$ and $\mathfrak{L}$ is conjugate to the pair of parallel lines K and L.* The direction of axes $\mathfrak{K}$ and $\mathfrak{L}$ (see pages 149 and 150) is the direction which lines K and L have in common; we shall call it the *direction of the pair of parallel lines K and L.*

We therefore have

THEOREM 20. *If two lines have two (distinct) directions in common, then they are identical.*

In investigations concerning the properties of the pair of parallel lines, the order of the points on the lines plays an important role. That is why instead considering a pair of parallel lines K and L, we shall, in general, consider the pair of axes $\mathfrak{K}$ and $\mathfrak{L}$ conjugate to them. In this connection it is convenient to employ some notions concerning lines for the case of axes. Thus, for example, we define the *distance $\varrho(a,\mathfrak{K})$ of a point a from an axis $\mathfrak{K}$* as the distance $\varrho(a,K)$ of point a from line K of axis $\mathfrak{K}$ and the *perpendicular projection of a point a upon axis $\mathfrak{K}$* as the perpendicular projection of point a upon line K of axis $\mathfrak{K}$. In dealing with a half-plane we shall sometimes regard as its boundary not the line K, but one of its two axis $\mathfrak{K}$ and $\mathfrak{K}^*$. We shall denote the half-plane having boundary $\mathfrak{K}$ and passing through point a by $\boldsymbol{HP}(\mathfrak{K}a)$.

We shall now investigate the properties of the pair of parallel lines K and L. In our considerations we take the order of points on both lines K and L as that induced by the pair of lines K and L.

We shall first investigate the distance of the points of one parallel axis from the other parallel axis. In Euclidean geometry this distance is constant. This is not the case in Bolyai-Lobachevskian geometry, as may be seen from the following:

LEMMA 1. *If $\mathfrak{K} \parallel \mathfrak{L}$, $\mathfrak{K} \neq \mathfrak{L}$, and $a \in \mathfrak{K}$, then for every number $\alpha > \varrho(a,\mathfrak{L})$ there exists a point $p \in \mathfrak{K}$ such that $\varrho(p,\mathfrak{L}) = \alpha$.*

PROOF. Let b denote the perpendicular projection of point a upon axis $\mathfrak{L}$ and let A be the half-line determined by point a on axis $\mathfrak{K}$ (Fig. 215).

Then the angle $\mathbf{H}(ab)A^*$ is obtuse and, by Theorem 60 of Chapter III, the distance of point $p \in a \cup A^*$ from axis $\mathfrak{L}$ is a continuous and unbounded function of p. Hence for some point $p \in a \cup A^*$ we have $\varrho(p,\mathfrak{L}) = \alpha$.

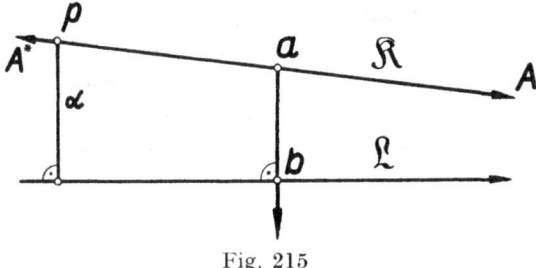

Fig. 215

A second lemma concerns the relation between two pairs of parallel axes.

LEMMA 2. *We assume that for $i = 1,2$ the conditions $\mathfrak{R}_i \parallel \mathfrak{L}_i$, $\mathfrak{R}_i \neq \mathfrak{L}_i$, $a_i, a_i' \in \mathfrak{R}_i$ and $a_i \prec a_i'$ hold. If $\varrho(a_1,a_1') = \varrho(a_2,a_2')$, then $\varrho(a_1,\mathfrak{L}_1) = \varrho(a_2,\mathfrak{L}_2)$ implies $\varrho(a_1',\mathfrak{L}_1) = \varrho(a_2',\mathfrak{L}_2)$.*

PROOF. Assume that $\varrho(a_1,a_1') = \varrho(a_2,a_2')$. Denoting by b_i and b_i', respectively, the perpendicular projections of points a_i and a_i' upon axis $\mathfrak{L}_i$ (Fig. 216) we have, by Theorem 22(I) of Chapter III,

$$\measuredangle\, a_1'a_1b_1 \equiv \measuredangle\, a_2'a_2b_2.$$

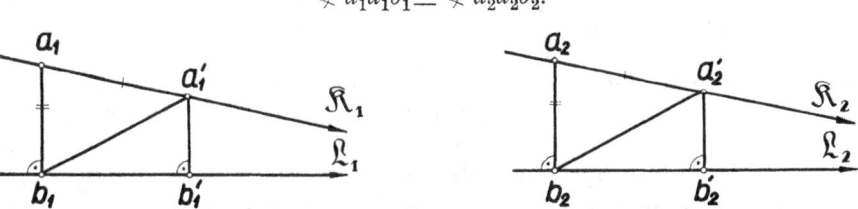

Fig. 216

Comparing the congruent triangles $a_1'a_1b_1$ and $a_2'a_2b_2$, and the congruent triangles $a_1'b_1b_1'$ and $a_2'b_2b_2'$ we obtain $a_1'b_1' \equiv a_2'b_2'$, that is $\varrho(a',\mathfrak{L}_1) = \varrho(a_2',\mathfrak{L}_2)$.

With the help of Lemmas 1 and 2 we prove

THEOREM 21. *Given two distinct parallel axes $\mathfrak{R}$ and $\mathfrak{L}$, let $\varphi(p) = \varrho(p,\mathfrak{L})$ for $p \in \mathfrak{R}$. Then φ is a continuous and decreasing function of point p and the range of φ coincides with the set of all positive numbers.*

PROOF. Let p be an arbitrary point on axis $\mathfrak{R}$ (Fig. 217). We take a point p' following p on $\mathfrak{R}$. Let q' be the perpendicular projection of p' upon axis $\mathfrak{L}$. Then the angle $pp'q$ is obtuse and, by Theorem 60 of Chapter III,

(i) the function φ is continuous at point p, and (ii) $\varphi(p) > \varphi(p')$, i.e. the function φ is decreasing.

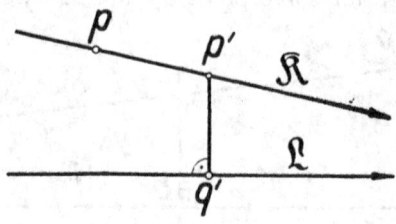

Fig. 217

Now let α be any real positive number. We seek on axis $\mathfrak{R}$ a point p such that $\varphi(p) = \alpha$. For this purpose we take any point $a \in \mathfrak{R}$. We may assume that $\varrho(a,\mathfrak{L}) \neq \alpha$. If $\alpha > \varrho(a,\mathfrak{L})$, then from Lemma 1 it immediately follows that there is a point $p \in \mathfrak{R}$ such that $\varrho(p,\mathfrak{L}) = \alpha$. Let us next assume that $\alpha < \varrho(a,\mathfrak{L})$. We now take any axis $\mathfrak{L}_1$ and a point p_1 such that $\varrho(p_1,\mathfrak{L}_1) = \alpha$ (Fig. 218). Through p_1 we produce an axis $\mathfrak{R}_1$ parallel to $\mathfrak{L}_1$. By Lemma 1 there is a point a_1 on axis $\mathfrak{R}_1$ such that

$$\varrho(a_1,\mathfrak{L}_1) = \varrho(a,\mathfrak{L}) > \alpha.$$

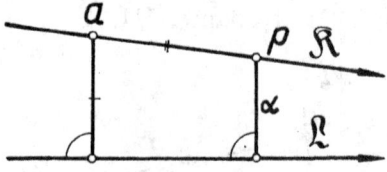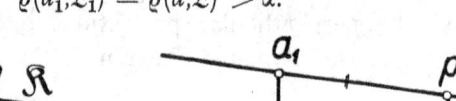

Fig. 218

Since we have already shown that the function φ is decreasing, then $a_1 \prec p_1$. Taking now a point p on axis $\mathfrak{R}$ in such a way that $a \prec p$ and $\varrho(a,p) = \varrho(a_1,p_1)$, we have, in accordance with Lemma 2,

$$\varphi(p) = \varrho(p,\mathfrak{L}) = \varrho(p_1,\mathfrak{L}_1) = \alpha.$$

In this way we have shown that function φ takes every real positive value. Since, in addition, φ is decreasing, then, by Theorem 44 of Chapter I, φ is continuous.

We shall still investigate the angle of parallelism for points $p \in \mathfrak{R}$ with respect to axis $\mathfrak{L}$.

THEOREM 22. *Given two distinct parallel axes $\mathfrak{R}$ and $\mathfrak{L}$, if point a_1 precedes point a_2 on $\mathfrak{R}$, then the angle of parallelism for point a_1 with respect to axis $\mathfrak{L}$ is smaller than the angle of parallelism for point a_2 with respect to axis $\mathfrak{L}$.*

PROOF. We assume that $a_1 \prec a_2$ on axis $\Re$ and let points b_1 and b_2 be the respective perpendicular projections of points a_1 and a_2 upon axis $\mathfrak{L}$. Then $b_1 \prec b_2$ on axis $\mathfrak{L}$. We denote by a_3 an arbitrary point on axis $\Re$ following point a_2 (Fig. 219). Then the angle of parallelism for point a_i $(i = 1,2)$ with respect to axis $\mathfrak{L}$ coincides with the angle $b_i a_i a_3$. We now have

(1) $$|\sphericalangle b_2 a_2 a_3| > |\sphericalangle a_1 b_2 a_2| + |\sphericalangle a_2 a_1 b_2|$$

and

$$\frac{\pi}{2} + |\sphericalangle a_1 b_2 a_2| = |\sphericalangle a_1 b_2 b_3| > \frac{\pi}{2} + |\sphericalangle b_1 a_1 b_2|;$$

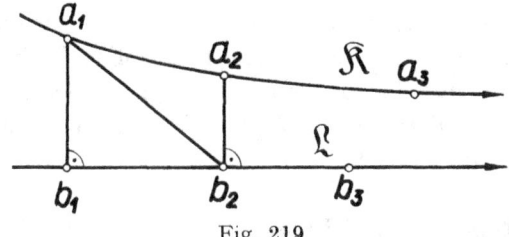

Fig. 219

from the latter we obtain

(2) $$|\sphericalangle a_1 b_2 a_2| > |\sphericalangle b_1 a_1 b_2|.$$

From formulas (1) and (2) it follows at once that

$$|\sphericalangle b_2 a_2 a_3| > |\sphericalangle b_1 a_1 b_2| + |\sphericalangle a_2 a_1 b_2| = |\sphericalangle b_1 a_1 a_3|.$$

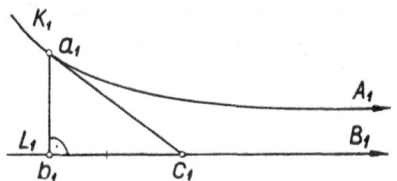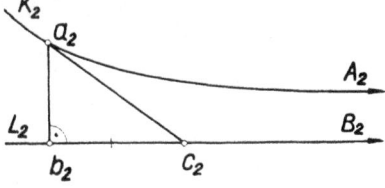

Fig. 220

In this way the theorem is proved.

Finally, we shall show that any two pairs of parallel lines are congruent to each other.

THEOREM 23. *Given lines K_1, K_2, L_1, L_2, if $K_1 \parallel L_1$, $K_2 \parallel L_2$, $K_1 \neq L_1$, and $K_2 \neq L_2$, then the figures $K_1 \cup L_1$ and $K_2 \cup L_2$ are congruent.*

PROOF. We assume that $K_i \parallel L_i$ and that $K_i \neq L_i$ $(i = 1,2)$. Let the pair of axes $\Re_i$ and $\mathfrak{L}_i$ be conjugate to the pair of lines K_i and L_i (Fig. 220). By Theorem 21, we can take on axis $\Re_i$ a point a_i such that

$$\varrho(a_1, \mathfrak{L}_1) = \varrho(a_2, \mathfrak{L}_2).$$

We denote by b_i the perpendicular projection of point a_i upon axis $\mathfrak{L}_i$; let B_i be the half-line determined on axis $\mathfrak{L}_i$ by point b_i. On half-line B_i we take a point c_i such that

$$b_1 c_1 \equiv b_2 c_2.$$

The function f:

$$f(a_1) = a_2, \ f(b_1) = b_2, \ f(c_1) = c_2,$$

realizes the congruence of triangles $a_1 b_1 c_1$ and $a_2 b_2 c_2$. By Theorems 101 and 102 of Chapter II the function f can be extended to the isometry g of the entire space S onto itself. This isometry maps half-line B_1 onto half-line B_2, half-line A_1 with origin a_1 and parallel to half-line B_1 onto half-line A_2 with origin a_2 and parallel to half-line B_2, line $K_1 \supset A_1$ onto line $K_2 \supset A_2$, and line $L_1 \subset B_1$ onto line $L_2 \supset B_2$. Thus, function g with the domain restricted to the set $K_1 \cup L_1$ realizes the congruence of the figures $K_1 \cup L_1$ and $K_2 \cup L_2$.

10. Pairs of Hyperparallel Lines

We shall prove the following:

LEMMA. *Given two hyperparallel lines K and L and a point $a \in K$, let A be any of the two half-lines determined by point a on line K. Then for any number $\alpha > 0$ there is a point $p \in A$ such that $\varrho(p, L) > \alpha$.*

PROOF. Let b denote the perpendicular projection of point a upon line L and let W denote the half-plane with boundary $\mathbf{L}(ab)$ containing half-line A (Fig. 221). Let $B = L \cap W$. From point a we produce a half-line $A_1 \parallel B$. Obviously $A_1 \subset W$. Let $K_1 = \mathbf{L}(A_1)$.

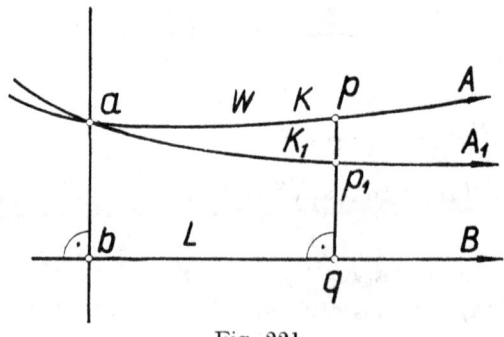

Fig. 221

By theorem 60 of Chapter III, we can find on half-line A a point p such that $\varrho(p, K_1) > \alpha$. We denote by q the perpendicular projection of point p upon line L. Since, as readily concluded from Theorem 19, half-line A and line L lie on opposite sides of line K_1, then line K_1 intersects the segment (pq) in some point p_1. We therefore have

$$\varrho(p,L) = \varrho(p,q) > \varrho(p,p_1) \geqslant \varrho(p,K_1) > \alpha,$$

which is what we had to prove.

THEOREM 24. *If lines K and L are hyperparallel, then there exists just one line M perpendicular to both these lines.*

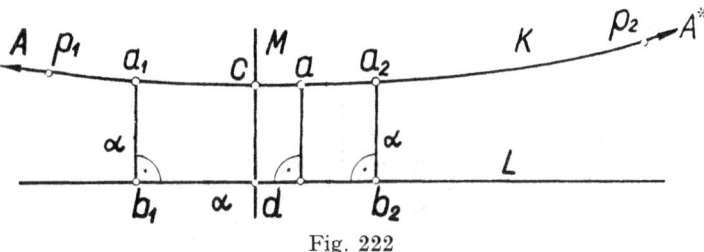

Fig. 222

PROOF. We pick a point a on line K and consider the decomposition $K = A \cup a \cup A^*$ (Fig. 222). According to the lemma we have just proved there are points $p_1 \in A$ and $p_2 \in A^*$ such that

$$\varrho(a,L) < \varrho(p_i,L) \quad \text{for} \quad i = 1,2.$$

Take a number α such that

$$\varrho(a,L) < \alpha < \varrho(p_i,L) \quad \text{for} \quad i = 1,2.$$

Since $\varrho(p,L)$ is a continuous function of the point $p \in K$ (see Theorem 59 of Chapter III), and segments $\langle ap_i \rangle$ are connected (see Theorem 2 of Chapter III), then, by Theorem 29 of Chapter I, there exist two (distinct) points $a_1 \in A$ and $a_2 \in A^*$ such that

$$\varrho(a_1,L) = \varrho(a_2,L) = \alpha.$$

We denote by b_1 and b_2 the perpendicular projections of points a_1 and a_2 upon line L. Then polygonal line (a_1,b_1,b_2,a_2) is a Saccheri quadrangle, from which it follows, by Theorem 83 of Chapter II, that the line M joining the midpoint c of the upper base $a_1 a_2$ and the midpoint d of the lower base $b_1 b_2$ is perpendicular to line K as well as to line L.

If there were to exist still another line M_1 also perpendicular both to lines K and L, then, denoting its intersection points with lines K and L by c_1 and d_1, respectively, we would obtain a rectangle (c,d,d_1,c_1), in contradiction to Theorem 4.

From Theorem 24 and 18 we obtain directly the following:

THEOREM 25. *In order that two distinct lines K and L be hyperparallel it is necessary and sufficient that they lie in one plane and that there exists a line M intersecting both of them perpendicularly.*

Let us consider two hyperparallel lines K and L. We denote by a_0 and b_0, respectively, the points of intersection of lines K and L with the only (by Theorem 24) line perpendicular to both lines K and L. Points a_0 and b_0 are therefore uniquely determined by the pair of hyperparallel lines K and L. Point a_0 will be called the *midpoint* of line K with respect to line L; similarly point b_0 will be called the *midpoint of line L with respect to line K*.

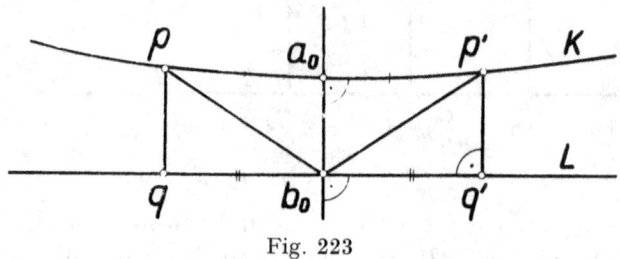

Fig. 223

THEOREM 26. *Given two hyperparallel lines K and L, let point a_0 be the midpoint of line K with respect to line L, and let $\varphi(p) = \varrho(p,L)$ for $p \in K$. Then for any two points $p,p' \in K$ we have: (i) $\varrho(a_0,p) = \varrho(a_0,p')$ implies $\varphi(p) = \varphi(p')$; (ii) the function φ is continuous, increasing, and unbounded on each closed half-line $a_0 \cup A$ determined by point a_0 on line K.*

PROOF. Let $p,p' \in K$ and $\varrho(a_0,p) = \varrho(a_0,p')$. We denote by b_0, q, and q', respectively, the perpendicular projections of points a_0, p, and p' upon line L (Fig. 223). In the congruent triangles a_0b_0p and a_0b_0p' we have

$$b_0p \equiv b_0p' \quad \text{and} \quad \sphericalangle a_0b_0p \equiv \sphericalangle a_0b_0p';$$

thus the right triangles pqb_0 and $p'q'b_0$ are congruent (see Theorem 75(I) of Chapter II), and specifically $pq \equiv p'q'$, that is, $\varphi(p) = \varphi(p')$. In this way the proof of the first part of the theorem is completed.

We now pass on to the second part of the theorem. On half line $a_0 \cup A$ the function φ is continuous, non-decreasing (cf. Note 1 following Theorem 60 of Chapter III), and, by the lemma proved at the beginning of this section, the function φ is unbounded on half-line $a_0 \cup A$. It therefore remains to prove that on half-line $a_0 \cup A$ there are no two distinct points p_1 and p_2 such that $\varphi(p_1) = \varphi(p_2)$. Indeed, if for some two distinct points $p_1,p_2 \in a \cup A$ we were to have $\varphi(p_1) = \varphi(p_2)$, then, denoting by q_1 and q_2, respectively, the perpendicular projections of points p_1 and p_2 upon line L, we would obtain a Saccheri quadrangle (p_1,q_1,q_2,p_2). As readily seen, the median of this quadrangle would be distinct from line a_0b_0, and would be perpendicular to lines K and L, in contradiction to Theorem 24. Thus the proof is completed.

11. Perpendicular Projection of a Line Upon a Line

We shall now prove the following:

THEOREM 27. *Consider on a plane P an acute angle AB with vertex a_0. The perpendicular projection of half-line A upon line $L \supset B$ is a segment (a_0b), where b is some point on half-line B. Half-line A is parallel to the line $M \subset P$ passing through point b and perpendicular to L.*

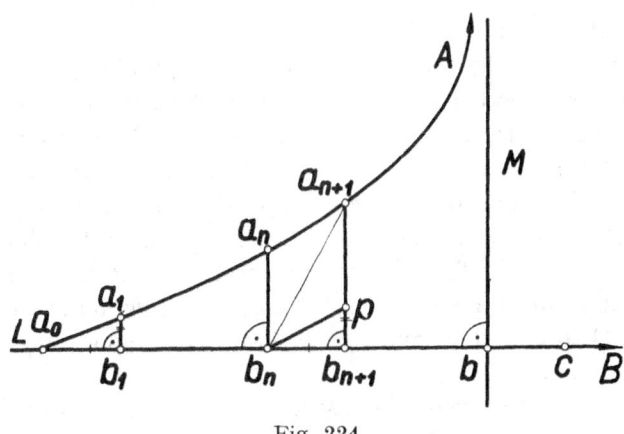

Fig. 224

PROOF. It is readily noted that the perpendicular projection of half-line A upon line L is included in half-line B. We shall now show that on half-line B there exist points which are not projections of any points on half-line A. To do this we take the infinite sequence of points (b_n) on half-line B (Fig. 224) defined by the condition $\varrho(a_0,b_n) = n$; let us suppose that each of the points b_n is the projection of some point $a_n \in A$. We shall show that for every natural n we then have

(1) $$\Delta(\langle a_0 a_n b_n \rangle) \geqslant n \cdot \Delta(\langle a_0 a_1 b_1 \rangle).$$

Indeed, this inequality holds for $n = 1$. Let us further assume that it holds for some $n \geqslant 1$. Since $a_1 b_1 < a_{n+1} b_{n+1}$, there then exists a point $p \in (a_{n+1} b_{n+1})$ such that $p b_{n+1} = a_1 b_1$. Triangles $a_0 a_1 b_1$ and $b_n p b_{n+1}$ are congruent, and therefore

(2) $$\Delta(\langle a_0 a_1 b_1 \rangle) = \Delta(\langle b_n p b_{n+1} \rangle).$$

Further, from Theorem 9 and formulas (1) and (2) it follows that

$$\Delta(\langle a_0 a_{n+1} b_{n+1}\rangle) = \Delta(\langle a_0 a_{n+1} b_n\rangle) + \Delta(\langle b_n a_{n+1} b_{n+1}\rangle)$$
$$> \Delta(\langle a_0 a_n b_n\rangle) + \Delta(\langle b_n p b_{n+1}\rangle)$$
$$\geqslant n \cdot \Delta(\langle a_0 a_1 b_1\rangle) + \Delta(\langle a_0 a_1 b_1\rangle) = (n+1) \cdot \Delta(\langle a_0 a_1 b_1\rangle).$$

Hence inequality (1) remains valid after the number n is replaced by the number $n+1$.

Since $\Delta(\langle a_0 a_1 b_1 \langle) > 0$, then we conclude from inequality (1) that the defect of triangles $\langle a_0 a_n b_n\rangle$ tends to infinity with n, in contradiction to the fact that the defect of the triangle cannot exceed the number π. Therefore, for sufficiently large n, the points b_n are not projections of any points of half-line A.

Let us now divide half-line B into two sets X_1 and X_2, where in set X_1 we include every point that is the projection of some point on half-line A, and in X_2 all the remaining points of half-line B. It is readily shown that (X_1, X_2) is a Dedekind cut of half-line B. Let b be the point determined by this cut. We shall show that $b \in X_2$. To do this, let us suppose, on the contrary, that $b \in X_1$. We denote by a that point on half-line A whose projection is point b. Then the projection b' of any point a' following point a on half-line A belongs to set X_2, in contradiction to the definition of set X_2. Therefore the projection of half-line A upon line L coincides with the segment $(a_0 b)$.

We now pass to the proof of the second part of the theorem. We denote by C the half-line of line M which has the origin b and lies on the same

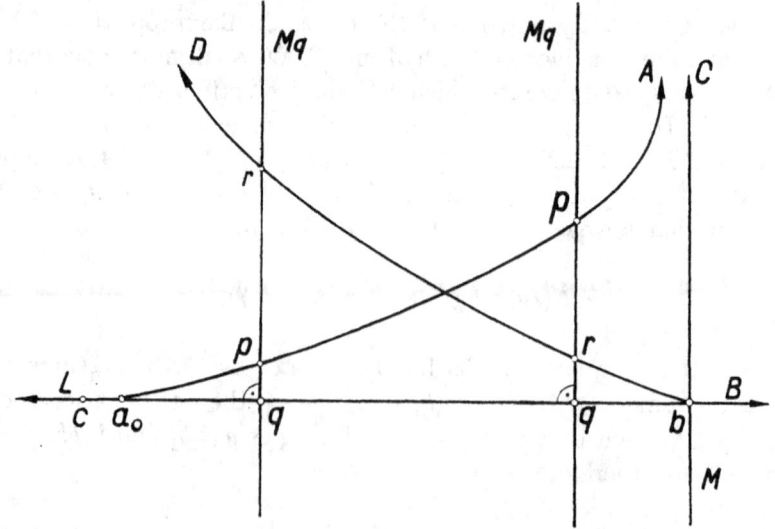

Fig. 225

side of line L as half-line A (Fig. 225). We shall show that $C \parallel A$. Let D be any half-line with origin b that lies between half-line ba and C. Since the angle $D\textbf{H}(ba_0)$ is acute, the projection of half-line D upon line L is a segment (bc) where $c \in \textbf{H}(ba_0)$. We take a point $q \in (bc) \cap (a_0b)$. The line $M_q \subset P$ passing through point q perpendicularly to line L intersects half-line A in some point p, and half-line D in some point r. We have

$$p = r \quad \text{or} \quad \textbf{B}(q,p,r) \quad \text{or} \quad \textbf{B}(q,r,p).$$

In each of these three cases it is readily seen that $A \cap D \neq 0$ (in the second case, by applying Pasch's Axiom to triangle bqr and half-line A; in the third case, by applying Pasch's Axiom to triangle a_0pq and half-line D). Therefore $C \parallel A$, and it thus follows that $A \parallel C$ and $A \parallel M$. This concludes the proof.

A direct conclusion from Theorem 27 is

THEOREM 28. *If lines K and L intersect in one point, but are not perpendicular to each other, then the perpendicular projection of line K upon line L is an open segment.*

Next we investigate the perpendicular projection of line K upon line L in case lines K and L are parallel or hyperparallel. We shall need the following:

LEMMA. *Consider, in a plane P, a right or obtuse angle AB with vertex a and a line L intersecting perpendicularly half-line B in some point b. The perpendicular projection of half-line A upon line L is a segment (bc), where A and c lie on the same side of line ab. Half-line A is parallel to line $M \subset P$ passing through point c and perpendicular to L.*

PROOF. Let W denote the half-plane with boundary $\textbf{L}(ab)$ which includes half-line A (Fig. 226) and let $L \cap W = C$. We produce, from point b, a half-line D parallel to half-line A. Then $D \subset W$. Angle DC is acute; therefore, by Theorem 27, there exists a point $c \in C$ such that segment (bc) is the projection of half-line D upon line L. Let the line $M \subset P$ pass through point c and be perpendicular to L. By Theorem 27 we have $D \parallel M$; thus $A \parallel M$. It remains to show that segment (bc) is also the projection of half-line A upon L.

Since $\textbf{B}(a,\textbf{L}(D),c)$, then $\textbf{B}(A,\textbf{L}(D),M)$. Hence $A \cap M = 0$, from which it readily follows that the projection of half-line A upon line L is included in half-line cb. Moreover, it is readily noted that this projection is included in half-line bc. Therefore the projection of half-line A upon line L is included in segment (bc).

Now let r be any point of segment (bc). Then the line $M_r \subset P$ produced through point r perpendicularly to line L intersects half-line D in some point q. The axis $\mathfrak{N}$ determined by half-line D is parallel to axis $\mathfrak{K}$ determined by half-line A. We denote by b_1 and q_1, respectively, the projections of points b and q upon $\mathfrak{K}$. Let

$$D_q = D - (bq\,\rangle = \mathbf{H}^*(qb), \quad B' = \mathbf{H}^*(qr).$$

Then

(1) $$D_q B' = \sphericalangle\ bqr < D\mathbf{H}(ba).$$

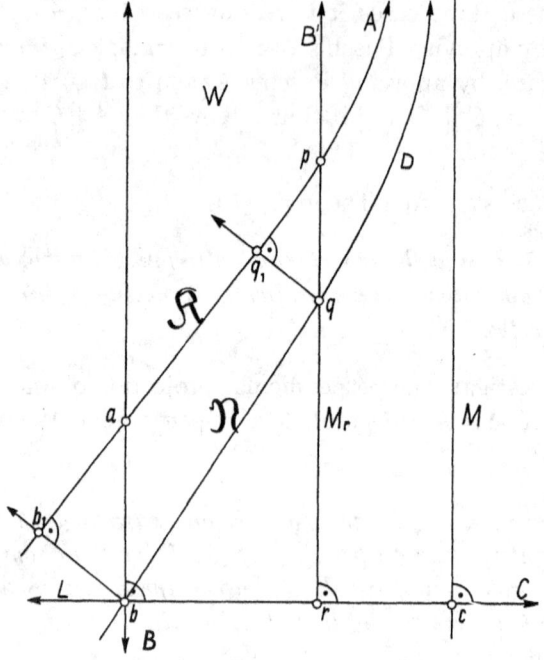

Fig. 226

Further, since AB is a right or obtuse angle, then $b_1 \in a \cup A^*$, and consequently

(2) $$D\mathbf{H}(ba) < D\mathbf{H}(bb_1).$$

Finally, since point b precedes q on axis $\mathfrak{N}$, we have, by Theorem 22,

(3) $$D\mathbf{H}(bb_1) < D_q\,\mathbf{H}(qq_1).$$

Formulas (1)–(3) imply that

$$D_q B' < D_q\mathbf{H}(qq_1).$$

It thus follows that half-line B' intersects axis $\mathfrak{K}$ in some point p following point q_1. Since $B' \subset M_r \subset W$, then $p \in A$, and point r is the projection of point p upon line L. In this way we have shown that each point r of segment (bc) is the projection of some point of half-line A. This concludes the proof of the lemma.

THEOREM 29. *If axes $\mathfrak{K}$ and $\mathfrak{L}$ are parallel and distinct, then the perpendicular projection of $\mathfrak{K}$ upon $\mathfrak{L}$ is a half-line of axis $\mathfrak{L}$.*

PROOF. We take any point a on axis $\mathfrak{K}$ (Fig. 227). We denote by b the perpendicular projection of a upon axis $\mathfrak{L}$. Let a determine a half-line A on $\mathfrak{K}$, and let b determine a half-line B on $\mathfrak{L}$. Then angle $\mathbf{H}(ab)A^*$ is obtuse, and by the Lemma, there is a point $c \in B^*$ such that segment (bc) is the projection of half-line A^* upon $\mathfrak{L}$.

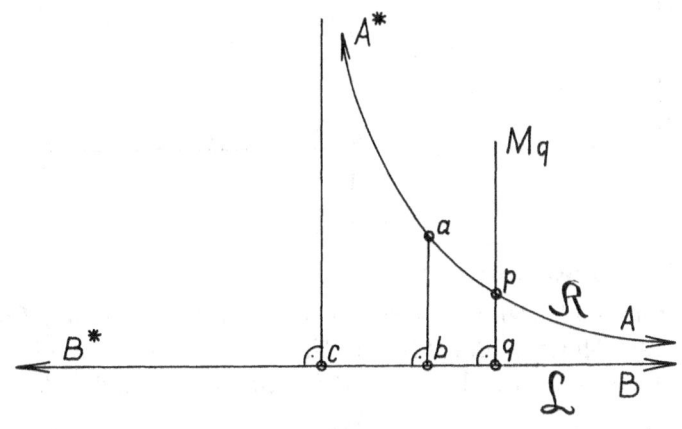

Fig. 227

The projection of half-line A upon axis $\mathfrak{L}$ is, as readily noted, included in half-line B. We take any point q on B. The size of the angle of parallelism for point c with respect to axis $\mathfrak{K}$ is, as readily noted, equal to $\pi/4$ (see the end of Lemma). Since q follows c on $\mathfrak{L}$, then the size of the angle of parallelism for q with respect to $\mathfrak{K}$ is greater than $\pi/4$ (see Theorem 22). As a result, line M_q produced (in the plane determined by axes $\mathfrak{K}$ and $\mathfrak{L}$) through point q perpendicularly to $\mathfrak{L}$ intersects $\mathfrak{K}$ in some point p, which, as easily seen, lies on half-line A. Therefore the entire half-line B is the projection of half-line A upon axis $\mathfrak{L}$. Consequently, the half-line $(cb\rangle \cup B$ of axis $\mathfrak{L}$ is the projection of the entire axis $\mathfrak{K}$ upon axis $\mathfrak{L}$.

THEOREM 30. *If lines K and L are hyperparallel then the perpendicular projection of K upon L is an open segment.*

PROOF. Let line M be the common perpendicular to lines K and L, and let $M \cap K = a$, $M \cap L = b$, $\mathbf{P}(KL) = W \cup \mathbf{L}(ab) \cup W^*$, $K \cap W = A$, $L \cap W = B$ (Fig. 228). Then $K \cap W^* = A^*$, $L \cap W^* = B^*$. By the Lemma, there exist points $c_1 \in B$ and $c_2 \in B^*$ such that segments (bc_1)

and (bc_2) are the respective projections of half-lines A and A^* upon line L. Since, moreover, point b is the projection of point a, then the projection of line K upon line L is the segment (c_1c_2).

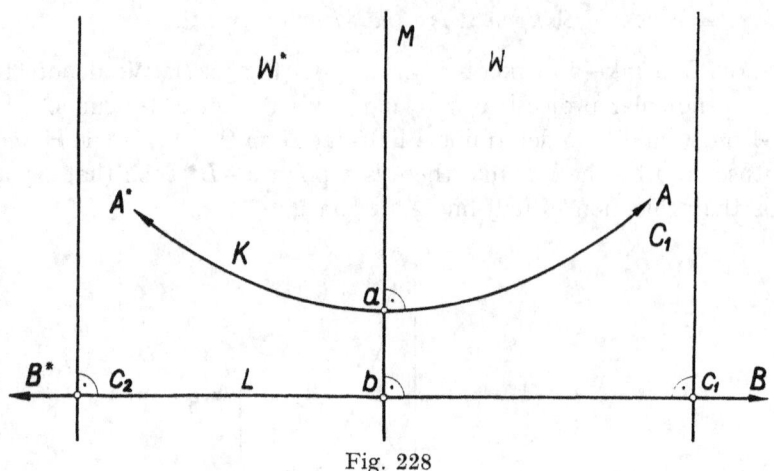

Fig. 228

12. The Line of Enclosure

Consider an angle AB with vertex a. We shall refer to a line M to which each of the half-lines A and B is parallel as a *line of enclosure* of the angle AB. In this case we shall also say that *line M encloses angle AB*.

THEOREM 31. *For every angle AB there exists just one line of enclosure M. Line M lies in the inner domain of angle AB.*

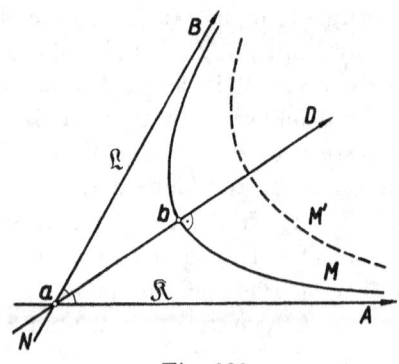

Fig. 229

PROOF. Let point a be the vertex of angle AB, and half-line D, its bisector (Fig. 229). Angle AD is acute; therefore the perpendicular projection of half-line A upon the line $N \supset D$ is a segment (ab), where $b \in D$. Through point b, in the plane of angle AB, we produce the line M perpendicular to N. By Theorem 27, we have $A \parallel M$. Hence $B \parallel M$

and M is the line of enclosure we are seeking. Since line M passes through point b lying in the inner domain of angle AB and does not intersect any of its sides, then it lies entirely in the inner domain of angle AB.

Let us now suppose that M' is any arbitrary line of enclosure of angle AB. Since $A \parallel M$ and $A \parallel M'$ then each of the lines M and M' has the direction of the axis $\Re$ determined by half-line A. Similarly, from $B \parallel M$ and $B \parallel M'$ it follows that each of the lines M and M' has the direction of the axis $\mathfrak{L}$ determined by half-line B. Hence lines M and M' have two (distinct) directions in common, and consequently they are identical (see Theorem 20).

We shall now show that the distance of vertex a from the line of enclosure M depends only on the size of angle AB.

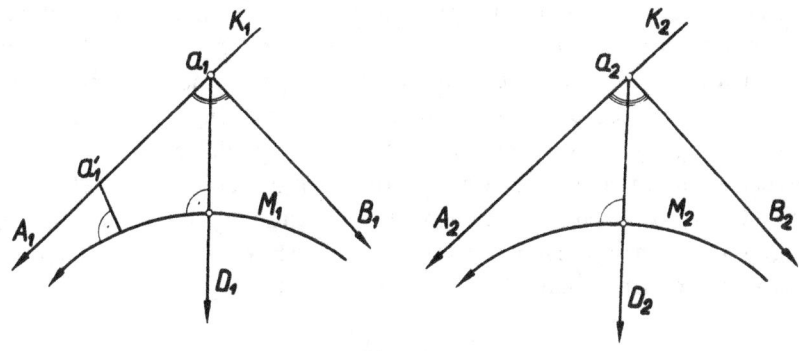

Fig. 230

THEOREM 32. *Let a line M_1 enclose an angle A_1B_1 with vertex a_1 and let a line M_2 enclose an angle A_2B_2 with vertex a_2. Then $A_1B_1 \equiv A_2B_2$ implies $\varrho(a_1, M_1) = \varrho(a_2, M_2)$.*

PROOF. Assume that

(1) $$A_1B_1 \equiv A_2B_2.$$

Lines $K_i \supset A_i$ and M_i $(i = 1,2)$ are parallel (Fig. 230). Let the pair of axes $\Re_i$ and $\mathfrak{M}_i$ be conjugate to the pair of lines K_i and M_i. We denote by D_i the bisector of angel A_iB_i. Then angle A_iD_i is the angle of parallelism for point a_i with respect to axis $\mathfrak{M}_i$, and from formula (1) it follows that

(2) $$A_1D_1 \equiv A_2D_2.$$

If $\varrho(a_1, \mathfrak{M}_1) \neq \varrho(a_2, \mathfrak{M}_2)$, then, by Theorem 21, there would exist on axis $\Re$ a point a_1' distict from point a_1 and such that $\varrho(a_1', \mathfrak{M}_1) = \varrho(a_2, \mathfrak{M}_2)$. Then the angle of parallelism for point a_1' with respect to axis $\mathfrak{M}_1$ would not be congruent to angle A_1D_1 (see Theorem 22) and simultaneously

it would be congruent to angle A_2D_2. This, however, would be in contra-
diction to formule (2). Hence $\varrho(a_1,\mathfrak{M}_1) = \varrho(a_2,\mathfrak{M}_2)$, that is, $\varrho(a_1,M_1) = \varrho(a_2,M_2)$.

13. Natural Basic Segment. Natural Measure of Segments

From Theorem 32 we conclude at once

THEOREM 33. *There exists just one free segment* $\mathfrak{a}$ *with the following
property: If AB is any right angle, point a—its vertex, line M—its line of
enclosure, and point b—the perpendicular projection of a upon M, then*
$ab \in \mathfrak{a}$.

We shall refer to the free segment $\mathfrak{a}$ defined in this way as the *(free)
natural basic segment* and denote it by letter $\mathfrak{r}$.

Thus, in Bolyai-Lobachevskian geometry, we have singled out one definite
free segment from the family of all free segments, in fact, free segment $\mathfrak{r}$.

By means of free segment $\mathfrak{r}$ (the natural basic segment) we may single
out one definite measure of segments. Indeed, by Theorem 35 of Chapter
III, there exists just one measure φ_0 of segments which assigns $\ln(\sqrt{2}+1)$
to every segment $ab \in \mathfrak{r}$. The choice of the number $\ln(\sqrt{2}+1)$ will be
justified in Section 30. We shall call the measure φ_0 the *natural* measure
of segments. Henceforth we shall use exclusively this measure. Therefore,
for any two distinct points a and b, we shall have

$$|ab| = \varphi_0(ab)$$

and

$$\varrho(a,b) = \varphi_0(ab).$$

We shall call the metric ϱ defined in this way the *natural metric* and its
value $\varrho(a,b)$ the *natural distance* between points a and b. We shall call
$|ab|$ the *natural length* of segment ab.

Let us consider the following two theorems on similitudes: Theorem 58
of Chapter III belonging to absolute geometry and stating that for any
similitude f of space **S** onto itself and for any angle AB

$$f(A)f(B) \equiv AB,$$

and Theorem 7 of Bolyai-Lobachevskian geometry stating that for any
similitude f of space **S** onto itself and for any segment ab

$$f(a)f(b) \equiv ab.$$

The proof of the first of these two theorems was based on the following
facts: (i) all primitive notions, and therefore all defined notions of
geometry, are invariants of similitudes; (ii) in absolute geometry, certain

free angle, in fact, the right free angle, was singled out from among all free angles.

The proof of the second theorem was based on a sufficient condition for the congruence of triangles (see Theorem 6), which holds only in Bolyai-Lobachevskian geometry. At present, in Bolyai-Lobachevskian geometry, from among all free segments one has been single dout (without using Theorem 7), in fact, the free segment $\mathfrak{r}$ (the natural basic segment). This permits Theorem 7 to be proved by means of a proof similar to that of Theorem 58 of Chapter III. More accurately, for this purpose we should supplement Theorem 56 of Chapter III by further points stating, in turn, that all notions necessary to define free segment $\mathfrak{r}$ and finally free segment $\mathfrak{r}$ itself are invariants of similitudes, and then prove for the measure of segments a theorem analogous to Theorem 57 of Chapter III on the measure of angles.

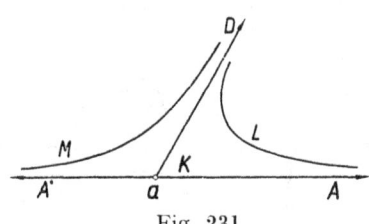

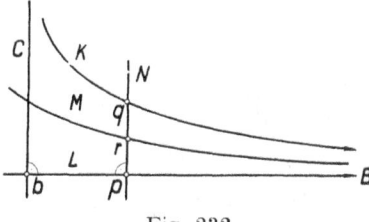

Fig. 231   Fig. 232

14. Three Parallel Lines

Three parallel lines K, L, M not having any direction in common may readily be found on a plane P. Indeed, let $K = A \cup a \cup A^*$ be any line in plane P and let $D \subset P$ be any half-line with origin a distinct from half-lines A and A^* (Fig. 231). If line L encloses angle AD, and line M encloses angle A^*D, then every two of the three lines K, L, M are parallel, and at the same time all three lines K, L, M do not have any direction in common.

On the other hand, we have

THEOREM 34. *If line M lies between parallel lines K and L, then it has the direction of the pair K and L.*

PROOF. Let the pair of parallel axes $\mathfrak{K}$ and $\mathfrak{L}$ be conjugate to the pair of parallel lines K and L (Fig. 232) and let half-line B of axis $\mathfrak{L}$ be the perpendicular projection of axis $\mathfrak{K}$ upon axis $\mathfrak{L}$ (see Theorem 29). If p is any point on half-line B, then the line N perpendicular to line L at point p intersects line K in some point q, and consequently, it intersects line M in some point $r \in (pq)$. Therefore the projection of line M upon line L includes half-line B, from which it follows, by Theorems 29 and 30, that $M \parallel L$, and that the direction of the pair of lines M and L coincides with the direction of axis $\mathfrak{L}$, i.e. with the direction of the pair of lines K and L.

15. Perpendicular Bisectors of the Three Sides of a Triangle

In Euclidean geometry it may readily be proved that the perpendicular bisectors of the three sides of a triangle intersect in a point. This is not the case in Lobachevskian geometry.

THEOREM 35. *The perpendicular bisectors of the three sides of a triangle abc (in plane abc) have either a common point or a common perpendicular line (in which case every two of them are hyperparallel), or a common direction (in which case every two of them are parallel).*

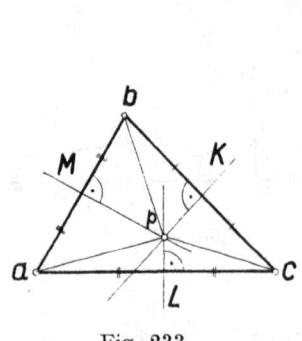

Fig. 233

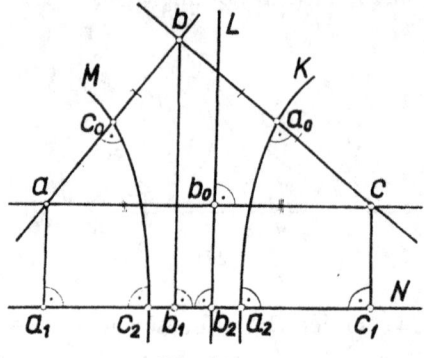

Fig. 234

PROOF. Consider any triangle abc and let lines K, L, M be the perpendicular bisectors of sides bc, ac, ab, respectively.

If lines K and L intersect in some point p (Fig. 233), then, by Theorem 81 of Chapter II,

$$\varrho(p,b) = \varrho(p,c) \quad \text{and} \quad \varrho(p,a) = \varrho(p,c).$$

It thus follows that $\varrho(p,a) = \varrho(p,b)$, which, by Theorem 81 of Chapter II, is equivalent to the condition $p \in M$. By permuting the vertices of triangle abc, we obtain the statement

(I) If a point p belongs to any two of the three perpendicular bisectors K, L, M, then point p belongs to each of the perpendicular bisectors K, L, M.

Let us next assume that lines K and L are hyperparallel. There then exists a line N perpendicular to each of them (Fig. 234). We shall show that line N is also perpendicular to line M. We denote by a_1, b_1, c_1, respectively, the perpendicular projections of vertices a, b, c upon line N.

Line N is, as readily noted, distinct from lines bc and ac; therefore, by Theorem 25, lines N and bc are hyperparallel (both are perpendicular to line K) and lines N and ac are also hyperparallel (both are perpendicular to line L). It thus follows that points a and b lie on the same side of line N, and that $bb_1 \equiv cc_1$ and $aa_1 \equiv cc_1$ (see Theorem 26); hence $aa_1 \equiv bb_1$. We thus readily conclude that $a_1 \neq b_1$ and that the polygonal line (a_1,a,b,b_1) is a Saccheri quadrangle. Therefore the line joining the midpoint c_0 of segment ab with the midpoint c_2 of segment a_1b_1 is perpendicular both to line ab and to line N. Consequently $\mathbf{L}(c_0c_2) = M$ and $M \perp N$. By permuting the vertices of triangle abc, we obtain the statement

(II) If a line N is perpendicular to any two of the three perpendicular bisectors K, L, M, then line N is perpendicular to each of the perpendicular bisectors K, L, M.

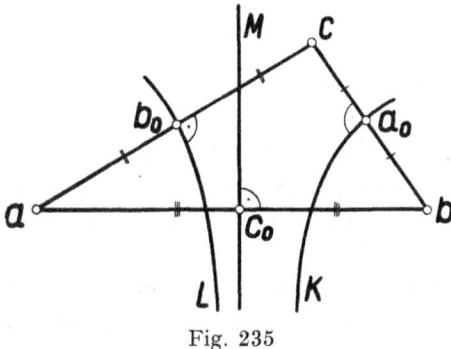

Fig. 235

Finally, let us assume that $K \parallel L$ (Fig. 235). By (I) and (II), we then have $K \parallel M$ and $L \parallel M$. Hence the assumptions regarding lines K, L, M are now symmetric and, without reducing the generality of the proof, we may assume that M is the perpendicular bisector of the longest side, i.e.

$$ab \geqslant ac \quad \text{and} \quad ab \geqslant bc.$$

Denote by a_0, b_0, c_0 the midpoints of sides bc, ac, ab, respectively. We have $\mathbf{B}(a,M,b)$. Further, $ab_0 \leqslant ac_0$ and $ba_0 \leqslant bc_0$; hence

$$\sim \mathbf{B}(a,M,b_0) \quad \text{and} \quad \sim \mathbf{B}(b,M,a_0).$$

It thus follows that $\mathbf{B}(a_0,M,b_0)$, and consequently $\mathbf{B}(K,M,L)$. By Theorem 34 it thus follows at once that lines K, L, M have a common direction

By constructing the appropriate triangles, the reader may readily show that each of the possibilities mentioned in Theorem 35 does occur.

16. The Lobachevskian Function Π

By Theorem 22(I) of Chapter III the size of the angle of parallelism for a point a with respect to an axis $\mathfrak{L}$ is, for any fixed measure of segments, a function of the distance of point a from axis $\mathfrak{L}$; in particular, it is a function of the natural distance $x = \varrho(a,\mathfrak{L})$ (see Section 13). We shall denote this function by the letter Π and call it the *(natural) Lobachevskian function*. We have

$$0 < \Pi(x) < \pi/2 \quad \text{for each} \quad x > 0.$$

THEOREM 36. Π *is a continuous and decreasing function of the positive variable x; the range of Π coincides with the open interval $(0,\pi/2)$. Furthermore,*

(I)
$$\Pi(\ln(\sqrt{2}+1)) = \frac{\pi}{4}$$

and

(II)
$$\lim_{x \to 0} \Pi(x) = \frac{\pi}{2} \quad \text{and} \quad \lim_{x \to \infty} \Pi(x) = 0.$$

PROOF. From Theorems 21 and 22 it follows at once that Π is a decreasing function, and from Theorem 31 it follows at once that Π takes on every value α of the interval $(0,\pi/2)$. Therefore function Π is continuous.

Formula (I) follows at once from the definition of the natural metric.

We shall now derive formula (II). Let (x_n) be any sequence of positive numbers tending to 0. For every number $\varepsilon > 0$ there exists a number $x > 0$ such that

$$\frac{\pi}{2} - \varepsilon < \Pi(x) < \frac{\pi}{2}.$$

Since for almost all n we have $0 < x_n < x$, then for almost all n we have

$$\frac{\pi}{2} - \varepsilon < \Pi(x) < \Pi(x_n) < \frac{\pi}{2};$$

Hence $\lim_{n \to \infty} \Pi(x_n) = \pi/2$, and consequently $\lim_{x \to 0} \Pi(x) = \pi/2$.

In a similar manner we may show that $\lim_{x \to \infty} \Pi(x) = 0$. Hence the proof of the theorem is completed.

We set $\Pi(0) = \pi/2$. Then function Π is defined in the set of all nonnegative numbers; by Theorem 36, it remains continuous and decreasing. It is sometimes convenient to extend the Lobachevskian function also to negative numbers by setting

$$\Pi(x) = \pi - \Pi(-x) \quad \text{for every} \quad x < 0.$$

Thus we obtain function Π defined in the set of all real numbers. By Theorem 36, we obtain immediately

THEOREM 37. *The Lobachevskian function Π is a continuous and decreasing function of the real variable x; the range of Π coincides with the open interval* $(0,\pi)$.

Far more difficult is the task of finding an analytical formula for $\Pi(x)$, and more specifically, of expressing the function Π in terms of elementary real functions. To find an analytical formula for $\Pi(x)$ is the central problem of Lobachevskian geometry. In particular, we use this formula when we derive the formula for the distance $\varrho(p,q)$ between two points p and q in terms of their coordinates (Section 30). The latter formula plays an essential role in our proof of categoricalness.

There are many different proofs of the analytical formula for $\Pi(x)$. In the proof which we shall present, use is made of the notion of the defect of an arbitrary plane figure (introduced in Section 7), and of the properties of a plane figure called the *infinite right triangle* and of a curve called the *horocycle*. That is why we shall take up the theory of the infinite triangle (Section 17) and the theory of the horocycle (Sections 18–24) before considering this proof. The analytical formula for $\Pi(x)$ will be given in Section 25.

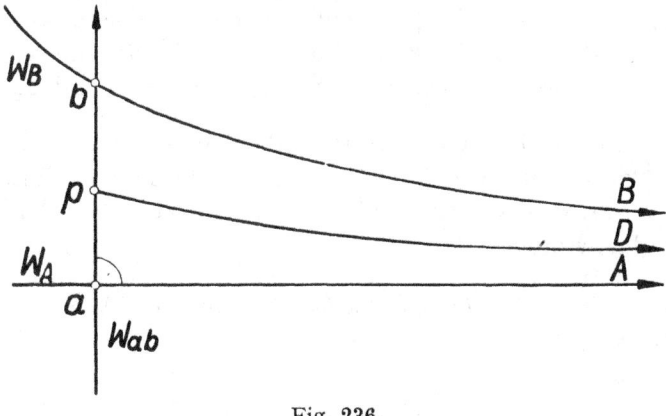

Fig. 236

17. The Infinite Right Triangle

Let there be given a segment ab and a half-line A with origin a. If $\mathsf{H}(ab)A$ is a right angle (Fig. 236), then the set-theoretical sum T of all closed half-lines $p \cup D$, where $p \in \langle ab \rangle$ and $D \parallel A$, will be called the *infinite (closed) right triangle* with *finite side ab* and *infinite side A*. We shall call the points a and b the *vertices*, and half-line B with origin b and parallel to half-line A the *hypotenuse* of infinite triangle T.

We denote by W_{ab} the half-plane with boundary $\mathsf{L}(ab)$ and including side A; we put $W_A = \mathsf{HP}(\mathsf{L}(A)b)$ and $W_B = \mathsf{HP}(\mathsf{L}(B)a)$. It may readily be proved that

(1) $$T = \bar{W}_{ab} \cap \bar{W}_A \cap \bar{W}_B.$$

We shall now give a sufficient condition for the congruence of two infinite right triangles:

THEOREM 38. *Given an infinite right triangle T_1 with finite side a_1b_1 and an infinite right triangle T_2 with finite side a_2b_2, if $a_1b_1 \equiv a_2b_2$, then $T_1 \equiv T_2$.*

PROOF. Let $a_1b_1 \equiv a_2b_2$. We assume that, in triangle T_i $(i = 1,2)$, the right angle is at vertex a_i. On the infinite side A_i we take a point c_i such that $a_1c_1 \equiv a_2c_2$. The function f:

$$f(a_1) = a_2, \quad f(b_1) = b_2, \quad f(c_1) = c_2,$$

realizing the congruence of triangles $a_1b_1c_1$ and $a_2b_2c_2$, can be extended, by Theorems 101 and 102 of Chapter II, to the isometry g of the whole space S onto itself. The function g maps segment $\langle a_1b_1 \rangle$ onto segment $\langle a_2b_2 \rangle$, half-line A_1 onto half-line A_2, and every closed half-line $p_1 \cup D_1$, where $p_1 \in \langle a_1b_1 \rangle$ and $D_1 \parallel A_1$, onto closed half-line $p_2 \cup D_2$, where $p_2 \in \langle a_2b_2 \rangle$ and $D_2 \parallel A_2$. Thus function g (with the domain restricted to set T_1) realizes the congruence of infinite triangles T_1 and T_2.

From Theorems 15 and 38 we infer that the defect Δ of an infinite right triangle T depends only on the natural length of its finite side ab; in other words, it is a function of the real variable $x = |ab|$.

We shall now prove

THEOREM 39. *If x is the natural length of the finite side of an infinite right triangle T, then*

$$\Delta(T) = \frac{\pi}{2} - \Pi(x).$$

PROOF. Let the segment ab, the half-line A with origin a, and the half-line B be, respectively, the finite side, the infinite side and the hypotenuse of an infinite triangle T. Then $x = |ab|$. Let E be any polygon included in infinite triangle T, and let points $p_1, p_2, \ldots, p_n$ be the vertices of some triangulation $\mathfrak{X}$ of polygon E. We denote by p_i', where $i = 1, 2, \ldots n$, the perpendicular projection of point p_i upon line $K \supset A$. Obviously, all points p_i' lie on half-line A; let point c be the last of them (Fig. 237). By Theorem 29, point c is the projection of some point $d \in B$. Let $W = \textbf{HP}(\textbf{L}(cd)a)$. It is readily noted (see formula (1)) that

$$E \subset T \cap (W \cap \textbf{L}(cd)) = \langle abc \rangle \cup \langle dbc \rangle.$$

We therefore have

$$\Delta(E) \leqslant \Delta(\langle abc \rangle) + \Delta(\langle dbc \rangle) = \pi - |\sphericalangle abc| - |\sphericalangle bca| - \frac{\pi}{2} +$$

$$+ \pi - |\sphericalangle dbc| - |\sphericalangle bcd| - |\sphericalangle cdb|.$$

Furthermore,

$$|\sphericalangle abc| + |\sphericalangle dbc| = \Pi(x), \quad |\sphericalangle bca| + |\sphericalangle bcd| = \frac{\pi}{2}, \quad |\sphericalangle cdb| > \frac{\pi}{2}.$$

Thus

$$\Delta(E) \leqslant \pi - \Pi(x) - |\sphericalangle cdb| < \frac{\pi}{2} - \Pi(x).$$

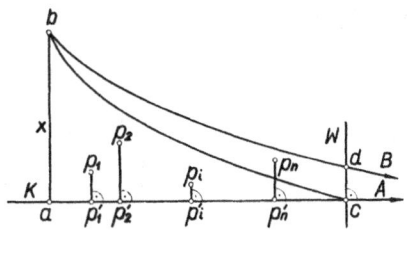

Fig. 237

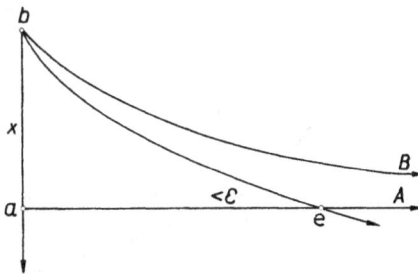

Fig. 238

Since the inequality $\Delta(E) < \dfrac{\pi}{2} - \Pi(x)$ holds for every polygon $E \in T$, then

$$\Delta(T) \leqslant \frac{\pi}{2} - \Pi(x).$$

It remains to show that for every $\varepsilon > 0$ there exists a polygon $E \subset T$ such that $\Delta(E) > \pi/2 - \Pi(x) - \varepsilon$. To do this we make use of Theorems 18, 44 and 39(I) of Chapter III and take a point e on half-line A (Fig. 238) such that

$$|\sphericalangle bea| < \varepsilon.$$

Since $B \parallel A$, then $\textbf{B}(\textbf{H}(ba), \textbf{H}(be), B)$. Taking triangle $\langle abe \rangle$ as polygon E, we have $E \subset T$ and

$$\Delta(E) = \Delta(\langle abe \rangle) = \pi - \frac{\pi}{2} - |\sphericalangle abe| - |\sphericalangle bea| > \frac{\pi}{2} - \Pi(x) - \varepsilon.$$

This concludes the proof.

18. The Horocycle

Let there be given any direction, i.e. any class $\mathcal{D}$ of parallel axes on a plane P (see Chapter II, Section 36). We define a horocycle *with respect to direction* $\mathcal{D}$ to be any figure H which has just one point in common with each axis $\mathfrak{N}$ of class $\mathcal{D}$ and which satisfies the following condition: for any two distinct points $a \in \mathfrak{R} \in \mathcal{D}$ and $b \in \mathfrak{L} \in \mathcal{D}$ on H,

(1)
$$\varrho(a,\mathfrak{L}) = \varrho(b,\mathfrak{R}).$$

Every axis $\mathfrak{R} \in \mathcal{D}$ will be called an *axis* of horocycle H, and the half-line A determined on axis $\mathfrak{R}$ by the point a in which $\mathfrak{R}$ intersects H will be called a *radius* of horocycle H. We shall show that the angle $\mathsf{H}(ab)A$ is always acute.

THEOREM 40. *Given in a plane P two distinct parallel axes $\mathfrak{R}$ and $\mathfrak{L}$, and points $a \in \mathfrak{R}$, $b \in \mathfrak{L}$, if $\varrho(a,\mathfrak{L}) = \varrho(b,\mathfrak{R})$ and A is the half-line determined on axis $\mathfrak{R}$ by point a, then the angle $\mathsf{H}(ab)A$ is acute.*

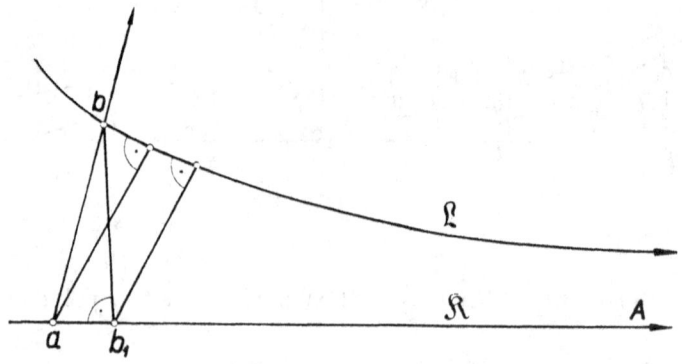

Fig. 239

PROOF. Let point b_1 be the perpendicular projection of point b upon axis $\mathfrak{R}$ (Fig. 239). Then

$$\varrho(b_1,\mathfrak{L}) < \varrho(b_1,b) = \varrho(b,\mathfrak{R}) = \varrho(a,\mathfrak{L}),$$

and therefore $b_1 \succ a$ on axis $\mathfrak{R}$, that is, $b_1 \in A$. Hence angle $\mathsf{H}(ab)A$ is acute.

We shall now show that in the definition of the horocycle condition (1) may be replaced by some other conditions:

THEOREM 41. *Given on a plane P two distinct parallel axes $\mathfrak{R}$ and $\mathfrak{L}$ of a direction $\mathcal{D}$, let A and B be the half-lines determined on $\mathfrak{R}$ and $\mathfrak{L}$ by* points $a \in \Re$ and $b \in \mathfrak{L}$, and let M be the perpendicular bisector (on plane P) of segment ab. Then the following three conditions are equivalent:

(I) $\varrho(a,\mathfrak{L}) = \varrho(b,\Re)$,
(II) $\textbf{HP}(ab)A \equiv \textbf{HP}(ba)B$,
(III) $\mathcal{D}$ is one of the two directions of line M.

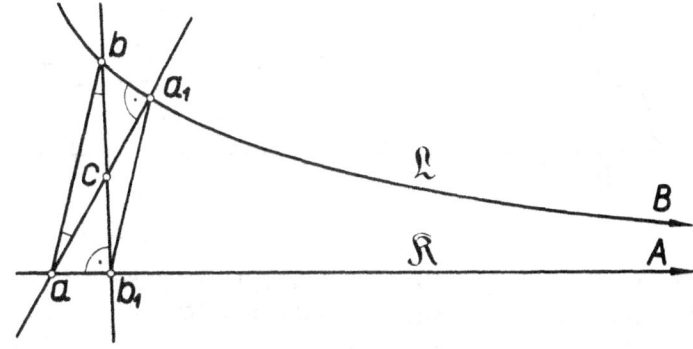

Fig. 240

PROOF. Let us assume that condition (I) is fulfilled. We denote by a the perpendicular projection of point a upon axis $\mathfrak{L}$, and by point b_1 the perpendicular projection of point b upon axis $\Re$ (Fig. 240). By Theorem 40,

2) $a_1 \in B$ and $b_1 \in A$,

and therefore

(3) $\sphericalangle a_1 a b_1 \equiv \sphericalangle b_1 b a_1$.

From formula (2) we further conclude that quadrangle (a,b,a_1,b_1) is convex, from which it follows that its diagonals $\textbf{L}(aa_1)$ and $\textbf{L}(bb_1)$ intersect in some point $c \in (aa_1) \cap (bb_1)$. Hence

$\sphericalangle acb_1 \equiv \sphericalangle a_1cb, \quad \sphericalangle cab_1 \equiv \sphericalangle cba_1, \quad \sphericalangle cb_1a \equiv \sphericalangle ca_1b,$

and applying Theorem 6 to triangles a_1bc and ab_1c, we there by conclude that $ca \equiv cb$, that is,

(4) $\sphericalangle a_1 ab \equiv \sphericalangle b_1 ba.$

Formula (II) follows from equalities (3) and (4).

We next assume that condition (II) is fulfilled. Let a_1 be the perpendicular projection of point a upon axis $\mathfrak{L}$ (Fig. 241). Then $a_1 \in B$. We take on half-line A a point b_1 such that $ab_1 \equiv ba_1$. Comparing congruent triangles abb_1 and baa_1 we conclude that angle ab_1b is a right angle and $bb_1 \equiv aa_1$, from which formula (I) follows.

In this way we have established the equivalence of conditions (I) and (II). We next show the equivalence of conditions (II) and (III).

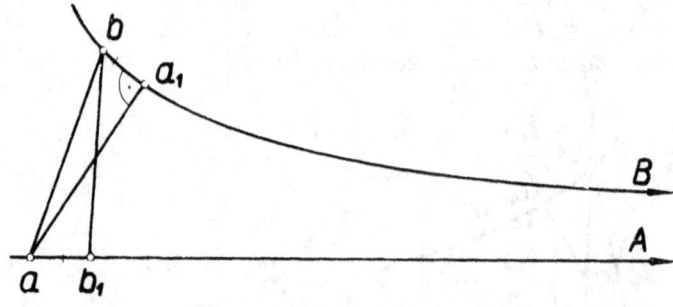

Fig. 241

We assume that condition (II) is fulfilled; let the pair of parallel axes $\Re$ and $\mathfrak{L}$ be conjugate to the pair of parallel lines K and L. If the perpendicular bisecter M of segment ab were to intersect line K in a point p (Fig. 242), then in congruent triangles acp and bcp we would have $\sphericalangle\, bap \equiv \sphericalangle\, abp$, from which it would follow that $p \in L$, in contradiction to the fact that lines K and L are disjoint. Hence $K \cap M = 0$. In the same way we may prove that $L \cap M = 0$. Therefore line M lies

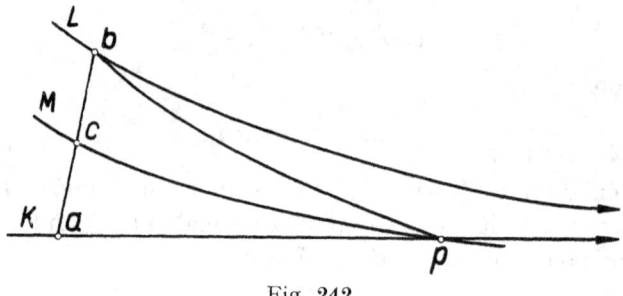

Fig. 242

between lines K and L, and consequently, by Theorem 34, it has the direction $\mathcal{D}$. We have thus derived formula (III).

Finally, we assume that condition (III) is fulfilled. Let axis $\mathfrak{M}$ of line M be parallel to axis $\Re$. Then $\mathfrak{M} \,||\, \mathfrak{L}$ and, since $ac \equiv bc$, formula (II) holds.

We shall now prove

THEOREM 42. *For any point a of a given plane P and for any direction $\mathcal{D}$ on P there is just one horocyle H with respect to $\mathcal{D}$ passing through a,*

PROOF. We denote by $\Re$ the axis of direction $\mathcal{D}$ passing through point a. If a horocycle with respect to $\mathcal{D}$ passes through point a, then it intersects every axis $\mathfrak{N} \neq \mathfrak{K}$ of direction $\mathcal{D}$ in the point $p_{\mathfrak{N}}$ uniquely determined by the equation

5) $$\varrho(p_{\mathfrak{N}},\mathfrak{K}) = \varrho(a,\mathfrak{N}).$$

It thus remains to show that the figure H formed by point a and all points $p_{\mathfrak{N}}$ for $\mathfrak{N} \in \mathcal{D} - \{\mathfrak{K}\}$ is a horocycle. To do this we take two distinct axes $\mathfrak{N}_1, \mathfrak{N}_2 \in \mathcal{D} - \{\mathfrak{K}\}$ and put $p_1 = p_{\mathfrak{N}_1}$, $p_2 = p_{\mathfrak{N}_2}$ (Fig. 243). Let M_i be the perpendicular bisector of segment ap_i ($i = 1,2$). By Theorem 41,

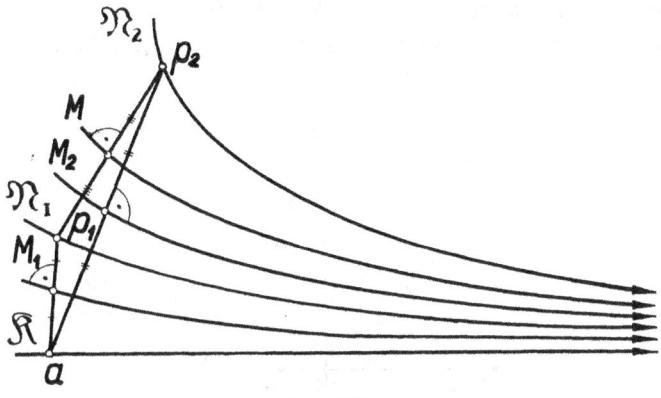

Fig. 243

we derive from equality (5) that $\mathcal{D}$ is one of the two directions of line M_i. Hence lines M_1 and M_2 are parallel. It thus follows, in particular, that points a, p_1, p_2 are not collinear, since if this were not the case, lines M_1 and M_2, would, by Theorem 25, be hyperparallel. Now let line M be the perpendicular bisector of side p_1p_2 of triangle ap_1p_2. By Theorem 35, $\mathcal{D}$ is one of the two directions of line M, from which it follows, by Theorem 41, that

$$\varrho(p_1,\mathfrak{N}_2) = \varrho(p_2,\mathfrak{N}_1).$$

From this last equality and from formula (5) we conclude that figure H is a horocycle.

NOTE. Since, on horocycle H, we have found three non-collinear points a, p_1, and p_2, then $P = \mathbf{P}(ap_1p_2)$, and hence horocycle H determines the plane P on which it lies. Since, further, the perpendicular bisectors M_1 of segment ap_1 and M_2 of segment ap_2 are parallel and $\mathcal{D}$ is their common direction, then horocycle H determines the direction $\mathcal{D}$ of its axes.

We shall now show that the horocycle is a homeomorphic image of a line. We shall need for the proof the following two lemmas:

LEMMA 1. *For every pair of numbers $\varepsilon, \lambda > 0$, there exists a number $\delta > 0$ such that if in a triangle abc the conditions*

$$|\sphericalangle a| < \delta \quad and \quad |ab| = |ac| = \lambda$$

are satisfied, then $|bc| < \varepsilon$.

PROOF. It is sufficient to carry out the proof for the case in which $\varepsilon < 2\lambda$, that is, when $\varepsilon/2 < \lambda$. We take any two perpendicular lines K and L intersecting each other in a point d (Fig. 244). Let $L = B \cup d \cup B^*$. We choose points $b' \in B$ and $c' \in B^*$ such that $\varrho(d, b') = \varrho(d, c') = \varepsilon/2$. Since $\varrho(b', p)$ is a continuous and, as readily noted, an unbounded function

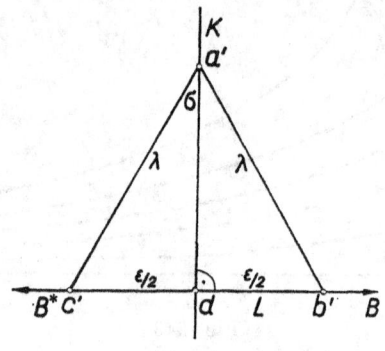

Fig. 244

of a point $p \in K$, and since $\lambda > \varrho(b', d)$, then there exists on line K a point a' such that $\varrho(a', b') = \lambda$. We have thus constructed an isosceles triangle $a'b'c'$ whose sides $a'b' \equiv a'c'$ have the length λ, and whose base $b'c'$ has the length ε. Taking $\delta = |\sphericalangle a'|$ and comparing triangle abc with triangle $a'b'c'$ we obtain, by Theorem 40 of Chapter II, the inequality $|bc| < \varepsilon$.

LEMMA 2. *For every pair of numbers $\varepsilon, \lambda > 0$ there exists a number $\delta > 0$ such that if in a triangle abc the conditions*

(6) $$|\sphericalangle a| < \delta, \quad ||ab| - \lambda| < \delta, \quad ||ac| - \lambda| < \delta$$

are satisfied, then $|bc| < \varepsilon$.

PROOF. Take a triangle abc and points $b_1 \in \mathbf{H}(ab)$ and $c_1 \in \mathbf{H}(ac)$ such that

(7) $$|ab_1| = |ac_1| = \lambda + \frac{\varepsilon}{3}.$$

By Lemma 1, there exists a number $\eta > 0$ such that

(8) $$\text{if } |\sphericalangle a| < \eta, \quad \text{then} \quad |b_1 c_1| < \frac{\varepsilon}{3}.$$

We shall show that the number

$$\delta = \min\left(\eta, \frac{\varepsilon}{3}\right)$$

satisfies the condition required in the lemma.

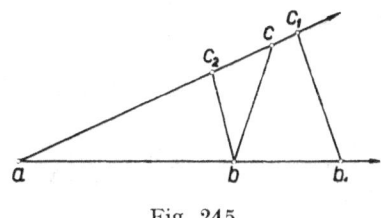

Fig. 245

Thus let us assume that in triangle abc condition (6) is fulfilled. Then

(9) $$|\sphericalangle a| < \eta$$

and

(10) $$\lambda - \frac{\varepsilon}{3} < |ab|, |ac| < \lambda + \frac{\varepsilon}{3}.$$

We may assume that $|ab| \leqslant |ac|$ (Fig. 245). Let c_2 denote a point on half-line ac such that

(11) $$ac_2 \equiv ab.$$

By formulas (7) and (10), we have on half-line ab

$$a \prec b \prec b_1,$$

and on half-line ac

$$a \prec c_2 \preceq c \prec c_1.$$

It is readily noted that $bc_2 < b_1c_1$. This, together with formulas (8) and (9), gives

$$|bc_2| < \frac{\varepsilon}{3},$$

which concludes the proof for the case in which $c = c_2$.

If $c \neq c_2$, then in triangle bcc_2 we have (see formulas (10) and (11))

$$|bc| < |bc_2| + |c_2c| < \frac{1}{3}\varepsilon + \frac{2}{3}\varepsilon = \varepsilon.$$

This completes the proof of Lemma 2.

We can now prove

THEOREM 43. *Given, on a plane P, a horocycle H with respect to a direction D and a point $a \in \mathfrak{K} \in \mathcal{D}$ on H, let A be the half-line determined on axis $\mathfrak{K}$ by point a and let W be one of the two half-planes determined on plane P by axis $\mathfrak{K}$. Then the function f correlating with every real number t a point $p = f(t)$ defined by the conditions*

(12)   *if $t = 0$,   then $p = a$;*

(13)   *if $t > 0$, then $p \in W$, $\mathbf{L}(ap)A = \Pi\left(\dfrac{t}{2}\right)$, $\varrho(a,p) = t$;*

(14)   *if $t < 0$, then $p \in W^*$, $\mathbf{L}(ap)A = \Pi\left(-\dfrac{t}{2}\right)$, $\varrho(a,p) = -t$;*

maps the set of real numbers homeomorphically onto horocycle H.

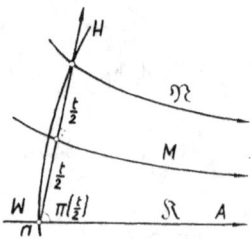

Fig. 246

PROOF. We shall first consider the function f for $t > 0$. Let $p = f(t)$. We produce through point p an axis $\mathfrak{N} \parallel \mathfrak{K}$ (Fig. 246). From the first equality in formula (13) it follows that the perpendicular bisector M of segment ap in plane P has the direction $\mathcal{D}$. By Theorem 41, we thus have $\varrho(a,\mathfrak{N}) = \varrho(p,\mathfrak{K})$, and hence $p \in H$. Since $p \in W$, then $p \in H \cap W$.

Conversely, let $p \in H \cap W$. We produce through point p an axis $\mathfrak{N} \parallel \mathfrak{K}$ and we put $t = \varrho(a,p)$. Since $p \in H$, then $\varrho(a,\mathfrak{N}) = \varrho(p,\mathfrak{K})$, and, by Theorem 41, the perpendicular bisector M of segment ap has the direction $\mathcal{D}$; consequently, $\mathbf{L}(ap)A = \Pi\left(\dfrac{t}{2}\right)$.

We have thus shown that function f maps the set of positive numbers onto the figure $H \cap W$. Since, in addition, $t_1 \neq t_2$ implies $f(t_1) \neq f(t_2)$, then f is a one-to-one transformation.

We have now come to the proof of the continuity. Let $p = f(t)$ for some $t > 0$. We take a number $t' > 0$ different from t and we put $p' = f(t')$

(Fig. 247). Because of Lemma 2, there exists a number $\delta > 0$ such that the inequalities

$$|\sphericalangle pap'| = |\Pi(\tfrac{1}{2}t') - \Pi(\tfrac{1}{2}t)| < \delta$$

and

$$|\varrho(a,p') - \varrho(a,p)| = |t' - t| < \delta$$

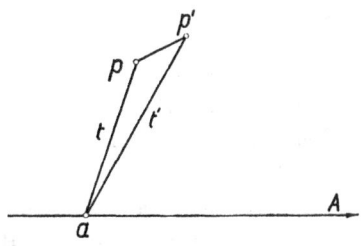

Fig. 247

imply the inequality $\varrho(p,p') < \varepsilon$. Since the function Π is continuous (see Theorem 36), we can choose along with the number δ a number $\eta > 0$ such that $\eta < \delta$ and that $|t' - t| < \eta$ implies $|\Pi(\tfrac{1}{2}t') - \Pi(\tfrac{1}{2}t)| < \delta$. Hence $|t' - t| < \eta$ implies $\varrho(p,p') < \varepsilon$. Thus we have proved the continuity of the function f for $t > 0$. From formula (13) it is seen that for the point $p \in H \cap W$ we have $f^{-1}(p) = \varrho(a,p)$, from which it follows at once that f^{-1} is continuous for $p \in H \cap W$. Thus function f maps the set of positive numbers homeomorphically onto the figure $H \cap W$.

Since our choice of half-plane W was quite arbitrary, the function f' defined by the formula

$$f'(t) = f(-t)$$

maps the set of positive numbers homeomorphically onto the figure $H \cap W^*$ and therefore function f maps the set of negative numbers homeomorphically onto the figure $H \cap W^*$.

Finally, we note that from the second equality in formula (13) and from the second equality in formula (14) it immediately follows that

$$\lim_{t \to 0} f(t) = a \quad \text{and} \quad \lim_{p \to a} f^{-1}(p) = 0,$$

from which we conclude, by (12), that function f homeomorphically maps the set of all real numbers onto the horocycle $H = H \cap W^* \cup a \cup H \cap W$, which is what we had to prove.

We shall call the function f defined, in Theorem 43, by the system of equalities (12)–(14) the *normal parametric representation* of horocycle H with *origin* a and *positive half-plane* W.

The normal parametric representation of the horocycle will be used in the proof of the following:

THEOREM 44. *Let $H_i \subset P_i$ (for $i = 1,2$) be a horocycle with respect to the direction $\mathcal{D}_i$ and let $a_i, b_i \in H_i$. If $\varrho(a_1, b_1) = \varrho(a_2, b_2)$, then there exists an isometry f mapping the plane P_1 onto the plane P_2 in such a way that $f(H_1) = H_2$, $f(a_1) = a_2$, and $f(b_1) = b_2$.*

PROOF. Assume that $\varrho(a_1, b_1) = \varrho(a_2, b_2)$. Let A_i $(i = 1,2)$ be the radius of horocycle H_i with origin a_i, and let $L_i = \mathbf{L}(A_i)$. If $a_i = b_i$, we denote by W_i either of the half-planes determined on plane P_i by line L_i; if $a_i \neq b_i$, we put $W_i = \mathbf{HP}(L_i b_i)$.

The isometry f described in the proof of Theorem 98 of Chapter II (with the help of half-lines A_1, A_2 and half-planes W_1, W_2) maps plane P_1 onto plane P_2. Let f_i be the normal parametric representation of horocycle H_i with origin a_i and positive half-plane W_i. As may readily be seen, for any real number t,

$$f(f_1(t)) = f_2(t).$$

Hence $f(H_1) = H_2$, $f(a_1) = a_2$, and $f(b_1) = b_2$.

From the theorem just proved we conclude, in particular, that any two horocycles are congruent. Furthermore, if in Theorem 44 we assume $H_1 = H_2$, then the property of the horocycle established there indicates a certain kind of homogeneity of horocycles.

19. Tangents and Secants of the Horocycle

From the normal parametric representation of the horocycle we may readily conclude

THEOREM 45. *Any line intersects any horocycle in at most two points.*

A line L intersecting horocycle H with respect to the direction $\mathcal{D}$ in two distinct points a_1 and a_2 (Fig. 248) is called a *secant* of horocycle H. We denote by A_i $(i = 1,2)$ the radius of horocycle H with origin a_i. The half-plane W with boundary L and containing radius A_i is called the *positive side* of secant L, and half-plane W^* complementary to W is called the *negative side* of secant L.

From the normal parametric representation of horocycle H it is seen that the only lines of plane $P \supset H$ which intersect horocycle H in just one point a are the line K of axis $\mathfrak{K} \in \mathcal{D}$ passing through point a and the line N perpendicular to line K at point a (Fig. 249). We shall call line N the *tangent* to horocycle H at point a. Instead of saying that horocycle

H is the horocycle *with respect* to the direction $\mathcal{D}$, we shall henceforth say that it is *orthogonal* to direction $\mathcal{D}$.

We shall now give a sufficient condition that two distinct tangents to horocycle H have a point in common.

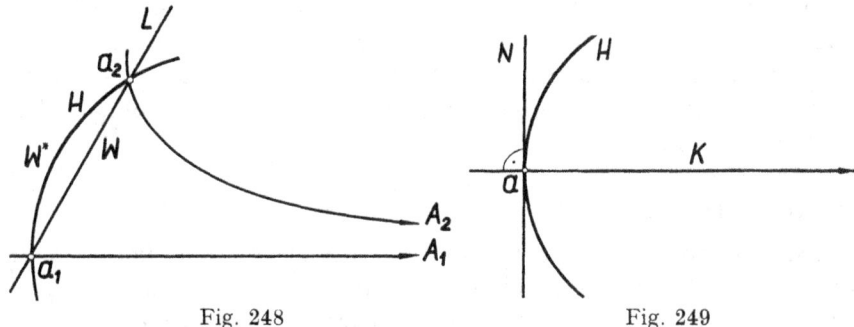

Fig. 248   Fig. 249

THEOREM 46. *Given two distinct points* a_1 *and* a_2 *on a horocycle* H, *if*

(1) $$\varrho(a_1,a_2) < 2\Pi^{-1}\left(\frac{\pi}{4}\right),$$

then the tangents N_1 *at point* a_1 *and* N_2 *at point* a_2 *to* H *intersect each other, and the intersection point* $p = N_1 \cap N_2$ *lies on the negative side of secant* a_1a_2.

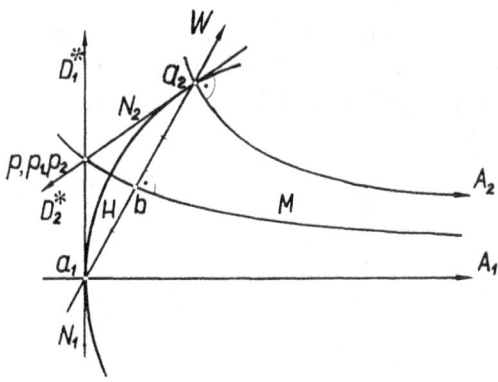

Fig. 250

PROOF. We denote by W^* the negative side of secant a_1a_2 and we put $D_i^* = N_i \cap W^*$ (for $i = 1,2$). Let A_i (for $i = 1,2$) denote the radius of horocycle H with origin a_i (Fig. 250). By Theorem 43 and formula (1) we have (see Theorem 36)

$$|\boldsymbol{H}(a_1a_2)A_1| = \Pi(\tfrac{1}{2}\varrho(a_1,a_2)) > \frac{\pi}{4},$$

from which we conclude that

$$\left|\textbf{H}(a_1a_2)D_1^*\right| = \frac{\pi}{2} - \left|\textbf{H}(a_1a_2)A_1\right| < \frac{\pi}{4}.$$

Hence half-line D_1^* intersects the perpendicular bisector M of segment a_1a_2 in some point p_1. In a similar way we prove that half-line D_2^* intersects line M in some point p_2. Denote by b the midpoint of segment a_1a_2; comparing the congruent right triangles ba_1p_1 and ba_2p_2 we get $bp_1 \equiv bp_2$, that is $p_1 = p_2$. Consequently, point $p = p_1 = p_2$ of half-plane W^* lies on each of the tangents N_1 and N_2.

20. Arc of the Horocycle

Since horocycle H is, by Theorem 43, a homeomorphic image of the set of real numbers, then any two distinct points on it, say a and b, determine just one closed arc with end points a and b included in horocycle H. We refer to this arc as the *arc ab* (of horocycle H) and denote it in formulas by $\smile ab$. Segment ab will be called the *chord* of arc ab. If f is any homeomorphism mapping the set of real numbers onto horocycle H and if $a = f(t_1)$ and $b = f(t_2)$, where $t_1 < t_2$, then arc ab is identical with the set of points $f(t)$ for $t_1 \leqslant t \leqslant t_2$, and its interior, with the set of points $f(t)$ for $t_1 < t < t_2$.

From the normal parametric representation of the horocycle H we readily get

THEOREM 47. *If $a \in \mathfrak{R} \in \mathcal{D}$ and $b \in \mathfrak{L} \in \mathcal{D}$ are two distinct points of a horocycle H orthogonal to the direction $\mathcal{D}$, then the interior of arc ab is identical to the part of horocycle H which lies between axes $\mathfrak{R}$ and $\mathfrak{L}$ (i.e. in the set $\textbf{HP}(\mathfrak{R}b) \cap \textbf{HP}(\mathfrak{L}a)$).*

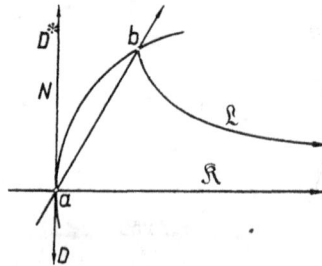

Fig. 251

By using the normal parametric representation we may also readily determine the position of arc ab with respect to the secant ab and the tangent to horocycle H at an end point of the arc (Fig. 251):

THEOREM 48. *Given two distinct points a and b on a horocycle H, if line N is tangent at point a to horocycle H and $D \subset N$ is the half-line with origin* *a, lying on the positive side of secant ab, then the interior of the arc ab is included in the inner domain of angle* **H**(ab)D*, *and the figure H — ⌣ab is included in the inner domain of angle* **H**(ab)D.

Finally, from the general properties of the arc and from Theorem 44 we obtain without difficulty:

THEOREM 49. *The arc a_1b_1 of a horocycle H_1 is congruent to the arc a_2b_2 of a horocycle H_2 if and only if the segment a_1b_1 is congruent to the segment a_2b_2.*

The arc of a horocycle is therefore characterized geometrically by the length of its chord.

21. Length of Arc of the Horocycle

Consider any arc *ab* of a horocycle *H*. Since every three points on horocycle *H* are non-collinear, then by the general definition of the length of the arc (see Introduction, end of Section 9) and by the triangle inequality we have

$$|\smile ab| > |ab|.$$

We shall now prove

THEOREM 50. *Every arc ab of a horocycle H has a finite length.*

PROOF. From the general properties of the arc (see Introduction, end of Section 9) it follows that any arc on a horocycle can be represented as the sum of a finite number of arcs whose chords have lengths smaller than $2\Pi^{-1}\left(\frac{\pi}{4}\right)$. Thus it is sufficient to prove the theorem for the case

$$\varrho(a,b) < 2\Pi^{-1}\left(\frac{\pi}{4}\right).$$

Then, by Theorem 46, the tangents to horocycle *H* at points *a* and *b* intersect one another in some point *c* lying on the negative side of secant *ab* (Fig. 252). Because of Theorem 48 the interior of arc *ab* lies

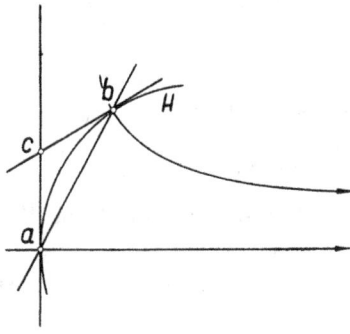

Fig. 252

simultaneously in the inner domains of angles bac and abc, that is, in triangle (abc). Therefore all vertices different from points a and b of any polygonal line Z inscribed in arc ab lie in triangle (abc). From Theorem 48 it further follows that polygonal line Z is convex. By Theorem 52 of Chapter II, we thus conclude that

$$|Z| < |ac| + |bc|.$$

Hence the length of polygonal lines Z inscribed in arc ab, and consequently also the length of arc ab, is bound from above by the number $|ac| + |bc|$ and therefore is finite.

We shall now prove

THEOREM 51. *For every horocycle H there exists a homeomorphic mapping f_0 of the set of real numbers onto H such that for any pair of real numbers $t_1 < t_2$ the arc $f_0(t_1)f_0(t_2)$ has the length $t_2 - t_1$.*

PROOF. Let f be any normal parametric representation of horocycle H. Putting

$$\varphi(u) = \text{Sign } u \cdot |\cup f(0)f(u)|,$$

we obtain a real function φ of real variable u, which, by the general properties of the arc (see Introduction, end of Section 9) is increasing and continuous, and, by Theorem 43, is bounded neither from below nor from above. It thus follows that function φ is a homeomorphism mapping the set of real numbers onto itself.

We put

$$f_0(t) = f\varphi^{-1}(t)$$

for every real t. If $t_1 < t_2$ and moreover $\varphi(u_1) = t_1$, $\varphi(u_2) = t_2$, then $u_1 < u_2$, and from

$$\varphi(u_1) = \text{Sign } u_1 \cdot |\cup f(0)f(u_1)| \quad \text{and} \quad \varphi(u_2) = \text{Sign } u_2 \cdot |\cup f(0)f(u_2)|$$

it follows at once that

$$|\cup f(u_1)f(u_2)| = \varphi(u_2) - \varphi(u_1), \quad \text{i.e.} \quad |\cup f_0(t_1)f_0(t_2)| = t_2 - t_1,$$

which is what we had to prove.

We shall refer to the homeomorhism f_0 as a *normalized parametric representation* of horocycle H.

We now take any arc a_1b_1 of horocycle H_1 and any arc a_2b_2 of horocycle H_2. Since congruent arcs have the same length, then from Theorem 49 it follows that

(1) $\qquad |a_1b_1| = |a_2b_2|$ implies $|\smile a_1b_1| = |\smile a_2b_2|$.

Let us now assume that

$$|\smile a_1b_1| = |\smile a_2b_2|.$$

From the existence of a normalized parametric representation of horocycle H_2 we conclude that besides point b_2 there exists just one point $b_2' \in H_2$ such that

$$|\smile a_2b_2'| = |\smile a_1b_1|.$$

On the other hand, there exist just two points $p,q \in H_2$ such that

(2) $\qquad |a_2p| = |a_2q| = |a_1b_1|,$

from which it follows, by formula (1), that

$$|\smile a_2p| = |\smile a_2q| = |\smile a_1b_1|.$$

Hence the pair of points p and q is identical with the pair of points b_2 and b_2', and consequently one of the two points p and q is identical with point b_2. It thus follows, by (2), that $|a_1b_1| = |a_2b_2|$. Thus we have proved

THEOREM 52. *The length of an arc a_1b_1 of a horocycle H_1 is equal to the length of an arc a_2b_2 of a horocycle H_2 if and only if $|a_1b_1| = |a_2b_2|$.*

With the help of Theorem 52 we shall prove

THEOREM 53. *Given an arc a_1b_1 on a horocycle H_1 and an arc a_2b_2 on a horocycle H_2, if $|a_1b_1| < |a_2b_2|$ then $|\smile a_1b_1| < |\smile a_2b_2|$.*

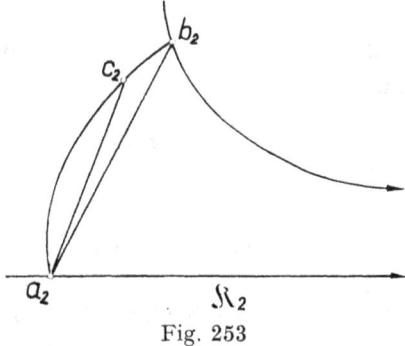

Fig. 253

PROOF. We assume that $|a_1b_1| < |a_2b_2|$. Let $\Re_2$ be the axis of horocycle H_2 passing through point a_2 (Fig. 253). Using the normal parametric representation of horocycle H_2 with origin a_2 and positive half-plane $\Re_2 b_2$ we find in the interior of arc a_2b_2 a point c_2 such that $|a_1b_1| = |a_2c_2|$. Then

$$|\smile a_1b_1| = |\smile a_2c_2| < |\smile a_2b_2|,$$

which concludes the proof.

By Theorem 52, the length of arc ab is a function of the length of chord ab. We shall show that the length of arc ab may be also treated as a function of the distance of point a from the axis $\mathfrak{L}$ passing through point b:

THEOREM 54. *Let* H_i *(where* $i = 1,2$*) be a horocycle orthogonal to a direction* $\mathcal{D}$ *and let* $a_i \in \mathfrak{R}_i \in \mathcal{D}_i$ *and* $b_i \in \mathfrak{L} \in \mathcal{D}_i$ *be two distinct points on horocycle* H_i. *If* $\varrho(a_1,\mathfrak{L}_1) = \varrho(a_2,\mathfrak{L}_2)$, *then* $|{\smile}a_1b_1| = |{\smile}a_2b_2|$.

PROOF. We assume that

(3) $$\varrho(a_1,\mathfrak{L}_1) = \varrho(a_2,\mathfrak{L}_2).$$

Let point a_i determine the half-line A_i on axis $\mathfrak{R}_i$ (Fig. 254) and let $W_2 = \boldsymbol{HP}(\mathfrak{R}_2b_2)$. We take a point $b_2' \in H_2 \cap W_2$ such that $|a_2b_2'| = |a_1b_1|$.

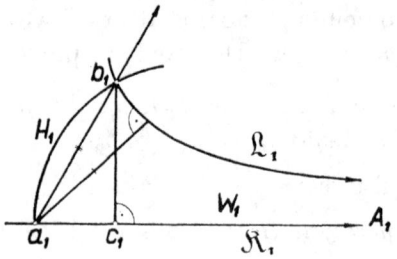 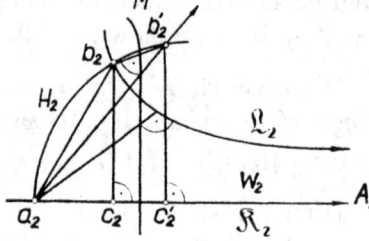

Fig. 254

Then $\boldsymbol{H}(a_1b_1)A_1 \equiv \boldsymbol{H}(a_2b_2')A_2$ (see the normal parametric representation of the horocycle); thus

$$\varrho(b_2',\mathfrak{R}_2) = \varrho(b_1,\mathfrak{R}_1)$$

which, together with formula (3), gives

$$\varrho(b_2',\mathfrak{R}_2) = \varrho(b_2,\mathfrak{R}_2).$$

We denote by c_2 and c_2' the perpendicular projections of points b_2 and b_2' upon axis $\mathfrak{R}_2$. If $b_2 \neq b_2'$, then polygonal line (b_2,c_2,c_2',b_2') would be a Saccheri quadrangle and consequently the perpendicular bisector M of chord b_2b_2' would intersect axis $\mathfrak{R}_2$, in contradiction to the fact that $\mathcal{D}$ is one of the two directions of line M. Hence $b_2 = b_2'$ and $|a_1b_1| = |a_2b_2|$. It thus follows, by Theorem 52, that $|{\smile}a_1b_1| = |{\smile}a_2b_2|$.

We take any horocycle H orthogonal to a direction $\mathcal{D}$, and on it two distinct points $a \in \mathfrak{R} \in \mathcal{D}$ and $b \in \mathfrak{L} \in \mathcal{D}$. Let c be the perpendicular projection of point a upon axis $\mathfrak{L}$. By Theorem 54, the length of arc

ab is a function of segment $[ac]$. If $[ac]$ coincides with the natural basic segment (see Section 13), we denote the length of arc ab by σ. In other words, arc ab has the length σ provided line L of axis $\mathfrak{L}$ is the line of enclosure of the right angle whose sides are the half-line A determined

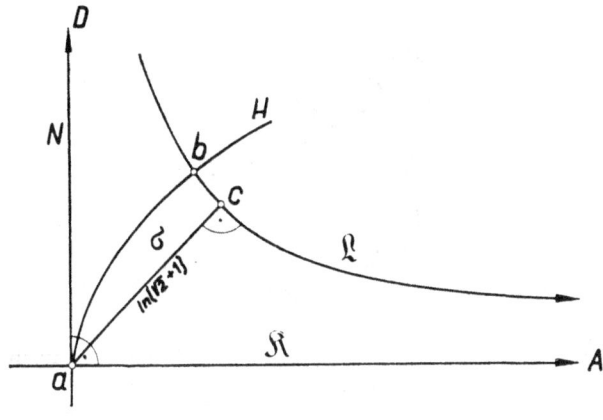

Fig. 255

by point a on axis $\mathfrak{R}$ and the half-line $D \subset \textbf{\textit{HP}}$ $(\mathfrak{R}b)$ with origin a of the tangent N to horocycle H at point a (Fig. 255).

22. Translations on a Plane. Length of Arc on a Translated Horocycle

Consider in a plane P a direction $\mathcal{D}$. Let x be any real number. For every point $p \in P$ there exists just one axis $\mathfrak{N} \in \mathcal{D}$ such that $p \in \mathfrak{N}$. We denote by $f(p)$ the point $p_1 \in \mathfrak{N}$ defined by the following conditions:

$$p_1 = p \quad \text{if} \quad x = 0,$$
$$p_1 \succ p \quad \text{and} \quad |p_1 p| = x \quad \text{if } x > 0,$$
$$p_1 \prec p \quad \text{and} \quad |p_1 p| = -x \quad \text{if } x < 0.$$

Function f so defined will be called the *translation on plane P in the direction $\mathcal{D}$ by x*. It is readily noted that we have the following

THEOREM 55. *The translations on a plane P in a direction $\mathcal{D}$ form a group of one-to-one transformations of plane P onto itself; if f is the translation by x, then f^{-1} is the translation by $-x$, and if f_1 is the translation by x_1 and f_2 is the translation by x_2, then $f_2 f_1$ is the translation by $x_1 + x_2$.*

We now take, in plane P, any horocycle H orthogonal to the direction $\mathcal{D}$ and apply the translation f in the direction $\mathcal{D}$ by x. We shall investigate the figure $f(H)$.

For the sake of discussion we assume that $x > 0$. Let $a \in \mathfrak{R} \in \mathcal{D}$ and $b \in \mathfrak{L} \in \mathcal{D}$ be any two distinct points on horocycle H (Fig. 256). Then $a\,f(a) \equiv b\,f(b)$ and $\measuredangle f(a)ab \equiv \measuredangle f(b)ba$ from which it follows that $a\,f(b) \equiv b\,f(a)$ and therefore $\measuredangle a\,f(a)f(b) \equiv \measuredangle b\,f(b)f(a)$. Because of

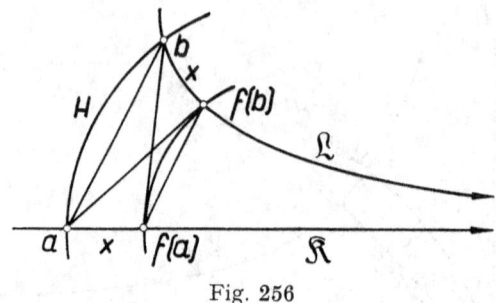

Fig. 256

Theorem 41 it thereby follows that $f(H)$ is a horocycle with respect to the direction $\mathcal{D}$. We therefore have

Theorem 56. *If f is a translation on a plane P in a direction $\mathcal{D}$ and if $H \subset P$ is a horocycle orthogonal to $\mathcal{D}$, then $f(H)$ is also a horocycle orthogonal to $\mathcal{D}$.*

Further, on the basis of Theorem 47 it may readily be shown that we have

Theorem 57. *Every translation f on a plane P in a direction $\mathcal{D}$ maps any arc ab of a horocycle $H \subset P$ orthogonal to $\mathcal{D}$ onto the arc $f(a)f(b)$ of the horocycle $f(H)$.*

We shall now investigate the relation between the length of arc ab on horocycle H and the length of arc $f(a)f(b)$ on horocycle $f(H)$. To do this we shall first prove the following two lemmas:

Lemma 1. *Given a real number x, let f_i be the translation on a plane P_i in a direction $\mathcal{D}_i$ by x and let $H_i \subset P_i$ be a horocycle orthogonal to D_i, where $i = 1,2$. Under these assumptions, $|\smile a_1 b_1| = |\smile a_2 b_2|$ implies $|\smile f_1(a_1)f_1(b_1)| = |\smile f_2(a_2)f_2(b_2)|$ for any arc $a_1 b_1$ on H_1 and any arc $a_2 b_2$ on H_2.*

Proof. From $|\smile a_1 b_1| = |\smile a_2 b_2|$ it follows that $|a_1 b_1| = |a_2 b_2|$ (Fig. 257). Considering, in succession, the congruent triangles $a_1 b_1 f_1(a_1)$, $a_2 b_2 f_2(a_2)$, and $f_1(a_1)f_1(b_1)b_1$, $f_2(a_2)f_2(b_2)b_2$, we obtain $|f_1(a_1)f_1(b_1)| = |f_2(a_2)f_2(b_2)|$, which implies $|\smile f_1(a_1)f_1(b_1)| = |\smile f_2(a_2)f_2(b_2)|$.

Lemma 2. *If f is the translation on a plane P in a direction $\mathcal{D}$ by $x > 0$, then the length of an arc ab of a horocycle $H \subset P$ orthogonal to $\mathcal{D}$ is greater than the length of the arc $f(a)f(b)$ of the horocycle $f(H)$.*

PROOF. Let c be the midpoint of chord ab (Fig. 258). Since the perpendicular bisector M of chord ab has the direction $\mathcal{D}$, then, as may readily be noted, it intersects segment $(f(a)f(b))$ in some point c'. Considering,

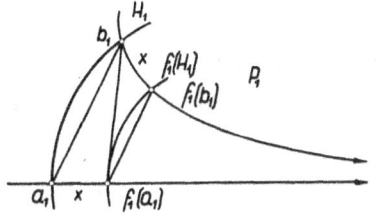 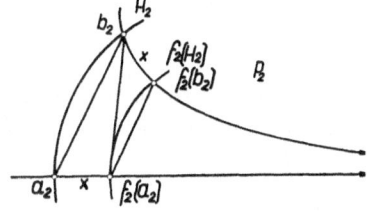

Fig. 257

in succession, congruent triangles $a f(a)c$, $b f(b)c$, and $c c'f(a)$, $c c'f(b)$, we conclude that angles $c c'f(a)$ and $c c'f(b)$ are right angles and that $f(a)c' \equiv f(b)c'$, from which it follows (see Theorem 21) that

$$\tfrac{1}{2}|ab| > \tfrac{1}{2}|f(a)f(b)|.$$

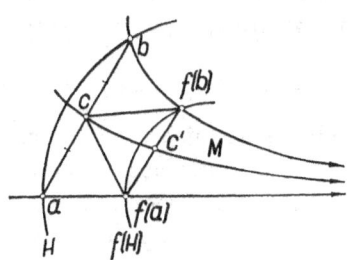

Fig. 258

Hence $|ab| > |f(a)f(b)|$, which implies, by Theorem 53, that $|\cup ab| > |\cup f(a)f(b)|$.

Now, let f be the translation on a plane P in a direction $\mathcal{D}$ by x and let H be a horocycle orthogonal to $\mathcal{D}$. We denote by g a normalized parametric representation of horocycle H. Putting, for $t > 0$,

$$\varphi(t) = |f(\cup g(0)g(t))| = |\cup f(g(0))f(g(t))|$$

we obtain a positive function φ of positive variable t. It is readily shown that function φ is additive. Indeed, we have

$$\varphi(t_1 + t_2) = |f(\cup g(0)g(t_1 + t_2))|$$
$$= |f(\cup g(0)g(t_1) \cup \cup g(t_1)g(t_1 + t_2))|$$
$$= |f(\cup g(0)g(t_1))| + |f(\cup g(t_1)g_1(t_1 + t_2))|.$$

But

$$|f(\smile g(0)g(t_1))| = \varphi(t_1),$$

and from $|\smile g(t_1)g(t_1 + t_2)| = |\smile g(0)g(t_2)|$ it follows, by Lemma 1 (for $H_1 = H_2 = H$), that

$$|f(\smile g(t_1)g(t_1 + t_2))| = |f(\smile g(0)g(t_2))| = \varphi(t_2).$$

By known arithmetical statements, φ as an additive function of real variable t with positive values has the form

$$\varphi(t) = \lambda \cdot t,$$

where λ is a positive constant.

Let us now take any arc ab of horocycle H and let $a = g(\alpha)$, $b = g(\beta)$, where $\alpha < \beta$. Then $|\smile ab| = \beta - \alpha$ and

$$|\smile f(a)f(b)| = |f(\smile ab)| = |f(\smile g(\alpha)g(\beta))|$$
$$= |f(\smile g(0)g(\beta - \alpha))| = \varphi(\beta - \alpha) = \lambda \cdot (\beta - \alpha).$$

The following relation thus holds between the length of arc ab of horocycle H and the length of arc $f(a)f(b)$ of horocycle $f(H)$:

$$|\smile f(a)f(b)| = \lambda \cdot |\smile ab|.$$

From Lemma 1 it follows at once that the constant λ depends only on the number x (but not on the choice of the plane P, the direction $\mathcal{D}$ and the horocycle H), and consequently $\lambda = \lambda(x)$.

Let us now consider the function λ for $x > 0$. From Lemma 2 we obtain

(1) $$0 < \lambda(x) < 1 \quad \text{for} \quad x > 0.$$

Denoting by f_i (for $i = 1,2$) the translation on plane P in the direction $\mathcal{D}$ by $x_i > 0$, and taking $f = f_2 f_1$, we have for any arc ab on a horocycle $H \subset P$ orthogonal to $\mathcal{D}$

$$\frac{|\smile f(a)f(b)|}{|\smile ab|} = \frac{|\smile f_2(f_1(a))f_2(f_1(b))|}{|\smile f_1(a)f_1(b)|} \cdot \frac{|\smile f_1(a)f_1(b)|}{|\smile ab|},$$

from which we get, by Theorem 55,

(2) $$\lambda(x_1 + x_2) = \lambda(x_1) \cdot \lambda(x_2).$$

By known arithmetical statements, λ as a function of real variable x satisfying conditions (1) and (2) has the form

$$\lambda(x) = e^{x/-\varkappa},$$

where $\varkappa$ is a positive constant.† In this way we have proved

THEOREM 58. *There exists a positive constant $\varkappa$ with the following property: If, on any plane P, we take any direction $\mathcal{D}$ and any horocycle H orthogonal to $\mathcal{D}$, then, for the translation f on plane P in the direction $\mathcal{D}$ by any x, the relation between the length of any arc ab on horocycle H and the length of the arc $f(a)f(b)$ on horocycle $f(H)$ is expressed by the formula*

$$|{\smile}f(a)f(b)| = e^{x/-\varkappa}\,|{\smile}ab|.$$

We shall calculate the numerical value of the constant $\varkappa$ in Section 25.

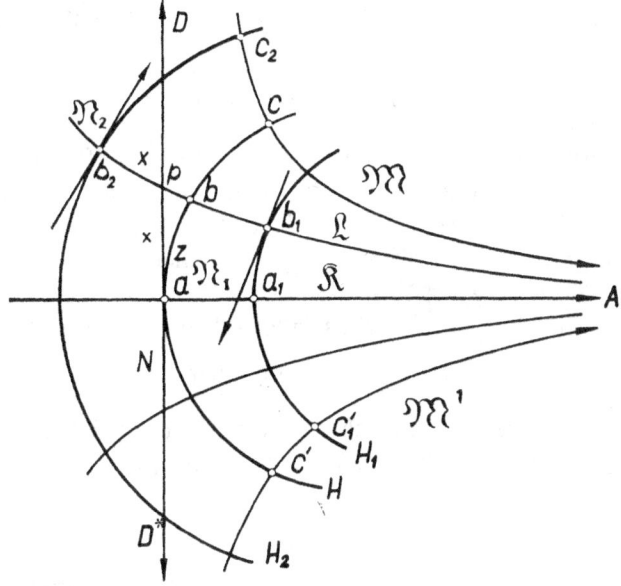

Fig. 259

23. Length of Arc of the Horocycle (Conclusion)

Consider a horocycle H orthogonal to a direction $\mathcal{D}$ and two points $a \in \mathfrak{K} \in \mathcal{D}$ and $b \in \mathfrak{L} \in \mathcal{D}$ on H (Fig. 259). We produce through point a the tangent N to horocycle H and we assume that N intersects axis $\mathfrak{L}$ in a point p. We shall now find the formulas expressing the length of segment bp and the length of arc ab as functions of the length of segment ap. We put

$$|ap| = x, \quad |bp| = y,$$
$$|{\smile}ab| = z.$$

---

† See e.g. Konrad KNOPP, *Theorie und Anwendungen der unendlichen Reihen*, Berlin 1924, p. 190.

Let A be the half-line determined on axis $\mathfrak{K}$ by point a; let $\mathfrak{N}$ be the axis determined on tangent N by half-line $D = \mathbf{H}(ap)$. We produce the line M enclosing right angle AD (see Theorem 31), and the line M' enclosing right angle AD^*; let

(1) $$\mathfrak{M} \parallel \mathfrak{K} \quad \text{and} \quad \mathfrak{M}^* \parallel \mathfrak{N}$$
and
(2) $$\mathfrak{M}' \parallel \mathfrak{K} \quad \text{and} \quad \mathfrak{M}'^* \parallel \mathfrak{N}^*,$$

where $\mathfrak{M}$, $\mathfrak{M}^*$ are the axes of line M and $\mathfrak{M}'$, $\mathfrak{M}'^*$ are the axes of line M'. Putting $c = H \cap M$ and $c' = H \cap M'$, we have (see end of Section 21)

(3) $$|\smile ac| = |\smile ac'| = \sigma.$$

It may readily be shown that $y < x$ (see Theorem 40). If we translate horocycle H in the direction $\mathcal{D}$ by $x - y$, we obtain a horocycle H_1 orthogonal to the direction $\mathcal{D}$. Let b_1 be the point of intersection of horocycle H_1 with axis $\mathfrak{L}$. Since $\varrho(p,b_1) = x$, then if we give the proper orientation to the tangent N_1 to horocycle H_1 at point b_1, we obtain an axis $\mathfrak{N}_1$ such that $\mathfrak{N}_1 \parallel \mathfrak{N}^*$. It thus follows from formula (2) that

$$\mathfrak{M}' \parallel \mathfrak{L} \quad \text{and} \quad \mathfrak{M}'^* \parallel \mathfrak{N}_1,$$

and therefore line M' encloses the right angle formed by the half-lines determined on axes $\mathfrak{L}$ and $\mathfrak{N}_1$ by point b_1. Hence, putting $c_1' = H_1 \cap M'$, we have

(4) $$|\smile b_1 c_1'| = \sigma.$$

Similarly, if we translate horocycle H in the direction $\mathcal{D}$ by $-(x + y)$, we obtain a horocycle H_2 orthogonal to $\mathcal{D}$. Let b_2 be the intersection point of horocycle H_2 and axis $\mathfrak{L}$. If we give the proper orientation to the tangent N_2 to horocycle H_2 at point b_2, we obtain an axis $\mathfrak{N}_2$ such that $\mathfrak{N}_2 \parallel \mathfrak{N}$. It thus follows from formula (1) that

$$\mathfrak{M} \parallel \mathfrak{L} \quad \text{and} \quad \mathfrak{M}^* \parallel \mathfrak{N}_2.$$

Hence, putting $c_2 = H_2 \cap M$ we again have

(5) $$|\smile b_2 c_2| = \sigma.$$

It is readily noted that the translation by $-(x - y)$ maps arc $b_1 c_1'$ onto arc bc' and that the translation by $x + y$ maps arc $b_2 c_2$ onto arc bc. From formulas (3)–(5), with the help of Theorem 58, we derive the following two equalities:

(6) $$\sigma + z = \sigma \cdot e^{(x - y)/\varkappa}, \quad \sigma - z = \sigma \cdot e^{-(x + y)/\varkappa}.$$

Adding the respective sides, we obtain

$$2\sigma = \sigma \cdot \left(e^{(x-y)/\varkappa} + e^{-(x+y)/\varkappa}\right);$$

hence

(7) $$e^{y/\varkappa} = \tfrac{1}{2}\left(e^{x/\varkappa} + e^{-x/\varkappa}\right),$$

that is,

$$y = \varkappa \cdot \ln \frac{e^{x/\varkappa} + e^{-x/\varkappa}}{2} = \varkappa \cdot \ln \cosh\left(\frac{x}{\varkappa}\right).$$

Substracting the respective sides of equalities (6), we obtain with the help of equality (7)

$$2z = \sigma \cdot \left(e^{x-y/\varkappa} - e^{-(x+y)/\varkappa}\right)$$

$$= 2\sigma \cdot \frac{e^{x/\varkappa} - e^{-x/\varkappa}}{e^{x/\varkappa} + e^{-x/\varkappa}} = 2\sigma \tanh\left(\frac{x}{\varkappa}\right).$$

We therefore have

THEOREM 59. *Given a horocycle H orthogonal to a direction* $\mathcal{D}$ *and two points* $a \in \mathfrak{R} \in \mathcal{D}$ *and* $b \in \mathfrak{L} \in \mathcal{D}$ *on H, if the tangent N to horocycle H at point a intersects axis* $\mathfrak{L}$ *in some point p, then*

(I) $$|pb| = \varkappa \cdot \ln \cosh\left(\frac{|ap|}{\varkappa}\right),$$

(II) $$|\smile ab| = \sigma \cdot \tanh\left(\frac{|ap|}{\varkappa}\right).$$

24. Sectors of the Horocycle

Given on a horocycle H an arc ab. We shall call the sum of the closed radii of horocycle H whose origins lie on arc ab (Fig. 260) the *sector of horocycle H based on arc ab*. By Theorem 17, the following theorem readily follows from this definition:

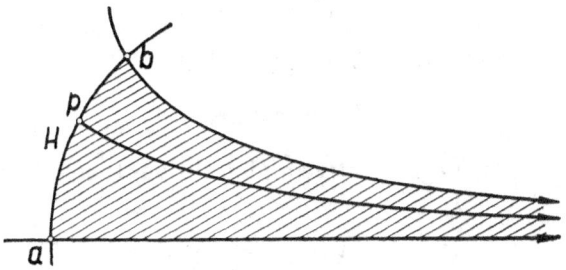

Fig. 260

THEOREM 60. *If p is any point of the interior of an arc ab of a horocycle H, then, denoting by F, F_1, F_2 the sectors based on arcs ab, ap, bp, respectively, we have $F = F_1 \cup F_2$ and*

$$\Delta(F) = \Delta(F_1) + \Delta(F_2).$$

If the length of an arc ab of a horocycle H is smaller than σ, then the tangent N to horocycle H at point a intersects in some point p the axis $\mathfrak{L}$ of horocycle H passing through point b (Fig. 261). Then sector F based on arc ab is included in the infinite right triangle T with vertices a and p, and, by Theorems 16 and 39, the defect of sector F is finite. Since

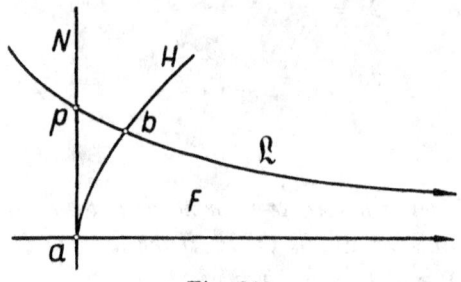

Fig. 261

from the existence of a normalized parametric representation of the horocycle (see Theorem 51) it follows that any arc of the horocycle can be represented as the sum of a finite number of arcs each of which has the length smaller than σ, we then have, by Theorem 60:

THEOREM 61. *Every sector of a horocycle has a finite defect.*

By employing Theorem 44 we may easily show that sectors of horocycles based on arcs of equal length are congruent. By Theorem 15, we thus conclude that the defect of sector F based on arc ab is a function of the positive variable $x = |\cup ab|$:

$$\Delta(F) = \varphi(x).$$

It follows immediately from Theorem 60 that

$$\varphi(x_1 + x_2) = \varphi(x_1) + \varphi(x_2);$$

and since the function φ is positive and additive, there exists a number $\alpha > 0$ such that

$$\varphi(x) = \alpha \cdot x \quad \text{for every } x > 0.$$

We thus have

THEOREM 62. *There exists a positive constant α such that for any horocycle H and for the sector F based on any arc ab of H we have*

$$\Delta(F) = \alpha \cdot |\smile ab|.$$

25. Formula of the Lobachevskian Function

We have now come to the determination of the analytical form of $\Pi(x)$ (cf. end of Section 16). The proof will run along the following lines: Let x be any real positive number. We take an infinite triangle T_x whose finite side ab has the natural length x. As we know, the defect of triangle T_x is some function of the variable x (see Section 17):

$$\Delta(T_x) = \psi(x).$$

By Theorem 39,

(1) $$\psi(x) = \frac{\pi}{2} - \Pi(x).$$

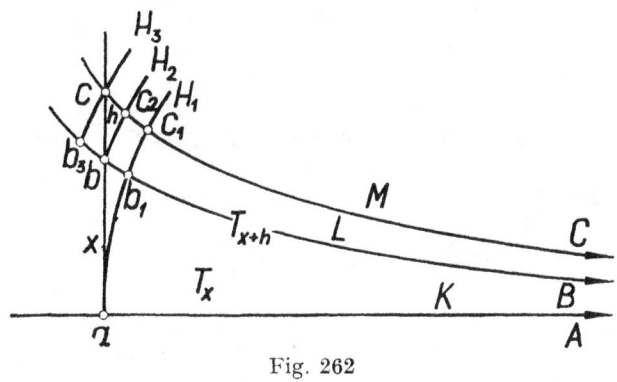

Fig. 262

Hence the function ψ has been defined by means of function Π. We now calculate the value $\psi(x)$ for the second time, this time with the help of some elementary real functions. If we compare both values for $\psi(x)$, we obtain an analytical formula for $\Pi(x)$ for $x > 0$. This formula proves to be correct for real numbers $x \leqslant 0$ as well.

We now pass on to the proof proper. Assume that in the infinite right triangle T_x the right angle is at vertex a, and let A be the infinite side, and B the hypotenuse, of triangle T_x (Fig. 262). Let us now increase x by a number h; for the sake of discussion let h be positive. We take a point c such that

$$\mathbf{B}(a,b,c) \quad \text{and} \quad |bc| = h.$$

Then the infinite right triangle T_{x+h} with finite side ac and infinite side A has the defect equal to $\psi(x+h)$. Let C be the hypotenuse of triangle T_{x+h}. Giving lines $K \supset A$, $L \supset B$, $M \supset C$, respectively, the orientations of half-lines A, B, C, we obtain parallel axes $\mathfrak{K}, \mathfrak{L}, \mathfrak{M}$; we denote their common direction by $\mathcal{D}$. We produce through points a, b, c the horocycles H_1, H_2, H_3 orthogonal to the direction $\mathcal{D}$ (see Theorem 42); let

$$b_1 = L \cap H_1, \ b_3 = L \cap H_3, \ c_1 = M \cap H_1, \ c_2 = M \cap H_2.$$

Denoting by F_2 the sector of horocycle H_2 based on arc bc_2, and by F_3 the sector of horocycle H_3 based on arc b_3c, we have, by Theorems 16 and 17,

$$(2) \qquad \Delta(F_2) \leqslant \psi(x + h) - \psi(x) \leqslant \Delta(F_3).$$

Using Theorem 59(I), we find that

$$|bb_1| = \varkappa \cdot \ln \cosh\left(\frac{x}{\varkappa}\right), \quad |cc_1| = \varkappa \cdot \ln \cosh\left(\frac{x+h}{\varkappa}\right).$$

By Theorem 58, we then have

$$|\smile bc_2| = |\smile b_1 c_1| \cdot e^{\frac{|bb_1|}{\varkappa}} = |\smile b_1 c_1| \cdot \cosh\left(\frac{x}{\varkappa}\right),$$

$$|\smile b_3 c| = |\smile b_1 c_1| \cdot e^{\frac{|cc_1|}{\varkappa}} = |\smile b_1 c_1| \cdot \cosh\left(\frac{x+h}{\varkappa}\right),$$

which, by Theorem 62, gives

$$(3) \qquad \Delta(F_2) = \alpha \cdot |\smile b_1 c_1| \cdot \cosh\left(\frac{x}{\varkappa}\right),$$

$$(4) \qquad \Delta(F_3) = \alpha \cdot |\smile b_1 c_1| \cdot \cosh\left(\frac{x+h}{\varkappa}\right).$$

From formulas (2)–(4) it follows that

$$\alpha \cdot \frac{|\smile b_1 c_1|}{h} \cdot \cosh\left(\frac{x}{\varkappa}\right) \leqslant \frac{\psi(x+h) - \psi(x)}{h} \leqslant \alpha \cdot \frac{|\smile b_1 c_1|}{h} \cdot \cosh\left(\frac{x+h}{\varkappa}\right),$$

from which we get

$$(5) \qquad \frac{d\psi}{dx} = \lim_{h \to 0} \frac{\psi(x+h) - \psi(x)}{h} = \alpha \cdot \cosh\left(\frac{x}{\varkappa}\right) \cdot \lim_{h \to 0} \frac{|\smile b_1 c_1|}{h},$$

if $\lim\limits_{h \to 0} \dfrac{|\smile b_1 c_1|}{h}$ exists. We have

$$|\smile b_1 c_1| = |\smile ac_1| - |\smile ab_1|,$$

and by formula (II) of Theorems 59 the length of arc ab_1 is a function χ of the variable x,

$$|\smile ab_1| = \sigma \cdot \tanh\left(\frac{x}{\varkappa}\right) = \chi(x).$$

Thus

$$\lim_{h \to 0} \frac{|\cup b_1 c_1|}{h} = \lim_{h \to 0} \frac{|\cup a c_1| - |\cup a b_1|}{h} = \lim_{h \to 0} \frac{\chi(x+h) - \chi(x)}{h} =$$

$$= \frac{d\chi}{dx} = \frac{\sigma}{\varkappa} \cdot \cosh^{-2}\left(\frac{x}{\varkappa}\right),$$

and substituting this value into formula (5), we finally obtain

$$\frac{d\psi}{dx} = \frac{\alpha\sigma}{\varkappa} \cdot \cosh^{-1}\left(\frac{x}{\varkappa}\right).$$

Since $\psi(0) = 0$, it thus follows that

(6) $$\psi(x) = \int_0^x \frac{\alpha\sigma}{\varkappa} \cdot \cosh^{-1}\left(\frac{t}{\varkappa}\right) dt = -2\alpha\sigma \left(\text{arc cot } e^{x/\varkappa} - \frac{\pi}{4}\right).$$

It still remains to calculate the constant $\alpha\sigma$. We have

(7) $$\lim_{x \to \infty} \text{arc cot } e^{x/\varkappa} = 0,$$

and from formula (I) and from Theorem 36 we get

(8) $$\lim_{x \to \infty} \psi(x) = \frac{\pi}{2}.$$

From formulas (6)–(8) we obtain $\frac{\pi}{2} = 2\alpha\sigma \cdot \frac{\pi}{4}$, that is, $\alpha\sigma = 1$. Therefore

(9) $$\psi(x) = -2 \text{ arc cot } e^{x/\varkappa} + \frac{\pi}{2}.$$

In this way the function ψ has been expressed in terms of the function arc cot and an exponential function.

Comparison of formulas (1) and (9) gives

(10) $$\Pi(x) = 2 \text{ arc cot } e^{x/\varkappa}$$

for all $x > 0$.

If $x < 0$, then

$$\Pi(x) = \pi - \Pi(-x) = 2\left(\frac{\pi}{2} - \text{arc cot } e^{-\frac{x}{\varkappa}}\right) = 2 \text{ arc cot } e^{\frac{x}{\varkappa}};$$

thus formula (10) also holds for all negative x. Since $\Pi(0) = \frac{\pi}{2}$, then formula (10) holds for $x = 0$ as well. Therefore formula (10) holds for all real x.

Formula (10) is equivalent to the formula

(11) $$x = \varkappa \cdot \ln \cot \left(\tfrac{1}{2}\Pi(x)\right).$$

It still remains to calculate the constant $\varkappa$. By Theorem 36, we have

$$\Pi\left(\ln\left(\sqrt{2}+1\right)\right) = \frac{\pi}{4}.$$

Thus, setting $x = \ln\left(\sqrt{2}+1\right)$ in formula (11), we obtain

$$\ln\left(\sqrt{2}+1\right) = \varkappa \cdot \ln \cot \frac{\pi}{8};$$

and since $\cot \dfrac{\pi}{8} = \sqrt{2}+1$, it follows that $\varkappa = 1$.

We therefore have

THEOREM 63. *For all real x*

$$\Pi(x) = 2 \operatorname{arc} \cot e^x$$

and

$$x = \ln \cot \left(\tfrac{1}{2}\Pi(x)\right).$$

26. The Functions sin Π and cos Π

From the equality

$$\Pi(-x) = \pi - \Pi(x)$$

we get at once

THEOREM 64. *For all real x*

$$\text{(I)} \quad \sin \Pi(-x) = \sin \Pi(x),$$

$$\text{(II)} \quad \cos \Pi(-x) = -\cos \Pi(x).$$

By Theorem 63,

$$\cot \tfrac{1}{2} \Pi(x) = e^x$$

from which we get

(1)
$$\sin \Pi(x) = \frac{2 \cot \tfrac{1}{2} \Pi(x)}{\cot^2 \tfrac{1}{2} \Pi(x) + 1} = \frac{2 e^x}{e^{2x}+1},$$

(2)
$$\cos \Pi(x) = \frac{\cot^2 \tfrac{1}{2} \Pi(x) - 1}{\cot^2 \tfrac{1}{2} \Pi(x) + 1} = \frac{e^{2x}-1}{e^{2x}+1}.$$

Further, we have

(3)
$$\frac{\sin \Pi(x) \sin \Pi(y)}{1 + \cos \Pi(x) \cos \Pi(y)} = \frac{4 e^{x+y}}{(e^{2x}+1)(e^{2y}+1) + (e^{2x}-1)(e^{2y}-1)}$$

$$= \frac{2 e^{x+y}}{e^{2(x+y)}+1} = \sin \Pi(x+y).$$

A similar calculation gives us

(4) $$\frac{\cos \Pi(x) + \cos \Pi(y)}{1 + \cos \Pi(x) \cos \Pi(y)} = \cos \Pi(x+y).$$

Formulas (3) and (4), together with Theorem 64, lead to the following:

THEOREM 65. *For any two real numbers x and y,*

(I) $$\sin \Pi(x \pm y) = \frac{\sin \Pi(x) \sin \Pi(y)}{1 \pm \cos \Pi(x) \cos \Pi(y)},$$

(II) $$\cos \Pi(x \pm y) = \frac{\cos \Pi(x) \pm \cos \Pi(y)}{1 \pm \cos \Pi(x) \cos \Pi(y)}.$$

27. Right Triangle

We consider a right triangle abc with the right angle c in a plane P. Let

$$\alpha = |\sphericalangle a| \quad \text{and} \quad \beta = |\sphericalangle b|.$$

We shall show that the following formulas hold:

(1) $$\Pi(|ab| + \Pi^{-1}(\beta)) + \alpha = \Pi(|ac|),$$
(2) $$\Pi(|ac|) + \alpha = \Pi(|ab| - \Pi^{-1}(\beta)).$$

Let $W = \mathbf{HP}(\mathbf{L}(ab)c)$. In order to prove formula (1) we take on line ab a point d (Fig. 263) such that

$$\mathbf{B}(a,b,d) \quad \text{and} \quad |bd| = \Pi^{-1}(\beta).$$

In half-plane W^*, we produce from point a a half-line A making with half-line ad an angle of size $\Pi(|ab| + \Pi^{-1}(\beta))$, and from point d a half-line D perpendicular to line ab. We then have

$$A \parallel D \quad \text{and} \quad D \parallel \mathbf{H}(cb).$$

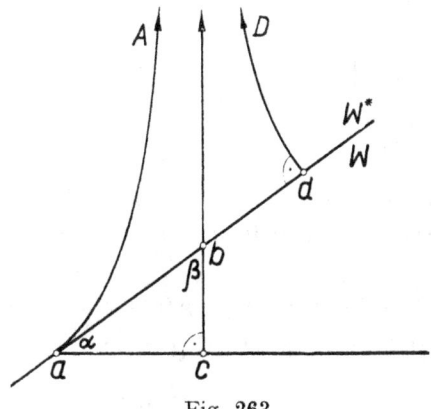

Fig. 263

Hence $A \parallel \mathbf{H}(cb)$, from which we get equality (1).

In order to prove formula (2) we take on line ab a point d (Fig. 264) such that

$$d \in \mathbf{H}(ba) \quad \text{and} \quad |bd| = \Pi^{-1}(\beta).$$

In half-plane W, we produce from point a a half-line A making with half-line ac an angle of size $\Pi(|ac|)$, and from point d a half-line D perpendicular to line ab. We then have

$$A \parallel \mathbf{H}(bc) \quad \text{and} \quad \mathbf{H}(bc) \parallel D,$$

from which it follows that

$$A \parallel D.$$

If $\mathbf{B}(a,d,b)$ (Fig. 264), equality (2) is obtained at once. If, however, $\mathbf{B}(b,a,d)$ (Fig. 265), then

$$\pi - (\Pi(|ac|) + \alpha) = \Pi(\Pi^{-1}(\beta) - |ab|) = \pi - \Pi(|ab| - \Pi^{-1}(\beta)),$$

from which we again obtain equality (2). Finally, it is readily noted that equality (2) also holds when $d = a$.

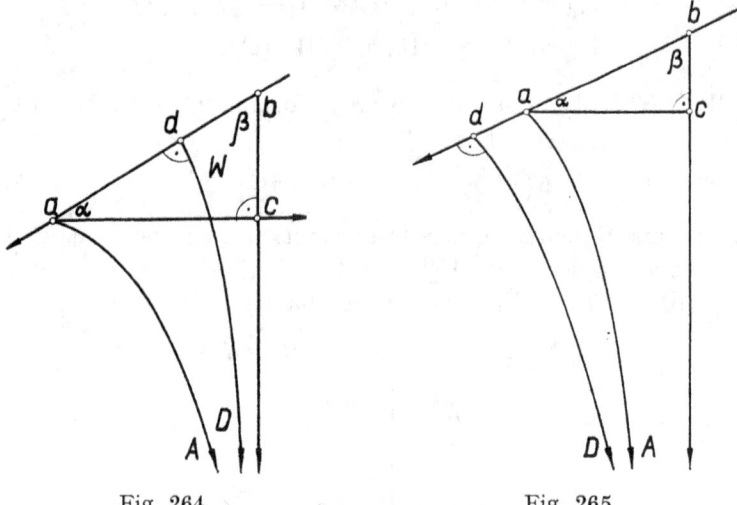

Fig. 264   Fig. 265

With the help of formulas (1) and (2) we shall express the lengths of all three sides of triangle abc in terms of the sizes α and β of its acute angles.

With this purpose in mind, we write equations (1) and (2) in the form

(3) $$\Pi(|ac|) = \Pi(|ab| + \Pi^{-1}(\beta)) + \alpha,$$

(4) $$\Pi(|ac|) = \Pi(|ab| - \Pi^{-1}(\beta)) - \alpha.$$

We then obtain

(5) $\cos \Pi(|ac|) = \cos \Pi(|ab| + \Pi^{-1}(\beta)) \cos \alpha - \sin \Pi(|ab| + \Pi^{-1}(\beta)) \sin \alpha,$

(6) $\cos \Pi(|ac|) = \cos \Pi(|ab| - \Pi^{-1}(\beta)) \cos \alpha + \sin \Pi(|ab| - \Pi^{-1}(\beta)) \sin \alpha.$

By Theorem 65(II), we have

(7) $$\cos \Pi(|ab| + \Pi^{-1}(\beta)) - \cos \Pi(|ab| - \Pi^{-1}(\beta))$$

$$= \frac{\cos \Pi(|ab|) + \cos \beta}{1 + \cos \Pi(|ab|) \cos \beta} - \frac{\cos \Pi(|ab|) - \cos \beta}{1 - \cos \Pi(|ab|) \cos \beta}$$

$$= \frac{2 (\cos \beta - \cos^2 \Pi(|ab|) \cos \beta)}{1 - \cos^2 \Pi(|ab|) \cos^2 \beta} = \frac{2 \sin^2 \Pi(|ab|) \cos \beta}{1 - \cos^2 \Pi(|ab|) \cos^2 \beta}$$

and

(8) $$\sin \Pi(|ab| + \Pi^{-1}(\beta)) + \sin \Pi(|ab| - \Pi^{-1}(\beta))$$

$$= \frac{\sin \Pi(|ab|) \sin \beta}{1 + \cos \Pi(|ab|) \cos \beta} + \frac{\sin \Pi(|ab|) \sin \beta}{1 - \cos \Pi(|ab|) \cos \beta} = \frac{2 \sin \Pi(|ab|) \sin \beta}{1 - \cos^2 \Pi(|ab|) \cos^2 \beta}.$$

Substracting the respective sides of equalities (6) and (5), we obtain, by formulas (7) and (8),

$$\sin^2 \Pi(|ab|) \cos \beta \cos \alpha = \sin \Pi(|ab|) \sin \beta \sin \alpha$$

that is

(9) $$\sin \Pi(|ab|) = \tan \alpha \tan \beta.$$

We now write equations (1) and (2) in the form

(10) $$\Pi(|ac|) - \alpha = \Pi(|ab| + \Pi^{-1}(\beta)),$$

(11) $$\Pi(|ac|) + \alpha = \Pi(|ab| - \Pi^{-1}(\beta)).$$

From formulas (10) and (11) we obtain, by Theorem 65(I),

(12) $$\sin \Pi(|ac|) \cos \alpha - \cos \Pi(|ac|) \sin \alpha = \frac{\sin \Pi(|ab|) \sin \beta}{1 + \cos \Pi(|ab|) \cos \beta},$$

(13) $$\sin \Pi(|ac|) \cos \alpha + \cos \Pi(|ac|) \sin \alpha = \frac{\sin \Pi(|ab|) \sin \beta}{1 - \cos \Pi(|ab|) \cos \beta}.$$

Adding the respective sides of equalities (12) and (13), we obtain, with the help of equality (9),

$$2 \sin \Pi(|ac|) \cos \alpha = \frac{2 \sin \Pi(|ab|) \sin \beta}{1 - \cos^2 \Pi(|ab|) \cos^2 \beta}$$

$$= \frac{2 \tan \alpha \tan \beta \sin \beta}{1 - (1 - \tan^2 \alpha \tan^2 \beta) \cos^2 \beta} = \frac{2 \sin \alpha \cos \alpha}{\cos \beta},$$

that is,

$$(14) \qquad \sin \Pi(|ac|) = \frac{\sin \alpha}{\cos \beta}.$$

By permutation of the vertices a and b, we obtain from formula (14) the formula

$$(15) \qquad \sin \Pi(|bc|) = \frac{\sin \beta}{\cos \alpha}.$$

From formulas (9), (14), and (15), the reader may readily derive the following two important formulas:

THEOREM 66. *If in a triangle abc the angle c is a right angle, then*

   (I) $\sin \Pi(|ab|) = \sin \Pi(|ac|) \sin \Pi(|bc|)$ *(the sine formula)*

and

   (II) $\cos \Pi(|ac|) = \cos \Pi(|ab|) \cos |\sphericalangle a|$ *(the cosine formula).*

28. Quadrangle with Three Right Angles

Let us consider a plane quadrangle (a,b,c,d) in which the angles a,b,c are right angles (Fig. 266), and let

$$x = |ab|, \ y = |bc|, \ z = |cd|, \ u = |bd|, \ \alpha = |\sphericalangle \ abd|, \ \beta = |\sphericalangle \ cbd|.$$

It is readily noted that quadrangle (a,b,c,d) is convex and therefore

$$(1) \qquad \alpha + \beta = \frac{\pi}{2}.$$

Further, applying the cosine formula (see Theorem 66) to triangles abd and bcd, we obtain

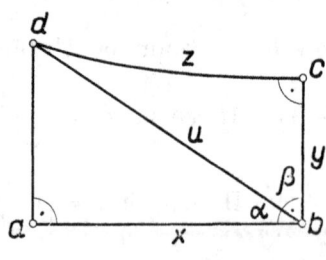

Fig. 266

$$(2) \qquad \cos \alpha = \frac{\cos \Pi(x)}{\cos \Pi(u)} \quad \text{and} \quad \cos \beta = \frac{\cos \Pi(y)}{\cos \Pi(u)}.$$

From formulas (1) and (2) it follows that

$$\cos^2 \Pi(u) = \cos^2 \Pi(x) + \cos^2 \Pi(y),$$

from which we obtain

(3) $$\sin^2 \Pi(u) = \sin^2 \Pi(y) - \cos^2 \Pi(x).$$

On the other hand, applying the sine formula (see Theorem 66) to triangle bcd, we obtain

(4) $$\sin^2 \Pi(u) = \sin^2 \Pi(y) \sin^2 \Pi(z).$$

From formulas (3) and (4) we conclude that $\cos \Pi(x) = \sin \Pi(y) \cos \Pi(z)$.

THEOREM 67. *If in a plane quadrangle (a,b,c,d) angles a, b, c are right angles, then*

$$\cos \Pi(|ab|) = \sin \Pi(|bc|) \cos \Pi(|cd|).$$

29. Rectangular Coordinates on a Plane

On a plane P, we take any arbitrary (but described in terms of the natural measure) rectangular coordinate system Ψ with axes $\mathfrak{R}_1$ and $\mathfrak{R}_2$ intersecting in a point a_0 (see Chapter III, Section 21). We shall seek the necessary and sufficient condition for a pair of numbers (ξ_1, ξ_2) to be the pair of coordinates (in the system Ψ) of a point p of plane P.

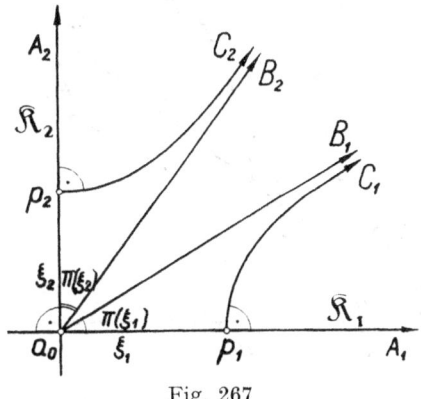

Fig. 267

Let us first assume that $\xi_1, \xi_2 > 0$. Let A_i (for $i = 1,2$) be the positive half-line of axis $\mathfrak{R}_i$ and let point $p_i \in A_i$ have the coordinate ξ_i on axis $\mathfrak{R}_i$ (Fig. 267). In the inner domain of angle A_1A_2, we produce from point p_i a half-line C_i perpendicular to axis $\mathfrak{R}_i$ and from point a_0 a half-line B_i parallel to C_i. Then

$$|A_iB_i| = \Pi(\xi_i) \quad \text{for } i = 1,2.$$

If the numbers ξ_1 and ξ_2 satisfy the inequality

$$\Pi(\xi_1) + \Pi(\xi_2) \leqslant \frac{\pi}{2},$$

then either $B_1 = B_2$ or both $\mathbf{B}(A_1,B_1,B_2)$ and $\mathbf{B}(A_2,B_2,B_1)$, from which it readily follows that half-lines C_1 and C_2 lie on opposite sides of line $L_i \supset B_i$, and therefore, they do not intersect. It thus follows that $\Psi(p) \neq (\xi_1,\xi_2)$ for every point $p \in P$.

Let us next assume that the numbers ξ_1 and ξ_2 satisfy the inequality

$$(1) \qquad\qquad \Pi(\xi_1) + \Pi(\xi_2) > \frac{\pi}{2}.$$

Then both $\mathbf{B}(A_1,B_2,B_1)$ and $\mathbf{B}(A_2,B_1,B_2)$ (Fig. 268). We produce from point a_0 any half-line B such that $\mathbf{B}(B_1,B,B_2)$. Then $\mathbf{B}(A_i,B,B_i)$ (for $i = 1,2$) and consequently half-line B intersects half-line C_i in some point q_i. If $q_1 = q_2$, then half-lines C_1 and C_2 intersect in the point $p = q_1 = q_2$. Let us next assume that $q_1 \neq q_2$ and let e.g. $\mathbf{B}(a_0,q_1,q_2)$. Then points p_2 and q_2 lie on opposite sides of line $M_1 \supset C_1$, from which it again follows that half-line C_1 intersects half-line C_2 in some point p. It is clear that $\Psi(p) = (\xi_1,\xi_2)$.

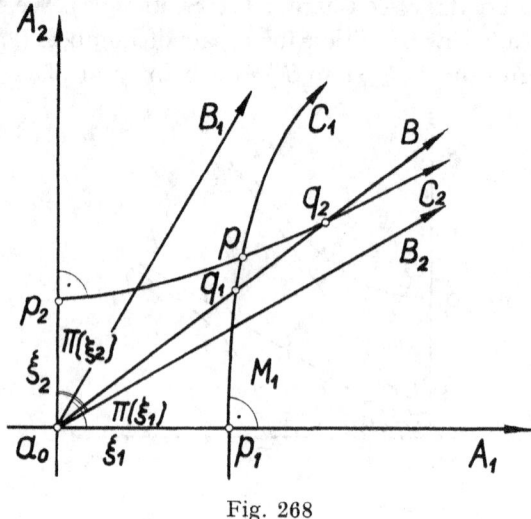

Fig. 268

In this way we have proved that

(I) For any system of rectangular coordinates Ψ and for any real numbers $\xi_1,\xi_2 > 0$, inequality (1) is a necessary and sufficient condition that $\Psi(p) = (\xi_1,\xi_2)$ for some point $p \in P$.

Inequality (1) can be given another form. We note that for any two numbers $0 < \alpha,\beta < \pi/2$ the inequality $\alpha + \beta > \pi/2$ is equivalent to the inequality

$$\sin^2\left(\frac{\pi}{2} - \alpha\right) < \sin^2 \beta,$$

which in turn is equivalent to the inequality

$$\cos^2 \alpha + \cos^2 \beta < 1.$$

From $\xi_1, \xi_2 > 0$ it follows that $0 < \Pi(\xi_1)$, $\Pi(\xi_2) < \pi/2$. Hence, for positive ξ_1 and ξ_2, condition (1) is equivalent to the condition

(2) $$\cos^2 \Pi(\xi_1) + \cos^2 \Pi(\xi_2) < 1.$$

Let us now consider two real numbers, ξ_1, ξ_2 such that $\xi_1 \cdot \xi_2 < 0$, say $\xi_1 > 0$ and $\xi_2 < 0$. Let Ψ' be the rectangular coordinate system with axes $\mathfrak{R}_1$ and $\mathfrak{R}_2^*$. Then for any point $p \in P$ we have

$$\Psi(p) = (\xi_1, \xi_2) \text{ if and only if } \Psi'(p) = (\xi_1, -\xi_2).$$

Hence, by applying (I) we conclude from the equivalence of formulas (1) and (2) that the inequality

(3) $$\cos^2 \Pi(\xi_1) + \cos^2 \Pi(-\xi_2) < 1$$

is a necessary and sufficient condition that $\Psi(p) = (\xi_1, \xi_2)$ for some point $p \in P$. But, by Theorem 64 (II), condition (3) is equivalent to condition (2).

Finally, we note that, since $\Pi(0) = \pi/2$, inequality (2) is satisfied by every pair of real numbers ξ_1 and ξ_2 at least one of which is equal to zero. On the other hand, for any real numbers ξ_1 and ξ_2 there exists, on axis $\mathfrak{R}_1$, a point p such that $\Psi(p) = (\xi_1, 0)$ and, on axis $\mathfrak{R}_2$, a point p such that $\Psi(p) = (0, \xi_2)$.

We therefore have the following general theorem:

THEOREM 68. *Let a rectangular coordinate system Ψ, described in terms of the natural measure, be fixed on a plane P. In order that a pair of real numbers (ξ_1, ξ_2) be a pair of coordinates of a point $p \in P$ in system Ψ it is necessary and sufficient that*

$$\cos^2 \Pi(\xi_1) + \cos^2 \Pi(\xi_2) < 1.$$

30. Beltrami Coordinates on a Plane. Formula for Distance

Given a plane P and, on it, a system of rectangular coordinates Ψ (described in terms of the natural measure) with axes $\mathfrak{R}_1$ and $\mathfrak{R}_2$ intersecting in point a_0. For any point $p \in P$ let $\Psi(p) = (\xi_1^p, \xi_2^p)$ and let

$$x_1^p = \cos \Pi(\xi_1^p), \ x_2^p = \cos \Pi(\xi_2^p).$$

The function Θ defined for every point $p \in P$ by the equality

$$\Theta(p) = (x_1^p, x_2^p)$$

will be called the *Beltrami coordinate system* on plane P and the numbers

x_1^p and x_2^p, *the first and second Beltrami coordinates* of point p in system Θ. By Theorems 68 and 37,

(I) The system Θ maps plane P in a one-to-one way onto the set of those pairs of real numbers (x_1, x_2) which satisfy the inequality $x_1^2 + x_2^2 < 1$, i.e. onto the Klein space $\mathbf{K}_2$.

We shall now express the distance between two points $a, b \in P$ in terms of their Beltrami coordinates:

$$x_1^a = \cos \Pi(\xi_1^a), \;\; x_2^a = \cos \Pi(\xi_2^a), \;\; x_1^b = \cos \Pi(\xi_1^b), \;\; x_2^b = \cos \Pi(\xi_2^b),$$

where

$$(\xi_1^a, \xi_2^a) = \Psi(a) \quad \text{and} \quad (\xi_1^b, \xi_2^b) = \Psi(b).$$

We denote by a_i and b_i the perpendicular projections of points a and b upon axis $\mathfrak{R}_i$ $(i = 1,2)$ and by c, the perpendicular projection of point a upon the line produced through point b and perpendicular to axis $\mathfrak{R}_1$ (Fig. 269). Let

$$\alpha = \varrho(a, a_1), \qquad \beta = \varrho(b, b_1),$$
$$\gamma = \varrho(c, b_1), \qquad \delta = \varrho(a, c),$$
$$\lambda = \varrho(b, c), \qquad \mu = \varrho(a, b).$$

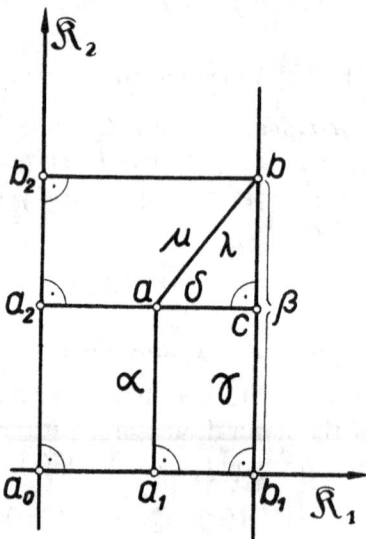

Fig. 269

For the sake of discussion we assume that

$$a, b \sim \in \mathfrak{R}_1, \; a, b \sim \in \mathfrak{R}_2, \quad \text{and} \quad a_1 \neq b_1.$$

Examining in succession quadrangles (a_2,a_0,a_1,a), (b_2,a_0,b_1,b), and (c,b_1,a_1,a), we obtain, by Theorems 67 and 64(II),

$$(1) \qquad \cos \Pi(\alpha) = \frac{\text{Sign}(\xi_2^a) \cdot \cos \Pi(\xi_2^a)}{\sin \Pi(\xi_1^a)},$$

$$(2) \qquad \cos \Pi(\beta) = \frac{\text{Sign}(\xi_2^b) \cdot \cos \Pi(\xi_2^b)}{\sin \Pi(\xi_1^b)},$$

$$(3) \qquad \cos \Pi(\gamma) = \sin \Pi(\xi_1^a - \xi_1^b) \cos \Pi(\alpha),$$

$$(4) \qquad \cos^2 \Pi(\xi_1^a - \xi_1^b) = \sin^2 \Pi(\gamma) \cos^2 \Pi(\delta).$$

We now calculate $\sin \Pi(\lambda)$. It is readily noted that

$$\text{if Sign}(\xi_2^a)\,\text{Sign}(\xi_2^b) = 1, \text{ then } \lambda = |\beta - \gamma|;$$
$$\text{if Sign}(\xi_2^a)\,\text{Sign}(\xi_2^b) = -1, \text{ then } \lambda = \beta + \gamma.$$

Hence, using Theorem 65(I), we have

$$(5) \qquad \sin \Pi(\lambda) = \frac{\sin \Pi(\beta) \sin \Pi(\gamma)}{1 - \text{Sign}(\xi_2^a)\,\text{Sign}(\xi_2^b) \cos \Pi(\beta) \cos \Pi(\gamma)}.$$

If $b \neq c$, then by applying the sine formula (see Theorem 66) to right triangle abc we obtain

$$(6) \qquad \sin \Pi(\mu) = \sin \Pi(\lambda) \sin \Pi(\delta).$$

It is readily noted that this formula also holds if $b = c$. Finally, we conclude from formulas (5) and (6) that

$$(7) \qquad \sin \Pi(\mu) = \frac{\sin \Pi(\beta) \sin \Pi(\gamma) \sin \Pi(\delta)}{1 - \text{Sign}(\xi_2^a)\,\text{Sign}(\xi_2^b) \cos \Pi(\beta) \cos \Pi(\gamma)}.$$

Using formulas (2)–(4) and Theorem 65(I), we obtain, in turn,

$$(8) \quad \sin^2 \Pi(\beta) = 1 - \cos^2 \Pi(\beta) = 1 - \frac{\cos^2 \Pi(\xi_2^b)}{1 - \cos^2 \Pi(\xi_1^b)} = \frac{1 - (x_1^b)^2 - (x_2^b)^2}{1 - (x_1^b)^2},$$

$$(9) \quad \sin^2\Pi(\gamma)\,\sin^2\Pi(\delta) = \sin^2\Pi(\gamma)\,(1 - \cos^2\Pi(\delta)) = \sin^2\Pi(\gamma) - \cos^2\Pi(\xi_1^a - \xi_1^b)$$

$$= \sin^2\Pi(\xi_1^a - \xi_1^b) - \cos^2\Pi(\gamma) = \sin^2\Pi(\xi_1^a - \xi_1^b)\,(1 - \cos^2\Pi(\alpha))$$

$$= \frac{\sin^2\Pi(\xi_1^b)\,(\sin^2\Pi(\xi_1^a) - \cos^2\Pi(\xi_2^a))}{(1 - \cos\Pi(\xi_1^a) \cos\Pi(\xi_1^b))^2} = \frac{(1 - (x_1^b)^2)\,(1 - (x_1^a)^2 - (x_2^a)^2)}{(1 - x_1^a x_1^b)^2},$$

$$(10) \qquad 1 - \text{Sign}(\xi_2^a)\,\text{Sign}(\xi_2^b) \cos\Pi(\beta) \cos\Pi(\gamma)$$

$$= 1 - \frac{\cos\Pi(\xi_2^b) \sin\Pi(\xi_1^a - \xi_1^b) \cos\Pi(\xi_2^a)}{\sin\Pi(\xi_1^b) \sin\Pi(\xi_1^a)}$$

$$= 1 - \frac{\cos\Pi(\xi_2^a) \cos\Pi(\xi_2^b)}{1 - \cos\Pi(\xi_1^a) \cos\Pi(\xi_1^b)} = \frac{1 - x_1^a x_1^b - x_2^a x_2^b}{1 - x_1^a x_1^b}.$$

From formulas (7)–(10) we at once derive the formula

$$(11) \qquad \sin \Pi(\mu) = \frac{\sqrt{1 - (x_1^a)^2 - (x_2^a)^2} \ \sqrt{1 - (x_1^b)^2 - (x_2^b)^2}}{1 - x_1^a x_1^b - x_2^a x_2^b}.$$

We may now readily calculate $\varrho(a,b)$. Indeed, by Theorem 63, we have

$$(12) \qquad \varrho(a,b) = \mu = \ln \cot\left(\tfrac{1}{2}\Pi(\mu)\right)$$

$$= \tfrac{1}{2}\ln \frac{1 + \cos\Pi(\mu)}{1 - \cos\Pi(\mu)} = \tfrac{1}{2}\ln \frac{1 + \sqrt{1 - \sin^2\Pi(\mu)}}{1 - \sqrt{1 - \sin^2\Pi(\mu)}}$$

and substituting the value for $\sin \Pi(\mu)$ given by formula (11) into formula (12) we obtain, after some simple calculations,

$$(13) \ \varrho(a,b) = \tfrac{1}{2}\ln \frac{(1 - x_1^a x_1^b - x_2^a x_2^b) + \sqrt{(x_1^a - x_1^b)^2 + (x_2^a - x_2^b)^2 - (x_1^a x_2^b - x_2^a x_1^b)^2}}{(1 - x_1^a x_1^b - x_2^a x_2^b) - \sqrt{(x_1^a - x_1^b)^2 + (x_2^a - x_2^b)^2 - (x_1^a x_2^b - x_2^a x_1^b)^2}}.$$

We have $\Theta(a) = x^a = (x_1^a, x_2^a) \in \mathbf{K}_2$ and $\Theta(b) = x^b = (x_1^b, x_2^b) \in \mathbf{K}_2$. Thus, comparing formula (13) with formula (9) on page 262 we see that

$$\varrho(a,b) = \rho_{\mathrm{K}}(\Theta(a), \Theta(b)),$$

where ρ_{K} is the metric in the Klein space $\mathbf{K}_2$. Together with (I) (on page 344) this gives

THEOREM 69. *Every plane P metrized by means of the natural metric ϱ is isometric with the Klein space $\mathbf{K}_2$.*

As may readily be shown, Theorem 69 of space Lobachevskian geometry is a consequence only of the plane axioms of the system (GBL_3). As a theorem of plane Lobachevskian geometry it takes the following form:

THEOREM 70. *Space (= plane) $\mathbf{S}$ metrized by means of the natural metric ϱ is isometric with the Klein space $\mathbf{K}_2$.*

With this we conclude our discussion of Bolyai-Lobachevskian geometry and pass on to the proof of categoricity.

31. Categoricity of Bolyai-Lobachevskian Geometry

We denote by (GBL_2) the axiom system of plane Bolyai-Lobachevskian geometry obtained from the axiom system (GA_2) of plane absolute geometry (see page 196) by adding Axiom BL (as formulated on page 197, line 16).

Let us consider any two models (M_1) and (M_2) for the axiom system (GBL_2). We metrize the set of M_i-points $(i = 1,2)$ by means of the natural metric ϱ_{M_i}. As such metric spaces the sets of M_i-points are both isometric with space $\mathbf{K}_2$ (see Theorem 70). Hence the space of M_1-points is isometric with the space of M_2-points; let function f establish this isometry. Proceeding similarly as in the proof of the categoricity of Euclidean geometry, we prove, with the help of Theorems 51 and 52 of Chapter III and Theorem 71 of Chapter I, that the isometry f establishes the isomorphism of models (M_1) and (M_2).

We therefore have

PROPOSITION 7. *The axiom system* (GBL_2) *of plane Lobachevskian geometry is categorical.*

A similar proof of categoricity can be made for space Lobachevskian geometry. To do so, the Beltrami coordinate system should be introduced in space S; then this system should be proved to map space S, in a one-to-one way, onto the set of those triples of real numbers (x_1, x_2, x_3) which satisfy the inequality $x_1^2 + x_2^2 + x_3^2 < 1$, i.e. onto the Klein space K_3; finally, it should be shown that the formula for the natural distance between two points of space S in Beltrami coordinates coincides with the formula for the distance between two points in space K_3.

Both Euclidean and Bolyai-Lobachevskian geometry turn out to be categorical theories; in other words both Euclidean geometry and Bolyai-Lobachevskian geometry, up to isomorphism, uniquely determine their models. There exists, however, an essential difference between the families of the models of these two geometries. The spaces of any two models of Euclidean geometry are isometric for any arbitrarily chosen proper metrics. On the other hand, in order that the spaces of two models of Bolyai-Lobachevskian geometry be isometric the proper metrics should be fixed in them in some suitable way, e.g. natural metrics in both spaces. For any arbitrarily chosen metrics, the spaces of two models of Bolyai-Lobachevskian geometry are only similar.

PART TWO

PROJECTIVE GEOMETRY

Introduction to Part Two

Primitive Notions and Axioms

As the primitive notions of projective geometry we take a set S, two classes $\mathfrak{GL}$ and $\mathfrak{PL}$ of subsets of S, and a four-termed relation D among elements of S.

Set S will be called *space*. Elements of set S will be called *points*, elements of class $\mathfrak{GL}$ — (*straight*) *lines*, elements of class $\mathfrak{PL}$ — *planes*. For points, lines, and planes we shall adopt the same notation as in Euclidean geometry. Also, the formulas $p \in L$ and $p \in P$ will read the same as in Euclidean geometry.

We shall refer to D as the relation of *division*. The formula $D(a,b;c,d)$ will read: *points a and b divide points c and d.*

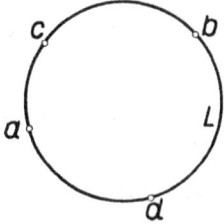

Fig. 270

The intuitive meaning of relation D is the following: The projective line topologically is identical with the circle. We take on projective line L four distinct points a, b, c, d (Fig. 270). By removing the points a and b projective line L is split into two arcs. If point c belongs to one of them, and point d to the other, then we say that points a and b divide points c and d.

The axioms of projective geometry split into three groups. The first group consists of ten axioms of incidence I'1–I'10; they determine the set-theoretical relations between points, lines, and planes. The second group consists of seven axioms of order O'1–O'7; they deal with the relation D. The Axiom of Continuity Co' constitutes the third group.

VII

Axioms of Incidence and Order

1. Axioms of Incidence

We take the same definition as in Euclidean geometry for three *collinear* or *non-collinear* points and four *coplanar* or *non-coplanar* points. Then the first nine axioms of incidence have the following form:

AXIOM I'1. *For any line L there exist (three) distinct points a, b, c such that $a,b,c \in L$.*

AXIOM I'2. *For any points a and b there exists at least one line L such that $a,b \in L$.*

AXIOM I'3. *If points a and b are distinct, then there exists at most one line L such that $a,b \in L$.*

AXIOM I'4. *For any plane P there exist (three) non-collinear points a, b, c such that $a,b,c \in P$.*

AXIOM I'5. *For any points a, b, c there exists at least one plane P such that $a,b,c \in P$.*

AXIOM I'6. *If points a, b, c are non-collinear, then there exists at most one plane P such that $a,b,c \in P$.*

AXIOM I'7. *For any line L and for any plane P, if there exist two distinct points a and b such that $a,b \in L$ and $a,b \in P$, then $L \subset P$.*

AXIOM I'8. *For any planes P and Q, if there exists a point a such that $a \in P$ and $a \in Q$, then there exists a point b distinct from a and such that $b \in P$ and $b \in Q$.*

AXIOM I'9. *There exist (four) non-coplanar points a, b, c, d.*

We see at once that the axioms of incidence I'2–I'9 of projective geometry coincide with the axioms of incidence I2–I9 of Euclidean geometry. On the other hand, the axiom of incidence I'1 is stronger than Axiom I1.

It thus follows at once:

PROPOSITION 1. *All the consequences of the system of axioms of incidence of Euclidean geometry are theorems of projective geometry.*

This concerns, in particular, all theorems from Sections 2–5 of Chapter I; we shall make use of them without referring each time to Proposition 1. Because of Theorem 6 of Chapter I two distinct points a and b determine a line, which we again denote by $L(ab)$. By Theorems 7, 8, and 9 three non-collinear points a, b, c, or a line K and a point $b \sim \in K$, or two distinct intersecting lines K and L determine a plane, which we again denote by $P(abc)$ or $P(Kb)$ or $P(KL)$, respectively.

The last axiom of incidence has the following form:

AXIOM I'10. *For any lines K and L, if there exists a plane P such that $K \subset P$ and $L \subset P$, then there exists a point c such that $c \in K$ and $c \in L$.*

Among the axioms of incidence of Euclidean geometry there is no axiom corresponding to Axiom I'10. Moreover, Axiom I'10 is not true in Euclidean geometry, since it states that any two lines lying on one plane have a point in common, which is in contradiction to Theorem 39 of Chapter II.

We shall refer to axioms I'1–I'4 and I'10 as *plane axioms* and to Axioms I'5–I'9 as *space axioms*.

2. Fundamental Existence Theorems

As we already mentioned, Axiom I'1 of projective geometry is stronger than the corresponding axiom I1 of Euclidean geometry. Indeed, Axiom I1 postulates the existence of only two distinct points on every line, while Axiom I'1 states that three distinct points lie on every line. We shall now investigate the consequences of this strengthening.

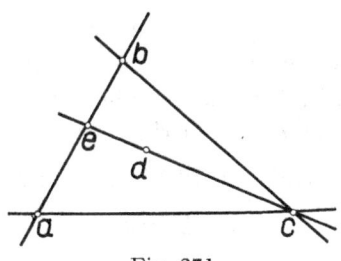

Fig. 271

THEOREM 1. *For any non-collinear points $a, b, c \in P$ there exists a point $d \in P$ non-collinear with any two of them.*

PROOF. On the basis of Axiom I'1 there is on line ab a point $e \neq a, b$ and there is on line ce a point $d \neq c, e$ (Fig. 271). Using Proposition 1, the reader may readily show that point d is not collinear with any two of the three points a, b, c.

We conclude from Theorem 1 and Axiom I4 (=I'4) that, on every plane P, there are four points a, b, c, d, no three of which are collinear. This is a strengthening of Axiom I4.

THEOREM 2. *For any non-coplanar points a, b, c, d there exists a point e non-coplanar with any three of the four points a, b, c, d.*

PROOF. By Theorem 1, there is a point p on plane abc non-collinear with any two of the three points a, b, c (Fig. 272), and by Axiom I'1, there is a point $e \neq p$, d on line pd. It may be readily shown that point e is non-coplanar with any three of the four points a, b, c, d.

From Theorem 2 and from Axiom I9 (=I'9) it follows that, in space S, there exist five points a, b, c, d, e, no four of which are coplanar. This is a strengthening of Axiom I9.

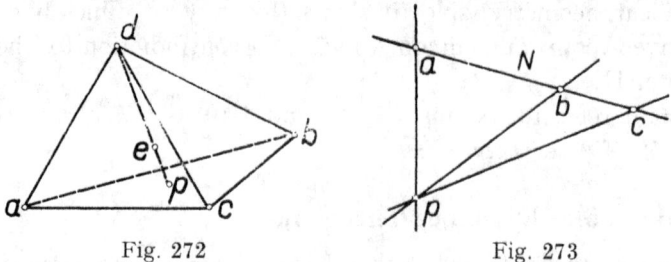

Fig. 272   Fig. 273

The first part of Theorem 12(II) of Chapter I can be strengthened as follows:

THEOREM 3. *For every point $p \in P$ there are three distinct lines $K, L, M \subset P$ passing through p.*

PROOF. By Theorem 12(I) of Chapter I, on plane P there is a line N such that $p \sim \in N$, and, by Axiom I'1, on line N there are three distinct points a, b, c (Fig. 273). We may readily show that lines $K = \mathsf{L}(pa)$, $L = \mathsf{L}(pb)$, and $M = \mathsf{L}(pc)$ are distinct and lie on plane P.

Next, we strengthen Theorem 12(V) of Chapter I:

THEOREM 4. *For any two lines K and L of plane P there exists a point $c \in P$ such that $c \sim \in K$ and $c \sim \in L$.*

PROOF. If $K = L$ the theorem follows at once from Theorem 12(V) of Chapter I. If $K \neq L$ (Fig. 274) then we take points $a \in K-L$ and $b \in L-K$. By Axiom I'1, a point $c \neq a$, b lies on line ab. It is clear that $c \sim \in K \cup L$.

Finally, we strengthen Theorem 12(VI) of Chapter I:

THEOREM 5. *For any two lines K and L on plane P there exists a line M which is skew to both line K and line L.*

PROOF. By Theorem 4, there exists a point $c \in P-(K \cup L)$ (Fig. 275), and, by Axiom I9 ($=$I'9), there exists a point $d \sim \in P$. Therefore line cd is skew to line K and to line L simultaneously.

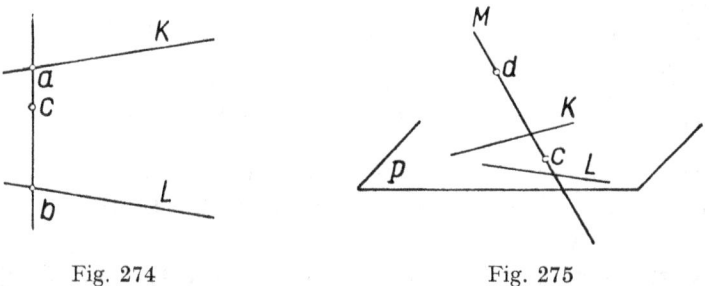

Fig. 274   Fig. 275

3. Intersections of Lines and Planes

In Section 2 we did not make use of Axiom I'10. We shall now show the simplest consequences of including this axiom.

From Axioms I'10 and I'3 we derive:

THEOREM 6. *Any two distinct lines on a given plane have just one point in common.*

We shall now prove

THEOREM 7. *Any two planes have a point in common.*

PROOF. Let us consider any two planes P_1 and P_2 and any two points $a \in P_1$ and $b \in P_2$. Through points a and b we produce any plane Q. Then there is in the set $P_1 \cap Q$ a point $a_1 \neq a$ and in the set $P_2 \cap Q$ a point $b_1 \neq b$. Lines aa_1 and bb_1 both lie in plane Q, and therefore, by Axiom I'10 they have some point c in common. We may readily show that point c lies on each of the planes P_1 and P_2.

With the help of Theorem 7 we strengthen Theorem 15 of Chapter I:

THEOREM 8. *The intersection of any two distinct planes is a line.*

Next, we shall prove

THEOREM 9. *Any line and any plane have a point in common.*

PROOF. Consider a line K and a plane P. If $K \subset P$ the theorem is obvious. We therefore assume that $K \sim \subset P$ and we choose any point $a \in P$. If $a \sim \in K$, we take the plane $Q = \mathbf{P}(Ka)$; by Theorem 8, the set $P \cap Q$ is a line. Lines K and $P \cap Q$ both lie in plane Q; by Axiom I'10 they have a point b in common. Obviously, $b \in K \cap P$.

Employing Theorem 9, we can strengthen Theorem 14 of Chapter I:

THEOREM 10. *If $L \sim \subset P$, then the line L and the plane P have just one point in common.*

The last theorem of this section deals with the intersection of three planes:

THEOREM 11. *Any three planes have at least one point in common.*

PROOF. Indeed, two of them include (according to Theorem 8) a line which intersects (according to Theorem 9) the third plane.

4. Central Projection upon a Line. Perspective and Projective Transformations

Let there be given a line L and a point $a \sim \in L$ in a plane P (Fig. 276) Consider any point $p \in P - a$. Line ap lies in plane P and, by Theorem 6,

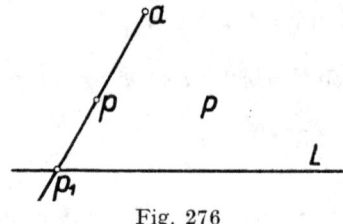

Fig. 276

intersects line L in just one point p_1. Point p_1 is called the *central projection of point p upon line L from center a*, and the function f correlating with point $p \in P - a$ the projection p_1 is called the *central projectivity upon line L from center a*.

Let $K \subset P - a$. The projectivity upon line L from center a with the domain restricted to line K is called the *perspective transformation of line K onto line L with center a*. By Theorem 6, we have at once

THEOREM 12. *For any plane P, any point $a \in P$ and any two lines $K, L \subset P - a$ the perspective transformation of K onto L with center a establishes a one to one correspondence between the points of K and the points of L.*

The superposition of a finite number of perspective transformations will be called a *projective transformation*.

We shall now prove

THEOREM 13. *If a_1, a_2, a_3 are three distinct points of line K, and b_1, b_2, b_3 are three distinct points of line L, then there exists a projective transformation f of line K onto line L such that $f(a_i) = b_i$ for $i = 1, 2, 3$.*

PROOF. We assume initially that $K \cap L = 0$ (Fig. 277). On line a_1b_1 we take a point $e_1 \neq a_1, b_1$. Then $e_1 \sim \in K$ and $e_1 \sim \in L(a_2b_1)$. Let f_1 be the perspective transformation of line K onto line a_2b_1 (in the plane $a_1a_2b_1$) with center e_1. Then

(1) $$f_1(a_1) = b_1, \quad f_1(a_2) = a_2, \quad f_1(a_3) = a_3',$$

where $a_3' = L(e_1a_3) \cap L(a_2b_1)$. Lines a_2b_2 and $a_3'b_3$ are distinct and both lie in plane $a_2b_1b_2$. We denote by e_2 their point of intersection. It is readily shown that $e_2 \sim \in L(a_2b_1)$ and $e_2 \sim \in L$. Let f_2 be the perspective transformation of line a_2b_1 upon line L (in plane $a_2b_1b_2$) with center e_2. Then

(2) $$f_2(b_1) = b_1, \quad f_2(a_2) = b_2, \quad f_2(a_3') = b_3.$$

Let us now take the projective transformation $f = f_2 f_1$ of line K upon line L. From formulas (1) and (2) it follows at once that $f(a_i) = b_i$ for $i = 1, 2, 3$.

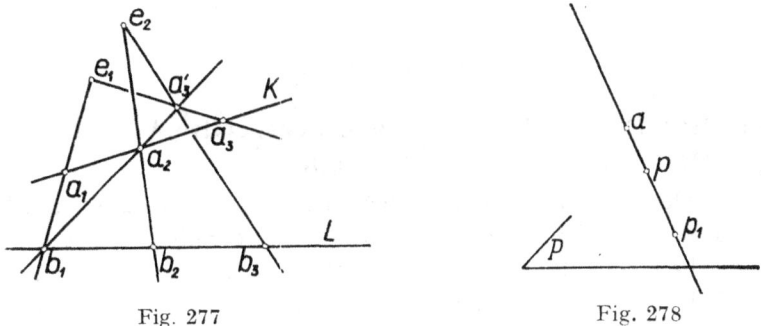

Fig. 277   Fig. 278

We next assume that $K \cap L \neq 0$. By Theorem 5, there exists a line M intersecting neither line K nor line L. We choose on line M three distinct points c_1, c_2, c_3. By what we have proved above, there exists a projective transformation f' of line K onto line M which maps the triple of points (a_1, a_2, a_3) onto the triple of points (c_1, c_2, c_3), and a projective transformation f'' of line M upon line L which maps the triple of points (c_1, c_2, c_3) onto the triple of points (b_1, b_2, b_3). The projective transformation $f = f''f'$ of line K onto line L maps the triple of points (a_1, a_2, a_3) onto the triple of points (b_1, b_2, b_3).

5. Central Projection upon a Plane

Consider in space S a plane P and a point $a \sim \in P$ (Fig. 278). By Theorem 10, for every point $p \neq a$ line ap intersects plane P in just one point p_1. Point p_1 is called the *central projection of point p upon plane P from center a*, and the function f correlating with point $p \in S - a$ the projection p_1 is called the *central projectivity upon plane P from center a*.

We shall prove

THEOREM 14. *If P, K $\subset$ S—a, then the projectivity f from center a upon plane P maps, in a one-to-one manner, line K onto some line of plane P.*

PROOF. We note that, on plane $Q = P(Ka)$, function f is the projectivity from center a upon line $L = P \cap Q$ (Fig. 279). To prove the theorem, it is then sufficient to make use of Theorem 12.

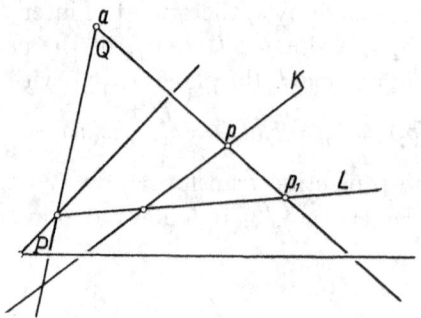

Fig. 279

6. Triangles. Perspective Center and Perspective Axis of Two Triangles. The Theorem of Desargues

The unordered triple of non-collinear points a, b, c will be called, as in Euclidean geometry, a *triangle*, and will be denoted by abc. Points a, b, c will be called the *vertices* of triangle abc. By the *sides* of triangle abc we shall now understand the lines $L(ab), L(ac), L(bc)$.

Let there be given two triangles $a_1 a_2 a_3$ and $b_1 b_2 b_3$ and a function f establishing a one-to-one correspondence between the vertices of triangle $a_1 a_2 a_3$ and the vertices of triangle $b_1 b_2 b_3$. We may assume that the vertices of triangle $b_1 b_2 b_3$ were denoted in such a way that

$$f(a_i) = b_i \quad \text{for} \quad i = 1, 2, 3.$$

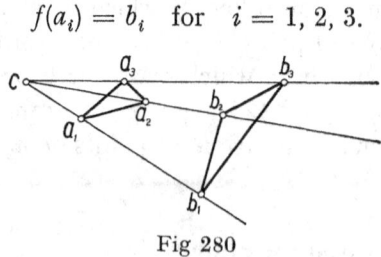

Fig 280

We shall say that point c is the *perspective center of triangles $a_1 a_2 a_3$ and $b_1 b_2 b_3$ for the correspondence f of vertices* (Fig. 280) if point c is collinear with each pair of corresponding vertices a_i, b_i ($i = 1, 2, 3$). We shall say that line M is the *perspective axis of triangles $a_1 a_2 a_3$ and $b_1 b_2 b_3$ for the correspondence f of vertices* if line M is *concurrent* with every pair of corresponding sides $L(a_i a_j)$ and $L(b_i b_j)$, i.e. $M \cap L(a_i a_j) \cap L(b_i b_j) \neq 0$ for every pair of indices $i \neq j$ $(i, j = 1, 2, 3)$ (Fig. 281).

It is readily noted that not every two triangles have a perspective center (or perspective axis) and that two triangles having a perspective center (or perspective axis) for some correspondence of their vertices may have no perspective center (or no perspective axis) for some other correspondence of their vertices.

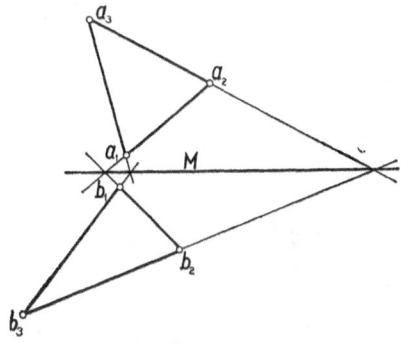

Fig. 281

One of the basic theorems of projective geometry is the *Desargues Theorem* (direct and converse) establishing the connection between the existence of the perspective center and the existence of the perspective axis of two triangles. The direct Desargues Theorem has the form:

THEOREM 15. *If two triangles $a_1 a_2 a_3$ and $b_1 b_2 b_3$ have a perspective axis for a correspondence f of their vertices, then they have a perspective center for the same correspondence f.*

The converse Desargues Theorem has the form:

THEOREM 16. *If two triangles $a_1 a_2 a_3$ and $b_1 b_2 b_3$ have a perspective center for a correspondence f of their vertices, then they have a perspective axis for the same correspondence f.*

In the proof of the Desargues Theorem we assume, without decreasing the generality of the proof, that $f(a_i) = b_i$ for $i = 1, 2, 3$. When referring, in the proof, to a perspective center or to a perspective axis of triangles $a_1 a_2 a_3$ and $b_1 b_2 b_3$, we shall always have in mind the perspective center or the perspective axis of these triangles for correspondence f.

In the proof of the Desargues Theorem, both direct and converse, we distinguish between two cases according to whether both triangles lie in one plane (plane case) or in two distinct planes (space case). We note that in the plane case as well we shall employ a space construction in the proof. We shall show in Section 15 of Chapter IX that the Desargues Theorem for the plane case cannot be derived from the plane axioms of incidence themselves.

PROOF OF THE DIRECT DESARGUES THEOREM IN THE SPACE CASE. Consider two triangles $a_1a_2a_3$ and $b_1b_2b_3$ having a perspective axis M, provided plane $P = \mathbf{P}(a_1a_2a_3)$ is distinct from plane $Q = \mathbf{P}(b_1b_2b_3)$. We shall seek the perspective center of triangles $a_1a_2a_3$ and $b_1b_2b_3$.

It is readily noted that the perspective axis M must lie both in plane P and in plane Q, and hence $M = P \cap Q$. The following two cases are now possible:

Case 1. $\mathbf{L}(a_ia_j) \neq \mathbf{L}(b_ib_j)$ for $1 \leqslant i < j \leqslant 3$ (Fig. 282). Since

$$\mathbf{L}(a_ia_j) \cap \mathbf{L}(b_ib_j) \cap M \neq 0,$$

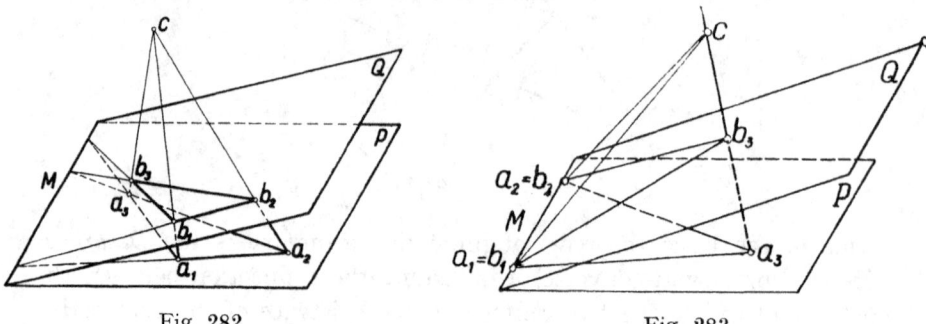

Fig. 282   Fig. 283

then lines $\mathbf{L}(a_ia_j)$ and $\mathbf{L}(b_ib_j)$ intersect one another, and consequently they determine a plane $P_{i,j}$. By Theorem 11, there exists a point $c \in P_{1,2} \cap P_{1,3} \cap P_{2,3}$. We shall show that point c is a perspective center of triangles $a_1a_2a_3$ and $b_1b_2b_3$.

In fact, it is readily noted that $P \neq Q$ implies $P_{1,2} \neq P_{1,3}$ and that points a_1 and b_1 lie on each of the planes $P_{1,2}$ and $P_{1,3}$. Therefore the product $P_{1,2} \cap P_{1,3}$ is a line containing points a_1, b_1, c. Point c is therefore collinear with points a_1 and b_1. In a similar manner, we may prove that point c is collinear with points a_2 and b_2 and with points a_3 and b_3. Therefore point c is the perspective center of triangles $a_1a_2a_3$ and $b_1b_2b_3$.

Case 2. One of the sides $\mathbf{L}(a_ia_j)$ of triangle $a_1a_2a_3$ coincides with the corresponding side $\mathbf{L}(b_ib_j)$ of triangle $b_1b_2b_3$. Let e.g. $\mathbf{L}(a_1a_2) = \mathbf{L}(b_1b_2)$ (Fig. 283). Then $M = \mathbf{L}(a_1a_2) = \mathbf{L}(b_1b_2)$, and therefore

$$\mathbf{L}(a_1a_3) \cap M = a_1 \quad \text{and} \quad \mathbf{L}(b_1b_3) \cap M = b_1.$$

Since $\mathbf{L}(a_1a_3) \cap \mathbf{L}(b_1b_3) \cap M \neq 0$, then $a_1 = b_1$. In a similar way, we prove that $a_2 = b_2$. Thus every point collinear with points a_3 and b_3 is a perspective center of triangles $a_1a_2a_3$ and $b_1b_2b_3$.

PROOF OF THE CONVERSE DESARGUES THEOREM IN THE SPACE CASE.
Consider two triangles $a_1a_2a_3$ and $b_1b_2b_3$ having a perspective center c, provided plane $P = \mathbf{P}(a_1a_2a_3)$ is distinct from plane $Q = \mathbf{P}(b_1b_2b_3)$. We shall show that the line $M = P \cap Q$ is a perspective axis of these triangles. We shall examine separately the following two cases:

Case 1. The perspective center c is distinct from each of the points a_i and b_i for $i = 1, 2, 3$ (Fig. 284). Then point c lies neither in plane P nor in plane Q. If e.g. we were to have $c \in P$, then $b_i \in \mathbf{L}(ca_i) \subset P$ for $i = 1, 2, 3$, and hence $P = Q$, contrary to the assumption. Therefore points c, a_i, a_j $(1 \leqslant i < j \leqslant 3)$ are non-collinear, and consequently they

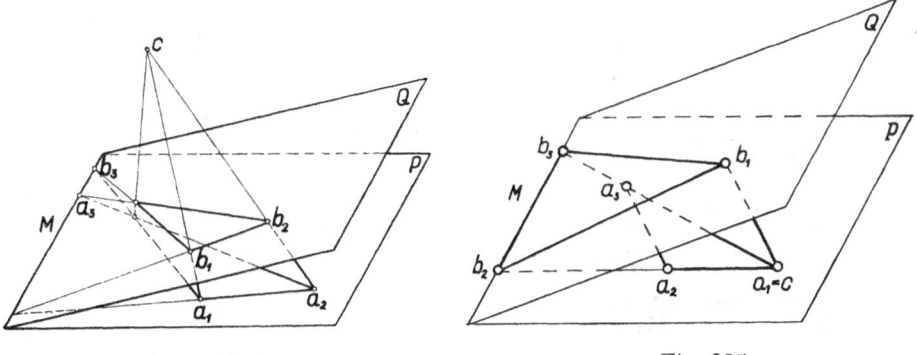

Fig. 284   Fig. 285

determine a plane $\mathbf{P}(ca_ia_j)$ which, as readily noted, is distinct from planes P and Q, contains points b_i and b_j, and intersects plane P along line a_ia_j and plane Q along line b_ib_j. It thus follows that lines a_ia_j and b_ib_j have some common point p lying on line $M = P \cap Q$, and therefore line M is a perspective axis of triangles $a_1a_2a_3$ and $b_1b_2b_3$.

Case 2. The perspective center c is one of the vertices a_i or b_i $(i = 1, 2, 3)$. Let e.g. $c = a_1$ (Fig. 285). Then

$$b_2 \in \mathbf{L}(a_1a_2) \subset P \quad \text{and} \quad b_3 \in \mathbf{L}(a_1a_3) \subset P$$

from which it follows that $M = P \cap Q = \mathbf{L}(b_2b_3)$. Hence

$$b_2 \in \mathbf{L}(a_1a_2) \cap \mathbf{L}(b_1b_2) \cap M \quad \text{and} \quad b_3 \in \mathbf{L}(a_1a_3) \cap \mathbf{L}(b_1b_3) \cap M,$$

and since lines a_2a_3 and b_2b_3 both lie in plane P, they have a common point; thus

$$\mathbf{L}(a_2a_3) \cap \mathbf{L}(b_2b_3) \cap M \neq 0.$$

In this way we have shown that line M is a perspective axis of triangles $a_1a_2a_3$ and $b_1b_2b_3$.

PROOF OF THE DIRECT DESARGUES THEOREM IN THE PLANE CASE.
Consider, in a plane P, two triangles $a_1a_2a_3$ and $b_1b_2b_3$ having a perspective
axis M. We shall seek a perspective center of these triangles. We shall
first examine three (special) cases.

Case 1. Perspective axis M coincides with one of the sides $L(a_ia_j)$ or
$L(b_ib_j)$ $(1 \leqslant i < j \leqslant 3)$. Let e.g. $M = L(a_1a_2)$ (Fig. 286). Then

$$L(a_1a_3) \cap M = a_1 \quad \text{and} \quad L(a_2a_3) \cap M = a_2,$$

and consequently $a_1 \in L(b_1b_3)$ and $a_2 \in L(b_2b_3)$. Hence point b_3 is a perspective center of triangles $a_1a_2a_3$ and $b_1b_2b_3$.

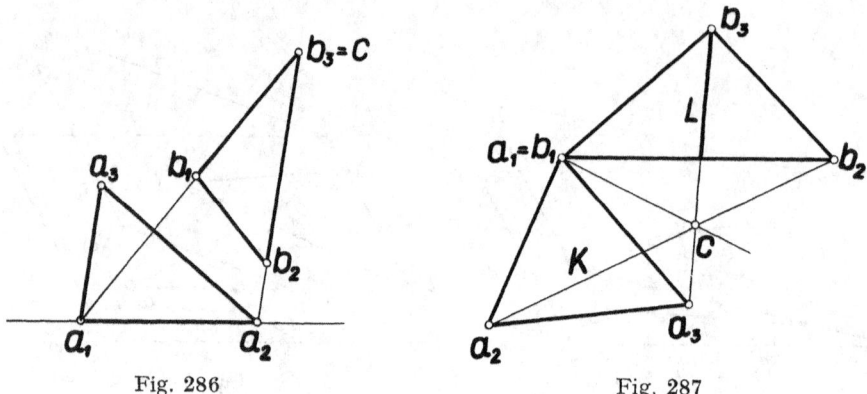

Fig. 286   Fig. 287

Case 2. Two corresponding vertices a_i and b_i coincide. Let e.g. $a_1 = b_1$
(Fig. 287). We produce, in plane P, a line K passing through points a_2
and b_2 and a line L passing through points a_3 and b_3. Lines K and L
have a common point c. Point c, being collinear with points a_2 and b_2
and with points a_3 and b_3, is, since $a_1 = b_1$, a perspective center of
triangles $a_1a_2a_3$ and $b_1b_2b_3$.

Case 3. One of the vertices a_i or b_i lies on perspective axis M. Let e.g.
$a_1 \in M$. Because of Case 1 we can assume additionally that $a_2, a_3 \sim \in M$.
Then lines $L(b_1b_2)$ and $L(b_1b_3)$ both pass through point a_1, and conse-
quently $a_1 = b_1$. Thus Case 3 reduces to Case 2.

Before passing on to the fourth (general) case we shall prove the following:

LEMMA. If c_1, c_2, c_3 are three distinct points of a line M lying on a
plane Q, then there exists, in the plane Q, a triangle $d_1d_2d_3$ such that
$d_1, d_2, d_3 \sim \in M$ and the line M intersects sides $L(d_1d_2)$, $L(d_1d_3)$, $L(d_2d_3)$, in
points c_3, c_2, c_1 respectively.

PROOF. We take any point $d_1 \in Q-M$, and, on line d_1c_3, we choose a
point d_2 distinct from points d_1 and c_3 (Fig. 288). Lines c_1d_2 and d_1c_2 are

distinct, let d_3 be their point of intersection. It is readily shown that triangle $d_1 d_2 d_3$ satisfies all the required conditions.

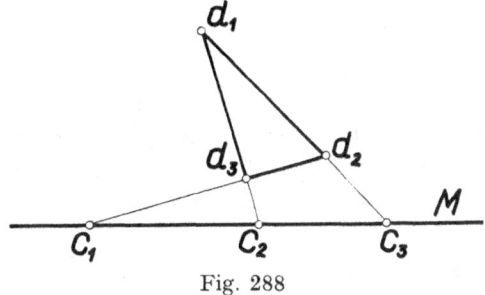

Fig. 288

Case 4. For $i = 1, 2, 3$ we have $a_i \sim \in M$, $b_i \sim \in M$ and $a_i \neq b_i$ (Fig. 289). From these assumptions it follows immediately that lines $\mathbf{L}(a_1 a_2)$, $\mathbf{L}(a_1 a_3)$, and $\mathbf{L}(a_2 a_3)$ intersect line M in three distinct points, which we denote respectively by c_3, c_2, and c_1. By Theorem 12(VII) of Chapter I we can find a plane $Q \neq P$ passing through line M. According to the Lemma, there exists a triangle $d_1 d_2 d_3 \subset Q - M$ such that its sides $\mathbf{L}(d_1 d_2)$, $\mathbf{L}(d_1 d_3)$, $\mathbf{L}(d_2 d_3)$ intersect line M in points c_3, c_2, c_1, respectively. Therefore

$$\mathbf{L}(a_1 a_2) \cap M = \mathbf{L}(b_1 b_2) \cap M = \mathbf{L}(d_1 d_2) \cap M = c_3,$$
$$\mathbf{L}(a_1 a_3) \cap M = \mathbf{L}(b_1 b_3) \cap M = \mathbf{L}(d_1 d_3) \cap M = c_2,$$
$$\mathbf{L}(a_2 a_3) \cap M = \mathbf{L}(b_2 b_3) \cap M = \mathbf{L}(d_2 d_3) \cap M = c_1,$$

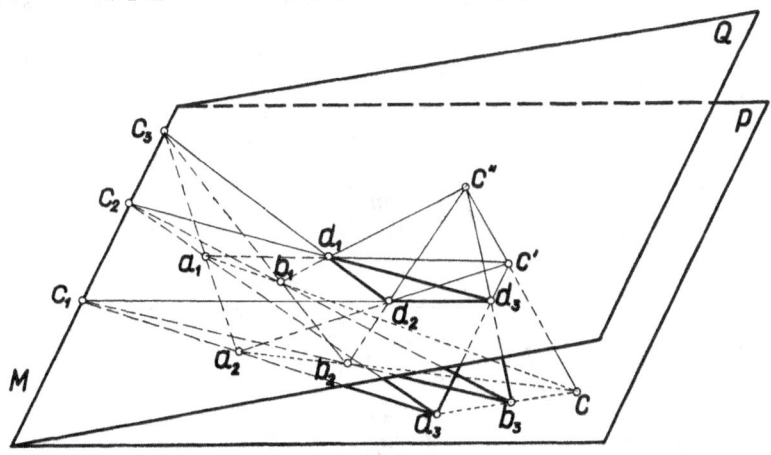

Fig. 289

and consequently line M is a perspective axis for triangles $a_1 a_2 a_3$ and $d_1 d_2 d_3$ for the correspondence f' of vertices given by the condition

$$f'(a_i) = d_i \quad \text{for} \quad i = 1, 2, 3,$$

as well as for triangles $b_1b_2b_3$ and $d_1d_2d_3$ for the correspondence f'' of vertices given by the condition

$$f''(b_i) = d_i \quad \text{for} \quad i = 1, 2, 3.$$

Hence, according the direct Desargues Theorem which we have already proved in the space case, there exists a perspective center c' for triangles $a_1a_2a_3$ and $d_1d_2d_3$ for the correspondence f' of vertices and a perspective center c'' for triangles $b_1b_2b_3$ and $d_1d_2d_3$ for the correspondence f'' of vertices.

If point c' were to lie in plane P, then, for each vertex $a_i \neq c'$, line $c'a_i$ would lie in plane P; thus $d_i \in P$ and finally $d_i \in P \cap Q$ in contradiction to the fact that line $M = P \cap Q$ does not pass through any of the points d_1, d_2, d_3. Consequently $c' \sim \in P$. In a similar way, we can show that $c'' \sim \in P$ and that $c', c'' \sim \in Q$. We note, further, that $c' \neq c''$, since otherwise line $c'd_1$ equal to line $c''d_1$ would pass through point a_1 and through point b_1, distinct by our assumption from point a_1, and thus would lie in plane P, in contradiction to the fact that $c' \sim \in P$.

According to Theorem 10, line $c'c''$ intersects plane P in some point c. We shall show that point c is a perspective center of triangles $a_1a_2a_3$ and $b_1b_2b_3$. Indeed, projection upon plane P from center c'' maps the collinear points c', a_i, d_i onto points c, a_i, b_i, respectively, from which it follows, by Theorem 14, that points c, a_i, b_i are collinear.

In this way we have concluded the proof of the direct Desargues Theorem.

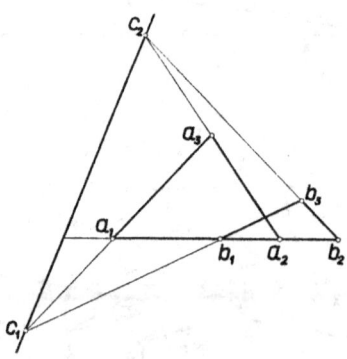

Fig. 290

PROOF OF THE CONVERSE DESARGUES THEOREM IN THE PLANE CASE. Consider, in a plane P, two triangles $a_1a_2a_3$ and $b_1b_2b_3$ having a perspective center c. We seek the perspective axis of these triangles. We shall first examine four (special) cases:

Case 1. One pair of corresponding sides $L(a_i a_j)$ and $L(b_i b_j)$ coincide. Let e.g. $L(a_1 a_2) = L(b_1 b_2)$ (Fig. 290). Let us consider the points

$$c_1 \in L(a_1 a_3) \cap L(b_1 b_3) \quad \text{and} \quad c_2 \in L(a_2 a_3) \cap L(b_2 b_3).$$

We see immediately that any line M passing through points c_1 and c_2 is a perspective axis of triangles $a_1 a_2 a_3$ and $b_1 b_2 b_3$.

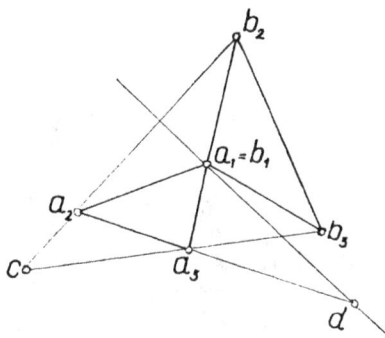

Fig. 291

Case 2. Two corresponding vertices a_i and b_i coincide. Let e.g. $a_1 = b_1$ (Fig. 291). Taking a point $d \in L(a_2 a_3) \cap L(b_2 b_3)$ we see immediately that the line M determined by points $a_1 = b_1$ and d is a perspective axis we are seeking.

Case 3. The perspective center c coincides with one of the vertices a_i or b_i. Let e.g. $c = a_1$ (Fig. 292). Then $b_2 \in L(a_1 a_2)$ and $b_3 \in L(a_1 a_3)$, and hence line $M = L(b_2 b_3)$ is a perspective axis of triangles $a_1 a_2 a_3$ and $b_1 b_2 b_3$.

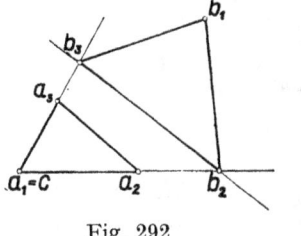

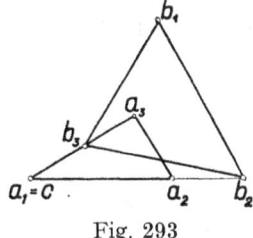

Fig. 292　　　　　　　　　　　　　Fig. 293

Case 4. One of the vertices a_i (or b_i) lies on one of the sides $L(b_i b_j)$ (or $L(a_i a_j)$) passing through the corresponding vertex b_i (or a_i). Let e.g. $b_3 \in L(a_1 a_3)$ (Fig. 293). Because of Cases 2 and 1, we may assume that

$$a_3 \neq b_3 \quad \text{and} \quad L(b_1 b_3) \neq L(a_1 a_3).$$

Then $b_1 \sim \in L(a_1 a_3)$, from which we conclude that lines $L(a_1 b_1)$ and

$L(a_3b_3) = L(a_1a_3)$ are distinct, and therefore have only one common point a_1. It thus follows that $c = a_1$, and Case 4 reduces to Case 3.

It still remains to examine the general case (Fig. 294):

Case 5. For $i \neq j$ $(i, j = 1, 2, 3)$

(1) $$a_i \sim \in L(b_ib_j) \quad \text{and} \quad b_i \sim \in L(a_ia_j)$$

and for $i = 1, 2, 3$

(2) $$c \neq a_i \quad \text{and} \quad c \neq b_i.$$

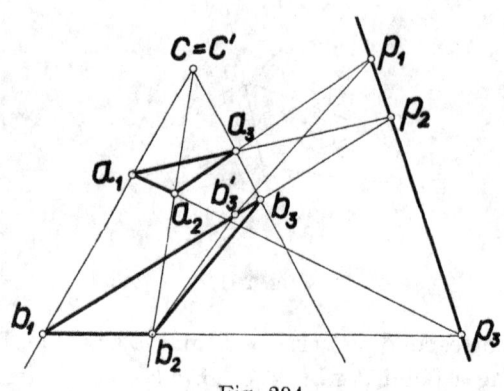

Fig. 294

The following conditions then hold:

(3) $$a_i \neq b_i,$$

(4) $$L(a_ia_j) \neq L(b_ib_j),$$

(5) $$L(a_ib_i) \neq L(a_jb_j) \quad \text{for} \quad i \neq j,$$

and, by formula (5),

(6) $$c = L(a_1b_1) \cap L(a_2b_2).$$

Let

(7) $$L(a_1a_2) \cap L(b_1b_2) = p_3 \quad \text{and} \quad L(a_1a_3) \cap L(b_1b_3) = p_2.$$

Because of formula (3) we have $p_2 \neq p_3$. We shall show that line $M = L(p_2p_3)$ is a perspective axis of triangle $a_1a_2a_3$ and $b_1b_2b_3$.

First of all, we note that, by formulas (1) and (7),

(8) $$M \neq L(a_2a_3) \quad \text{and} \quad M \neq L(b_1b_2).$$

We put

(9) $$L(a_2a_3) \cap M = p_1.$$

Because of formulas (1), (7), and (9)

(10) $$p_1 \neq b_2 \quad \text{and} \quad p_1 \neq p_3.$$

Line $b_2 p_1$ is obviously distinct from line $b_1 b_3$. Let $\mathbf{L}(b_2 p_1) \cap \mathbf{L}(b_1 b_3) = b_3'$. Then

(11) $$p_1 \in \mathbf{L}(b_2 b_3').$$

We shall show that $b_3' \neq b_1$. Indeed, if $b_3' = b_1$, then, by formulas (9) and (11) we would have $p_1 \in \mathbf{L}(b_1 b_2) \cap M$; since, by formula (7), we also have $p_3 \in \mathbf{L}(b_1 b_2) \cap M$, then, because of formula (10), we would have $\mathbf{L}(b_1 b_2) = M$, in contradiction to (8). Hence $b_3' \neq b_1$ and we can write

(12) $$\mathbf{L}(b_1 b_3) = \mathbf{L}(b_1 b_3').$$

From $b_3' \neq b_1$ it further follows that points b_1, b_2, b_3' are non-collinear. It is readily shown, using formulas (7), (9), (11), (12) that line M is a perspective axis of triangles $a_1 a_2 a_3$ and $b_1 b_2 b_3'$ for the correspondence f' of vertices given by the equalities

$$f'(a_1) = b_1, \; f'(a_2) = b_2, \; f'(a_3) = b_3'.$$

Thus, on the basis of the direct Desargues Theorem (already proved) there exists a perspective center c' of triangles $a_1 a_2 a_3$ and $b_1 b_2 b_3'$ for the correspondence f' of vertices. Since, by formula (5), $\mathbf{L}(a_1 b_1) \neq \mathbf{L}(a_2 b_2)$, it must be that $c' = \mathbf{L}(a_1 b_1) \cap \mathbf{L}(a_2 b_2)$, and, by formula (6), point c' coincides with the perspective center c of triangles $a_1 a_2 a_3$ and $b_1 b_2 b_3$.

By formulas (2) and (1), we have $c \neq a_3$ and $\mathbf{L}(a_3 c) \neq \mathbf{L}(b_1 b_3)$. From the equality $c' = c$ and from formula (12) it therefore follows that

$$b_3 = \mathbf{L}(a_3 c) \cap \mathbf{L}(b_1 b_3) = \mathbf{L}(a_3 c') \cap \mathbf{L}(b_1 b_3) = b_3'.$$

Thus $b_3 = b_3'$ and therefore line M is a perspective axis of triangles $a_1 a_2 a_3$ and $b_1 b_2 b_3$. This concludes the proof of the converse Desargues Theorem.

7. Axioms of Order

We shall now give the axioms of order characterizing the primitive relation $\mathbf{D}$. We recall that the formula $\mathbf{D}(a,b;c,d)$ reads: *points a and b divide points c and d.*

AXIOM O'1. *If $\mathbf{D}(a,b;c,d)$, then points a,b,c,d are collinear and distinct.*

AXIOM O'2. *If $\mathbf{D}(a,b;c,d)$, then $\mathbf{D}(c,d;a,b)$ and $\mathbf{D}(b,a;c,d)$.*

AXIOM O'3. *If $\mathbf{D}(a,b;c,d)$, then $\sim \mathbf{D}(a,c;b,d)$.*

AXIOM O'4. *If points a,b,c,d are collinear and distinct, then $\mathbf{D}(a,b;c,d)$ or $\mathbf{D}(a,c;b,d)$ or $\mathbf{D}(a,d;b,c)$.*

AXIOM O'5. *If points a, b, c are collinear and distinct, then there exists a point d such that* $\mathbf{D}(a,b;c,d)$.

AXIOM O'6. *For any collinear and distinct points a, b, c, d, e, if* $\sim\mathbf{D}(a,b;c,d)$ *and* $\sim\mathbf{D}(a,b;c,e)$, *then* $\sim\mathbf{D}(a,b;d,e)$.

AXIOM O'7. *If f is a perspective transformation of a line K onto a line L then* $\mathbf{D}(a,b;c,d)$ *implies* $\mathbf{D}(f(a),f(b);f(c),f(d))$ *for any four points* $a,b,c,d \in K$.

The sense of Axioms O'1–O'6 is clear. We supplement Axiom O'6 as follows:

THEOREM 17. *For any collinear and distinct points a, b, c, d, e*
if $\mathbf{D}(a,b;c,d)$ *and* $\mathbf{D}(a,b;c,e)$, *then* $\sim\mathbf{D}(a,b;d,e)$.

PROOF. We assume that

(1) $$\mathbf{D}(a,b;c,d) \quad \text{and} \quad \mathbf{D}(a,b;c,e).$$

Then, by Axiom O'2, we also have

(2) $$\mathbf{D}(b,a;c,d) \quad \text{and} \quad \mathbf{D}(b,a;c,e).$$

With the help of Axiom O'3, we conclude from formula (1) that

$$\sim\mathbf{D}(a,c;b,d) \quad \text{and} \quad \sim\mathbf{D}(a,c;b,e),$$

which, together with Axiom O'6, gives

(3) $$\sim\mathbf{D}(a,c;d,e).$$

In the same way, we obtain from formula (2)

(4) $$\sim\mathbf{D}(b,c;d,e).$$

By Axiom O'2, we conclude from formulas (3) and (4) that

$$\sim\mathbf{D}(d,e;c,a) \quad \text{and} \quad \sim\mathbf{D}(d,e;c,b),$$

which, together with Axioms O'6 and O'2, gives $\sim\mathbf{D}(a,b;d,e)$.

Axiom O'7 states that the relation of division is an invariant of perspective transformations. Hence the following theorem follows immediately:

THEOREM 18. *If f is a projective transformation of line K onto line L, then* $\mathbf{D}(a,b;c,d)$ *implies* $\mathbf{D}(f(a),f(b);f(c),f(d))$ *for any four points* $a,b,c,d \in K$.

8. Segments. Open Segments

An unordered pair of two distinct points a and b will be called, as in Euclidean geometry, a *segment* and will be denoted by ab. In Euclidean geometry, segment ab uniquely determines the open segment ab consisting of points lying between points a and b. In projective geometry, instead of the relation of betweenness, we have the relation of division. With help of this relation, we introduce the notion of two complementary open segments with end points a and b. We shall do this in a way analogous to that used in Euclidean geometry, where, employing relation **B**, we introduced the notion of two complementary half-lines with origin a.

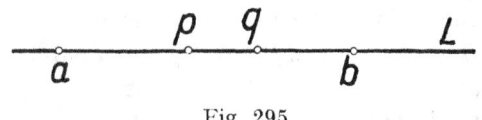

Fig. 295

We fix on a line L two distinct points a and b. Let us investigate the relation R that occurs between points $p, q \in L - \{a,b\}$ if and only if $\sim \mathbf{D}(a,b;p,q)$, that is, if points a and b do not divide points p and q (Fig. 295). By Axiom O'1 we have $\sim \mathbf{D}(a,b;p,p)$ for every point $p \in L - \{a,b\}$, and therefore relation R is reflexive. Next, from Axiom O'2 it follows immediately that $\sim \mathbf{D}(a,b;p,q)$ implies $\sim \mathbf{D}(a,b;q,p)$ for any two points $p, q \in L - \{a,b\}$, and hence relation R is symmetric. Finally, from Axioms O'2 and O'6 it follows that $\sim \mathbf{D}(a,b;p,q)$ and $\sim \mathbf{D}(a,b;q,r)$ implies $\sim \mathbf{D}(a,b;p,r)$, and hence relation R is transitive.

Relation R, as an equivalence relation, determines the partition of the set $L - \{a,b\}$ into equivalence classes. We shall show that there are just two such classes. Thus we take any point $p_0 \in L$ distinct from points a and b. By Axiom O'5, there exists a point q_0 such that

$$(1) \qquad\qquad \mathbf{D}(a,b;p_0,q_0).$$

Obviously, points p_0 and q_0 belong to two different equivalence classes. Now let r be any point of the set $L - \{a,b\}$. If we were to have simultaneously $\mathbf{D}(a,b;p_0,r)$ and $\mathbf{D}(a,b;q_0,r)$, then, by Theorem 17, we would have $\sim \mathbf{D}(a,b;p_0,q_0)$ in contradiction to (1). Therefore it must be that

$$\sim \mathbf{D}(a,b;p_0,r) \quad \text{or} \quad \sim \mathbf{D}(a,b;q_0,r),$$

and hence point r belongs either to the equivalence class containing point p_0 or to the one containing point q_0.

We therefore have

THEOREM 19. *For any two distinct points a and b of a line L, the set*

$L—\{a,b\}$ can be uniquely represented as the sum of two non-empty and disjoint sets

$$L—\{a,b\} = I_1 \cup I_2$$

satisfying the following three conditions:

(i) if $p,q \in I_1$, then $\sim\mathbf{D}(a,b;p,q)$,

(ii) if $p,q \in I_2$, then $\sim\mathbf{D}(a,b;p,q)$,

(iii) if $p \in I_1$ and $q \in I_2$, then $\mathbf{D}(a,b;p,q)$.

Each of the two sets I_1 and I_2 will be called an *open segment with end points a and b*. Hence, in projective geometry, segment ab determines not one open segment, but the pair of open segments with common end points a and b. The two open segments I_1 and I_2 are said to be *complementary* to one another. If I is any open segment, then the segment complementary to I will be denoted by I^*. The open segment I is uniquely determined by its end points a and b and one of its points c. We shall denote it by $(a\text{-}c\text{-}b)$. We therefore have

$$p \in (a\text{-}c\text{-}b) \quad \text{if and only if} \quad \sim\mathbf{D}(a,b;c,p)$$

and

$$p \in (a\text{-}c\text{-}b)^* \quad \text{if and only if} \quad \mathbf{D}(a,b;c,p).$$

By adding to open segment $(a\text{-}c\text{-}b)$ its end points a and b, we obtain the *closed segment* $\langle a\text{-}c\text{-}b \rangle$.

9. Properties of Open Segments

Let us consider three non-collinear points a, b, c. Each pair of those points determines two complementary open segments. We shall now consider the relations which hold between the six open segments formed in this way.

THEOREM 20. *Let a, b, c be three non-collinear points. For any segments $(a\text{-}c_0\text{-}b)$ and $(a\text{-}b_0\text{-}c)$ there exists, among the two complementary open segments with end points b and c, a segment $(b\text{-}a_0\text{-}c)$ such that every line L intersecting segments $(a\text{-}c_0\text{-}b)$ and $(a\text{-}b_0\text{-}c)$ also intersects segment $(b\text{-}a_0\text{-}c)$.*

PROOF. We denote by a_0 the intersection point of lines bc and b_0c_0 (Fig. 296). Obviously $a_0 \neq b$ and $a_0 \neq c$. We shall show that every line L intersecting both segments $(a\text{-}c_0\text{-}b)$ and $(a\text{-}b_0\text{-}c)$ also intersects segment $(b\text{-}a_0\text{-}c)$. We may assume that $L \neq \mathbf{L}(b_0c_0)$. Let

(1) $L \cap (a\text{-}c_0\text{-}b) = p$ and $L \cap (a\text{-}b_0\text{-}c) = q$

and

$$L \cap \mathbf{L}(bc) = r \quad \text{and} \quad L \cap \mathbf{L}(b_0c_0) = r_0.$$

We therefore have to prove that $r \in (b\text{-}a_0\text{-}c)$, that is, $\sim D(b,c;a_0,r)$. Let us examine four possible cases, according to the position of point r_0.

Case 1. $r_0 \in L(bc)$. Then $r_0 = a_0 = r$, and, because of Axiom O'1, we have $\sim D(b,c;a_0,r)$.

Case 2. $r_0 \in L(ab)$ (Fig. 297). Then the perspective transformation of line ac onto line bc from center r_0 maps points a, b_0, q, c onto points b, a_0, r, c, respectively, and, by Axiom O'7, we conclude from $\sim D(a,c;b_0q)$ (see formula (1)) that $\sim D(b,c;a_0r)$.

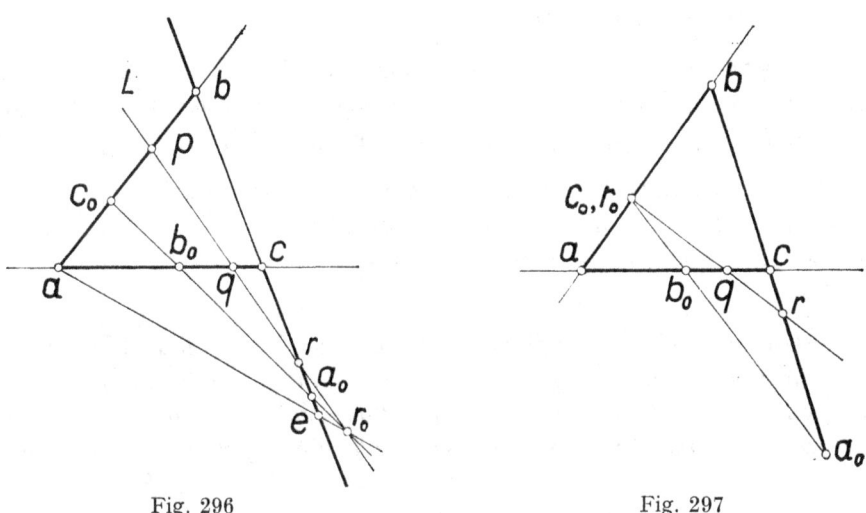

Fig. 296   Fig. 297

Case 3. $r_0 \in L(a,c)$. The proof is similar to that of Case 2.

Case 4. Point r_0 does not lie on any of the lines ab, ac, bc (Fig. 296). Let e be the intersection point of lines ar_0 and bc. The perspective transformation of line ab onto line bc from center r_0 maps the quadruple (a,c_0,p,b) onto the quadruple (e,a_0,r,b), and the perspective transformation of line ac onto line bc from center r_0 maps the quadruple (a,b_0,q,c) onto the quadruple (e,a_0,r,c). By Axiom O'7, we conclude from $\sim D(a,b;c_0,p)$ and $\sim D(a,c;b_0,q)$ (see formula (1)) that $\sim D(e,b;a_0r)$ and $\sim D(e,c;a_0r)$, which implies that $\sim D(b,c;a_0,r)$.

Thus the proof is complete.

As a direct conclusion from Theorem 20 we have

THEOREM 21. *Let a, b, c be three non-collinear points. For any segments $(a\text{-}c_0\text{-}b)$ and $(a\text{-}b_0\text{-}c)$ there exists, among the two complementary segments with end points b and c, just one segment $(b\text{-}a_0\text{-}c)$ such that no line L intersects each of the three segments $(a\text{-}c_0\text{-}b)$, $(a\text{-}b_0\text{-}c)$, and $(b\text{-}a_0\text{-}c)$.*

10. The Triple of Open Sides of a Triangle

Consider a triangle abc. Each of the three segments ab, ac, and bc determines two complementary open segments, so that by choosing one segment from each such pair we obtain eight systems, each consisting of three open segments with end points a and b, a and c, b and c. By Theorem 21, some of these systems satisfy the condition that no line L intersects each of the three open segments of the system; we shall refer to any such system as the *system of open sides* of triangle abc. From Theorem 21 we obtain at once

THEOREM 22. *If a triple of open segments* $\{I_1, I_2, I_3\}$ *is a system of open sides of a triangle* abc, *then the remaining systems of open sides of triangle* abc *are the triples* $\{I_1, I_2^*, I_3^*\}$, $\{I_1^*, I_2, I_3^*\}$, *and* $\{I_1^*, I_2^*, I_3\}$.

Hence every triangle abc has four systems of open sides.

By such a definition of the system of open sides, Pasch's Theorem, known from Euclidean geometry, applies to triangles of projective geometry:

THEOREM 23. *Given in a plane* P *a triangle* abc *and a line* L. *If segments* $(a\text{-}c_0\text{-}b)$, $(a\text{-}b_0\text{-}c)$, *and* $(b\text{-}a_0\text{-}c)$ *constitute a system of open sides of triangle* abc, *and line* L *intersects side* $(a\text{-}c_0\text{-}b)$ *and does not pass through any of the vertices* a, b, c, *then line* L *also intersects at least one of the two sides* $(a\text{-}b_0\text{-}c)$ *and* $(b\text{-}a_0\text{-}c)$.

PROOF. If line L intersecting side $(a\text{-}c_0\text{-}b)$ were to intersect line ac in some point of segment $(a\text{-}b_0\text{-}c)^*$, and line bc in some point of segment $(b\text{-}a_0\text{-}c)^*$, then segments $(a\text{-}c_0\text{-}b)$, $(a\text{-}b_0\text{-}c)^*$, $(b\text{-}a_0\text{-}c)^*$, and thus, by Theorem 22, also segments $(a\text{-}c_0\text{-}b)$, $(a\text{-}b_0\text{-}c)$, $(b\text{-}a_0\text{-}c)$ would not constitute a system of open sides of triangle abc.

11. Model for the Euclidean Axioms of Incidence and Order in Projective Geometry. Proper Points, Lines, and Planes

Let us denote by (IO) the system composed of the axioms of incidence and order of Euclidean geometry. We shall now construct a model for the axiom system (IO) in projective geometry, or more precisely, in that part based only on the axioms of incidence and order. This will allow us to employ some of the theorems of Euclidean geometry for the purposes of projective geometry.†

Let us fix in space S of projective geometry an arbitrary plane which we shall call an *improper plane* or *plane at infinity* and denote it by P_∞.

---

† We are modelling ourselves here, in principle, on the book Vysšaya geometriya by N. V. EFIMOV, Moscow-Leningrad 1949 (in Russian).

We shall call points of plane P_∞ *improper points* or *points at infinity*, and lines of plane P_∞—*improper lines* or *lines at infinity*. The remaining points, lines, and planes of space **S** will be called *proper points, lines* and *planes*. It is readily noted that:

LEMMA 1. *Every proper line contains just one improper point.*

LEMMA 2. *The set of improper points of any proper plane is an (improper) line.*

From Lemma 1 it follows immediately, by Axiom I'1, that

LEMMA 3. *There are (at least) two distinct proper points on every proper line.*

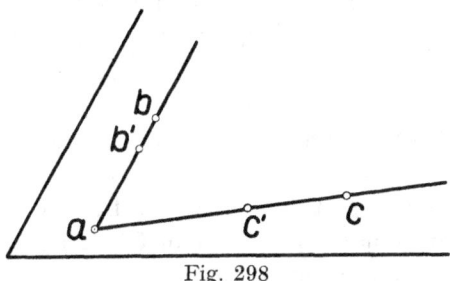

Fig. 298

We shall next show that:

LEMMA 4. *There are three non-collinear proper points on every proper plane.*

By Axiom I'4, given a proper plane P, we have three noncollinear points a, b, c on it (Fig. 298); by Lemma 2, one of them, say point a, is certainly a proper one. Because of Lemma 3, there is on proper line ab a proper point $b' \neq a$, and on proper line ac a proper point $c' \neq a$. Proper points a, b', c' obviously lie on plane P and are non-collinear.

In a similar way, it follows from Axiom I'9 that

LEMMA 5. *There exist four non-coplanar proper points.*

We now come to the construction of a model for the axiom system (IO) in projective geometry. This model will be called model (E) (from the name EUCLID). We define E-*points* as proper points of space **S**, E-*lines* as the sets of proper points of proper lines, and finally, E-*planes* as the sets of proper points of proper planes. The E-*betweenness relation*, which, just as the betweenness relation in Euclidean geometry, will be denoted by **B**, is defined in the following manner: For any E-points a, b, c

$$\mathbf{B}(a,b,c) \text{ if and only if } \mathbf{D}(a,c;b,p_\infty),$$

where p_∞ is the point at infinity of line ac.

From the definition of the relation **B** we get

THEOREM 24. *For any arbitrarily fixed plane at infinity, the E-segment* (ac) *is identical with the open segment with end points a and c not containing the improper point* p_∞ *of line ac, that is, with the segment* $(a\text{-}p_\infty\text{-}c)^*$.

We shall now show that the interpretation

(E)   E-points, E-lines, E-planes, **B**

is a model for the axiom system (IO) in projective geometry.

It is readily shown that the interpretation (E) satisfies all the axioms of incidence of Euclidean geometry. For Axioms I1, I4, I9 this follows from Lemmas 3, 4, and 5, respectively; for Axioms I2, I3, I5, I6, I7, directly from Axioms I'2, I'3, I'5, I'6, I'7, respectively, of projective geometry; for Axiom I8, from Axiom I'8 with the help of Axiom I'7 and Lemma 3.

We now come to the axioms of order of Euclidean geometry. From Axioms of Order O'1–O'5 of projective geometry it follows at once that interpretation (E) satisfies Axioms O1–O6. Let us now suppose that interpretation (E) does not satisfy Axiom O7. Then for some E-points a, b, c, d we would simultaneously have

(1) $\sim \textbf{\textit{D}}(a,c\,;b,p_\infty)$,

(2) $\textbf{\textit{D}}(b,d\,;c,p_\infty)$,

(3) $\sim \textbf{\textit{D}}(a,d\,;b,p_\infty)$.

But by Axioms O'2, and O'6 we conclude from (1) and (3) that $\textbf{\textit{D}}(c,d\,;b,p_\infty)$, which, because of Axioms O'2 and O'3, contradicts (2). Hence interpretation (E) satisfies Axioms O7. In a similar way it may be shown that interpretation (E) satisfies Axiom O8.

It now remains to verify Axiom O9, or its equivalent, the Pasch Theorem (Theorem 78 of Chapter I). Because of Theorem 23, for this purpose it is sufficient to prove

THEOREM 25. *For any arbitrarily fixed plane at infinity, if points a, b, c are proper and non-collinear, then the E-segments* (ab), (ac), (bc) *form a system of open sides of triangle abc.*

PROOF. In the contrary case, segments (ab), (ac), (bc)* would, by Theorem 21, form a system of open sides of triangle abc. Then, because of Theorem 22, segments (ab)*, (ac)*, (bc)* would also form a system of open sides of triangle abc. But this is impossible, since from Theorem 24 it follows that the improper line of plane abc intersects each of the three segments (ab)*, (ac)*, (bc)*.

Henceforth, instead of saying that interpretation (E) satisfies theorem T of Euclidean Geometry, we shall say that *theorem* T *is satisfied in* E-*space* $S-P_\infty$. The fact that interpretation (E) is a model for the system of axioms (IO) can now be expressed as

PROPOSITION 2. *If in space* S *we fix in any arbitrary way the plane at infinity* P_∞, *then in* E-*space* $S-P_\infty$ *there are satisfied all those theorems of Euclidean geometry which are consequences of the axioms of incidence and order only.*

It is readily noted, on the basis of Theorem 12(VIII) of Chapter I, that, given any plane P and any line L on it, we can choose a plane at infinity P_∞ in such a way that plane P is a proper plane and line L, an improper line. In other words, plane P can be treated as a proper plane with the line at infinity L.

Similarly, given any line L and any point p on it we can choose a plane at infinity P_∞ (see Theorem 12(IV) of Chapter I) in such a way that L is a proper line and p an improper point. In other words, line L can be treated as a proper line with the point at infinity p. This helps us to employ (through the intermediary of Proposition 2) the theorems of Euclidean geometry resulting from the axioms of the systems (IO) in the proofs of theorems of projective geometry.

As an example we shall prove in this way

THEOREM 26. *Every open segment* I *contains infinitely many points.*

PROOF. Segment I lie on a line L. We regard L as a proper line with the point at infinity $p_\infty \in I^*$. Then, by Theorem 24, open segment I is an open E-segment of E-line $L-p_\infty$, and hence, by Theorem 19 of Chapter I, it contains infinitely many points.

12. Ordinary Triangles

In Section 10 we have showed that any triangle abc has four systems of open sides. Triangle abc with a system of open sides $\{I_1, I_2, I_3\}$ will be called an *ordinary triangle* and will be denoted by $abc\{I_1, I_2, I_3\}$. More precisely, ordinary triangle $abc\{I_1, I_2, I_3\}$ is an ordered pair consisting of triangle abc and a system of its open sides $\{I_1, I_2, I_3\}$.

By the *boundary of ordinary triangle* $abc\{I_1, I_2, I_3\}$ we shall understand the point set $F = abc \cup I_1 \cup I_2 \cup I_3$. By the *inner domain of ordinary triangle* $abc\{I_1, I_2, I_3\}$ we shall understand the set of points $p \in \mathbf{P}(abc)$ satisfying the following condition: Every line $L \subset \mathbf{P}(abc)$ passing through point p intersects boundary F of triangle $abc\{I_1, I_2, I_3\}$ in just two points. The inner domains of ordinary triangles will also be called *open triangles*.

If we fix, in any way, a plane at infinity P_∞ then, by Theorem 24, every E-open E-segment (ab) of E-space $\mathbf{S}-P_\infty$ is an open segment of space $\mathbf{S}$. We shall also show that every E-open E-triangle abc of E-space $\mathbf{S}-P_\infty$ is an open triangle of space $\mathbf{S}$:

THEOREM 27. *For any arbitrarily fixed plane at infinity, an E-triangle* (abc) *coincides with the inner domain of the ordinary triangle* $abc\{I_1,I_2,I_3\}$, *where* $I_1=(ab)$, $I_2=(ac)$, $I_3=(bc)$.

PROOF. Let L_∞ be the improper line of plane $P=\mathbf{P}(abc)$. Further, let $F=abc\cup I_1\cup I_2\cup I_3$. Then F is the E-boundary of E-triangle (abc) as well as the boundary of ordinary triangle $abc\{I_1,I_2,I_3\}$, and

(1) $$F\subset P-L_\infty.$$

Hence the inner domain G of ordinary triangle $abc\{I_1,I_2,I_3\}$ consists of only proper points; for, if an improper point p_∞ were to belong to set G, then the line L_∞ passing through point p_∞ would intersect boundary F, in contradiction to (1). Therefore

(2) $$G\subset P-L_\infty.$$

By formulas (1) and (2) the inner domain G of ordinary triangle $abc\{I_1,I_2,I_3\}$ can be characterized as the set of E-points $p\in P-L_\infty$ such that every E-line $L\in P-L_\infty$ passing through p intersects set F in just two E-points. By Theorem 81 of Chapter I, it thus follows that G is the E-tringle (abc).

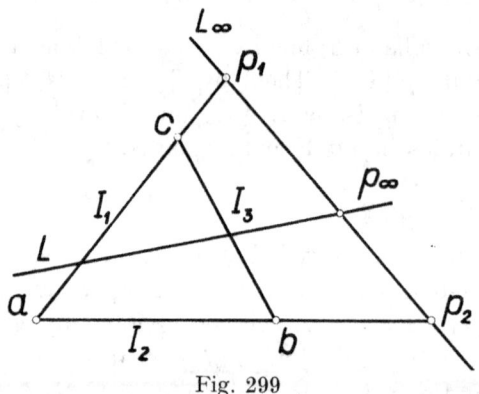

Fig. 299

We shall now investigate the intersection of an open triangle and a line:

THEOREM 28. *For any ordinary triangle* $abc\{I_1,I_2,I_3\}$ *and any line L on a given plane P the intersection of L with the inner domain G of triangle* $abc\{I_1,I_2,I_3\}$ *is either an empty set or an open segment.*

PROOF. Let us take points $p_1\in I_1^*$ and $p_2\in I_2^*$ and let us regard the line $L_\infty=\mathbf{L}(p_1p_2)$ as the improper line of plane P (Fig. 299). Line L_∞

does not pass through any of the points a, b, c and, by Pasch's Theorem (Theorem 23), it does not intersect any of the open segments I_1, I_2, I_3. Hence segments I_1, I_2, I_3 are E-sides of E-triangle (abc), from which it follows, by Theorem 27, that G coincides with E-triangle (abc). Therefore

$$G \subset P - L_\infty.$$

Thus, if line L and set G have a point in common, then L is a proper line. Let $L \cap G \neq 0$ and $L \cap L_\infty = p_\infty$. Since $L \cap G = (L - p_\infty) \cap G$, the set $(L - p_\infty) \cap G$ coincides, by Theorem 82 of Chapter I, with some E-open E-segment of E-line $L - p_\infty$, that is, with some open segment of line L.

13. Topology on the Line

Let L be any line. Let us denote by $\mathfrak{U}$ the class of all open segments included in line L. Class $\mathfrak{U}$ satisfies, the following three conditions:

(I) *For any two distinct points $p, q \in L$ there exist open segments $U_1, U_2 \in \mathfrak{U}$ such that $p \in U_1$, $q \in U_2$ and $U_1 \cap U_2 = 0$.*

PROOF. It suffices to take on line L any point $a \neq p, q$ and any point b such that $\mathbf{D}(p, q; a, b)$. Then segments $U_1 = (a\text{-}p\text{-}b)$ and $U_2 = (a\text{-}q\text{-}b)$ satisfy the required conditions.

Condition (I) could have been proved in another way by making use of a similar condition proved for the E-line in Section 11 of Chapter I. If we regard line L as a proper line with the improper point $p_\infty \neq p, q$, then, by Theorem 28 of Chapter I, there exist on E-line $L - p_\infty$ two E-open E-segments U_1' and U_2' such that $p \in U_1'$, $q \in U_2'$, and $U_1' \cap U_2' = 0$. By Theorem 24, we have $U_1, U_2 \in \mathfrak{U}$.

(II) *If p and a are two distinct points on line L and $p \in U \in \mathfrak{U}$, then there exists an open segment $U' \subset U$ such that $p \in U'$, but $a \sim \, \in U'$.*

PROOF. We regard line L as a proper line with the point at infinity $p_\infty \in U^* - a$ (by Theorem 26 such a point always exists). Then points p and a are E-points, and segment U is an E-open E-segment of E-line $L - p_\infty$. By Theorem 28 of Chapter I, there thus exist E-open E-segments U_1 and U_2 such that $p \in U_1$, $a \in U_2$ and $U_1 \cap U_2 = 0$, and an E-open E-segment $U' \subset U \cap U_1$ such that $p \in U'$. Obviously $U' \subset U$ and $a \sim \, \in U'$.

(III) *If $U_1, U_2 \in \mathfrak{U}$ and $p \in U_1 \cap U_2$, then there exists an open segment $U_0 \in \mathfrak{U}$ such that $p \in U_0$ and $U_0 \subset U_2$.*

PROOF. We regard line L as a proper line with the point at infinity $p_\infty \neq p$. By condition (II), there exist on E-line $L - p_\infty$ two E-open

E-segments $U_0' \subset U_1$ and $U_2' \subset U_2$ such that $p \in U_1' \cap U_2'$; hence there exists an E-open E-segment $U_0 \subset U \ \cap U \ \subset U_1' \cap U_2'$ such that $p \in U_0$.

From conditions (I) and (III) we obtain

THEOREM 29. *The class* $\mathfrak{U}$ *of all open segments of a line* L *constitutes a base of neighborhoods for line* L.

In this way line L becomes a topological space. Let us fix on line L the point at infinity p_∞. By Theorem 28 of Chapter I the class of all E-open E-segments of E-line $L - p_\infty$ is the base of neighborhoods for E-line $L - p_\infty$ and as such determines on E-line $L - p_\infty$ a topology which we shall call E-*topology*. The names of the topological notions in E-topology will be preceded by the prefix "E". If $p \in L - p_\infty$, then, by condition (II), in every neighborhood of point p there is included an E-neighborhood of p, and, by Theorem 24, every E-neighborhood of p is simultaneously a neighborhood of p. From this it is readily concluded that, as applied to E-space $L - p_\infty$, the topological notions of space L (based on the notion of the neighborhood) coincide with the analogous notions of E-space $L - p_\infty$ (based on the notion of the E-neighborhood). If e.g. $p \in L - p_\infty$ and $F \subset L - p_\infty$, then point p is a point of accumulation of set F if and only if it is an E-point of accumulation of set F; if $F \subset L - p_\infty$, then set F is connected if and only if it is E-connected; if the function f defined in the set $L - p_\infty$ takes on the values belonging to this set, then function f is continuous at point $p \in L - p_\infty$ if and only if it is E-continuous at point p.

14. Topology on the Plane

Let P be any plane. We denote by $\mathfrak{U}$ the class of all open triangles included in plane P. Proceeding similarly as in the preceding section where we established a topology on the line, we can show, with the help of Theorems 84 and 86 of Chapter I and Theorem 27, that class $\mathfrak{U}$ satisfies the following three conditions:

(I) *For any two distinct points* $p,q \in P$ *there exist open triangles* $U_1, U_2 \in \mathfrak{U}$ *such that* $p \in U_1$, $q \in U_2$ *and* $U_1 \cap U_2 = 0$.

(II) *Let* K *be any line in plane* P. *If* $p \in P - K$ *and* $p \in U \in \mathfrak{U}$, *then there exists an open triangle* $U' \subset U$ *such that* $p \in U'$ *and* $K \cap U' = 0$.

(III) *If* $U_1, U_2 \in \mathfrak{U}$ *and* $p \in U_1 \cap U_2$, *then there exists an open triangle* $U_0 \in \mathfrak{U}$ *such that* $p \in U_0$ *and* $U_0 \subset U_1 \cap U_2$.

From conditions (I) and (III) we obtain

THEOREM 30. *The class* $\mathfrak{U}$ *of all open triangles of a plane P constitutes a base of neighborhoods for plane P.*

In this way plane P becomes a topological space. Let us fix on plane P the line at infinity L_∞. By Theorem 84 of Chapter I, the class of all E-open E-triangles of E-plane $P-L_\infty$ constitutes a base of neighborhoods for E-plane $P-L_\infty$ and as such determines on E-plane $P-L_\infty$ a topology which we shall call E-*topology*. The names of the topological notions in E-topology will be preceded by the prefix "E". If $p \in P-L_\infty$, then, by condition (II), in every neighborhood of point p there is included an E-neighborhood of p, and, by Theorem 27, every E-neighborhood of p is simultaneously a neighborhood of p. From this we conclude that, as applied to space $P-L_\infty$, the topological notions of space P coincide with the analogous notions of E-space $P-L_\infty$.

The topology on plane P induces a topology on every line $L \subset P$. We note that the topology introduced in this way on line L coincides with the topology previously defined (in Section 13). Indeed, by Theorem 28, if line L intersects a neighborhood $U \in \mathfrak{U}$, then $L \cap U$ is an open segment, and, as readily noted, every open segment of line L can be regarded as the intersection of line L with a neighborhood $U \in \mathfrak{U}$.

15. Quadrangles

The unordered quadruple of coplanar points a, b, c, d, no three of which are collinear, will be called a (plane) *quadrangle* and will be denoted by *abcd*. Points a, b, c, d will be called the *vertices*, and the six lines determined by pairs of different vertices will be called the *sides of quadrangle abcd*

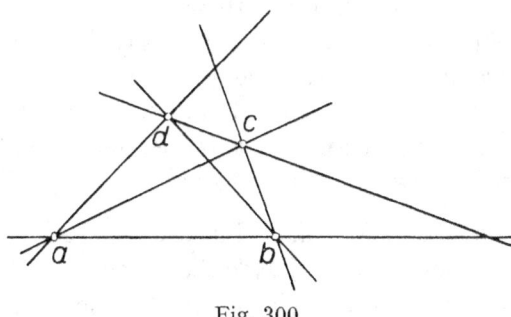

Fig. 300

(Fig. 300). We shall say that *two sides are opposite* if together they contain all the vertices of the quadrangle. Quadrangle *abcd* therefore has three pairs of opposite sides: $\mathbf{L}(ab)$ and $\mathbf{L}(cd)$, $\mathbf{L}(ac)$ and $\mathbf{L}(bd)$, $\mathbf{L}(ad)$ and $\mathbf{L}(bc)$. Each pair of opposite sides intersects in a point called a *diagonal point* of quadrangle *abcd*.

THEOREM 31. *The three diagonal points of a quadrangle are non-collinear.*

PROOF. The diagonal points of a quadrangle $abcd$ lying in a plane P coincide with the points

$$p = \mathbf{L}(ab) \cap \mathbf{L}(cd), \quad q = \mathbf{L}(ac) \cap \mathbf{L}(bd), \quad r = \mathbf{L}(ad) \cap \mathbf{L}(bc).$$

Obviously, points p and r are distinct, and each of them is distinct from each of the vertices of quadrangle $abcd$ (Fig. 301). It should be shown that point q does not lie on line pr.

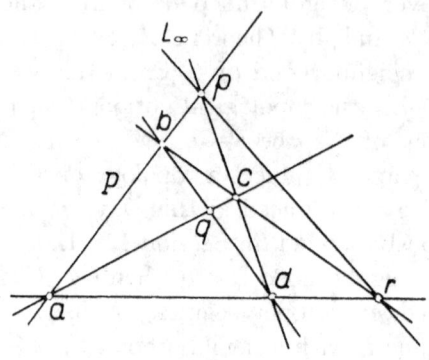

Fig. 301

Let us regard P as a proper plane with the line at infinity $L_\infty = \mathbf{L}(pr)$. We note that line L_∞ does not pass through any of the vertices of quadrangle $abcd$, since otherwise it would coincide simultaneously with one of the two sides $\mathbf{L}(ab)$ or $\mathbf{L}(cd)$ and with one of the two sides $\mathbf{L}(ad)$ or $\mathbf{L}(bc)$, which is impossible because of the non-collinearity of each three of the vertices of the quadrangle. Hence points a, b, c, d are proper points, i.e. E-points of E-plane $P - L_\infty$. It may readily be shown that E-quadrangle (a,b,c,d) is E-convex, and therefore (because of Theorem 93(I) of Chapter I) its E-diagonals ac and bd intersect in an E-point. Since $\mathbf{L}(ac) \cap \mathbf{L}(bd) = q$, hence q is a proper point, and consequently $q \sim \in \mathbf{L}(pr)$, which was what we were to prove.

By Theorem 31 the three diagonal points of a quadrangle form a triangle. We shall call the sides of this triangle the *diagonals* of the quadrangle.

By fixing a pair of opposite sides of quadrangle $abcd$, say $\mathbf{L}(ac)$ and $\mathbf{L}(bd)$, we determine a diagonal point $q = \mathbf{L}(ac) \cap \mathbf{L}(bd)$. Hence point q is a function of four (ordered) vertices:

$$q = f(a,b,c,d).$$

Then $p = f(a,c,b,d)$ and $r = f(a,b,d,c)$. We shall show that function f is continuous on the entire plane P. To show this, we fix points a, b, c, d

and—as in the proof of Theorem 31—we choose the line $L = \mathbf{L}(pr)$ as the line at infinity. Then E-quadrangle (a,b,c,d) is E-convex, and, by Theorem 96 of Chapter I, there exist E-neighborhoods U_a, U_b, U_c, U_d of E-points a, b, c, d, respectively, such that if

$$a' \in U_a, \quad b' \in U_b, \quad c' \in U_c, \quad d' \in U_d,$$

then E-points a',b',c',d' are distinct and E-quadrangle (a',b',c',d') is also E-convex. In the E-neighborhoods U_a, U_b, U_c, U_d (as the domains of the first, second, third, and fourth arguments, respectively, of f), the function f coincides with the function f' correlating to each E-convex E-quadrangle (a',b',c',d') lying on E-plane $P-L_\infty$ its E-diagonal E-point q'. By Theorem 96 of Chapter I, f' is an E-continuous function of points a,b,c,d; hence f is a continuous function of points a,b,c,d. We therefore have

THEOREM 32. *Every diagonal point of a quadrangle abcd on a plane P is a continuous function of the four vertices a, b, c, d.*

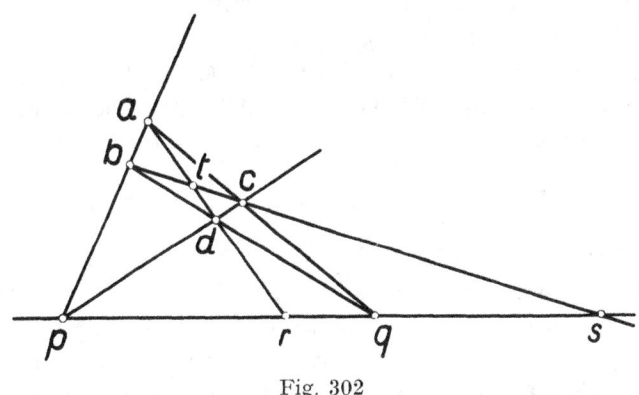

Fig. 302

16. Harmonic Quadruples

We say that *points p and q divide points r and s harmonically*—in symbols, $\mathbf{D}_h(p,q;r,s)$—if there exists a quadrangle *abcd* for which points p and q are diagonal points, and points r and s are the intersection points of the diagonal pq and the two sides of quadrangle *abcd* which do not pass through points p and q (Fig. 302). We then also say that the quadruple of points (p,q,r,s) is *harmonic* and that this harmonicity is *realized* by quadrangle *abcd*. From the definition of the relation $\mathbf{D}_h$ and from Theorem 31 we obtain at once

THEOREM 33. *If $\mathbf{D}_h(p,q;r,s)$, then points p, q, r, s are collinear and distinct.*

We shall now prove

THEOREM 34. *If* $\mathbf{D}_h(p,q;r,s)$, *then* $\mathbf{D}(p,q;r,s)$.

PROOF. Let the harmonicity of the quadruple (p,q,r,s) be realized by a quadrangle *abcd* (Fig. 302) and let

$$p = \mathbf{L}(ab) \cap \mathbf{L}(cd), \quad q = \mathbf{L}(ac) \cap \mathbf{L}(bd), \quad r = \mathbf{L}(pq) \cap \mathbf{L}(ad),$$

$$s = \mathbf{L}(pq) \cap \mathbf{L}(bc).$$

We denote the third diagonal point of quadrangle *abcd* by t; hence

$$t = \mathbf{L}(ad) \cap \mathbf{L}(bc).$$

Let f_1 be the perspective transformation from center a of diagonal $\mathbf{L}(pq)$ onto side $\mathbf{L}(bc)$, and f_2, the perspective transformation from center d of side $\mathbf{L}(bc)$ onto diagonal $\mathbf{L}(pq)$. Transformation f_1 maps the quadruple (p,q,r,s) onto the quadruple (b,c,t,s), and transformation f_2 maps the quadruple (b,c,t,s) onto the quadruple (q,p,r,s). Therefore, the projective transformation $f = f_2 f_1$ of diagonal $\mathbf{L}(pq)$ onto itself maps the quadruple (p,q,r,s) onto the quadruple (q,p,r,s).

Points p, q, r, s are, by Theorem 33, collinear and distinct, and therefore, because of Axioms O'4, O'3, and O'2, only one of the following three formulas holds:

$$\mathbf{D}(p,q;r,s) \quad \text{or} \quad \mathbf{D}(p,r;q,s) \quad \text{or} \quad \mathbf{D}(p,s;q,r).$$

We shall show that neither the second nor the third formula holds Indeed, by Theorem 18, transformation f preserves the relation of division, and hence it would follow from $\mathbf{D}(p,r;q,s)$ that $\mathbf{D}(q,r;p,s)$, and from $\mathbf{D}(p,s;q,r)$, that $\mathbf{D}(q,s;p,r)$. But one or the other would contradict Axiom O'3. Thus $\mathbf{D}(p,q;r,s)$, which is what we had to prove.

By the above theorem, the relation $\mathbf{D}_h$ is included in the relation $\mathbf{D}$. In the coming sections we shall show that $\mathbf{D}_h$ preserves some properties of $\mathbf{D}$. In particular, this concerns the properties formulated in Axioms O'2, O'5, and O'7.

17. Permutations of Harmonic Quadruples

THEOREM 35. *If* $\mathbf{D}_h(p,q;r,s)$, *then* $\mathbf{D}_h(q,p;r,s)$ *and* $\mathbf{D}_h(r,s;p,q)$.

PROOF. From the definition of the relation $\mathbf{D}_h$ it follows at once that the order of points p and q, as well as r and s, does not play any role; in particular, $\mathbf{D}_h(p,q;r,s)$ implies $\mathbf{D}_h(q,p;r,s)$.

We shall now show that $\mathbf{D}_h(p,q;r,s)$ implies $\mathbf{D}_h(r,s;p,q)$. Let *abc* be a quadrangle realizing the harmonicity of the quadruple (p,q,r,s) (Fig. 303) and let

$$p = \mathbf{L}(ab) \cap \mathbf{L}(cd), \quad q = \mathbf{L}(ac) \cap \mathbf{L}(bd), \quad r = \mathbf{L}(pq) \cap \mathbf{L}(ad)$$
$$s = \mathbf{L}(pq) \cap \mathbf{L}(bc).$$

Further, let

$$\mathbf{L}(ad) \cap \mathbf{L}(bc) = t \quad \text{and} \quad \mathbf{L}(as) \cap \mathbf{L}(br) = e.$$

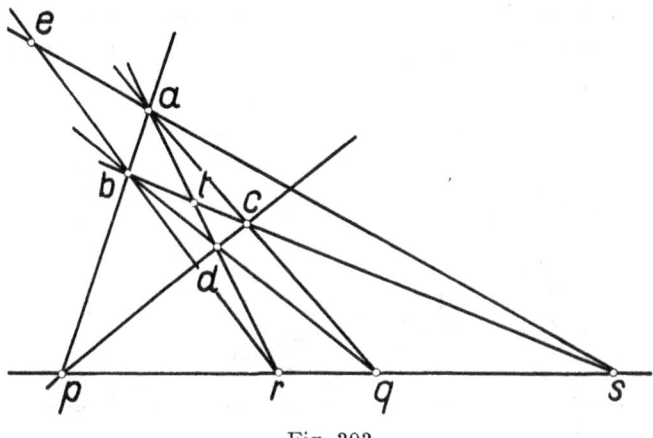

Fig. 303

It may readily be shown that no three of the points a, t, e, b are collinear and therefore points a, t, e, b form a quadrangle $ateb$. We shall show that the quadrangle $ateb$ realizes the harmonicity of the quadruple (r,s,p,q). We have

$$\mathbf{L}(at) \cap \mathbf{L}(eb) = r, \quad \mathbf{L}(ae) \cap \mathbf{L}(tb) = s, \quad \mathbf{L}(rs) \cap \mathbf{L}(ab) = p.$$

It thus remains to show that $\mathbf{L}(rs) \cap \mathbf{L}(te) = q$, that is, that points q, e, t are collinear. To show this, we note that line pq is a perspective axis of triangles abe and cdt for correspondence f of vertices given by the equalities

$$f(a) = c, \; f(b) = d, \; f(e) = t.$$

According to the direct Desargues Theorem, triangles abe and cdt therefore have a perspective center for correspondence of vertices f. This center must coincide with the intersection point of lines ac and bd, that is, with point q. Hence point q is collinear with points e and t, which is what we had to prove.

If $\mathbf{D}_h(p,q;r,s)$, then we also say that *segment pq is harmonically conjugate to segment rs*. From Theorem 35 it follows that this relation between segments is symmetric.

18. The Fourth Harmonic Point

Axiom O′5 says that for any three distinct collinear points a, b, c there exists a point d such that $\mathbf{D}(a,b;c,d)$, and on the basis of the remaining axioms it is readily concluded that there are infinitely many such points. The situation is quite different for the relation $\mathbf{D}_h$. In order to show this, we shall first prove the following:

LEMMA. *For any three distinct points p,q,r on a given line L there exists a point s such that $\mathbf{D}_h(p,q;r,s)$, where the harmonicity of the quadruple (p,q,r,s) is realized by some quadrangle $abcd$ lying on a given plane P passing through L.*

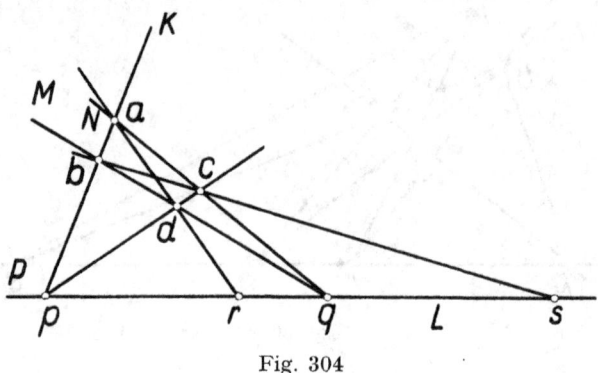

Fig. 304

PROOF. Employing Theorem 3, let us produce in plane P through point p a line K distinct from line L, and through point q two distinct lines M and N, each of which is distinct from line L (Fig. 304). Let

$$a = K \cap N, \ b = K \cap M, \ d = M \cap \mathbf{L}(ar), \ c = N \cap \mathbf{L}(pd),$$

and let

$$s = L \cap \mathbf{L}(bc).$$

From the construction it follows that points a, b, c, d lie in plane P and are vertices of the quadrangle $abcd$, for which the points $p = \mathbf{L}(ab) \cap \mathbf{L}(cd)$ and $q = \mathbf{L}(ac) \cap \mathbf{L}(bd)$ are the diagonal points, and the points r and s are the intersection points of diagonal $L = \mathbf{L}(pq)$ and sides $\mathbf{L}(ad)$ and $\mathbf{L}(bc)$. Hence $\mathbf{D}_h(p,q;r,s)$, and point s proves to be the point we are seeking.

We now come to the theorem proper:

THEOREM 36. *For every three distinct collinear points p, q, r there exists just one point s such that $\mathbf{D}_h(p,q;r,s)$.*

PROOF. Because of the lemma we have just proved, it is sufficient to show that there exists at most one such point s. Let us denote by L the line on which points p, q, r lie (Fig. 305). Let us assume, further, that the formulas

$$\boldsymbol{D}_h(p,q;r,s_1) \quad \text{and} \quad \boldsymbol{D}_h(p,q;r,s_2)$$

hold simultaneously; and for $i = 1,2$ let the harmonicity of the quadruple (p,q,r,s_i) be realized by the quadrangle $a_i b_i c_i d_i$ lying in a plane P_i. We

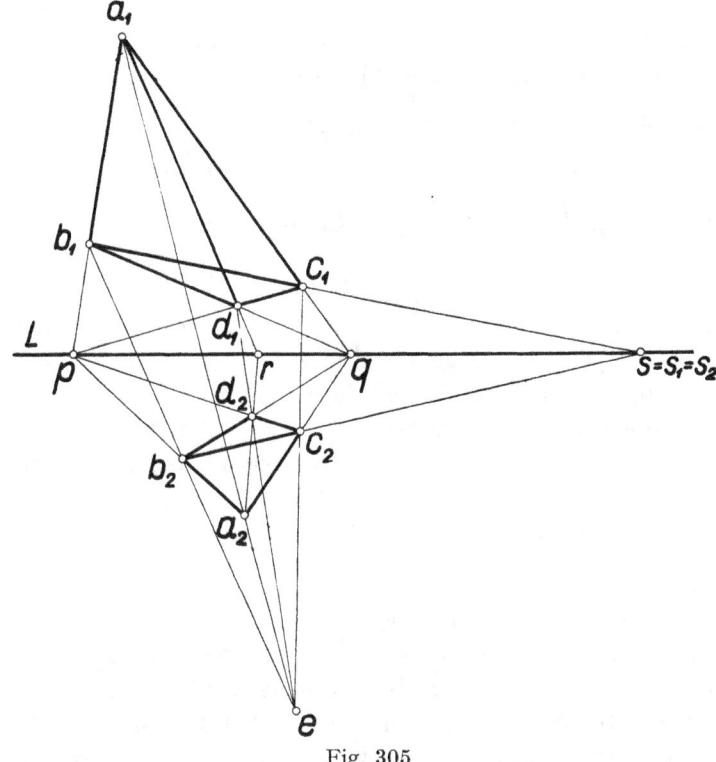

Fig. 305

have to show that $s_1 = s_2$. We may assume, without diminishing the generality of the proof, that

(1) $p = \boldsymbol{L}(a_1 b_1) \cap \boldsymbol{L}(c_1 d_1), \quad q = \boldsymbol{L}(a_1 c_1) \cap \boldsymbol{L}(b_1 d_1),$

$\quad\quad\; r = L \cap \boldsymbol{L}(a_1 d_1), \quad\quad\;\; s = L \cap \boldsymbol{L}(b_1 c_1),$

(2) $p = \boldsymbol{L}(a_2 b_2) \cap \boldsymbol{L}(c_2 d_2), \quad q = \boldsymbol{L}(a_2 c_2) \cap \boldsymbol{L}(b_2 d_2),$

$\quad\quad\; r = L \cap \boldsymbol{L}(a_2 d_2), \quad\quad\; s_2 = L \cap \boldsymbol{L}(b_2 c_2).$

Let the function f transform vertices a_1, b_1, c_1, d_1 of quadrangle $a_1 b_1 c_1 d_1$ onto the vertices a_2, b_2, c_2, d_2 of quadrangle $a_2 b_2 c_2 d_2$, respectively. From conditions (1) and (2) it follows at once that line L is, for correspondence f, a perspective axis for triangles

$$a_1b_1d_1 \quad \text{and} \quad a_2b_2d_2$$

as well as for triangles

$$a_1c_1d_1 \quad \text{and} \quad a_2c_2d_2.$$

From the direct Desargues Theorem it thus follows that for correspondence f there exists a perspective center e for triangles $a_1b_1d_1$ and $a_2b_2d_2$ and a perspective center e' for triangles $a_1c_1d_1$ and $a_2c_2d_2$.

We shall now examine two possible cases:

Case 1. $P_1 \neq P_2$ (Fig. 305). Then $P_1 \cap P_2 = L$, and, since

$$a_1b_1c_1d_1 \subset P_1 - L \quad \text{and} \quad a_2b_2c_2d_2 \subset P_2 - L,$$

we have $a_1 \neq a_2$, $d_1 \neq d_2$, $\mathbf{L}(a_1a_2) \neq \mathbf{L}(d_1d_2)$, and

(3) $$\mathbf{L}(a_1b_1) \neq \mathbf{L}(a_2b_2), \quad \mathbf{L}(a_1c_1) \neq \mathbf{L}(a_2c_2).$$

Hence $e = \mathbf{L}(a_1a_2) \cap \mathbf{L}(d_1d_2) = e'$, and therefore point e is, for correspondence f, a perspective center also for triangles $a_1b_1c_1$ and $a_2b_2c_2$. Using now the converse Desargues Theorem, we thereby conclude that, for correspondence f, there exists a perspective axis for triangles $a_1b_1c_1$ and $a_2b_2c_2$. By formulas (1), (2), and (3), this axis must coincide with line L. Since

$$L \cap \mathbf{L}(b_1c_1) = s_1 \quad \text{and} \quad L \cap \mathbf{L}(b_2c_2) = s_2,$$

then $s_1 = s_2$.

This concludes the proof for Case 1.

Case 2. $P_1 = P_2$. Let us produce through line L a plane P distinct from plane P_1, and therefore distinct also from plane P_2. On the basis of the Lemma there exists a point s such that $\mathbf{D}_h(p,q;r,s)$, where the harmonicity of the quadruple (p,q,r,s) is realized by a quadrangle $abcd$ lying in plane P. Because of Case 1, we conclude that $s = s_1$ and $s = s_2$, from which it follows that $s_1 = s_2$.

This concludes the proof for Case 2.

By the theorem which we have just proved, every triple (p,q,r) composed of distinct points of a line L uniquely determines a point $s \in L$ such that (p,q,r,s) is a harmonic quadruple. We shall refer to s as the *fourth harmonic point* for the triple of points (p,q,r). Hence the fourth harmonic point s is a function of the points $p,q,r \in L$:

$$s = s(p,q,r).$$

We shall now show that the function s is continuous for every three distinct points $p,q,r \in L$. To show this, let us take any plane P including line L and let us fix on it any two points $a \in P-L$ and

$$e \in P-(L \cup L(ap) \cup L(aq))$$

(Fig. 306). By Theorem 32, the point $b = b(p,q) = L(ap) \cap L(eq)$, as a diagonal point of quadrangle $apeq$, is a continuous function of vertices p and q, and point $d = d(p,q,r) = L(ar) \cap L(bq)$, as a diagonal point of quadrangle $arbq$, is a continuous function of vertices r, $b(p,q)$, and q,

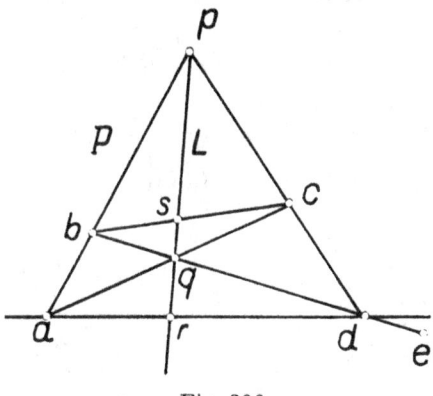

Fig. 306

and therefore, indirectly, of points p, q, r. By applying again Theorem 32 to the quadrangle $pdaq$, we conclude that its diagonal point $c = c(p,q,r) = L(pd) \cap L(aq)$ is a continuous function of vertices p, $d(p,q,r)$, and q, and therefore, indirectly, of points p, q, r. Finally, let us consider quadrangle $abcd$. Since

$$p = L(ab) \cap L(cd), \qquad q = L(ac) \cap L(bd),$$
$$r = L(pq) \cap L(ad),$$

then for the fourth harmonic point s it must be that

$$s = s(p,q,r) = L(pq) \cap L(bc).$$

We thus conclude that point s is a diagonal point of quadrangle $pqbc$ and consequently is a continuous function of vertices p,q, $b(p,q)$ and $c(p,q,r)$, and therefore, indirectly, of points p, q, r. We thus have

THEOREM 37. *For three distinct points p,q,r on a given line L the fourth harmonic point s for the triple (p,q,r) is a continuous function of points p,q,r.*

19. Perspective Transformations of Harmonic Quadruples

THEOREM 38. *If f is a perspective transformation of a line K onto a line L from a center $a \sim \in K \cup L$, then, for any four points $p,q,r,s \in K$ $\mathbf{D}_h(p,q;r,s)$ implies $\mathbf{D}_h(f(p),f(q);f(r),f(s))$.*

PROOF. If $K = L$, then f is the identity transformation and the theorem is obvious. Let us therefore assume that $K \neq L$. We shall consider separately two possible cases:

Case 1. The intersection point of lines K and L is one of the points p,q,r,s. Because of Theorem 35, we may assume that this point coincides with point p (Fig. 307). Then $f(p) = p$. We denote by b the point of intersection of lines ap and $qf(r)$. For quadrangle $abf(q)f(r)$ we have

$$p = \mathbf{L}(ab) \cap \mathbf{L}(f(q)f(r)), \qquad q = \mathbf{L}(af(q)) \cap \mathbf{L}(bf(r)),$$

$$r = K \cap \mathbf{L}(af(r)).$$

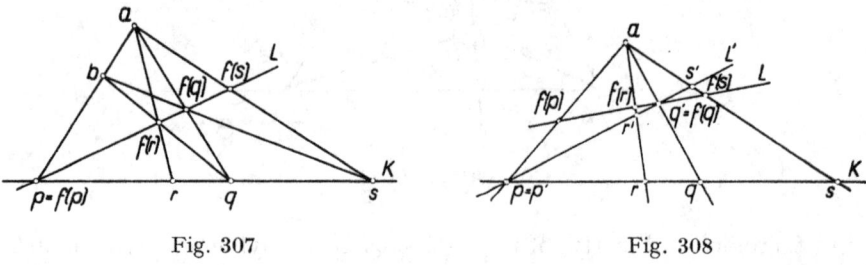

Fig. 307   Fig. 308

It thus follows that the intersection point of diagonal K and side $\mathbf{L}(bf(q))$ coincides with the fourth harmonic point for the triple (p,q,r), which, since $\mathbf{D}_h(p,q;r,s)$, gives

$$s = K \cap \mathbf{L}(bf(q)).$$

Let us now consider quadrangle $basq$. We have

$$\mathbf{L}(ba) \cap \mathbf{L}(sq) = f(p), \qquad \mathbf{L}(bs) \cap \mathbf{L}(aq) = f(q),$$
$$L \cap \mathbf{L}(bq) = f(r), \qquad L \cap \mathbf{L}(as) = f(s),$$

and consequently $\mathbf{D}_h(f(p),f(q);f(r),f(s))$.

Case 2. The intersection point of lines K and L is distinct from each of the points p,q,r,s (Fig. 308). In particular, we then have $p \neq f(q)$. Let $L' = pf(q)$. It is readily shown that $a \sim \in L'$. The perspective transformation f' of line K onto line L' from center a maps the points p, q, r, s, respectively, onto the points

$$p' = p, \quad q' = f(q), \quad r' = L' \cap \mathbf{L}(ar), \quad s' = L' \cap \mathbf{L}(as).$$

Since $p \in L'$, then we have to do with Case 1. Hence

(1) $$\mathbf{D}_h(p',q';r',s').$$

The perspective transformation f'' of line L' onto line L from center a maps the points p',q',r',s', respectively, onto the points $f(p),f(q),f(r),f(s)$. Since $q' \in L$, then again we have to do with Case 1, and from formula (1) it follows that $\mathbf{D}_h(f(p),f(q);f(r),f(s))$.

This completes the proof.

20. Continuity of the Central Projectivity

Given in a plane P a line L and a point $a \sim \in L$, let f be the projectivity from center a upon line L. We shall show that function f is continuous in the set $P-a$.

First, we shall show the continuity at a point $p \sim \in L$ (Fig. 309). Let $p_1 = f(p)$. We fix on line L two distinct points $b_1,b_2 \neq p_1$, and we

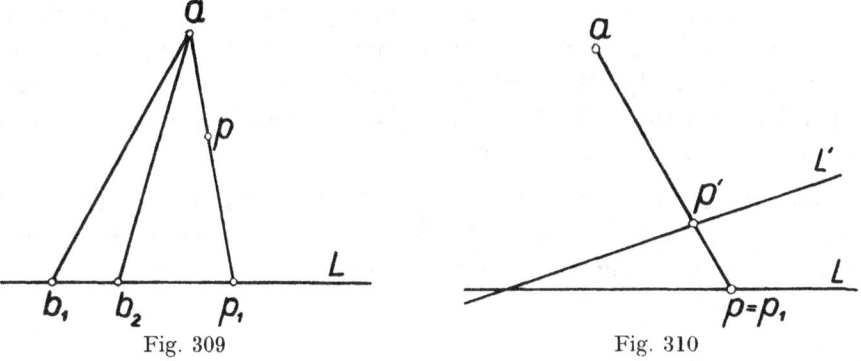

Fig. 309   Fig. 310

take a neighborhood U of point p such that $U \cap \mathbf{L}(ab_1) = 0$ and $U \cap \mathbf{L}(ab_2) = 0$. Then $p_1 = \mathbf{L}(ap) \cap \mathbf{L}(b_1b_2)$ is a diagonal point of quadrangle apb_1b_2, from which it follows, by Theorem 32, that point p_1 is a continuous function of point p (in neighborhood U).

We now assume that $p \in L$ (Fig. 310). Then $p_1 = f(p) = p$. Let $L' \subset P$ be any line not passing through either point a or point p. We denote by f' the projectivity from center a upon line L' and we put $p'=\mathbf{L}(ap) \cap L'$. Then $p' \sim \in L$ and $p_1 = f(p')$; also $p \sim \in L'$ and $p' = f'(p)$. By what has been proved above, it is seen that point p_1 is a continuous function of point p', and point p' is a continuous function of point p. Hence, indirectly, point p_1 is a continuous function of point p.

We have thus proved

THEOREM 39. *For any line L and any point $a \sim \in L$ on a given plane P the projectivity f upon L from center a is a continuous function of the point $p \in P-a$.*

21. Midpoint of a Segment

We now fix an arbitrarily chosen plane which in the succeeding sections will be used as the plane at infinity and will be denoted by Q_∞.

Let p and q be any two distinct proper points. We denote by r_∞ the point at infinity of line pq. According to Theorem 36, there exists just one point s which is the fourth harmonic for the triple of points (p,q,r_∞). We shall call point s the *midpoint of points p and q*. From Theorem 35 it follows that the midpoint s of points p and q is simultaneously the midpoint of points q and p, and therefore we can also call it the *midpoint of segment pq*. With the help of Theorems 33, 34, and 37, we obtain

THEOREM 40. *Consider a proper line L with the point at infinity r_∞, and two distinct proper points $p,q \in L$. The midpoint s of segment pq lies on the E-line $L-r_\infty$ between points p and q and is a continuous ($=$E-continuous) function of points p and q.*

By Theorem 35, it follows from $\mathbf{D}_h(p,q;r_\infty,s)$ that $\mathbf{D}_h(r_\infty,s;p,q)$. From this and from Theorem 36 we conclude that not only segment pq uniquely determines its midpoint s, but also points s and p uniquely determine a point q such that s is the midpoint of segment pq. On the basis of Theorems 33 and 37 we obtain

THEOREM 41. *Let L be a proper line with the point at infinity r_∞, and let s and p be two distinct proper points on L. The point q such that point s is the midpoint of segment pq lies on E-line $L-r_\infty$ and is a continuous ($=$E-continuous) function of points s and p.*

22. Natural Net

By a *natural net* we understand every sequence of proper points

$$(a_0,a_1,a_2,\ldots,a_{n-1},a_n,a_{n+1},\ldots)$$

such that a_n is the midpoint of the segment $a_{n-1}a_{n+1}$ for any natural n.

We choose any two distinct proper points p and q. It is readily noted that there exists just one natural net (a_n) such that $a_0 = p$ and $a_1 = q$. All terms a_n of this net are, by Theorem 41, proper points of the proper line $L = \mathbf{L}(pq)$. We denote by a_∞ the improper point of line L. From the definition of the natural net (a_n) it follows, by Theorem 40, that on E-line $L-a_\infty$ we have

$$\mathbf{B}(a_{n-1},a_n,a_{n+1}) \text{ for } n = 1, 2,\ldots$$

Thus it is readily shown, by induction, that if we order E-line $L-a_\infty$ from point a_0 to point a_1, then $m < n$ implies $a_m \prec a_n$ for any integers $m,n \geqslant 0$. We therefore have

THEOREM 42. *For any two distinct proper points p and q there exists just one natural net (a_n) such that $a_0 = p$ and $a_1 = q$. The values a_n of this net are proper points of the line $L = \mathbf{L}(pq)$. If a_∞ is the improper point of line L and E-line $L - a_\infty$ is ordered from p to q, then net (a_n) is an increasing sequence.*

We shall prove one more important property of the natural net:

THEOREM 43. *If a sequence (a_n) is a natural net, then for every natural number n point a_n is the midpoint of segment $a_0 a_{2n}$.*

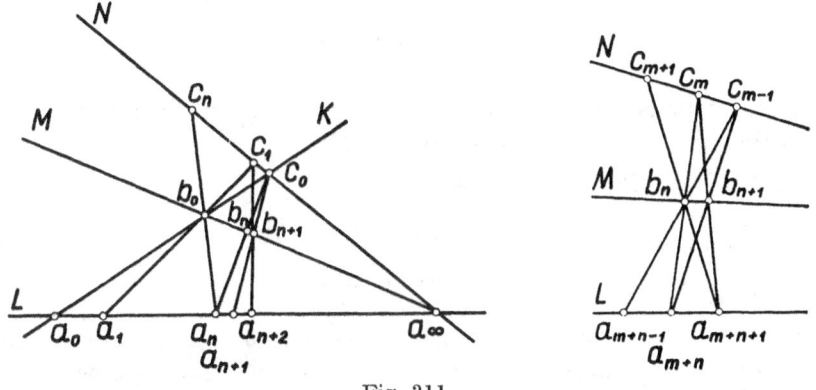

Fig. 311

PROOF. Let $L = \mathbf{L}(a_0 a_1)$ and let a_∞ be the improper point of line L (Fig. 311). Take any plane P including line L. In plane P we produce (basing ourselves on Theorem 3) through point a_0 any line K distinct from line L, and through point a_∞ any two distinct lines M and N, each of which is distinct from line L. Let

(1) $$b_0 = K \cap M, \quad c_0 = K \cap N.$$

We set

(2) $$b_n = \mathbf{L}(c_0 a_n) \cap M, \quad c_n = \mathbf{L}(b_0 a_n) \cap N$$
$$(n = 1, 2, \ldots).$$

From formulas (1) and (2) it follows at once that

(3) $$\text{points } c_0, b_n, a_n \text{ are collinear}$$
$$(n = 0, 1, 2, \ldots).$$

We shall show that also

(4) $$\text{points } c_1, b_n, a_{n+1} \text{ are collinear}$$
$$(n = 0, 1, 2, \ldots).$$

We employ induction with respect to n. The collinearity of points c_1, b_0, a_1

follows immediately from formula (2). Let us now assume that for some $n \geqslant 0$ points c_1, b_n, a_{n+1} are collinear. We consider quadrangle $c_0 b_{n+1} c_1 b_n$. With the help of the inductive promise and formulas (3) and (2), we obtain

$$\mathbf{L}(c_0 b_{n+1}) \cap \mathbf{L}(c_1 b_n) = a_{n+1}, \qquad \mathbf{L}(c_0 c_1) \cap \mathbf{L}(b_{n+1} b_n) = a_\infty,$$

$$L \cap \mathbf{L}(c_0 b_n) = a_n,$$

and, since from the definition of a net it follows that point a_{n+2} is the fourth harmonic for the triple of points (a_{n+1}, a_∞, a_n), we have

$$L \cap \mathbf{L}(b_{n+1} c_1) = a_{n+2}.$$

We thus conclude that points c_1, b_{n+1}, and a_{n+2} are collinear.
    In this way we have proved formula (4).
    We shall now prove that, generally,

(5)   points c_m, b_n, a_{m+n} are collinear

$$(m,n = 0, 1, 2, \ldots).$$

We employ induction with respect to m. By (3) and (4) the theorem holds for $m = 0$ and $m = 1$ (for every $n \geqslant 0$). We now assume that for some $m \geqslant 1$ points c_{m-1}, b_n, a_{m+n-1}, as well as points c_m, b_n, a_{m+n}, are collinear for every $n \geqslant 0$. With this assumption we shall show that points c_{m+1}, b_n, a_{m+n+1} are collinear for every $n \geqslant 0$. To do this, let us investigate for any $n \geqslant 0$ the quadrangle $a_{m+n} b_n a_{m+n+1} b_{n+1}$. Because of our inductive promise we have

$$\mathbf{L}(a_{m+n} b_n) \cap \mathbf{L}(a_{m+n+1} b_{n+1}) = c_m,$$

$$\mathbf{L}(a_{m+n} a_{m+n+1}) \cap \mathbf{L}(b_n b_{n+1}) = a_\infty,$$

$$N \cap \mathbf{L}(a_{m+n} b_{n+1}) = c_{m-1},$$

and, since for every $i \geqslant 0$ point c_i is the projection of point a_i from center $b_0 \sim \in L \cup N$ upon line N, then, by Theorem 38, point c_{m+1} is the fourth harmonic for the triple of points (c_m, a_∞, c_{m-1}), and therefore

$$N \cap \mathbf{L}(b_n a_{m+n+1}) = c_{m+1}.$$

Hence points c_{m+1}, b_n, and a_{m+n+1} are collinear. This completes the proof of formula (5).
    In order to show that point a_n is the center of points a_0 and a_{2n}, it is sufficient to note that, because of formulas (2) and (5), quadrangle $c_0 b_n c_n b_0$ realizes the harmonicity of the quadruple $(a_n, a_\infty, a_0, a_{2n})$.

23. Integral Net

By an *integral net* we shall understand every function f of the integral variable n, whose values $f(n)$ are proper points and such that for every integer n the point $f(n)$ is the midpoint of the segment $f(n-1)f(n+1)$.

We take two arbitrary distinct and proper points p and q. One may readily construct an integral net f such that $f(0) = p$ and $f(1) = q$. To do so, we take (employing Theorem 42) two natural nets (a_n) and (b_n) defined by the conditions

$$a_0 = p, \ a_1 = q \quad \text{and} \quad b_0 = q, \ b_1 = p,$$

and we define the function f in the following way:

$$f(n) = a_n \qquad \text{for} \quad n \geqslant 0,$$

$$f(n) = b_{-n+1} \quad \text{for} \quad n < 0.$$

Since, for every integer n, points $f(n-1), f(n)$, and $f(n+1)$ are three successive terms either of the natural net (a_n) or of the natural net (b_n), then point $f(n)$ is the midpoint of segment $f(n-1)f(n+1)$ and function f is an integral net.

Let us now assume that also for an integral net g we have $g(0) = p$ and $g(1) = q$. The sequences

$$f(0), f(1), \ldots, f(n), \ldots \quad \text{and} \quad f(1), f(0), \ldots, f(-n+1), \ldots$$

and

$$g(0), g(1), \ldots, g(n), \ldots \quad \text{and} \quad g(1), g(0), \ldots, g(-n+1), \ldots$$

are, as readily noted, natural nets. By Theorem 42, it thereby follows that $f(n) = g(n)$ for every integer n.

Hence the function f constructed above is the only integral net satisfying the conditions $f(0) = p$ and $f(1) = q$. From Theorem 42 it follows at once that all values $f(n)$ of net f are proper points of the proper line $L = \mathbf{L}(pq)$.

We denote by a_∞ the improper point of line L, and we order E-line $L - a_\infty$ from point p to point q. Then, by Theorem 42, for any natural numbers m and n it follows from $m < n$ that $a_m \prec a_n$ and $b_m \prec b_n$. It is thus readily concluded that for any integers m and n it follows from $m < n$ that $f(m) < f(n)$.

We therefore have

THEOREM 44. *For any two distinct proper points p and q there exists just one integral net f such that $f(0) = p$ and $f(1) = q$. The values $f(n)$ of this net are proper points of the line $L = \mathbf{L}(pq)$. If a_∞ is the improper* *point of line L and E-line L—a_∞ is ordered from p to q, then net f is an increasing function.*

We shall prove one more important property of the integral net.

THEOREM 45. *If f is an integral net, then for every two distinct integers m and n whose sum is even, $f\left(\dfrac{m+n}{2}\right)$ is the midpoint of segment f(m)f(n).*

PROOF. Assume that $m+n$ is an even number and let $m < n$. The sequence (c_k) defined by the formula

$$c_k = f(m + k) \ \text{ for } \ k = 0, 1, 2,\ldots$$

is a natural net. The number $n - m$ is positive and even, we may therefore represent it in the form $n - m = 2l$ where l is a natural number.

We now have

$$c_0 = f(m), \quad c_l = f\left(\frac{m+n}{2}\right), \quad c_{2l} = f(n).$$

Making use of Theorem 43, we conclude that $f\left(\dfrac{m+n}{2}\right)$ is the midpoint of segment $f(m)f(n)$.

24. Dyadic Net

As in Chapter II, by dyadic numbers we shall understand rational numbers of the form $n/2^k$, where n is any integer and k a non-negative integer. By a *dyadic net* we shall understand every function f of the dyadic variable w whose values $f(w)$ are proper points and such that for every two distinct dyadic numbers w_1 and w_2 the point $f\left(\dfrac{w_1 + w_2}{2}\right)$ is the midpoint of the segment $f(w_1)f(w_2)$.

Let us take two distinct proper points p and q. We construct a dyadic net f such that $f(0) = p$ and $f(1) = q$. To do this, we define a sequence of proper points

$$(q_k) \qquad\qquad q_0, q_1, \ldots, q_k, \ldots$$

in the following way: $q_0 = q$, and for $k \geqslant 0$ point q_{k+1} is the midpoint of segment pq_k. Employing Theorem 44 we now construct a sequence of functions

$$(f_k) \qquad\qquad f_0, f_1, \ldots, f_k, \ldots,$$

where f_k is the integral net defined by the conditions

$$f_k(0) = p \quad \text{and} \quad f_k(1) = q_k.$$

It is readily verified that

(1) $$f_k(n) = f_{k+1}(2n)$$

for every non-negative k and for every integer n. Indeed, the function f'_{k+1} defined by the formula

$$f'_{k+1}(n) = f_{k+1}(2n) \quad \text{for} \quad n = 1, 2, \ldots$$

is, by Theorem 45, an integral net, and from the definition of the sequences (q_k) and (f_k) it follows that

$$f'_{k+1}(0) = 0 \quad \text{and} \quad f'_{k+1}(1) = q_k.$$

Hence, by Theorem 44, $f_k = f'_{k+1}$.

Iterating formula (1) we obtain

(2) $$f_k(n) = f_{k+l}(2^l n)$$

for any non-negative integers k and l and for any integer n.

We now define the function f in the following way:

$$\text{if} \quad w = \frac{n}{2^k}, \quad \text{then} \quad f(w) = f_k(n).$$

Because of formula (2), the value $f(w)$ of function f does not depend on the choice of numbers n and k, but only on the dyadic number w.

It is verified directly that

$$f(0) = f_0(0) = p \quad \text{and} \quad f(1) = f_0(1) = q.$$

We shall now show that function f is a dyadic net. Indeed, any two distinct dyadic numbers w_1 and w_2 may be represented in the form

$$w_1 = \frac{2n_1}{2^k} \quad \text{and} \quad w_2 = \frac{2n_2}{2^k},$$

where n_1 and n_2 are two different integers and $k \geqslant 0$. Then

$$f(w_1) = f_k(2n_1), \; f(w_2) = f_k(2n_2), \; f\left(\frac{w_1 + w_2}{2}\right) = f_k(n_1 + n_2),$$

and, by Theorem 45, point $f\left(\dfrac{w_1 + w_2}{2}\right)$ is the midpoint of segment $f(w_1)f(w_2)$.

Let us now consider any dyadic net g such that $g(0) = p$ and $g(1) = q$. The dyadic nets f and g with domains restricted to the set of integers are integral nets, from which it follows that $f(n) = g(n)$ for every integer n. From the definition of the dyadic net we conclude, further, that, for any two different dyadic numbers w_1 and w_2, it follows from $f(w_1) = g(w_1)$ and $f(w_2) = g(w_2)$ that

$$f\left(\frac{w_1 + w_2}{2}\right) = g\left(\frac{w_1 + w_2}{2}\right),$$

from which we readily conclude that $f(w) = g(w)$ for every dyadic number w.

In this way we have shown that the function constructed above is the only dyadic net satisfying the conditions $f(0) = p$ and $f(1) = q$. Since all terms of the sequence (q_k) lie on the line $L = \mathbf{L}(pq)$, then from Theorem 44 it follows at once that all the values $f(w)$ of net f are proper points of line L. Let us denote by a_∞ the improper point of line L, and let us order E-line $L - a_\infty$ from point p to point q. We shall show that $w < v$ implies $f(w) \prec f(v)$ for any dyadic numbers w and v. Indeed, if $w < v$, then, having reduced w and v to a common denominator, we have for some $k \geqslant 0$ the equalities

$$w = \frac{m}{2^k} \quad \text{and} \quad v = \frac{n}{2^k},$$

where $m < n$. Then

$$f(w) = f_k(m) \quad \text{and} \quad f(v) = f_k(n),$$

and, since it may readily be shown by induction that E-half-line pq_k has the same orientation as E-half-line pq, then, by Theorem 44, we have $f(w) \prec f(v)$.

We therefore have

THEOREM 46. *For any two distinct proper points p and q there exists just one dyadic net f such that $f(0) = p$ and $f(1) = q$. The values $f(w)$ of net f are proper points of line $L = \mathbf{L}(pq)$. If a_∞ is the improper point of line L and E-line $L - a_\infty$ is ordered from p to q, then the net f is an increasing function.*

VIII

Axiom of Continuity

1. Axiom of Continuity

The last (third) group of axioms of projective geometry consists of the *Axiom of Continuity*.

AXIOM Co'. *Given two arbitrary non-empty point sets X and Y and a point c, if there exists a point a such that*

(1) $$p \in X \text{ and } q \in Y \qquad \text{implies } \mathbf{D}(a,q\,;p,c),$$

then there exists a point b such that

(2) $$p \in X - b \text{ and } q \in Y - b \text{ implies } \mathbf{D}(p,q\,;b,c).$$

From the Axiom of Continuity we at once obtain

THEOREM 1. *For any arbitrarily chosen plane at infinity P_∞ and for any two non-empty E-point sets X and Y, if there exists an E-point a such that*

(3) $$p \in X \text{ and } q \in Y \qquad \text{implies } \mathbf{B}(a,p,q),$$

then there exist an E-point b such that

(4) $$p \in X - b \text{ and } q \in Y - b \text{ implies } \mathbf{B}(p,b,q).$$

PROOF. If there exists an E-point a such that the condition (3) holds then sets X and Y an E-point a lie on a proper line L. Let c be the point at infinity of L. Then (3) implies (1). Thus, by Axiom Co', there exists a point b such that formula (2) is satisfied. Obviously b is an E-point and formula (2) implies formula (4).

By Theorem 1 model (E) satisfies the Axiom of Continuity of Euclidean geometry (see page 151). Thus Proposition 2 of Chapter VII (page 373) can be strengthened in the following way:

PROPOSITION 3. *If in space $\mathbf{S}$ we fix in any arbitrary way the plane at infinity P_∞, then in E-space $\mathbf{S} - P_\infty$ there are satisfied all the theorems of Euclidean geometry which are consequences of the axioms of incidence, order, and continuity only.*

In particular, in any arbitrary E-space $\mathbf{S} - P_\infty$ all the theorems of

Euclidean geometry from Section 1 of Chapter III are satisfied (see Note at the end of Section 1). Thus, for example, from Theorem 2 of Section 1 we conclude that every E-open E-segment is E-connected, and therefore is connected. Since any open segment I becomes, for the suitably chosen plane at infinity P_∞, an E-open E-segment, then we have

THEOREM 2. *Every open segment is a connected set.*

2. Dyadic Net (Conclusion)

In this and the succeeding sections we shall understand by the plane at infinity the plane Q_∞, fixed in Section 21 of Chapter VII, with which we introduced, in turn, the notion of the midpoint of a pair of proper points and the notions of the natural net, the integral net, and the dyadic net. We shall now prove the following important property of the dyadic net:

THEOREM 3. *If the values $f(w)$ of a dyadic net f are (proper) points of a proper line L with the point at infinity a_∞, then the range F of the net f is dense (= E-dense) on E-line $L-a_\infty$.*

PROOF. We order E-line $L-a_\infty$ from point $f(0)$ to point $f(1)$. Then it suffices to show that for any two points $a, b \in L-a_\infty$ such that $a \prec b$, there exists a dyadic number w satisfying the condition

(1) $a \prec f(w) \prec b$.

Let us suppose, on the contrary, that there exist two points a and b belonging to E-line $L-a_\infty$ such that $a \prec b$ and that condition (1) is satisfied for no dyadic number w. We then have one of the following three cases:

Case 1. For every dyadic number w we have $f(w) \precsim a$.
Case 2. For every dyadic number w we have $f(w) \succsim b$.
Case 3. There exist dyadic numbers w_1 and w_2 such that

$$f(w_1) \precsim a \prec b \precsim f(w_2).$$

In Case 1 we divide E-line $L-a_\infty$ into two sets X_1 and X_2 by including in set X_1 every point $p \in L-a_\infty$ such that for some dyadic number w we have $p \precsim f(w)$, and in set X_2 all the remaining points of E-line $L-a_\infty$. From the definition of set X_1 it follows at once that

(2) $F \subset X_1$.

The division $L-a_\infty = X_1 \cup X_2$ is a Dedekind cut, since $f(0) \in X_1, b \in X_2$, and every point of set X_1 precedes every point of set X_2. From Theorem 1 of Chapter III it therefore follows (through the intermediary of Proposition 3) that there exists a proper point c_0 determined by this cut, i.e. such that for every point $p_1 \in X_1$ and for every point $p_2 \in X_2$ we have

$$p_1 \leqq c_0 \leqq p_2.$$

Point c_0 is not a value of net f. For, if for some dyadic number w_0 we were to have $c_0 = f(w_0)$, then c_0 would have to be, by Theorem 46 of Chapter VII, less than $f(w)$ for any dyadic number $w > w_0$, in contradiction to formula (2). By Theorem 33 of Chapter I point c_0 is an E-point of accumulation of set X_1. Hence in any E-open E-segment $(rs) \subset L - a_\infty$ containing point c_0 (Fig. 312) there is a point p of set X_1. This point p

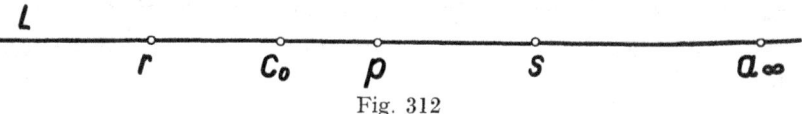

Fig. 312

precedes (or coincides with) a point $f(w)$ which, in turn, by formula (2), precedes point c_0. Hence $f(w) \in (rs)$. As a result, on E-line $L - a_\infty$, in any E-neighborhood of point c_0, there is a point $f(w) \neq c_0$, and consequently

(3) c_0 is an E-point of accumulation of set F.

Point c_0 is distinct from point $f(0)$. Let us now take a proper point d_0 such that point c_0 is the midpoint of segment $f(0)d_0$. From Theorem 40 of Chapter VII it follows that

(4) $f(0) \prec c_0 \prec d_0.$

E-half-line $U_0 = \mathbf{H}(c_0 d_0) \subset L - a_\infty$ is an E-neighborhood of point d_0. By Theorem 41 of Chapter VII, there exists an E-neighborhood U of point c_0 not containing point $f(0)$ and satisfying the condition that for every point $c \in U$ the point d such that c is the midpoint of segment $f(0)d$ belongs to E-neighborhood U_0 of point d_0. By formula (3), there exists a dyadic number w such that $f(w) \in U$. Hence point $f(2w)$ belongs to E-half-line $c_0 d_0$, which, by formula (4), gives $c_0 \prec f(2w)$ contrary to formula (2). Therefore Case 1 is impossible.

In Case 2 we divide E-line $L - a_\infty$ into sets X_1 and X_2 by including in set X_2 every point $p \in L - a_\infty$ such that for some dyadic number w we have $f(w) \leqq p$, and in set X_2—all the remaining points of E-line $L - a_\infty$. An argument quite similar to that used in Case 1 again leads us to a contradiction.

In Case 3 we effect two divisions of E-line $L - a_\infty$. First, we divide E-line $L - a_\infty$ into sets X_1 and X_2 by including in set X_1 every point $p \in L - a_\infty$ such that for some dyadic number w we have $p \leqq f(w) \leqq a$, and in set X_2 all the remaining points of E-line $L - a_\infty$. Then, we divide the E-line into sets X_1' and X_2' by including in set X_2' every point $p \in L-a_\infty$ such that for some dyadic number w we have $b \leq f(w) \leq p$, and in set X_1' all the remaining points of E-line $L-a_\infty$. It is readily verified that each of the divisions made is a Dedekind cut of E-line $L-a_\infty$. Let cut (X_1, X_2) determine the point $c_0 \in L-a_\infty$ and cut (X_1', X_2'), the point $c_0' \in L-a_\infty$. On the basis of Theorem 33 of Chapter I, we can readily show that

$$c_0 \leq a \prec b \leq c_0'.$$

From the definition of set X_1 it follows that if $c_0 \neq a$, then no point $f(w)$ of set F belongs to E-segment $(c_0 a)$; and from the definition of set X_2' it follows that if $c_0' \neq b$, then no point $f(w)$ of set F belongs to E-segment $\langle b c_0' \rangle$. Hence

(5) $(c_0 c_0') \cap F = 0.$

Further, proceeding similarly to Case 1, we show that

(6)   if $c_0 \sim \in F$, then c_0 is an E-point of accumulation of set F,

(7)   if $c_0' \sim \in F$, then c_0' is an E-point of accumulation of set F.

Now let d_0 denote the midpoint of segment $c_0 c_0'$. We then have

$$c_0 \prec d_0 \prec c_0'.$$

By Theorem 40 of Chapter VII, there exist for the E-neighborhood $U_0 = (c_0 c_0')$ of point d_0 two disjoint E-neighborhoods, U of point c_0 and U' of point c_0', such that for any two points $c \in U$ and $c' \in U'$ the midpoint of segment cc' belongs to E-neighborhood U_0. Because of formulas (6) and (7), there exist two dyadic numbers w and w' such that $f(w) \in U$ and $f(w') \in U'$. The point $f\left(\dfrac{w + w'}{2}\right)$ belongs to E-neighborhood U_0, in contradiction to formula (5). In this way we have shown that Case 3 also is impossible. This completes the proof of the theorem.

3. Real Net

By a *real net* we understand every continuous function f of a real variable x whose values $f(x)$ are proper points and such that for every two distinct real numbers x_1 and x_2 the point $f\left(\dfrac{x_1 + x_2}{2}\right)$ is the midpoint of segment $f(x_1)f(x_2)$.

Let us choose any two proper points a_0 and a_1. With the help of the Axiom of Continuity we shall construct a real net f such that $f(0) = a_0$

and $f(1) = a_1$. We proceed in the following manner: Basing ourselves on Theorem 46 of Chapter VII, we take a dyadic net f_0 determined by the conditions $f_0(0) = a_0$ and $f_0(1) = a_1$. Its values $f_0(w)$ are proper points of the proper line $L = \mathbf{L}(a_0 a_1)$. Let a_∞ be the improper point of line L. We order E-line $L - a_\infty$ from point a_0 to point a_1. We now correlate with every real number x the division of E-line $L - a_\infty$ into two sets $A_1(x)$ and $A_2(x)$, provided those points q for which there exist a dyadic number w such that

$$w \leqslant x \quad \text{and} \quad q \lneqq f_0(w),$$

belong to set $A_1(x)$, and the remaining points belong to set $A_2(x)$. It is readily noted, by making use of the fact that f is an increasing function (Theorem 46 of Chapter VII), that the sets $A_1(x)$ and $A_2(x)$ constitute a Dedekind cut of E-line $L - a_\infty$. Let $p \in L - a_\infty$ be the point determined by this cut. We put $f(x) = p$.

From the definition of the function f it readily follows that

(1)　　　　　　　$f(w) = f_0(w)$ for every dyadic number w.

Further, we have

(I) The values $f(x)$ of function f are points of E-line $L - a_\infty$, and $x_1 < x_2$ implies $f(x_1) \prec f(x_2)$.

Indeed, for any two real numbers $x_1 < x_2$ there exist two dyadic numbers w_1 and w_2 such that

$$x_1 < w_1 < w_2 < x_2,$$

which implies

$$f(x_1) \lneqq f_0(w_1) \prec f_0(w_2) \lneqq f(x_2).$$

We shall now show that the values of the function f fill the set $L - a_\infty$. To show this, given any point $p \in L - a_\infty$, we divide the set Θ of dyadic numbers into two sets $\Theta_1^{(p)}$ and $\Theta_2^{(p)}$ defined in the following way:

$$w \in \Theta_1^{(p)} \text{ if } f_0(w) \lneqq p, \quad \text{and} \quad w \in \Theta_2^{(p)} \text{ if } f_0(w) \succ p.$$

By Theorem 43 of Chapter I and Theorem 3 neither of the sets $\Theta_1^{(p)}$ and $\Theta_2^{(p)}$ is empty, and each number w_1 of set $\Theta_1^{(p)}$ is less than each number w_2 of set $\Theta_2^{(p)}$. Hence $(\Theta_1^{(p)}, \Theta_2^{(p)})$ is a Dedekind cut of set Θ, and consequently there exists just one real number x satisfying the condition

$$w_1 \leqslant x \leqslant w_2 \text{ for every } w_1 \in \Theta_1^{(p)} \text{ and for every } w_2 \in \Theta_2^{(p)}.$$

Basing ourselves on Theorem 3 we may readily show that $f(x) = p$.

Since, in addition, by (I), the function f maps different real numbers onto distinct points of E-line $L-a_\infty$, then:

(II) Function f establishes a one-to-one correspondence between real numbers and points of E-line $L-a_\infty$.

We shall now show that

(III) Function f is a real net, and $f(0) = a_0$, $f(1) = a_1$.

By (I), function f is monotonic. To prove the continuity of f it is therefore sufficient, by (II), to show that E-line $L-a_\infty$ is connected. This follows immediately from the fact that E-line $L-a_\infty$ is E-connected (see Theorem 3 from Chapter III). From (I) it also follows that the function f^{-1} is monotonic, which, along with the connectivity of the set of real numbers, gives the continuity of the function f^{-1}.

Further, from formula (1) we infer that $f(0) = a_0$ and $f(1) = a_1$.

Let us now consider two different numbers x_1 and x_2. There exist sequences of dyadic numbers $(w_{1,n})$ and $(w_{2,n})$ such that

$$\lim_{n \to \infty} w_{1,n} = x_1, \quad \lim_{n \to \infty} w_{2,n} = x_2,$$

and

$$w_{1,n} \neq w_{2,n} \text{ for } n = 0, 1,\ldots$$

Because of the continuity of function f we thereby obtain

$$\lim_{n \to \infty} f(w_{1,n}) = f(x_1) \text{ and } \lim_{n \to \infty} f(w_{2,n}) = f(x_2).$$

By applying formula (1) we conclude that the point $f\left(\dfrac{w_{1,n} + w_{2,n}}{2}\right)$ is the midpoint of segment $f(w_{1,n})f(w_{2,n})$. On the other hand, by (II), there exists a real number x_0 such that $f(x_0)$ is the midpoint of segment $f(x_1)f(x_2)$. Using Theorem 40 of Chapter VII, we thus conclude that

$$\lim_{n \to \infty} f\left(\frac{w_{1,n} + w_{2,n}}{2}\right) = f(x_0),$$

which, by the continuity of function f^{-1}, gives

$$\lim_{n \to \infty} \frac{w_{1,n} + w_{2,n}}{2} = x_0,$$

from which we obtain $x_0 = \dfrac{x_1 + x_2}{2}$. Hence $f\left(\dfrac{x_1 + x_2}{2}\right)$ is the midpoint of segment $f(x_1)f(x_2)$.

Let us now assume that a real net g also satisfies the conditions $g(0) = a_0$ and $g(1) = a_1$. The real nets f and g with domains restricted to the set of dyadic numbers are dyadic nets, from which, by Theorem 46 of Chapter VII, it follows that

(2) $f(w) = g(w)$ for every dyadic number w.

Let x be any real number. We pick a sequence of dyadic numbers (w_n) tending to x. Owing to the continuity of nets f and g we have

$$\lim_{n \to \infty} f(w_n) = f(x) \quad \text{and} \quad \lim_{n \to \infty} g(w_n) = g(x),$$

which, by (2), gives $f(x) = g(x)$ for every real number x. Hence $f = g$. This, together with (I) — (III), gives

THEOREM 4. *For any two distinct proper points a_0 and a_1 there exists just one real net f such that $f(0) = a_0$ and $f(1) = a_1$. The net f establishes a biunique and bicontinuous correspondence between the real numbers and the proper points of the line $L = \mathbf{L}(a_0 a_1)$. If a_∞ is the improper point of line L and E-line $L - a_\infty$ is ordered from a_0 to a_1, then net f is an increasing function.*

4. Cartesian Coordinates on a Line

Consider a proper line L with the point at infinity a_∞. We pick on L two distinct proper points a_0 and a_1; let f be the real net defined by the conditions $f(0) = a_0$ and $f(1) = a_1$. We shall call the function $\Phi = f^{-1}$ the *Cartesian coordinate system on line L with the pair of basic points* (a_0, a_1). Hence $\Phi(a_0) = 0$ and $\Phi(a_1) = 1$. By Theorem 4, the Cartesian coordinate system Φ assigns to every point $p \in L - a_\infty$ a real number $x^p = \Phi(p)$. We shall call this number x^p the *Cartesian coordinate* of point p in system Φ. From the properties of net f we obtain immediately the corresponding properties of coordinate system Φ.

THEOREM 5. *Any Cartesian coordinate system Φ on a proper line L with the point at infinity a_∞ establishes a biunique and bicontinuous correspondence between E-line $L - a_\infty$ and the Cartesian space $\mathbf{C}_1$. The system Φ maps the midpoint of any two distinct points $p, q \in L - a_\infty$ onto the Cartesian midpoint of the points $\Phi(p)$, $\Phi(q) \in \mathbf{C}_1$, and if E-line $L - a_\infty$ is ordered from point $\Phi^{-1}(0)$ to point $\Phi^{-1}(1)$, then system Φ is an increasing function.*

5. Cartesian Coordinates on a Plane

Consider a proper plane P with the line at infinity K_∞. We choose on P any two distinct proper lines K_1 and K_2 intersecting one another in a proper point a_0 (Fig. 313). We denote by $a_{\infty,i}$ (for $i = 1,2$) the improper point of line K_i. By Theorem 1 of Chapter VII, there exists a point

$c \in P - K_\infty - K_1 - K_2$. Let c_1 be the projection of point c from center $a_{\infty,2}$ upon line K_1, and c_2—the projection of point c from center $a_{\infty,1}$ upon line K_2. Then

$$c_1 = K_1 \cap \mathbf{L}(a_{\infty,2} c), \qquad c_2 = K_2 \cap \mathbf{L}(a_{\infty,1} c).$$

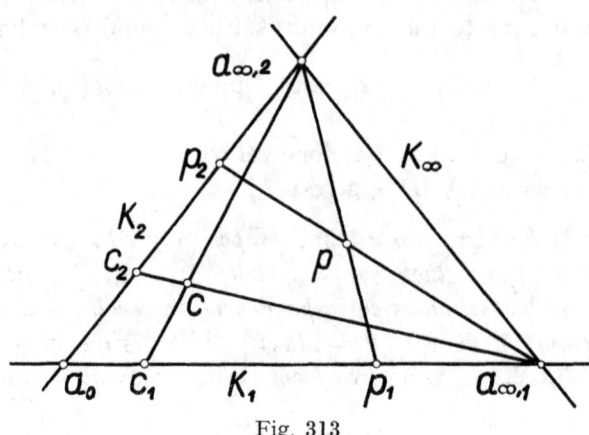

Fig. 313

Obviously, points c_1 and c_2 are both proper and distinct from point a_0. We fix on line K_i $(i = 1,2)$ the Cartesian coordinate system Φ_i with the pair of basic points (a_0, c_i).

Take any proper point p of plane P. Let us denote by p_1 its projection from center $a_{\infty,2}$ upon line K_1, by p_2 its projection from center $a_{\infty,1}$ upon line K_2, and let

$$x_i^p = \Phi_i(p_i) \text{ for } i = 1, 2.$$

The real numbers x_1^p and x_2^p correlated in this way with the point $p \in P - K_\infty$ will be called the *first and second Cartesian coordinates* of point p, and the function Φ defined by the condition

$$\Phi(p) = (x_1^p, x_2^p)$$

will be called the *Cartesian coordinate system on plane P with the pair of axes†* (K_1, K_2) *and with the basic point c.*

For a fixed line at infinity K_∞ the coordinate system Φ is uniquely determined by the pair of its axes (K_1, K_2) and by the basic point c.

The coordinate system Φ correlates with every proper point $p \in P$ an ordered pair of real numbers (x_1^p, x_2^p). Conversely, if an arbitrary pair of real numbers (x_1, x_2) is given, then there exists on plane P just one proper point p such that $\Phi(p) = (x_1, x_2)$. For, by Theorem 5, there exists just

---

† We have used here an imprecise terminology. Proper axes are E-lines $K_1 - a_{\infty,1}$ and $K_2 - a_{\infty,2}$ oriented by half-lines $a_0 c_1$ and $a_0 c_2$, respectively.

one point $p_i \in L_i - a_\infty$ (for $i = 1,2$) with the coordinate $\Phi_i(p_i) = x_i$, and since points $a_{\infty,1}$ and $a_{\infty,2}$ are distinct, then lines $L(a_{\infty,2}p_1)$ and $L(a_{\infty,1}p_2)$ intersect in just one proper point p. Hence the coordinate system Φ establishes a one-to-one correspondence between points of E-plane $P - K_\infty$ and pairs of real numbers. We shall show that the correspondence Φ is bicontinuous.

Since the first coordinate $x_1^p = \Phi_1(p_1)$ of point p is, by Theorem 5, a continuous function of point p_1, and point p_1, as the projection of point p from center $a_{\infty,2}$ upon line K_1, is, by Theorem 39 of Chapter VII, a continuous function of point p, then, indirectly, the coordinate x_1^p is a continuous function of point p. In a similar way it can be proved that the second coordinate x_2^p of point p is a continuous function of point p. Hence Φ is continuous.

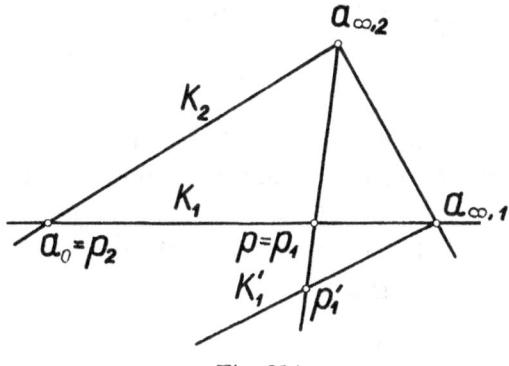

Fig. 314

In order to prove that the function Φ^{-1} is continuous, we shall first of all show that point p is a continuous function of its projections p_1 and p_2.

This is obvious if $p \in P - (K_\infty \cup K_1 \cup K_2)$, since p is then a diagonal point of the quadrangle $a_{\infty,2}p_1 a_{\infty,1}p_2$. Let us now assume that $p \in K_1 - a_0$ (Fig. 314); then $p_1 = p$ and $p_2 = a_0$. We fix any proper line K_1' distinct from line K_1 and passing through the point at infinity $a_{\infty,1}$; let $p_1' = L(a_{\infty,2}p) \cap K_1'$. Point p, as a diagonal point of quadrangle $a_{\infty,2}p_1' a_{\infty,1}p_2$ is a continuous function of points p_1' and p_2, and point p_1', as the projection of point p from center $a_{\infty,2}$ upon line K_1' is a continuous function of point p_1. Hence, indirectly, point p is a continuous function of points p_1 and p_2. In a similar manner we prove this for points $p \in K_2 - a_0$. Let us assume, finally, that $p = a_0$ (Fig. 315). Then $p = p_1 = p_2$. We fix (for $i = 1,2$) any proper line $K_i' \neq K_i$ passing through the point at infinity $a_{\infty,i}$. We set $p_1' = L(a_{\infty,2}p) \cap K_1'$ and $p_2' = L(a_{\infty,1}p) \cap K_2'$. Then point p is a continuous function of points p_1' and p_2', point p_i' (for $i = 1,2$)

is a continuous function of point p_i, and therefore, indirectly, point p is a continuous function of points p_1 and p_2.

By Theorem 5, point p_i (for $i = 1,2$) is a continuous function of the coordinate x_i^p, and hence, indirectly, point p is a continuous function of its coordinaties x_1^p and x_2^p, that is, Φ^{-1} is a continuous function.

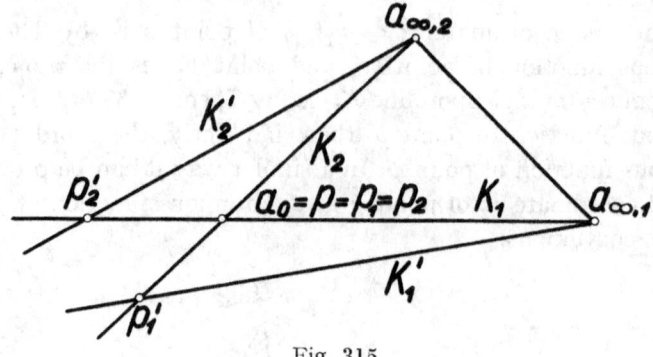

Fig. 315

In conclusion, we shall establish a relation between the coordinates of two distinct proper points p and q and the coordinates of their midpoint r. If $x_1^p = x_1^q$, then line pq passes through point $a_{\infty,2}$, and consequently all points of line pq, in particular point r, have the first coordinate equal

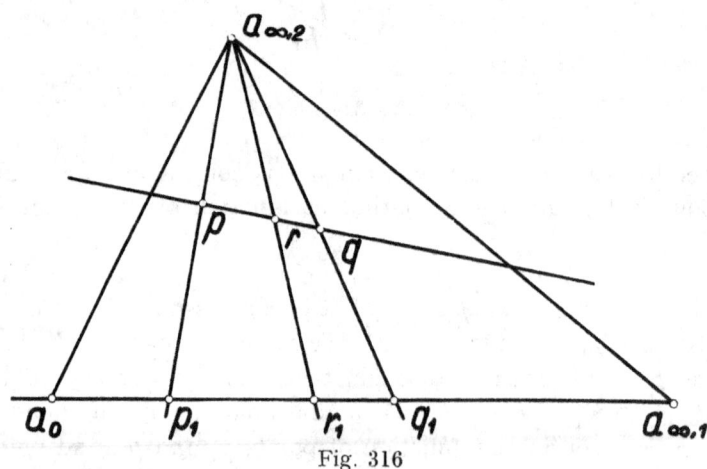

Fig. 316

to $x_1^p = \frac{1}{2}(x_1^p + x_1^q)$. Let us next assume that $x_1^p \neq x_1^q$. Then $a_{\infty,2} \sim \in L(pq)$ (Fig. 316). We denote by p_1, q_1, r_1, respectively, the projections of points p, q, r from center $a_{\infty,2}$ upon line K_1. Then $p_1 \neq q_1$ and, by Theorem 38 of Chapter VII, point r_1 is the midpoint of segment p_1q_1. Hence, by Theorem 5, we have $x_1^r = \frac{1}{2}(x_1^p + x_1^q)$. In a similar way, it is proved that $x_2^r = \frac{1}{2}(x_2^p + x_2^q)$.

Thus we have

THEOREM 6. *Any Cartesian coordinate system Φ on a proper plane P with the line at infinity K_∞ establishes a biunique and bicontinuous correspondence between E-plane P and the two-dimensional Cartesian space $\mathbf{C}_2$. The system Φ maps the midpoint of any two distinct points $p,q \in P - K_\infty$ onto the Cartesian midpoint of the points $\Phi(p),\Phi(q) \in \mathbf{C}_2$.*

6. Equation of the Set of Proper Points of a Proper Line on a Plane

Consider a proper plane P and on it a Cartesian coordinate system Φ. Let $L \subset P$ be any proper line on plane P. We choose on line L two distinct proper points a and b; let Ψ be the Cartesian coordinate system on line L with the pair of basic points (a,b). We shall now investigate the relation which occurs, for a proper point $p \in L$, between its Cartesian coordinate x in system Ψ and its pair of Cartesian coordinates x_1^p and x_2^p in system Φ. To do this, we consider the functions

(1) $$x_i(t) = (1-t)x_i^a + t x_i^b \quad \text{for} \quad i = 1, 2,$$

and

(2) $$f(t) = \Phi^{-1}(x_1(t), x_2(t)).$$

We shall show that

(3) $$\Psi^{-1}(t) = f(t) \quad \text{for every real } t;$$

in other words we shall show that if a proper point p of line L has in system Ψ the coordinate t, then it has in system Φ (as a proper point of plane P) the coordinates $x_1(t)$ and $x_2(t)$.

From the definition of the Cartesian coordinate system on a line it follows that the function Φ^{-1} is a real net and $\Phi^{-1}(0) = a$, $\Psi^{-1}(1) = b$. On the other hand, we find, directly, that $f(0) = a$ and $f(1) = b$. Hence, by Theorem 4, in order to prove equality (3) it is sufficient to show that function f is also a real net. From Theorem 6 it follows that function f in a continuous way maps real numbers onto proper points of plane P.

We shall now show that for any two different real numbers t_1 and t_2 the point $f\left(\dfrac{t_1 + t_2}{2}\right)$ is the midpoint of segment $f(t_1)f(t_2)$. Indeed, by Theorem 6, the ith coordinate $(i = 1, 2)$ of the midpoint of segment $f(t_1)f(t_2)$ is identical with the number $\frac{1}{2}(x_i(t_1) + x_i(t_2))$. Then

$$\tfrac{1}{2}(x_i(t_1) + x_i(t_2)) = \tfrac{1}{2}((1-t_1)x_i^a + t_1 x_i^b + (1-t_2)x_i^a + t_2 x_i^b) =$$

$$= \left(1 - \frac{t_1 + t_2}{2}\right)x_i^a + \frac{t_1 + t_2}{2}x_i^b = x_i\left(\frac{t_1 + t_2}{2}\right).$$

In this way we have proved formula (3).

The points $\Psi^{-1}(t)$, for t running over the real numbers, form the set of proper points of line $L = \mathbf{L}(ab)$. On the other hand, the pairs of real numbers $\Phi\Psi^{-1}(t)$, for t running over the real numbers, form, by formulas (1)–(3), the Cartesian line determined in Cartesian space $\mathbf{C}_2$ by the points $\Phi(a) = (x_1^a, x_2^a)$ and $\Phi(b) = (x_1^b, x_2^b)$. We therefore have

THEOREM 7. *Any Cartesian coordinate system Φ on a proper plane P maps the set of proper points of any proper line $L \subset P$ onto a line of Cartesian space $\mathbf{C}_2$.*

7. Cartesian Coordinates in Space

In space $\mathbf{S}$ we choose three proper lines K_1, K_2, K_3 intersecting one another in a proper point a_0 and not lying in one plane (Fig. 317). We

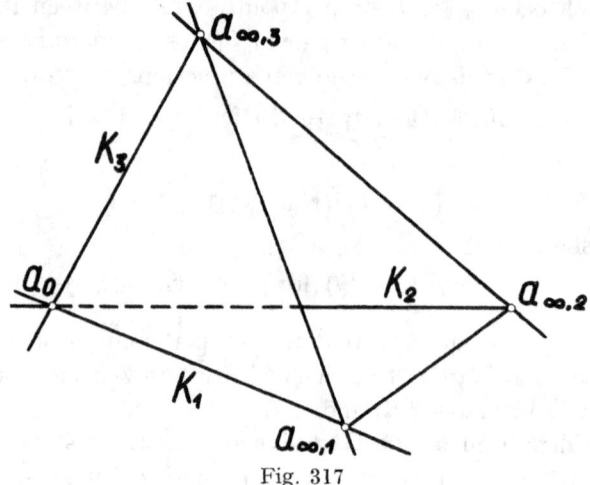

Fig. 317

denote by $a_{\infty,i}$ (for $i = 1,2,3$) the improper point of line K_i. It is clear that points $a_{\infty,1}, a_{\infty,2}, a_{\infty,3}$ are non-collinear. We put $P_{i,j} = \mathbf{P}(K_i K_j)$ for $1 \leqslant i < j \leqslant 3$. By Theorem 2 of Chapter VII, there exists a point

$$(1) \qquad\qquad c \in \mathbf{S} - Q_\infty - (P_{1,2} \cup P_{1,3} \cup P_{2,3}).$$

For any proper point p we now define the proper planes

$$Q_{p,1} = \mathbf{P}(p a_{\infty,2} a_{\infty,3}), \qquad Q_{p,2} = \mathbf{P}(p a_{\infty,1} a_{\infty,3}),$$

$$Q_{p,3} = \mathbf{P}(p a_{\infty,1} a_{\infty,2}).$$

Since the improper point $a_{\infty,i}$ of line K_i ($i = 1,2,3$) does not lie on the improper line of plane $Q_{p,i}$ then $K_i \sim \subset Q_{p,i}$ and the intersection point of line K_i and plane $Q_{p,i}$ is a proper point. Let, in particular,

$$c_i = K_i \cap Q_{c,i} \quad \text{for } i = 1, 2, 3.$$

From formula (1) it follows that $c_i \neq a_0$. We fix on line K_i the Cartesian coordinate system Φ_i with the pair of basic points (a_0, c_i). Let point p be any proper point of space $\mathbf{S}$. We put

$$p_i = K_i \cap Q_{p,i} \text{ for } i = 1, 2, 3$$

and

$$z_i^p = \Phi_i(p_i) \text{ for } i = 1, 2, 3.$$

We call the real numbers x_1^p, x_2^p, x_3^p correlated in this way with the point $p \in \mathbf{S} - Q_\infty$ the *first, second, and third Cartesian coordinates* of the point p, and we call the function Φ defined by the condition

$$\Phi(p) = (x_1^p, x_2^p, x_3^p)$$

the *Cartesian coordinate system in space* $\mathbf{S}$ *with the triple of axes*† (K_1, K_2, K_3) *and the basic point c*. For the fixed plane at infinity Q_∞ the coordinate system Φ is uniquely determined by its triple of axes (K_1, K_2, K_3) and by the basic point c.

The coordinate system Φ correlates with every proper point $p \in \mathbf{S}$ an ordered triple of real numbers (x_1^p, x_2^p, x_3^p). Conversely, if any triple of real numbers (x_1, x_2, x_3) is given, then there exists just one proper point p of space $\mathbf{S}$ such that $\Phi(p) = (x_1, x_2^!, x_3)$. For, by Theorem 5, there exists just one point $p_i \in K_i - a_{\infty,i}$ (for $i = 1, 2, 3$) with the coordinate $\Phi_i(p_i) = x_i$, and the planes

$$\mathbf{P}(p_1 a_{\infty,2} a_{\infty,3}), \quad \mathbf{P}(p_2 a_{\infty,1} a_{\infty,3}), \quad \mathbf{P}(p_3 a_{\infty,1} a_{\infty,2})$$

are readily seen to have just one point, and in fact a proper point, in common. We therefore have

THEOREM 8. *Any Cartesian coordinate system* Φ *in space* $\mathbf{S}$ *establishes a one-to-one correspondence between the proper points of space* $\mathbf{S}$ *and the points of three-dimensional Cartesian space* $\mathbf{C_3}$.

From the definition of the Cartesian coordinate system Φ in space $\mathbf{S}$ it follows immediately that for any proper point p

(2)   if $\Phi(p) = (x_1^p, x_2^p, x_3^p)$, then $\Phi_i(p_i) = x_i^p$ for $i = 1, 2, 3$.

Now let the triple of numbers (i, j, k) be any permutation of the triple of indices $(1, 2, 3)$ satisfying the condition $i < j$. For any proper point p we denote by $p_{i,j}$ its projection from center $a_{\infty,k}$ upon plane $P_{i,j}$. We therefore have

(3) $$p_{i,j} = P_{i,j} \cap \mathbf{L}(a_{\infty,k} \, p)$$

† See footnote on page 402.

and, in particular, $c_{i,j} = P_{i,j} \cap L(a_{\infty,k}\, c)$. Thus from formula (1) it follows that $c_{i,j} \in P_{i,j} - L(a_{\infty,i}\, a_{\infty,j}) - K_i - K_j$.

We denote by $\Phi_{i,j}$ the Cartesian coordinate system fixed on plane $P_{i,j}$ by the pair of axes (K_i, K_j) and the basic point $c_{i,j}$. We shall show that for any proper point p

(4)   if $\Phi(p) = (x_1^p, x_2^p, x_3^p)$, then $\Phi_{i,j}(p_{i,j}) = (x_i^p, x_j^p)$ for $1 \leqslant i < j \leqslant 3$.

Indeed, we obtain formula (4) at once from the following equalities which result from formula (3):

$$K_i \cap P(a_{\infty,j}\, a_{\infty,k}\, p) = K_i \cap P_{i,j} \cap P(a_{\infty,j}\, a_{\infty,k}\, p) = K_i \cap L(a_{\infty,j}\, p_{i,j})$$

and

$$K_j \cap P(a_{\infty,i}\, a_{\infty,k}\, p) = K_j \cap P_{i,j} \cap P(a_{\infty,i}\, a_{\infty,k}\, p) = K_j \cap P(a_{\infty,i}\, p_{i,j}).$$

These equalities hold for any proper point p and, in particular, for basic point c.

8. Equation of the Set of Proper Points of a Proper Line in Space

In space S we fix a Cartesian coordinate system Φ with axes K_1, K_2, K_3 and with basic point c. We keep the same notation as in the preceding section. Let $L \subset S$ be any proper line with the point at infinity p_∞. Point p_∞ does not lie on at least one of the lines $a_{\infty,1}a_{\infty,2}$, $a_{\infty,1}a_{\infty,3}$, $a_{\infty,2}a_{\infty,3}$, say line $a_{\infty,1}a_{\infty,2}$. Then, for any two proper points $q, r \in L$,

(1) $q \neq r$ implies $x_3^q \neq x_3^r$.

Take on line L any two distinct proper points a and b. Let

$$L_{1,3} = L(a_{1,3}b_{1,3}) \quad \text{and} \quad L_{2,3} = L(a_{2,3}b_{2,3}).$$

We take any proper point p of line L. By Theorem 14 of Chapter VII, we have $p_{1,3} \in L_{1,3}$ and $p_{2,3} \in L_{2,3}$. Employing formula (4) from the preceding section and Theorem 7, we conclude that for some two real numbers t_1 and t_2

$$x_1^p = (1-t_1)x_1^a + t_1 x_1^b \quad \text{and} \quad x_3^p = (1-t_1)x_3^a + t_1 x_3^b$$

and

$$x_2^p = (1-t_2)x_2^a + t_2 x_2^b \quad \text{and} \quad x_3^p = (1-t_2)x_3^a + t_2 x_3^b.$$

From the above we have

$$(1-t_1)x_3^a + t_1 x_3^b = (1-t_2)x_3^a + t_2 x_3^b, \text{ that is, } (t_2-t_1)x_3^a = (t_2-t_1)x_3^b.$$

Since, by formula (1), we have $x_3^a \neq x_3^b$, then $t_2 = t_1$. Thus we have shown that for any proper point $p \in L$ there exists a real number t $(t = t_1 = t_2)$ such that

(2) $$x_i^p = (1-t)x_i^a + tx_i^b \quad \text{for} \quad i = 1, 2, 3.$$

Conversely, let us assume that the coordinates x_1^p, x_2^p, x_3^p of some point p are given by formula (2), where t is a real number. By Theorem 7 and formula (4) of Section 7, we have $p_{1,3} \in L_{1,3}$ and $p_{1,2} \in L_{1,2}$, and therefore, by Theorem 14 of Chapter VII, point $p_{1,3}$ is the projection from center $a_{\infty,2}$ upon plane $P_{1,3}$ of some proper point $q \in L$, and point $p_{2,3}$ is the projection from center $a_{\infty,1}$ upon plane $P_{2,3}$ of some proper point $r \in L$. We therefore have

(3) $$x_1^q = x_1^p \quad \text{and} \quad x_3^q = x_3^p$$

and

(4) $$x_2^r = x_2^p \quad \text{and} \quad x_3^r = x_3^p,$$

as a result of which $x_3^q = x_3^r$. From this it follows, by formulas (1), (3), and (4), that $q = r = p$, and consequently $p \in L$.

Points $x^p = (x_1^p, x_2^p, x_3^p) \in \mathbf{C}_3$, where coordinate x_i^p is defined by formula (2), constitute, for t running over the real numbers, a line of Cartesian space $\mathbf{C}_3$ (see Chapter IV, Section 2). From Theorem 8 it follows that by a suitable choice of proper points a and b, (and hence, indirectly, of line $L = \mathbf{L}(ab)$), every line of Cartesian space $\mathbf{C}_3$ may be obtained in this way.

We therefore have

THEOREM 9. *Any Cartesian coordinate system Φ in space* **S** *maps the sets of proper points of proper lines of* **S** *onto lines of Cartesian space* $\mathbf{C}_3$.

9. Equation of the Set of Proper Points of a Proper Plane in Space

Consider in space **S** a Cartesian coordinate system Φ. Let $P \subset \mathbf{S}$ be any proper plane. We choose on plane P three non-collinear proper points a, b, c; let

$$\Phi(a) = (x_1^a, x_2^a, x_3^a), \quad \Phi(b) = (x_1^b, x_2^b, x_3^b), \quad \Phi(c) = (x_1^c, x_2^c, x_3^c).$$

For any real numbers λ, μ, ν we denote by $f(\lambda, \mu, \nu)$ the point $p = \Phi^{-1}(x_1^p, x_2^p, x_3^p)$, where for $i = 1, 2, 3$

$$x_i^p = \lambda x_i^a + \mu x_i^b + \nu x_i^c.$$

Let F be the set of points $f(\lambda, \mu, \nu)$ for $\lambda + \mu + \nu = 1$. Thus, set F is identical with that subset of space **S** whose Φ-image coincides with the Cartesian plane determined in space $\mathbf{C}_3$ by Cartesian points $\Phi(a), \Phi(b), \Phi(c)$ (see Chapter IV, Section 2).

We shall show that every proper point of plane P belongs to set F. Indeed, if p is a proper point of line ab, then, according to Theorem 9, there exists a real number t such that

$$x_i^p = (1 - t)x_i^a + tx_i^b \quad \text{for } i = 1, 2, 3,$$

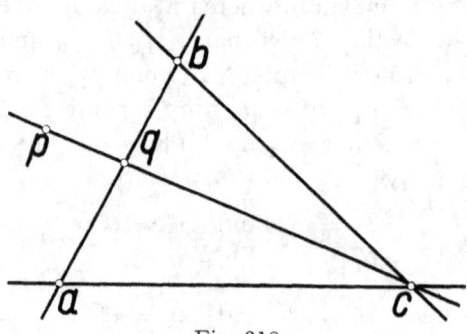

Fig. 318

and consequently $p = f(1-t, t, 0) \in F$. We proceed similarly when p is a proper point of line ac or of line bc. Let us next assume that p is a proper point of plane P not lying on any of the lines ab, bc, ac (Fig. 318). Then points a, b, c, p are vertices of quadrangle $abcp$, whose diagonal points

$$\mathbf{L}(ab) \cap \mathbf{L}(cp), \quad \mathbf{L}(ac) \cap \mathbf{L}(bp), \quad \mathbf{L}(ap) \cap \mathbf{L}(bc)$$

are, by Theorem 31 of Chapter VII, non-collinear. Hence at least one of them must be a proper point. Let, for example, $q = \mathbf{L}(ab) \cap \mathbf{L}(cp)$ be a proper point. Then there exists real numbers t and u such that

$$x_i^q = (1-t)x_i^a + tx_i^b \qquad (i = 1, 2, 3)$$
and
$$x_i^p = (1-u)x_i^c + ux_i^q \qquad (i = 1, 2, 3),$$

and consequently $p = f(u(1-t), ut, 1-u) \in F$. In this way, we have shown that the set of proper points of plane P is included in set F.

We shall now show that every proper point of set F is a proper point of plane P. Thus, let $p = f(\lambda, \mu, \nu)$, where $\lambda + \mu + \nu = 1$. One of the numbers λ, μ, ν, say ν, is different from 1. By Theorem 9, the point

$$q = f\left(\frac{\lambda}{1-\nu}, \frac{\mu}{1-\nu}, 0\right)$$

is a proper point of line ab included in plane P. Further, for $i = 1, 2, 3$, we have

$$x_i^p = (1-\nu)\left(\frac{\lambda}{1-\nu}x_i^a + \frac{\mu}{1-\nu}x_i^b\right) + \nu x_i^c = (1-\nu)x_i^q + \nu x_i^c.$$

Hence point p is a proper point of line qc included in plane P. In this way, we have shown that every point p of set F is a proper point of plane P.

Hence we have shown that the set of proper points of plane P coincides with set F.

The set $\Phi(F)$ constitutes a plane of Cartesian space $\mathbf{C}_3$ (see Chapter IV, Section 2). From Theorem 8 it follows that by a suitable choice of non-collinear proper points a, b, c (and hence, indirectly, of plane $P = \mathbf{P}(abc)$) every plane of Cartesian space $\mathbf{C}_3$ may be obtained in this way.

We therefore have

THEOREM 10. *Any Cartesian coordinate system Φ in space $\mathbf{S}$ maps the sets of proper points of proper planes of $\mathbf{S}$ on to planes of Cartesian space $\mathbf{C}_3$.*

10. Projective Coordinates in Space

Let us fix in space $\mathbf{S}$ a Cartesian coordinate system Φ with basic point c and with axes K_1, K_2, K_3 intersecting in a proper point a_0.

Let p be any proper point of space $\mathbf{S}$ and let $\Phi(p) = (x_1^p, x_2^p, x_3^p)$. Every quadruple of real numbers (x_0, x_1, x_2, x_3) proportional to the quadruple $(1, x_1^p, x_2^p, x_3^p)$, that is, such that for some real $\lambda \neq 0$ we have $x_0 = \lambda$ and $x_i = \lambda x_i^p$ for $i = 1, 2, 3$, will be called a *projective coordinate system of proper point p.*

Now let $p = p_\infty$ be any improper point of space $\mathbf{S}$. We produce through p_∞ any proper line M and we pick on it any two distinct proper points a and b. Let $\Phi(a) = (x_1^a, x_2^a, x_3^a)$ and $\Phi(b) = (x_1^b, x_2^b, x_3^b)$. Every quadruple of real numbers (x_0, x_1, x_2, x_3) proportional to the quadruple

$$(0, x_1^a - x_1^b, x_2^a - x_2^b, x_3^a - x_3^b),$$

that is, such that $x_0 = 0$ and $x_i = \lambda(x_i^a - x_i^b)$ for some real $\lambda \neq 0$ and $i = 1, 2, 3$, will be called a *projective coordinate system of improper point p.*

We shall show that this definition does not depend on the choice of line M and of points a and b. To do this, we choose any line N passing through point p_∞. If $M = N$, then Cartesian lines $\Phi(M - p_\infty)$ and $\Phi(N - p_\infty)$ (see Theorem 9) are identical. If, however, $M \neq N$, then $M \cap N = p_\infty$, and, by Theorems 10 and 9, the coordinate system Φ maps the set of proper points of plane MN onto some plane of Cartesian space $\mathbf{C}_3$, and the sets of proper points of lines M and N onto disjoint lines $\Phi(M - p_\infty)$ and $\Phi(N - p_\infty)$ of this plane. Hence in each of these two cases, for any two distinct proper points $c = \Phi^{-1}(x_1^c, x_2^c, x_3^c)$ and $d = \Phi^{-1}(x_1^d, x_2^d, x_3^d)$ of line N, the triple of numbers $(x_1^c - x_1^d, x_2^c - x_2^d, x_3^c - x_3^d)$ is proportional to the triple of numbers $(x_1^a - x_1^b, x_2^a - x_2^b, x_3^a - x_3^b)$ and therefore the quadruple $(0, x_1^c-x_1^d, x_2^c-x_2^d, x_3^c-x_3^d)$ is proportional to the quadruple $(0, x_1^a-x_1^b, x_2^a-x_2^b, x_3^a-x_3^b)$.

The function Ψ assigning to every (proper or improper) point p of space S the class of all systems (x_0, x_1, x_2, x_3) of the projective coordinates of p will be called the *system of projective coordinates in space S with the triple of axes (K_1, K_2, K_3) and with basic point c*. For the fixed plane at infinity Q_∞ the system Ψ is uniquely determined by the triple of axes (K_1, K_2, K_3) and by the basic point c.

The projective coordinate system Ψ maps the points of space S onto the classes of quadruples of real numbers $(x_0, x_1, x_2, x_3) \neq (0,0,0,0)$ such that two quadruples belong to one class if and only if they are proportional. In other words, system Ψ maps the points p of S onto the points $\Psi(p)$ of three-dimensional projective space P_3 (see Chapter IV, Section 6). We shall now show that $p \neq q$ implies $\Psi(p) \neq \Psi(q)$ for any points p and q in S. This is seen at once in case one of the two points p and q is proper, and the other improper, and it follows immediately from Theorem 8 if both points p and q are proper. If two distinct points $p = p_\infty$ and $q = q_\infty$ are both improper, then lines $L_1 = L(a_0 p_\infty)$ and $L_2 = L(a_0 q_\infty)$ are distinct, and hence sets L_1-p_∞ and L_2-q_∞ are also distinct; consequently, system Φ maps them onto two distinct Cartesian lines $\Phi(L_1-p_\infty)$ and $\Phi(L_2-q_\infty)$ intersecting in point $\Phi(a_0) = (0,0,0)$. If, therefore, $a \in L_1-p_\infty-a_0$, $b \in L_2-q_\infty-a_0$ and $\Phi(a) = (x_1^a, x_2^a, x_3^a)$, $\Phi(b) = (x_1^b, x_2^b, x_3^b)$, then the triples of numbers (x_1^a, x_2^a, x_3^a) and (x_1^b, x_2^b, x_3^b) are not proportional, and therefore the quadruples $(0, x_1^a, x_2^a, x_3^a) \in \Phi(p)$ and $(0, x_1^b, x_2^b, x_3^b) \in \Psi(q)$ also are not proportional. Hence $\Psi(p) \neq \Psi(q)$.

In conclusion, we show that every point x of projective space P_3 is a Ψ-image of some point p of space S. Indeed, if point x contains a quadruple of numbers of the form $(1, x_1, x_2, x_3)$ then $p = \Phi^{-1}(x_1, x_2, x_3)$, and if x contains a quadruple of numbers of the form $(0, x_1, x_2, x_3)$, then p is the improper point of the line $a_0 \Phi^{-1}(x_1, x_2, x_3)$.

In this way we have proved

THEOREM 11. *Any projective coordinate system Ψ in space S establishes a one-to-one correspondence between space S and three-dimensional projective space P_3.*

11. Equations of the Plane and of the Line in Space

Consider in space S a projective coordinate system Ψ. Let P be any plane. We shall prove that there exist real numbers $\alpha_0, \alpha_1, \alpha_2, \alpha_3$, not all equal to zero, such that plane P coincides with the set of those points p whose coordinates $x_0^p, x_1^p, x_2^p, x_3^p$ satisfy the equation

(1) $$\alpha_0 x_0 + \alpha_1 x_1 + \alpha_2 x_2 + \alpha_3 x_3 = 0.$$

If P is the improper plane, it is sufficient to take $\alpha_0 = 1$ and $\alpha_1 = \alpha_2 = \alpha_3 = 0$. We then obtain the equation $x_0 = 0$, which is the equation of the improper plane of projective space $\boldsymbol{P}_3$ (see Chapter IV, Section 6).

We next assume that P is a proper plane. Let Φ be the Cartesian coordinate system in space $\boldsymbol{S}$ with the same triple of axes and with the same basic point as system Ψ. By Theorem 10, there exist real numbers $\alpha_0, \alpha_1, \alpha_2, \alpha_3$, not all equal to zero, such that the set of proper points of plane P coincides with the set of those proper points $p \in \boldsymbol{S}$ whose Cartesian coordinates x_1^p, x_2^p, x_3^p (in system Φ) satisfy the equation

$$\alpha_0 + \alpha_1 x_1 + \alpha_2 x_2 + \alpha_3 x_3 = 0,$$

that is, whose projective coordinates (in system Ψ) $\lambda, \lambda x_1^p, \lambda x_2^p, \lambda x_3^p$, where $\lambda \neq 0$, satisfy equation (1).

We shall now show that the projective coordinates of an improper point $p \in \boldsymbol{S}$ satisfy equation (1) if and only if $p \in P$. To do this, we take any proper point $a \in P$ with projective coordinates $1, x_1^a, x_2^a, x_3^a$. Then

$$(2) \qquad \alpha_0 + \alpha_1 x_1^a + \alpha_2 x_2^a + \alpha_3 x_3^a = 0.$$

Let p be any improper point. We choose a proper point $b \neq a$ on line ap with projective coordinates $1, x_1^b, x_2^b, x_3^b$. If $p \in P$, then $b \in P$, and consequently

$$(3) \qquad \alpha_0 + \alpha_1 x_1^b + \alpha_2 x_2^b + \alpha_3 x_3^b = 0.$$

Subtracting equality (3) from equality (2), we see that the numbers $0, x_1^a - x_1^b, x_2^a - x_2^b, x_3^a - x_3^b$, that is, the projective coordinates of point p, satisfy equation (1). If, however, $p \sim \in P$, then $b \sim \in P$, and consequently

$$(4) \qquad \alpha_0 + \alpha_1 x_1^b + \alpha_2 x_2^b + \alpha_3 x_3^b \neq 0.$$

Subtracting inequality (4) from equality (2) we see that the projective coordinates $0, x_1^a - x_1^b, x_2^a - x_2^b, x_3^a - x_3^b$ of point p do not now satisfy equation (1).

In this way, we have shown that point p lies on proper plane P if and only if its coordinates (in system Ψ) satisfy equation (1). Equation (1) is the equation of a proper plane in projective space $\boldsymbol{P}_3$ (see Chapter IV, Section 6). Further, from Theorem 10 it follows that by a suitable choice of proper plane P we can obtain in this way the equation of every proper plane of projective space $\boldsymbol{P}_3$. We therefore have:

THEOREM 12. *Any projective coordinate system Ψ in space $\boldsymbol{S}$ maps the planes $P \subset \boldsymbol{S}$ onto the planes $\Psi(P)$ of projective space $\boldsymbol{P}_3$.*

Any line L may be represented as the intersection $L = P_1 \cap P_2$ of two distinct planes P_1 and P_2. By Theorems 12 and 11, we have

$$\Psi(L) = \Psi(P_1) \cap \Psi(P_2),$$

where $\Psi(P_1)$ and $\Psi(P_2)$ are two distinct planes of projective space $\mathbf{P}_3$. Hence $\Psi(L)$ is a line of space $\mathbf{P}_3$ (see Chapter IV, Section 6). On the basis of Theorem 12, we can, by a suitable choice of two distinct planes P_1 and P_2 (and hence, indirectly, of line $L = P_1 \cap P_2$) obtain, as $\Psi(L)$, any line of projective space $\mathbf{P}_3$. We therefore have

THEOREM 13. *Any projective coordinate system Ψ in space $\mathbf{S}$ maps the lines $L \subset \mathbf{S}$ onto the lines $\Psi(L)$ of projective space $\mathbf{P}_3$.*

In a similar manner, by means of Theorems 11–13, the reader may readily prove

THEOREM 14. *Given in space $\mathbf{S}$ a projective coordinate system Ψ and in a plane $P \subset \mathbf{S}$ a line L and a point $a \sim \in L$, if point p_1 is the projection of any point $p \in P$—a from center a upon line L, then, in projective space $\mathbf{P}_3$, the point $\Psi(p_1)$ is the projection of the point $\Psi(p) \in \Psi(P) - \Psi(a)$ from center $\Psi(a) \in \Psi(P) - \Psi(L)$ upon the line $\Psi(L) \subset \Psi(P)$.*

12. The Relation of Division and the Cross Ratio

In Section 11 we showed that the analytical analogues of the line and the plane of space $\mathbf{S}$ are the line and the plane of projective space $\mathbf{P}_3$. The analytical analogue of the relation of division is connected with the cross ratio of four collinear points in projective space $\mathbf{P}_3$ (here and in what follows, see Chapter IV Section 6):

THEOREM 15. *If Ψ is a projective coordinate system in space $\mathbf{S}$, then for any four distinct collinear points $a,b,c,d \in \mathbf{S}$ we have $\mathbf{D}(a,b; c,d)$ if and only if the cross ratio $(\Psi(a),\Psi(b);\Psi(c),\Psi(d))$ of points $\Psi(a)$, $\Psi(b)$, $\Psi(c)$, $\Psi(d)$ of projective space $\mathbf{P}_3$ is negative.*

PROOF. Let K_1 be the first axis of coordinate system Ψ. We consider three proper points a_0, b_0, c_0 of axis K_1 given by the conditions

$$(1,0,0,0) \in \Psi(a_0), \quad (1,2,0,0) \in \Psi(b_0), \quad (1,1,0,0) \in \Psi(c_0);$$

for any point $p \in K_1 - a_0$ we choose the coordinate system $(x^p,1,0,0)$ and we put

$$\varphi(p) = (\Psi(a_0),\Psi(b_0);\Psi(c_0),\Psi(p)).$$

By applying the general formula for the cross ratio (in space $\mathbf{P}_3$), we find that

(1) $$\varphi(p) = 2x^p - 1,$$

in particular,

(2) $$\varphi(c_0) = 2 \cdot 1 - 1 = 1,$$

and for the improper point a_∞ of line K_1

(3) $$\varphi(a_\infty) = 2 \cdot 0 - 1 = -1.$$

We shall now show that the function φ is a continuous function of point $p \in K_1 - a_0$. By formula (1), it is sufficient to show that the co-ordinate x^p is a continuous function of point p. We denote by Φ_1 the Cartesian coordinate system on line K_1 with basic points a_0 and c_0, and we order E-line $K_1 - a_\infty$ from point a_0 to point c_0. If point p is a proper point, then

$$x^p = \frac{1}{\Phi_1(p)},$$

and from the continuity of function Φ_1 (see Theorem 5) it follows that coordinate x^p is a continuous function of point p. Now let $p = a_\infty$. Then $x_p = 0$. Consider any number $\varepsilon > 0$. Let $q = \Phi_1^{-1}(-1/\varepsilon)$ and $r = \Phi_1^{-1}(1/\varepsilon)$. Then, by Theorem 5, we have $q \prec r$. Consider the neighborhood $U = (q-a_\infty-r)$ of point a_∞. If a proper point s belongs to neighborhood U, then on E-line $K_1 - a_\infty$ either $s \prec q$ or $s \succ r$. In the first case $\Phi_1(s) < \Phi_1(q)$, that is, $1/x^s < -1/\varepsilon$, from which it follows that

$$-\varepsilon < x^s < 0.$$

In the second case $\Phi_1(s) > \Phi_1(r)$, that is, $1/x^s > 1/\varepsilon$, from which it follows that

$$0 < x^s < \varepsilon.$$

Hence for every point s of neighborhood U we have $|x^s| < \varepsilon$, which shows that also at point $p = a_\infty$ coordinate x^p is a continuous function of point p. In this manner the continuity of function φ is proved.

By Theorem 2, segments $(a_0-c_0-b_0)$ and $(a_0-a_\infty-b_0)$ are both connected sets. Further, since function φ takes on the value 0 only for point $p = b_0$, then, because of the continuity of function φ, from formulas (2) and (3) it follows at once that

(4) $$\varphi(p) > 0 \text{ for } p \in (a_0-c_0-b_0),$$

(5) $$\varphi(p) < 0 \text{ for } p \in (a_0-a_\infty-b_0).$$

Let us now consider any four distinct points a, b, c, d of a line L. By Theorem 13 of Chapter VII, there exists a projective transformation f of line L onto line K_1 such that

$$f(a) = a_0, \quad f(b) = b_0, \quad f(c) = c_0.$$

Let $f(d) = d_0$. Since in projective space $\mathbf{P}_3$ the cross ratio is an invariant of perspective projectivities, then, by Theorem 14,

$$(\Psi(a), \Psi(b); \Psi(c), \Psi(d)) = (\Psi(a_0), \Psi(b_0); \Psi(c_0), \Psi(d_0)) = \varphi(d_0).$$

Let us assume that $\sim \mathbf{D}(a,b;c,d)$. Then, by Theorem 18 of Chapter VII, we have $\sim \mathbf{D}(a_0,b_0;c_0,d_0)$, that is, $d_0 \in (a_0 - c_0 - b_0)$, and consequently, by formula (4), we have $\varphi(d_0) > 0$. Let us next assume that $\mathbf{D}(a,b;c,d)$. Then, by Theorem 18 of Chapter VII,

(6) $\mathbf{D}(a_0,b_0;c_0,d_0).$

Furthermore, since point c_0 is the midpoint of segment $a_0 b_0$ (see Theorem 5), we have $\mathbf{D}_h(a_0,b_0;c_0,a_\infty)$, from which it follows, by Theorem 34 of Chapter VII, that

(7) $\mathbf{D}(a_0,b_0;c_0,a_\infty).$

From formulas (6) and (7) we obtain $\sim \mathbf{D}(a_0,b_0;a_\infty,d_0)$; hence

$$d_0 \in (a_0 - a_\infty - b_0)$$

and by formula (5) we have $\varphi(d_0) < 0$.

Thus the theorem is proved.

With this we conclude our treatment of projective geometry. The fact that in space $\mathbf{S}$ we have introduced projective coordinates constitutes, together with Theorems 11–13 and 15, a sufficient basis for the proof of the categoricity of projective geometry.

IX

Models of Projective Geometry

1. Problems of Consistency and Categoricity

In this chapter we shall consider the problems of the consistency and categoricity as regards projective geometry, in fact, space projective geometry based on the system of primitive notions

points, lines, planes, D

and on the system of axioms

$$(GP_3) \qquad\qquad I'1-I'10,\ O'1-O'7,\ Co',$$

and plane projective geometry based on the system of primitive notions

points, lines, D

and on the system of axioms

$$(GP_2) \qquad\qquad I'1-I'4,\ I'10,\ O'1-O'7,\ Co'.$$

Axioms I'4 and I'10 take the following form in the plane case:

AXIOM I'4. *There exist (three) non-collinear points a,b,c.*

AXIOM I'10. *For any lines K and L there exists a point c such that $c \in K$ and $c \in L$.*

In Section 2 we shall prove that the axiom system (GP_3) is consistent by constructing an arithmetic model (P) of (GP_3) formed by some notions of analytic geometry of three-dimensional projective space $\boldsymbol{P_3}$. In constructing this model we shall base ourselves on the analytic geometry given in Section 6 of Chapter IV. (We have already made use of the considerations of that section in constructing the Klein model.) In Section 3 we shall prove that the axiom system (GP_3) is categorical by showing that each of its models is isomorphic with the model (P) referred to above. In Section 6 we shall consider the question of the categoricity of plane projective geometry; an essential point here will be the construction of the so-called *Hilbert model*. In Sections 4 and 5 we shall discuss those notions of analytic geometry which are used in constructing the Hilbert model and which were not sufficiently discussed in Chapter IV.

2. Model (P). Consistency of Projective Geometry

We define P-*points*, P-*lines*, and P-*planes* respectively as points, lines, and planes of projective space $\boldsymbol{P}_3$. The relation of P-*division* $\boldsymbol{D}_P$ is defined in the following way: For any P-points a, b, c, d,

$$\boldsymbol{D}_P(a,b;c,d)$$

if and only if P-points a, b, c, d are P-collinear, distinct, and

$$(a,b;c,d) < 0.$$

(The definition of the cross ratio is on page 236.)

PROPOSITION 4. *The interpretation*

(P)   P-*points*, P-*lines*, P-*planes*, $\boldsymbol{D}_P$

is a model for the axiom system (GP$_3$).

PROOF. From Statements 40, 41 and 46–51 † it follows at once that interpretation (P) satisfies all the axioms of incidence (see pages 354 and 355).

We now pass on to the axioms of order (see page 365). From the definition of the relation $\boldsymbol{D}_P$ it follows at once that interpretation (P) satisfies Axiom O'1. From Statement 44 it readily follows that interpretation (P) also satisfies Axioms O'2 and O'3. We now come to Axiom O'4. Let there be given collinear and distinct P-points a, b, c, d. Let $(a,b;c,d) = \alpha$. Then $\alpha \neq 0,1$ and

$$(a,c;b,d) = 1 - \alpha \quad \text{and} \quad (a,d;b,c) = 1 - \frac{1}{\alpha}.$$

We have

$$\alpha(1-\alpha)\left(1-\frac{1}{\alpha}\right) = -(1-\alpha)^2 < 0.$$

Hence at least one of the numbers α, $1-\alpha$, $1 - \frac{1}{\alpha}$ is negative. Therefore interpretation (P) satisfies Axiom O'4. From Statement 45 it follows that interpretation (P) also satisfies Axiom O'5. We next consider Axiom O'6. Let us take any collinear and dictinct P-points a, b, c, d, e. It is readily calculated that

$$\frac{(a,b;c,e)}{(a,b;c,d)} = (a,b;d,e).$$

Hence, if $(a,b;c,e) > 0$ and $(a,b;c,d) > 0$, then also $(a,b;d,e) > 0$. We have thus verified Axiom O'6. The last axiom of order, Axiom O'7, is verified as follows: By Statement 56, every P-perspective transformation

---

† All the statements referred to in this proof are from Chapter IV.

(i.e. perspective transformation in projective space $\boldsymbol{P}_3$) f of P-line K onto P-line L can be extended to a linear transformation of the entire space $\boldsymbol{P}_3$ onto itself. Hence, for any four distinct P-points a, b, c, d of P-line K we have

$$(a,b;c,d) = (f(a),f(b);f(c),f(d)),$$

from which it immediately follows that

$$\boldsymbol{D}_\mathrm{P}(a,b;c,d), \text{ implies } \boldsymbol{D}_\mathrm{P}(f(a),f(b);f(c),f(d)).$$

It still remains to verify the Axiom of Continuity (see page 399) Let X and Y be two arbitrary non-empty sets of P-points and let c be an arbitrary P-point. Assume that a P-point a satisfies the condition

(1)   if $p \in X$ and $q \in Y$,then $\boldsymbol{D}_\mathrm{P}(a,q;p,c)$.

Then $a \neq c$, $a,c \sim \in X \cup Y$, and sets X and Y and P-points a and c lie on a P-line L with the parametric equation

$$x\,(\lambda,\mu) = [\lambda a_0 + \mu c_0, \lambda a_1 + \mu c_1, \lambda a_2 + \mu c_2, \lambda a_3 + \mu c_3].$$

For any $p \in X$ and any $q \in Y$ let

$$p = x\,(1,\lambda_q), \qquad q = x\,(1,\lambda_q).$$

Then, by (1), for any $p \in X$ and any $q \in Y$

(2) $$(a,q;p,c) = \frac{\lambda_p}{\lambda_p - \lambda_q} < 0.$$

From the above inequality it readily follows that for $p \in X$ the parameter λ_p is either always positive or always negative. Let us suppose e.g. that $\lambda_p > 0$ for $p \in X$. Then, by (2)

if $p \in X$ and $q \in Y$, then $\lambda_p < \lambda_q$,

and from the continuity of the set of real numbers it follows that there exists a number λ_b such that

(3)   if $p \in X$ and $q \in Y$, then $\lambda_p \leqslant \lambda_b \leqslant \lambda_q$.

We set $b = x\,(1,\lambda_b)$. Then, by (3), for any $p \in X - b$ and $q \in Y - b$ we have

$$(p,q;b,c) = \frac{\lambda_b - \lambda_p}{\lambda_b - \lambda_q} < 0.$$

Thus $\boldsymbol{D}_\mathrm{P}(p,q;b,c)$.

We have thus shown that interpretation (P) satisfies all the axioms of the system (GP$_3$).

From Proposition 4 we obtain

PROPOSITION 5. *If the arithmetic of real numbers is consistent, then the axiom system* (GP$_3$) *of space projective geometry is consistent.*

The axiom system (GP$_2$) of plane projective geometry is also consistent and the proof of the consistency of system (GP$_2$) is carried out in the same way as the proof of the consistency of system (GP$_3$). Thus the plane interpretation

(P)   P-points, P-lines, $\boldsymbol{D}_P$

is constructed, where P-points and P-lines are points and lines of two-dimensional projective space $\boldsymbol{P}_2$ and for any P-points a,b,c,d the formula $\boldsymbol{D}_P(a,b;c,d)$ is defined just as in the space case (see page 418). Proceeding precisely in the same way as in the proof of Proposition 4 we may prove that plane interpretation (P) is a model for the system (GP$_2$).

3. Categoricity of Space Projective Geometry

We have to prove that every two models for the axiom system (GP$_3$) are isomorphic. To do this it is sufficient to show that any arbitrary model for the axiom system (GP$_3$)

(M)   M-points, M-lines, M-planes, $\boldsymbol{D}_M$

is isomorphic with the model

(P)   P-points, P-lines, P-planes, $\boldsymbol{D}_P$.

which we constructed in the preceding section.

We choose in M-space an M-system of projective coordinates Ψ. We shall show that transformation Ψ establishes an isomorphism between models (M) and (P).

Indeed, by Theorems 11, 13, and 12 of Chapter VIII, function Ψ maps, in a one-to-one manner, the set of M-points onto the set of P-points, the class of M-lines onto the class of P-lines, and the class of M-planes onto the class of P-planes; and, by Theorem 15 of Chapter VIII, we have $\boldsymbol{D}_M(a,b;c,d)$ if and only if $\boldsymbol{D}_P(\Psi(a),\Psi(b);\Psi(c),\Psi(d))$, for any four M-points a, b, c, d.

We have thus proved

PROPOSITION 6. *The axiom system* (GP$_3$) *of space projective geometry is categorical.*

4. The Ellipse. Some Theorems Concerning the Circle and the Ellipse

We say that a curve $E \subset \mathbf{P}_2$ is an *ellipse* if there exists an affine transformation f of space $\mathbf{P}_2$ onto itself (see page 240) which maps the circle S with the center $(0,0)$ and radius 1 onto curve E. Thus the ellipse is a particular case of the conic. Circle S consists only of proper points and therefore so does the ellipse. Thus the ellipse can be investigated as a curve of Cartesian space $\mathbf{C}_2$.

Let an affine transformation f (in $\mathbf{C}_2$) transform circle S onto an ellipse E. We refer to the f-image G^i of the inner domain of circle S as the *inner domain* of ellipse E, and to the f-image G^o of the outer domain of circle S as the *outer domain* of ellipse E. By Statement 32 of Chapter IV the inner and the outer domains of the ellipse are affine notions.

Since affine transformations are homeomorphisms and preserve lines, half-lines, and half-planes, then Statements 32–37 of Chapter IV concerning the circle apply also to the ellipse.

In space $\mathbf{C}_2$ circle S is given by the algebraic equation

$$x_1^2 + x_2^2 = 1,$$

or by the parametric equation

$$x(\varphi) = (\cos \varphi, \sin \varphi) \quad \text{for} \quad 0 \leqslant \varphi < 2\pi.$$

The affine transformation f given by the formula

$$f(x_1, x_2) = (\alpha x_1, \beta x_2) \quad \text{for} \quad \alpha, \beta > 0$$

maps the circle S onto the ellipse E given by the algebraic equation

$$\frac{x_1}{\alpha^2} + \frac{x_2}{\beta^2} = 1$$

or by the parametric equation

$$x(\varphi) = (\alpha \cos \varphi, \beta \sin \varphi) \quad \text{for} \quad 0 \leqslant \varphi < 2\pi.$$

Assume now, that $\alpha > \beta$. It follows from the parametric equation that the distance of point $x(\varphi)$ of ellipse E from the point $(0,0)$ is equal to the number $\sqrt{\alpha^2 \cos^2 \varphi + \beta^2 \sin^2 \varphi}$ and therefore it is always $\leqslant \alpha$. Hence the distance between any two points of ellipse E is always $\leqslant 2\alpha$; furthermore, $\rho(x(0), x(\pi)) = 2\alpha$. Thus the diameter of ellipse E coincides with the number 2α; it is readily shown that the diameter of the figure $G^i \cup E$ also coincides with the number 2α.

We shall now prove a few special theorems dealing with the ellipse and the circle.

STATEMENT 1. *If a circle S with center (x_0, y_0) intersects an ellipse E given by the equation*

$$\frac{x^2}{\alpha^2} + \frac{y^2}{\beta^2} = 1 \quad (\alpha \geq \beta)$$

in at least three distinct points, then

(1) $$|x_0| \leq \frac{\alpha^2 - \beta^2}{\alpha} \quad \text{and} \quad |y_0| \leq \frac{\alpha^2 - \beta^2}{\beta}.$$

PROOF. Ellipse E has the parametric equation

$$x(\varphi) = (\alpha \cos \varphi, \beta \sin \varphi), \quad \text{where} \quad 0 \leq \varphi < 2\pi.$$

Let

$$x_i = x(\varphi_i) \quad \text{for} \quad i = 1, 2, 3$$

be the three distinct intersection points of circle S and ellipse E. Then

$$(x_0 - \alpha \cos \varphi_1)^2 + (y_0 - \beta \sin \varphi_1)^2 = (x_0 - \alpha \cos \varphi_2)^2 + (y_0 - \beta \sin \varphi_2)^2,$$

$$(x_0 - \alpha \cos \varphi_1)^2 + (y_0 - \beta \sin \varphi_1)^2 = (x_0 - \alpha \cos \varphi_3)^2 + (y_0 - \beta \sin \varphi_3)^2,$$

that is

$$2\alpha x_0(\cos \varphi_2 - \cos \varphi_1) + 2\beta y_0(\sin \varphi_2 - \sin \varphi_1)$$
$$= \alpha^2(\cos^2\varphi_2 - \cos^2\varphi_1) + \beta^2(\sin^2\varphi_2 - \sin^2\varphi_1),$$
$$2\alpha x_0(\cos \varphi_3 - \cos \varphi_1) + 2\beta y_0(\sin \varphi_3 - \sin \varphi_1)$$
$$= \alpha^2(\cos^2\varphi_3 - \cos^2\varphi_1) + \beta^2(\sin^2\varphi_3 - \sin^2\varphi_1).$$

The determinant of this system of equations has the form

$$\Delta = 4\alpha\beta \begin{vmatrix} \cos \varphi_2 - \cos \varphi_1 & \sin \varphi_2 - \sin \varphi_1 \\ \cos \varphi_3 - \cos \varphi_1 & \sin \varphi_3 - \sin \varphi_1 \end{vmatrix}$$

$$= 16\alpha\beta \sin \frac{\varphi_1 - \varphi_2}{2} \sin \frac{\varphi_2 - \varphi_3}{2} \sin \frac{\varphi_3 - \varphi_1}{2} \neq 0.$$

The numerator Δ_x of the determinant formula for x_0 has the form

$$\Delta_x = 2\beta \begin{vmatrix} \alpha^2(\cos^2\varphi_2 - \cos^2\varphi_1) + \beta^2(\sin^2\varphi_2 - \sin^2\varphi_1) & \sin\varphi_2 - \sin\varphi_1 \\ \alpha^2(\cos^2\varphi_3 - \cos^2\varphi_1) + \beta^2(\sin^2\varphi_3 - \sin^2\varphi_1) & \sin\varphi_3 - \sin\varphi_1 \end{vmatrix}$$

$$= 2\beta(\beta^2 - \alpha^2)(\sin\varphi_2 - \sin\varphi_1)(\sin\varphi_3 - \sin\varphi_1) \begin{vmatrix} \sin\varphi_2 + \sin\varphi_1 & 1 \\ \sin\varphi_3 + \sin\varphi_1 & 1 \end{vmatrix}$$

$$= 2\beta(\alpha^2 - \beta^2)(\sin\varphi_1 - \sin\varphi_2)(\sin\varphi_2 - \sin\varphi_3)(\sin\varphi_3 - \sin\varphi_1)$$

$$= 16\beta(\alpha^2 - \beta^2)\cos\frac{\varphi_1 + \varphi_2}{2}\cos\frac{\varphi_2 + \varphi_3}{2}\cos\frac{\varphi_3 + \varphi_1}{2}\sin\frac{\varphi_1 - \varphi_2}{2}\sin\frac{\varphi_2 - \varphi_3}{2}\sin\frac{\varphi_3 - \varphi_1}{2}.$$

From this we get

$$x_0 = \frac{\Delta_x}{\Delta} = \frac{\alpha^2 - \beta^2}{\alpha}\cos\frac{\varphi_1 + \varphi_2}{2}\cos\frac{\varphi_2 + \varphi_3}{2}\cos\frac{\varphi_3 + \varphi_1}{2}.$$

Similarly we find

$$y_0 = \frac{\Delta_y}{\Delta} = \frac{\beta^2 - \alpha^2}{\beta}\sin\frac{\varphi_1 + \varphi_2}{2}\sin\frac{\varphi_2 + \varphi_3}{2}\sin\frac{\varphi_3 + \varphi_1}{2}.$$

From this it follows immediately that the inequalities (1) are satisfied.

STATEMENT 2. *Given an ellipse E with the equation*

$$\frac{x^2}{\alpha^2} + \frac{y^2}{\beta^2} = 1$$

and a circle S, if the radius of circle S is greater than the number

(2) $$\omega(E) = \alpha + \frac{\alpha^2 - \beta^2}{\alpha\beta}\sqrt{\alpha^2 + \beta^2},$$

then circle S has at most two points in common with ellipse E.

PROOF. If circle S were to intersect ellipse E in three distinct points, then, according to Statement 1, for the distance between every point p of ellipse E and the center c of circle S we would have (Fig. 319)

$$\rho(p,c) < \alpha + \sqrt{\left(\frac{\alpha^2 - \beta^2}{\alpha}\right)^2 + \left(\frac{\alpha^2 - \beta^2}{\beta}\right)^2} = \alpha + \frac{\alpha^2 - \beta^2}{\alpha\beta}\sqrt{\alpha^2 + \beta^2}$$

and therefore the radius of circle S would be smaller than the number $\omega(E)$.

STATEMENT 3. *Given an ellipse E with the equation*

$$\frac{x^2}{\alpha^2} + \frac{y^2}{\beta^2} = 1,$$

if a circle S with radius $\lambda > \omega(E)$ has two points p and q in common with ellipse E, then one of the open arcs determined on circle S by points p and q lies in the outer domain of ellipse E, and the other—in the inner domain of ellipse E.

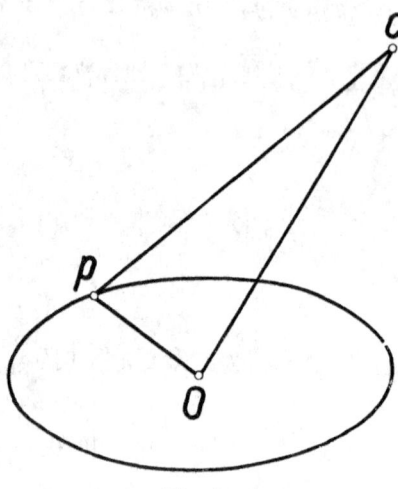

Fig. 319

PROOF. We denote by G^i and G^o, respectively, the inner and outer domains of ellipse E (Fig. 320). Let us assume that a circle S with radius

(2) $\lambda > \omega(E)$

has two points p and q in common with ellipse E. Let W_1 and W_2 be the half-planes determined by line pq. From Statement 2 it follows that $S \cap E = \{p,q\}$; consequently, each of the two arcs $J_1 = S \cap W_1$ and $J_2 = S \cap W_2$ lies either entirely in domain G^i or entirely in domain G^o. From formulas (1) and (2) it follows that the diameter of circle S is greater than the number 2α, that is, greater than the diameter of domain $G^i \cup E$. Hence the center a of circle S does not lie on line pq, and, if we assume that $a \in W_1$, arc J_1 lies in domain G^o. Let c and M denote the midpoint and perpendicular bisector, respectively, of segment pq. Then $c \in G^i$ and line M intersects arc J_2 in some point s. If arc J_2 were included in domain G^o, then some point r of ellipse E would lie on segment (cs). It is readily noted that the radius λ' of the circle S' determined by points p, q, r would be greater than the radius λ of circle S, and therefore greater than the number $\omega(E)$, contrary to the fact that circle S' would intersect ellipse E in three distinct points p, q, r. Hence arc J_2 lies in domain G^i.

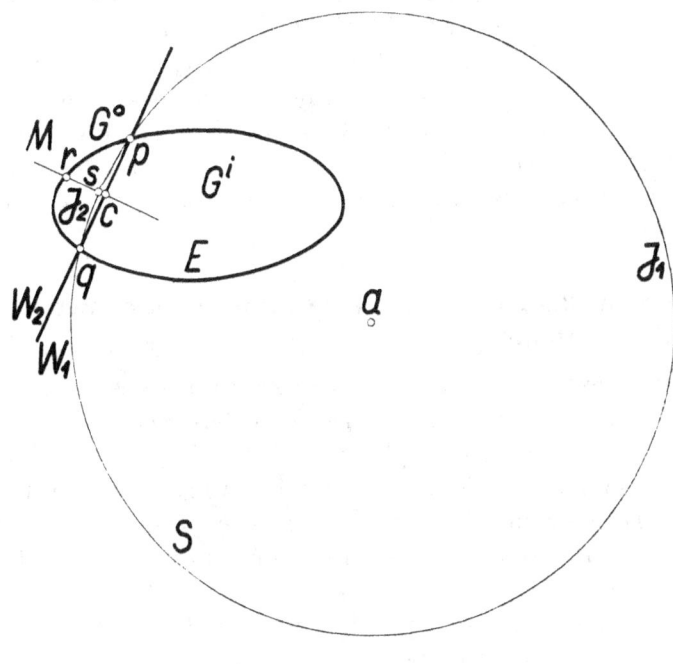

Fig. 320

5. The Limit of a Sequence of Circles

In this section we shall understand by circles the curves of plane $\mathbf{C}_2$ defined by equations of the form

(1) $\alpha_0 (x_1^2 + x_2^2) + 2\alpha_1 x_1 + 2\alpha_2 x_2 + \alpha_3 = 0$,   where $\alpha_1^2 + \alpha_2^2 > \alpha_0 \alpha_3$.

In the case $\alpha_0 = 0$ this equation determines a line (which we can regard as a circle with infinite radius); in the case $\alpha_0 \neq 0$ this equation determines a circle, in the usual meaning of this word, with the center $\left(-\dfrac{\alpha_1}{\alpha_0}, -\dfrac{\alpha_2}{\alpha_0} \right)$ and radius $\sqrt{\dfrac{\alpha_1^2 + \alpha_2^2 - \alpha_0 \alpha_3}{\alpha_0^2}}$.   Each of the circles S so understood splits the plane into two domains; in the case $\alpha_0 = 0$ these are two half-planes with boundary S, and in the case $\alpha_0 \neq 0$ these are the inner and outer domains of circle S. We say that figures $F_1, F_2 \subset \mathbf{C}_2$ lie on opposite sides of circle S if one of them lies in one of these domains and the second in the other.

The coefficients $\alpha_0, \alpha_1, \alpha_2, \alpha_3$ of equation (1), determined uniquely, up to a constant factor, by circle S will be called the *coordinates* of circle S. It is clear that by correlating with circle S the point $[\alpha_0,\alpha_1,\alpha_2,\alpha_3]$ of projective space $\boldsymbol{P}_3$, we obtain a one-to-one correspondence between circles in $\boldsymbol{C}_2$ and the points $[\alpha_0,\alpha_1,\alpha_2,\alpha_3]$ of space $\boldsymbol{P}_3$ satisfying the condition $\alpha_1^2 + \alpha_2^2 > \alpha_0\alpha_3$. This fact permits the introduction of the notion of a limit for a sequence of circles. If $\alpha_{0,n}, \alpha_{1,n}, \alpha_{2,n}, \alpha_{3,n}$ are the coordinates of a circle S_n $(n = 1,2,\ldots)$, then we say that *the sequence of circles* (S_n) *tends to the circle* S with coordinates $\alpha_0, \alpha_1, \alpha_2, \alpha_3$, if and only if in space $\boldsymbol{P}_3$ the sequence of points $([\alpha_{0,n}, \alpha_{1,n}, \alpha_{2,n}, \alpha_{3,n}])$ tends to the point $[\alpha_0, \alpha_1, \alpha_2, \alpha_3]$ (The topology in $\boldsymbol{P}_3$ was defined in the Introduction, Section 10.)

6. Problem of Categoricity of Plane Projective Geometry. The Hilbert Model

It might be expected that (GP_2) (see page 417) is a categorical axiom system of plane projective geometry. This, however, is not the case. To prove this, two non-isomorphic models for the axiom system (GP_2) should be found. With this purpose in mind we shall employ the direct Desargues Theorem in the plane case, and we shall construct for the axiom system (GP_2) two models such that one of the models satisfies this theorem, but not the other.

The first of these models is just the plane model (P) described at the end of Section 2. We shall prove

PROPOSITION 7. *The model*

(P)   P-*points*, P-*lines*, $\boldsymbol{D}_{\mathrm{P}}$

for the axiom system (GP_2) *satisfies the direct and converse Desargues Theorem.*

PROOF. Let us take two P-triangles $a_0a_1a_2$ and $b_0b_1b_2$. We shall show that for the correspondence f of the P-vertices given by the formula

$$f(a_i) = b_i \quad \text{for} \quad i = 0, 1, 2,$$

P-triangles $a_0a_1a_2$ and $b_0b_1b_2$ have a P-perspective axis if and only if they have a P-perspective center. We shall examine separately two cases.

Case 1. Two corresponding P-vertices coincide; let e.g. $a_0 = b_0$. Then the P-line M passing through P-point a_0 and through any P-point belonging to the intersection of P-lines a_1a_2 and b_1b_2 (by Statement 43 of Chapter IV, this intersection is non-empty) is a P-perspective axis; and P-point c, P-collinear with P-points a_1 and b_1 and with P-points a_2 and b_2 (by Statement 43 of Chapter IV such a P-point c always exists), is a P-perspective center of P-triangles $a_0a_1a_2$ and $b_0b_1b_2$ for correspondence f of P-vertices.

Case 2. For $i = 0,1,2$ we have $a_i \neq b_i$. Employing Statement 53 of Chapter IV we may apply a suitable linear transformation to space P_2 and reduce this case to the special case

$$a_0 = [1,0,0], \quad a_1 = [0,1,0], \quad a_2 = [0,0,1].$$

We put

$$b_i = [b_{i0}, b_{i1}, b_{i2}] \quad \text{for} \quad i = 0, 1, 2$$

and

$$\Delta = \begin{vmatrix} b_{00} & b_{01} & b_{02} \\ b_{10} & b_{11} & b_{12} \\ b_{20} & b_{21} & b_{22} \end{vmatrix}.$$

Let (i,j,k) be any permutation of the triple $(0,1,2)$. Then

P-line $a_i a_j$ has the equation $\quad x_k = 0$,

P-line $b_i b_j$ has the equation $\quad \begin{vmatrix} x_0 & x_1 & x_2 \\ b_{i0} & b_{i1} & b_{i2} \\ b_{j0} & b_{j1} & b_{j2} \end{vmatrix} = 0$;

P-line $a_i b_i$ has the equation $\quad b_{ij} x_k - b_{ik} x_j = 0$.

P-triangles $a_0a_1a_2$ and $b_0b_1b_2$ have, for the correspondence f of their P-vertices, a P-perspective axis M if and only if there exist three real numbers $\alpha_0, \alpha_1, \alpha_2$ (coefficients in the equation $\alpha_0 x_0 + \alpha_1 x_1 + \alpha_2 x_2 = 0$ of P-line M), not vanishing simultaneously, such that

$$\alpha_i \cdot \begin{vmatrix} b_{ii} & b_{ik} \\ b_{ji} & b_{jk} \end{vmatrix} + \alpha_j \cdot \begin{vmatrix} b_{ij} & b_{ik} \\ b_{jj} & b_{jk} \end{vmatrix} = 0 \quad \text{for } k = 0, 1, 2,$$

that is, if and only if

$$\begin{vmatrix} \begin{vmatrix} b_{00} & b_{02} \\ a_{10} & b_{12} \end{vmatrix} & \begin{vmatrix} b_{01} & b_{02} \\ b_{11} & b_{12} \end{vmatrix} & 0 \\ 0 & \begin{vmatrix} b_{11} & b_{10} \\ b_{21} & b_{20} \end{vmatrix} & \begin{vmatrix} b_{12} & b_{10} \\ b_{22} & b_{20} \end{vmatrix} \\ \begin{vmatrix} b_{20} & b_{21} \\ b_{00} & b_{01} \end{vmatrix} & 0 & \begin{vmatrix} b_{22} & b_{21} \\ b_{02} & b_{01} \end{vmatrix} \end{vmatrix} = 0.$$

P-triangles $a_0a_1a_2$ and $b_0b_1b_2$ have, for correspondence f of their P-vertices, a P-perspective center if and only if

$$\Delta_2 = \begin{vmatrix} 0 & -b_{02} & b_{01} \\ b_{12} & 0 & -b_{10} \\ -b_{21} & b_{20} & 0 \end{vmatrix} = 0.$$

As may be readily shown, we have $\Delta_1 = \Delta \cdot \Delta_2$. Since $\Delta \neq 0$, then $\Delta_1 = 0$ if and only if $\Delta_2 = 0$. Hence interpretation (P) satisfies Desargues' Theorem.

Henceforth, in denoting the notions of the model (P), we shall omit the prefix (or index) "P". Thus, instead of "P-line", we shall write simply "line", instead of "$\mathbf{D_P}$", we shall write simply "$\mathbf{D}$", and so on.

The second model for the axiom system (GP_2) originates with Hilbert and will therefore be called the *Hilbert model*. It is obtained by a modification of model (P).

We take in Cartesian space $\mathbf{C_2} \subset \mathbf{P_2}$ the ellipse E_0 with the equation

$$x^2 + 4y^2 = 1.$$

The function ω (see formula (2) on page 423) takes, for ellipse E_0, the value

$$\omega(E_0) = 1 + \frac{3}{4}\sqrt{5} < \frac{11}{4}$$

and therefore every circle S passing through point $a_0 = (7,0)$ and intersecting ellipse E_0 has the radius

(1) $$\lambda > \frac{1}{2}(7-1) = 3 > \omega(E_0).$$

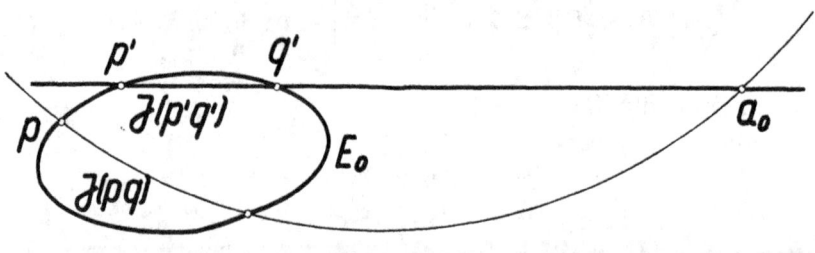

Fig. 321

It will be convenient to regard lines of space $\mathbf{P_2}$ which pass through point a_0 as special cases of circles. Then through any two distinct points p and q of ellipse E_0 just one circle $S(pq)$ can be produced to pass through point a_0 (Fig. 321). From Statement 3 and from formula (1) it follows that just one of the open arcs with end points p and q of circle $S(pq)$ lies in the inner domain of ellipse E_0. We shall denote this arc by $J(pq)$.

STATEMENT 4. *Let p_1, q_1, p_2, q_2 be distinct points of ellipse E_0. The arcs $J(p_1q_1)$ and $J(p_2q_2)$ intersect in a point if and only if the segments (p_1q_1) and (p_2q_2) intersect in a point; and the intersection point r of the arcs $J(p_1q_1)$ and $J(p_2q_2)$, if it exists, is unique and depends in a continuous way on points p_1, q_1, p_2, q_2.*

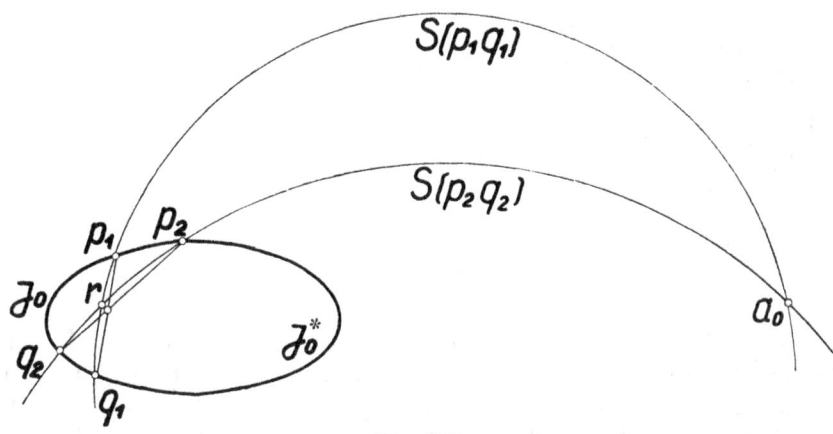

Fig. 322

PROOF. Points p_1 and q_1 determine on ellipse E_0 two open arcs J_0 and J_0^* (Fig. 322). These arcs, as readily noted, lie on opposite sides of circle $S(p_1q_1)$. If arcs $J(p_1q_1)$ and $J(p_2q_2)$ intersect in some point r, then points a_0 and r determine on circle $S(p_2q_2)$ two open arcs which lie on opposite sides of circle $S(p_1q_1)$ and both intersect ellipse E_0. It thus follows that one of the two points p_2 and q_2 lies on arc J_0 and the other, on arc J_0^*. Hence points p_2 and q_2 lie on different sides of line p_1q_1 and consequently the segment (p_2q_2) intersects line p_1q_1 in some point which, as may readily be shown, belongs to segment (p_1q_1).

Conversely, if segments (p_1q_1) and (p_2q_2) have a common point, then points p_2 and q_2 lie on different sides of line p_1q_1, and hence one of them lies on arc J_0 and the other, on arc J_0^*. Hence points p_2 and q_2 lie on opposite sides of circle $S(p_1q_1)$. As a result, arc $J(p_2q_2)$ must intersect circle $S(p_1q_1)$ in some point r which, as readily noted, belongs to arc $J(p_1q_1)$. There is only one such point, since circles $S(p_1q_1)$ and $S(p_2q_2)$ also intersect in point a_0. That point r depends continuously on points p_1, q_1, p_2, q_2 follows immediately from the analytical formulas for its coordinates.

Let L be any line of space $\mathbf{P}_2$. We denote by L° the set of points which is formed as follows:

(i) If line L has at most one point in common with ellipse E_0, then $L^\circ = L$.

(ii) If line L has two points p and q in common with ellipse E_0, then $L° = (L - (pq)) \cup J(pq)$, provided the segment (pq) lies in the inner domain of ellipse E_0 (Fig. 323).

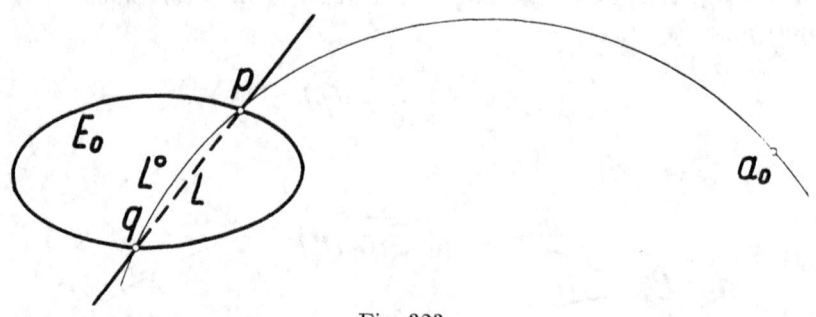

Fig. 323

It is clear that the point set $L°$ is homeomorphic with line L. This homeomorphism is fixed e.g. by the function g_L defined, in case (i), as the identity transformation, and, in case (ii), on set $L° - J(pq)$ as the identity transformation and on arc $J(pq)$ as the perpendicular projectivity upon (Cartesian) line pq.

From $L_1 \neq L_2$ it obviously follows that $L_1° \neq L_2°$ and, as a direct conclusion from Statement 4, we obtain the following:

STATEMENT 5. *For any lines L_1 and L_2 of space* $\mathbf{P_2}$, *we have* $L_1° \cap L_2° \neq 0$ *if and only if* $L_1 \cap L_2 \neq 0$.

We shall now prove

STATEMENT 6. *For any two distinct points a and b of space* $\mathbf{P_2}$ *there exists just one line L such that points a and b belong to set $L°$.*

PROOF. We denote by G^i and G^o, respectively, the inner and outer domains of ellipse E_0. It is readily seen that the theorem holds when points a and b both belong either to the set $G^i \cup E_0$ or to the set $G^o \cup E_0$.

It therefore remains to prove the theorem for the case one of the points, say a, lies in domain G^i and the other, say b, in domain G^o (Fig. 324).

Let us produce from point b the tangents L_1 and L_2 to ellipse E_0. Let $s_1 = L_1 \cap E_0$ and $s_2 = L_2 \cap E_0$, and let S_j $(j = 1,2)$ denote the circle passing through point a_0 and tangent at point s_j to line L_j. We denote by $\langle s_1 s_2 \rangle$ that closed segment with end points s_1 and s_2 whose interior is included in domain G^i; to each number $1 \leqslant \alpha \leqslant 2$ we correlate a point $s_\alpha \in \langle s_1 s_2 \rangle$ such that

(2) $$\varrho(s_1, s_\alpha) = (\alpha - 1) \cdot \varrho(s_1, s_2).$$

Obviously, point s_α is a continuous function of the parameter α. We

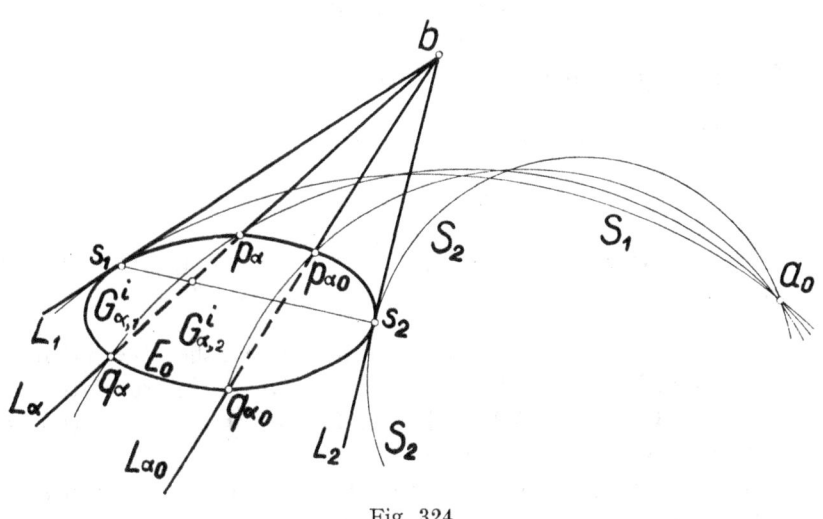

Fig. 324

denote by p_α and q_α the intersection points of line bs_α and ellipse E_0. From the equations of circles $S(p_\alpha q_\alpha)$, S_1, and S_2 it follows that as α tends to j ($j = 1,2$) circle $S(p_\alpha q_\alpha)$ tends to circle S_j.

For $1 < \alpha < 2$ arc $J(p_\alpha q_\alpha)$ divides domain G^i into two point sets $G^i_{\alpha,1}$ and $G^i_{\alpha,2}$ such that the boundary of set $G^i_{\alpha,j}$ (for $j = 1,2$) contains the point s_j. Since, as readily follows from Statement 2, circle S_j has only point s_j in common with the set $G^i \cup J_0$, then, as α tends to j, every point belonging to the boundary of set $G^i_{\alpha,j}$ tends to point s_j. Therefore, as α tends to j, the diameter of the boundary of set $G^i_{\alpha,j}$, and consequently the diameter of set $G^i_{\alpha,j}$ itself, tends to zero. Since $G^i_{\alpha,1}$ and $G^i_{\alpha,2}$ are open sets, for α sufficiently close to 1 we have $a \in G^i_{\alpha,2}$ and for α sufficiently close to 2 we have $a \in G^i_{\alpha,1}$. As a result, for some $1 < \alpha_0 < 2$ we have $a \in J(p_{\alpha_0} q_{\alpha_0})$. There exists only one such number α_0, since for $\alpha \neq \alpha_0$ segments $(p_{\alpha_0} q_{\alpha_0})$ and $(p_\alpha q_\alpha)$ are disjoint and hence, by Statement 4, arcs $J(p_{\alpha_0} q_{\alpha_0})$ and $J(p_\alpha q_\alpha)$ are disjoint.

The Hilbert model will be called here *model* (H). We define H-*points* as points of the projective space $\mathbf{P}_2$ and H-*lines* as the sets of points L°, where L is any line of space $\mathbf{P}_2$. For any four arbitrary points a, b, c, d we set

$$\mathbf{D}_{\mathrm{H}}(a,b\,;c,d)$$

if and only if a, b, c, d lie on some H-line L° and

$$\mathbf{D}(g_L(a), g_L(b); g_L(c), g_L(d)),$$

i.e.

$$(g_L(a), g_L(b); g_L(c), g_L(d)) < 0.$$

(The function g_L is defined on page 434.)

PROPOSITION 8. *The interpretation*

(H)   H-*points*, H-*lines*, $\mathbf{D}_H$

is a model for the axiom system (GP$_2$), *but does not satisfy the direct Desargues Theorem.*

PROOF. It is evident that interpretation (H) satisfies the axioms of incidence I'1 and I'4 and, by Statements 6 and 5, interpretation (H) also satisfies Axioms I'2, I'3, and I'10. Hence interpretation (H) is a model for the axioms of incidence.

We now consider the axioms of order. It is at once seen that interpretation (H) satisfies Axioms O'1—O'6. It remains to show that Axiom O'7 is also satisfied.

Let f be the H-perspective transformation of H-line K° upon H-line L° from H-center s; for H-points $a, b, c, d \in K^\circ$ let $\mathbf{D}_H(a, b; c, d)$. We have to prove that $\mathbf{D}_H(f(a), f(b); f(c), f(d))$.

Let us first assume that the intersection point r of H-lines K° and L° lies either on ellipse E_0 or in its outer domain (Fig. 325). H-points r and s determine some H-line M_0°. We produce from point s an arbitrary line N and we put $s_1 = N \cap K$, $s_2 = N \cap L$. We denote by $\langle s_1 s_2 \rangle$ that segment with end points s_1 and s_2 which does not contain point s, and with the numbers $1 \leqslant \alpha \leqslant 2$ we correlate, in a continuous way, points s_α of segment $\langle s_1 s_2 \rangle$, e.g. by means of formula (2).

We put

$$M_\alpha = \mathbf{L}(r s_\alpha);$$

then

(3) $M_1 = K, \quad M_2 = L.$

Let p_α and q_α be the intersection points of line M_α and the ellipse E_0. As follows from the analytical formulas for their coordinates, points p_α and q_α are continuous functions of parameter α.

Let us now denote the H-lines joining H-point s with points a, b, c, d, respectively, by $N_a^\circ, N_b^\circ, N_c^\circ, N_d^\circ$, and let $a_\alpha, b_\alpha, c_\alpha, d_\alpha$ be, respectively, intersection points of the H-lines $N_a^\circ, N_b^\circ, N_c^\circ, N_d^\circ$ with H-line M_α°. Points $a_\alpha, b_\alpha, c_\alpha, d_\alpha$ are distinct and, according to Statement 4, are continuous functions of points p_α and q_α, and consequently also of parameter α. From this it follows that points $g_{M_\alpha}(a_\alpha)$, $g_{M_\alpha}(b_\alpha)$, $g_{M_\alpha}(c_\alpha)$, $g_{M_\alpha}(d_\alpha)$ are distinct and are continuous functions of parameter α. Hence the cross ratio

$$\varphi(\alpha) = (g_{M_\alpha}(a_\alpha), g_{M_\alpha}(b_\alpha); g_{M\alpha}(c_\alpha), g_{M_\alpha}(d_\alpha))$$

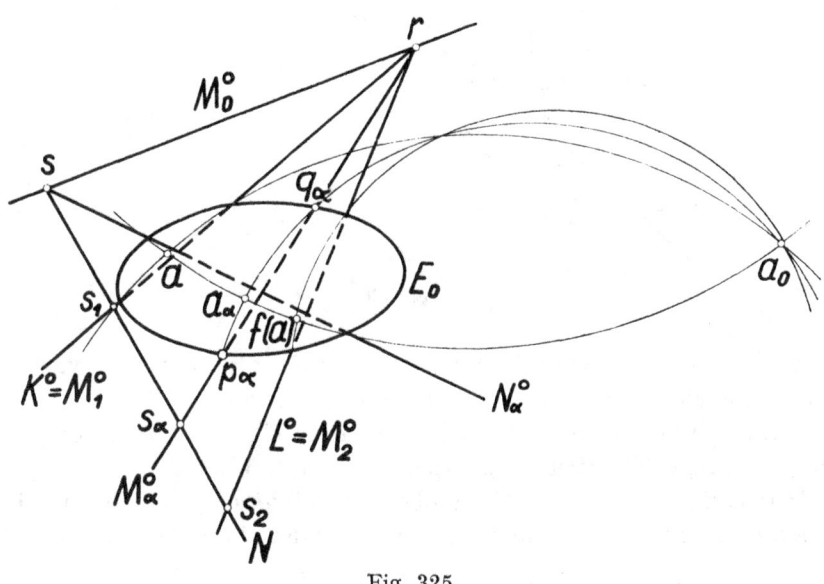

Fig. 325

is a continuous function of parameter α. From $\mathbf{D}_H(a,b;c,d)$ it follows that

$$\mathbf{D}(g_{M_1}(a), g_{M_1}(b); g_{M_1}(c), g_{M_1}(d))$$

and therefore $\varphi(1) < 0$. Since the number $\varphi(\alpha)$ is always different from zero, then from the continuity of function φ and the inequality $\varphi(1) < 0$ it follows that the number $\varphi(\alpha)$ is always negative and, in particular, $\varphi(2) < 0$. Therefore

$$\mathbf{D}(g_{M_2}(f(a)), g_{M_2}(f(b)); g_{M_2}(f(c)), g_{M_2}(f(d))),$$

which, together with formula (3), gives $\mathbf{D}_H(f(a), f(b); f(c), f(d))$.

If the intersection point r of H-lines K° and L° lies in the inner domain of ellipse E_0 (Fig. 326), then in the outer domain of ellipse E_0 we pick points $r_1 \in K$ and $r_2 \in L$ in such a way that line $L_1 = \mathbf{L}(r_1 r_2)$ does not pass through point s. Let f' and f'' be the H-perspective transformations from center s, respectively, of H-line K° upon H-line L_1° and of H-line L_1° upon H-line L°. Then $f = f'' f'$ and, by what was proved above, $\mathbf{D}_H(a,b;c,d)$ implies $\mathbf{D}_H(f'(a), f'(b); f'(c), f'(d))$ and $\mathbf{D}_H(f'(a), f'(b); f'(c), f'(d)$ implies $\mathbf{D}_H(f(a), f(b); f(c), f(d))$.

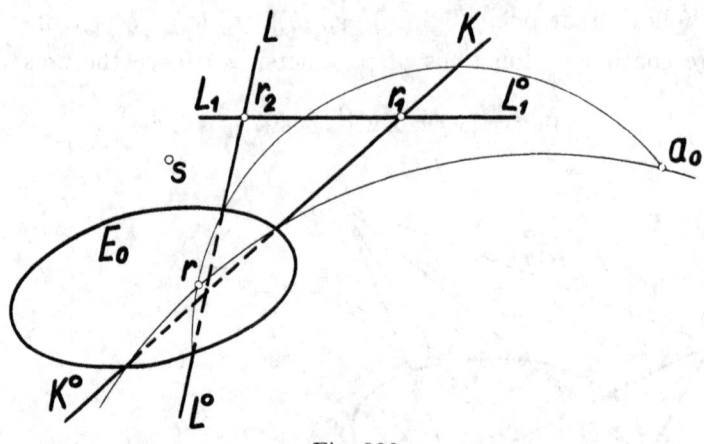

Fig. 326

Thus we have shown that interpretation (H) also satisfies Axiom O'7 and is therefore a model for the axioms of order.

Finally, from the definition of the relation $\mathbf{D}_H$ it follows at once that interpretation (H) satisfies also the Axiom of Continuity.

It remains to show that the Hilbert model does not satisfy the direct Desargues Theorem. To show this we consider the points

$$p_1 = \left(\frac{3}{5}, \frac{2}{5}\right),\ q_1 = \left(-\frac{3}{5}, -\frac{2}{5}\right),\ p_2 = \left(0, \frac{1}{2}\right),\ q_2 = \left(0, -\frac{1}{2}\right),\ c = (0,0)$$

and the lines

$$L_1 = \mathbf{L}(p_1 q_1),\quad L_2 = \mathbf{L}(p_2 q_2),\quad L_3 = \mathbf{L}(c a_0),$$

and we choose on line $L_i (i = 1,2,3)$ two distinct points a_i and b_i in such a way that points a_1, a_2, a_3 form a triangle $a_1 a_2 a_3$, points b_1, b_2, b_3 — a triangle $b_1 b_2 b_3$, and the sides of triangles $a_1 a_2 a_3$ and $b_1 b_2 b_3$ lie in the outer domain of ellipse E_0 (Fig. 327). Then point c is the perspective center of triangles $a_1 a_2 a_3$ and $b_1 b_2 b_3$ for the correspondence g of their vertices given by the formula

$$g(a_i) = b_i \quad \text{for} \quad i = 1, 2, 3,$$

and, according to the converse Desargues Theorem (see Proposition 7), there exists for these triangles a perspective axis M. Then, as readily seen, H-line $M°$ is a H-perspective axis for H-triangles $a_1 a_2 a_3$ and $b_1 b_2 b_3$. We shall show that no H-perspective center exists for these H-triangles. To do so it is sufficient to show that the intersection point c' of arcs $J(p_1 q_1)$ and $J(p_2 q_2)$ does not lie on Cartesian line L_3. Circles $S(p_1 q_1)$, $S(p_2 q_2)$ and line L_3 have, respectively, the equations

$$x^2 + y^2 - \frac{1212}{175}\, x + \frac{1818}{175}\, y - \frac{13}{25} = 0,$$

$$x^2 + y^2 - \frac{195}{28}\, x - \frac{1}{4} = 0,$$

$$y = 0.$$

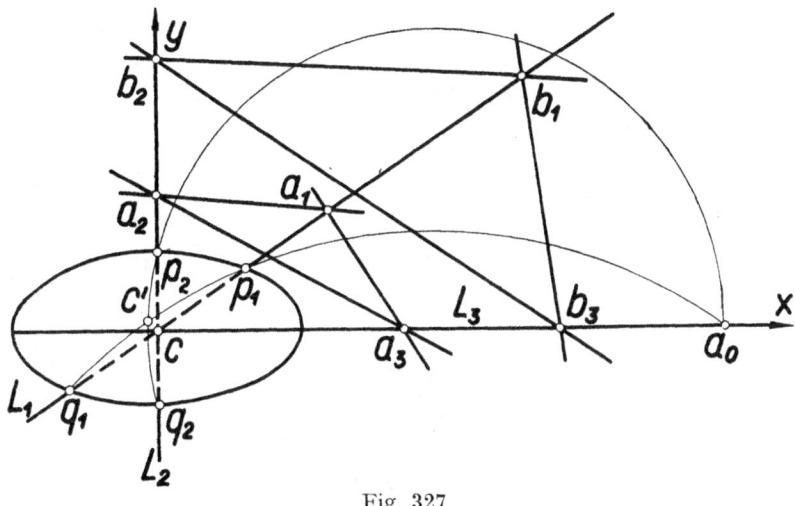

Fig. 327

The common points of circles $S(p_1 q_1)$ and $S(p_2 q_2)$ lie on the Cartesian line with the equation

$$\left(-\frac{1212}{175} + \frac{195}{28}\right) x + \frac{1818}{175}\, y - \frac{13}{25} + \frac{1}{4} = 0.$$

But this line intersects line L_3 only in the point $a_0 = (7,0)$ lying in the outer domain of ellipse E_0, and therefore distinct from point c'.

Thus we have shown that the direct Desargues Theorem (in the plane case) is independent of the axiom system (GP_2) and hence it becomes clear why in the proof of this theorem (Theorem 15 of Chapter VII) we had to use a space construction.

At the same time we have proved

PROPOSITION 9. *The system* (GP_2) *is not a categorical axiom system for plane projective geometry.*

If to the system (GP_2) we add the direct Desargues Theorem, then the axiom system thus extended is a categorical system. For, the direct Desargues Theorem implies the converse Desargues Theorem (see the proof on pages 362–365) and once we have the direct and converse

Desargues Theorems, we may then introduce projective coordinates on the plane in the same way as we did in space (the Desargues Theorem is necessary only to develop the theory of harmonic quadruples), and for the plane prove theorems analogous to Theorems 11, 13, 15, on which the proof of categoricity is based.

Index of Geometrical Symbols

Index